Forest Protection and Utilization

Forest Protection and Utilization

Shenaz Sultana Ahmed
Plant Pathology, College of Agriculture,
Assam Agricultural University, Jorhat, Assam, India

Yamini Baghel
Silviculture and Agroforestry
College of Forestry and Research Station,
Mahatma Gandhi University of Horticulture and Forestry,
Sankara, Patan, Durg, Chhattisgarh, India

Sheetanshu Gupta
Plant Biochemistry, College of Food Technology,
Central Agriculture University, Imphal, Manipur, India

CRC Press
Taylor & Francis Group
Boca Raton London New York

CRC Press is an imprint of the
Taylor & Francis Group, an **informa** business

-EPH-
Elite Publishing House

First published 2027
by CRC Press
4 Park Square, Milton Park, Abingdon, Oxon, OX14 4RN

and by CRC Press
2385 NW Executive Center Drive, Suite 320, Boca Raton FL 33431

CRC Press is an imprint of Informa UK Limited

British Library Cataloguing-in-Publication Data
A catalogue record for this book is available from the British Library

Print edition not for sale in India

ISBN: 9781041372080 (hbk)
ISBN: 9781041378822 (pbk)
ISBN: 9781042011063 (ebk)

DOI: 10.1201/9781042011063

Typeset in A Caslon Pro
by Elite Publishing House

-EPH-

Contents

About the Authors

Dr. Shenaz Sultana Ahmed holds an M.Sc. (Agri.) and Ph.D. (Agri.) in Plant Pathology from the College of Agriculture, Assam Agricultural University, Jorhat, Assam, India. She was awarded the UGC Ishan Uday Fellowship during her B.Sc. (Agri.) program. In her Master's, she received the *Prof. Debeswar Pathak Memorial Gold Medal* and the *Lt. S. Bora Memorial CP Award* for her academic excellence and research contributions. Her outstanding work in plant pathology earned her the *Best Master's Thesis Award* from the Agricultural and Environmental Technology Development Society (AETDS). She is also a qualified *ASRB-NET* scholar. Dr. Ahmed was honored with the prestigious *INSPIRE Fellowship* by the Department of Science and Technology, Government of India, an award granted to the top 1% of Life Science scholars. Her research interests span Plant and Forest Pathology, Food Security, Climate Change, and Sustainable Agriculture. She has authored 11 research papers, 5 book chapters, 2 popular articles, and 2 books, and has received 15 presentation awards at various national and international conferences.

Dr. Yamini Baghel is a committed Assistant Professor at College of Forestry and Research Station, Mahatma Gandhi University of Horticulture and Forestry in Sankara, Patan, Durg, Chhattisgarh. She earned her Ph.D. in Forestry in 2022, showcasing her dedication to advancing the field. Dr. Baghel holds both a Master's and Bachelor's degree in Forestry & Environmental Sciences from Guru Ghasidas Vishwavidyalaya, Bilaspur, Chhattisgarh.

Her academic and research interests are robust, focusing on forest Silviculture, Traditional Agroforestry, Carbon Mitigation, and the Indigenous Tribes of central India. With significant expertise in teaching and research, Dr. Baghel has successfully presented her work at numerous national and international conferences. Her extensive research contributions have been published in several peer-reviewed journals and book chapters, delving into vital topics such as species diversity, nutrient cycling, the role of minor forest products in sustainable livelihoods, urban forests, and Indigenous Traditional Knowledge.

Dr. Baghel's exceptional academic performance was recognized with a Gold Medal during her B.Sc. program. Her ongoing work is instrumental in shaping the discourse surrounding agroforestry, floral diversity, and ethnobotany and their crucial role in climate change mitigation, particularly in the context of India's diverse natural resources and conservation.

Dr. Sheetanshu Gupta is currently working as Assistant Professor (Plant Biochemistry) at the College of Food Technology, Central Agriculture University, Imphal, Manipur. Earlier, he was Examiner of Patent and Design (Gazetted A Officer) in the Ministry of Commerce at the IPO office in Delhi. He was an alumnus of G.B. Pant University of Agriculture & Technology, Pantnagar, Uttarakhand, India from the last 13 years. Dr Gupta completed his Ph D (Biochemistry) in 2011 from G.B. Pant University of Agriculture & Technology, Pantnagar, Uttarakhand, India He cleared ICAR ARS NET (Plant Biochemistry and Plant Physiology) and CSIR NET. Dr. Gupta gained valuable experience in the areas of environmental problems, aeroponics and hydroponics, and protein biochemistry. His work showcased his expertise in applying innovative farming techniques and exploring novel approaches to plant protection. Dr. Gupta researched on tissue culture hydroponics and mushroom farming, contributing to the advancement of these fields. Dr. Gupta's academic qualifications, research experience, and expertise in biochemistry and biotechnology make him a valuable asset in the field. With his passion for teaching and research, dedication to innovative farming techniques, and student advocacy, Dr. Gupta continues to contribute significantly to the advancement of agricultural and biochemical sciences with more than 30 published books, more than 250 chapters, and nearly 20 Research/ Review papers.

Chapter - 1

Basics of Forest Pathology and Entomology

1.1 Definition and Historical Evolution of Forest Pathology and Entomology

Foundations of Forest Pathology and Entomology

Forest pathology and entomology are two critical pillars of forest health science that provide insights into the diverse biological, ecological, and environmental processes that influence the well-being of forest ecosystems. Forest pathology primarily addresses diseases affecting trees and forests, focusing on biological and abiotic agents that cause damage or stress. These agents range from fungi, bacteria, and viruses to environmental factors like pollution and climate variability. In parallel, forest entomology studies the intricate roles of insects within forest ecosystems, highlighting their dual nature as beneficial agents of pollination and decomposition and as destructive pests capable of devastating large forested areas.

Forest pathology examines the cycles of disease, from infection to recovery, emphasizing the interplay between pathogens, host trees, and environmental conditions. This understanding aids in identifying the factors that predispose forests to disease outbreaks and helps in devising mitigation strategies. Similarly, forest entomology delves into insect biology, behavior, and interactions with plants, recognizing the critical roles insects play in nutrient cycling, seed dispersal, and forest succession. However, it also addresses the challenges posed by pest infestations, which can lead to significant economic losses and ecological imbalances. The foundations of forest pathology and entomology are rooted in the observation of natural processes and human interventions in forest ecosystems. These disciplines serve not only as tools for diagnosing and managing forest health issues but also as frameworks for understanding broader ecological patterns. Their integration into forest management strategies underscores their importance in balancing timber production, biodiversity conservation, and ecosystem services. In essence, forest pathology and entomology contribute to the sustainable management of forests by providing the scientific basis for addressing current challenges and anticipating future threats.

Historical Context: From Traditional Knowledge to Modern Science

The evolution of forest pathology and entomology as scientific disciplines mirrors humanity's growing understanding of forest ecosystems and their vital role in sustaining life. In ancient times, the health of

DOI: 10.1201/9781042011063-1

forests was perceived through the lens of traditional and indigenous knowledge systems. These systems, developed over millennia, were rooted in a deep connection to nature and an intuitive understanding of the relationships between trees, insects, and diseases. For example, early forest users recognized symptoms of decline, such as wilting or dieback, and attributed these phenomena to environmental changes, spiritual imbalances, or the influence of unseen forces.

In many traditional societies, forest health management practices were based on empirical observations passed down through generations. Techniques such as selective tree harvesting, controlled burning, and the application of herbal treatments reflected a practical approach to mitigating forest diseases and pests. For instance, indigenous peoples in tropical regions often observed that certain tree species were more resistant to disease and pests, leading to their selective cultivation. These practices, while lacking the scientific rigor of modern approaches, underscored the critical role of forests in providing resources and maintaining ecological balance.

The transition from traditional knowledge to systematic scientific inquiry began in the early modern period, spurred by the increasing economic importance of forests. In Europe, the rise of forestry as a profession during the 17th and 18th centuries marked the beginning of organized efforts to study forest health. Foresters documented tree diseases and insect infestations with growing precision, recognizing their impacts on timber production and forest sustainability. The establishment of forestry schools and research institutions provided a platform for the formalization of forest pathology and entomology as distinct fields of study. The 19th century was a transformative era for forest pathology. The pioneering work of Robert Hartig, a German scientist widely regarded as the "Father of Forest Pathology," laid the foundation for understanding tree diseases as biological phenomena. Hartig's groundbreaking studies on wood decay and root rot revealed the role of fungal pathogens in tree mortality, challenging earlier notions that attributed disease solely to environmental factors. His detailed investigations, published in works such as *Diseases of Trees* (1874), established a scientific framework for diagnosing and managing forest diseases. Hartig's contributions also highlighted the importance of studying the interaction between pathogens, host trees, and environmental conditions, a concept that remains central to forest pathology today.

During the same period, forest entomology emerged as a vital discipline, driven by the need to address insect outbreaks that threatened forest resources. The 19th century saw the rise of entomological studies focusing on the taxonomy, behavior, and ecology of forest insects. Scientists such as Jean-Henri Fabre in France and Charles Valentine Riley in the United States made significant contributions to understanding insect life cycles and their impacts on forests. These studies provided the basis for developing early pest management strategies, including the use of natural predators, chemical controls, and cultural practices to mitigate insect damage. The 20th century witnessed the convergence of forest pathology and entomology with advances in related scientific fields such as microbiology, genetics, and ecology. The discovery of fungal spores as reproductive units, combined with the development of molecular techniques for identifying pathogens, revolutionized forest pathology, enabling more precise diagnoses and targeted interventions. Similarly, advancements in insect physiology and behavior enriched the field of forest entomology, leading to the development of pheromone traps, biological control agents, and integrated pest management (IPM) approaches.

Technological innovations have further propelled the study of forest pathology and entomology into the modern era. Remote sensing, geographic information systems (GIS), and drone technology have transformed the way forest health is monitored and managed, allowing scientists to detect disease outbreaks and pest infestations at landscape scales. Molecular tools such as polymerase chain reaction (PCR) and next-generation sequencing have enhanced the ability to identify pathogens and insects with high accuracy, facilitating rapid responses to emerging threats. The historical evolution of forest pathology and entomology underscores their critical role in addressing the challenges of forest management in a rapidly changing world. From their roots in traditional knowledge to their current status as multidisciplinary scientific fields, these disciplines have continually adapted to the growing complexity of forest ecosystems and the demands of human society. As global forests face unprecedented pressures from climate change, invasive species, and habitat loss, integrating forest pathology and entomology into sustainable management practices remains essential for ensuring the health and resilience of these vital ecosystems.

1.2 Early Understanding of Forest Diseases and Insects

Cultural and Indigenous Perspectives on Forest Health

The early understanding of forest diseases and insects was deeply rooted in cultural and indigenous traditions that often framed forest health through a holistic lens, intertwining ecological, spiritual, and practical knowledge. Indigenous peoples around the world have long coexisted with forests, relying on them not only as a source of material resources but also as a foundation for cultural identity, spiritual beliefs, and communal well-being. These communities observed the natural rhythms of the forest and developed intricate systems of knowledge that recognized the interdependence of all its elements, including trees, insects, pathogens, and climatic factors.

In many indigenous cultures, forests were perceived as sacred entities inhabited by spirits or deities responsible for maintaining ecological balance. Signs of forest decline, such as tree dieback or pest infestations, were often interpreted as manifestations of spiritual displeasure or cosmic imbalance. This spiritual interpretation led to ritualistic practices aimed at appeasing forest deities or restoring harmony to the land. For example, in various tropical forest regions, ceremonies involving offerings, chants, or symbolic acts of purification were conducted to address perceived disturbances in forest health.

Indigenous knowledge systems were also grounded in empirical observations and adaptive practices that effectively managed forest resources. Communities in tropical rainforests, for instance, recognized the varying susceptibility of tree species to disease and pests. This awareness guided selective harvesting practices, the cultivation of resistant species, and the maintenance of diverse forest stands to minimize the spread of pathogens and pests. Similarly, some groups used natural repellents derived from plants or applied ash and oils to protect trees from insect damage, demonstrating a nuanced understanding of pest control long before the advent of chemical pesticides.

The intimate relationship between indigenous peoples and forests allowed them to identify subtle changes in forest health and attribute specific symptoms to environmental factors or biological agents. For example, the presence of certain fungi, discoloration of leaves, or unusual insect activity was recognized as an indicator of underlying stressors. This knowledge was often transmitted orally through stories, songs, and communal teachings, ensuring that successive generations inherited the tools to manage and sustain forest ecosystems. Although these practices lacked the systematic methodology of modern science, they were remarkably effective in addressing the challenges of forest health within the context of traditional livelihoods. The contribution of indigenous and cultural perspectives to the early understanding of forest diseases and insects highlights the value of integrating traditional knowledge with scientific approaches. Many of these practices remain relevant today, offering insights into sustainable forest management, biodiversity conservation, and ecosystem resilience. By recognizing and respecting the ecological wisdom embedded in cultural traditions, modern forestry can foster a more inclusive and holistic approach to addressing forest health challenges.

Early Observations and Misconceptions About Pathogens and Pests

As human societies transitioned from subsistence economies to organized forestry and agriculture, their understanding of forest diseases and insects began to evolve, albeit often constrained by misconceptions and limited scientific knowledge. Early observations of tree decline and pest outbreaks were typically anecdotal and lacked the empirical rigor that characterizes modern forest health sciences. Nevertheless, these observations laid the groundwork for the eventual development of forest pathology and entomology as systematic disciplines.

In ancient civilizations, forest health issues were frequently attributed to supernatural forces or divine intervention. For example, the Greeks and Romans believed that tree dieback was caused by malevolent spirits or the wrath of gods displeased by human actions. This worldview often resulted in ritualistic solutions, such as the consecration of forests, sacrifices to deities, or the avoidance of certain trees deemed cursed or sacred. While these practices reflected a deep respect for nature, they were rooted in mythological frameworks that offered limited practical insight into the biological causes of forest decline. During the Middle Ages, the understanding of forest diseases and pests was further influenced by folklore and superstition. In Europe, the appearance of fungi on tree trunks or the sudden defoliation of a forest stand was sometimes interpreted as an omen of impending disaster. Pest outbreaks, particularly those involving locusts or caterpillars, were often seen as divine punishment or the work of witches and sorcerers. These beliefs hindered the development of scientific approaches to forest health, as they prioritized spiritual explanations over empirical investigation. Despite these misconceptions, early foresters and land managers made important practical observations that contributed to the gradual accumulation of forest health knowledge. For instance, the relationship between environmental conditions and disease outbreaks was noted in various regions. Prolonged periods of drought, waterlogging, or unseasonal weather were often linked to increased tree mortality or pest activity. Similarly, the role of forest density and tree species composition in influencing disease and pest dynamics was recognized, leading to rudimentary forms of forest management aimed at reducing vulnerability.

The Renaissance and early modern period marked a turning point in the understanding of forest diseases and insects, as scientific inquiry began to challenge traditional misconceptions. The invention of the microscope in the 17th century enabled the observation of microorganisms, providing the first glimpse into the world of fungi, bacteria, and other potential disease agents. This technological advancement paved the way for the identification of pathogens as biological entities rather than supernatural phenomena. At the same time, early entomologists began classifying and studying insects in greater detail, recognizing their life cycles, feeding behaviors, and ecological roles. One notable development during this period was the realization that some insects acted as vectors for disease, transmitting pathogens from one tree to another. For example, bark beetles were observed to infest weakened trees, creating entry points for fungal pathogens that caused wood decay. This understanding highlighted the interconnected nature of forest health issues and underscored the need for integrated approaches to disease and pest management. While early observations of forest diseases and insects were often hindered by misconceptions, they also reflected humanity's enduring curiosity and reliance on forests. The transition from mythological interpretations to empirical investigation marked a critical step in the evolution of forest health sciences. By building on the insights of traditional knowledge and early observations, modern forest pathology and entomology have developed into robust scientific disciplines capable of addressing the complex challenges of forest ecosystems in the contemporary era.

1.3 Milestones in the Development of Forest Health Sciences

The Emergence of Forest Pathology as a Scientific Discipline

The establishment of forest pathology as a scientific discipline marks a pivotal chapter in the broader narrative of forest health sciences. Its emergence was driven by the need to understand and address tree diseases that threatened forest resources, timber industries, and ecological stability. While early forest users recognized symptoms of disease and employed intuitive remedies, the scientific foundation of forest pathology began to take shape in the 19th century with the work of pioneering researchers.

One of the most influential figures in the history of forest pathology is Robert Hartig, a German scientist often referred to as the "**Father of Forest Pathology**." Hartig's seminal research in the mid-19th century transformed the understanding of tree diseases by identifying fungi as causal agents of wood decay and root rot. His groundbreaking publication, *Diseases of Trees* (1874), provided the first systematic account of forest diseases, linking observable symptoms to specific pathogens. Hartig's meticulous studies demonstrated that tree diseases were not merely the result of environmental stress but were often driven by biological agents interacting with host trees under specific conditions. His work established the concept of host-pathogen interactions, a cornerstone of modern forest pathology.

Hartig's contributions extended beyond diagnosis to the practical management of tree diseases. He emphasized the importance of preventive measures, such as thinning dense forests to reduce humidity and the risk of fungal infections. His recommendations for controlling root diseases, including the removal of infected stumps and roots, remain relevant in contemporary forestry practices. Hartig's legacy continues to shape the field, with his methods serving as a foundation for the development of diagnostic tools, disease modeling, and forest health management strategies.

The late 19th and early 20th centuries witnessed further advancements in forest pathology, driven by technological innovations and interdisciplinary collaboration. The invention of the light microscope allowed researchers to study the morphology of fungal spores, bacterial cells, and other microscopic agents with unprecedented detail. This breakthrough enabled the identification of new pathogens and the classification of existing ones, enhancing the precision of disease diagnosis. Concurrently, advances in microbiology and plant physiology deepened the understanding of how pathogens invaded trees, disrupted physiological processes, and spread within forest ecosystems. One notable milestone during this period was the recognition that abiotic stressors contribute to tree diseases. Researchers began investigating how drought, pollution, frost, and soil degradation predisposed trees to infection by weakening their defenses. This holistic perspective underscored the importance of environmental management in maintaining forest health, leading to integrated approaches that combined silviculture, pathology, and ecological principles. The 20th century saw the expansion of forest pathology into global contexts, as researchers addressed diseases affecting diverse forest types, from temperate woodlands to tropical rainforests. The discovery and management of diseases such as Dutch elm disease, caused by the fungal pathogen *Ophiostoma ulmi*, demonstrated the devastating impact of invasive pathogens on native tree populations. Efforts to combat such diseases included the development of resistant tree varieties, fungicidal treatments, and quarantine measures to prevent the spread of pathogens across regions. These efforts exemplified the application of forest pathology in protecting biodiversity, supporting timber industries, and conserving ecosystems.

Advances in Forest Entomology and Their Applications

Forest entomology, a complementary field to forest pathology, has undergone remarkable development over the past two centuries, addressing the dual role of insects as both beneficial agents and destructive pests. The systematic study of forest insects began in the 18th and 19th centuries, catalyzed by the economic and ecological impacts of pest outbreaks. Early entomologists sought to classify insect species, understand their life cycles, and determine their interactions with forest trees, laying the groundwork for the scientific discipline of forest entomology.

One of the earliest milestones in forest entomology was the classification and documentation of economically significant forest pests. Entomologists identified bark beetles, defoliators, borers, and sap-sucking insects as major threats to timber production and forest health. For instance, the European spruce bark beetle (*Ips typographus*) was recognized as a key driver of tree mortality in coniferous forests, particularly during periods of drought or stress. Understanding the behavior and population dynamics of such pests enabled foresters to predict and mitigate outbreaks, reducing economic losses and preserving forest resources. The 19th century also witnessed the development of pest control strategies that combined ecological insights with practical interventions. Biological control emerged as a promising approach, involving the introduction of natural predators, parasitoids, and pathogens to regulate pest populations. For example, the use of predatory beetles to control bark beetle infestations demonstrated the potential of harnessing ecological interactions for sustainable pest management. This era also saw the introduction of mechanical and cultural control methods, such as trapping, pruning, and sanitation logging, which targeted specific pest species without relying on chemical interventions.

The 20th century marked a turning point in forest entomology, driven by advances in technology and ecological theory. The development of pheromone traps revolutionized pest monitoring and management, allowing foresters to detect early signs of infestation and implement targeted control measures. Pheromones, chemical signals used by insects for communication, were synthesized and deployed to disrupt mating behaviors or attract pests to traps. This method proved highly effective in managing species such as the pine shoot beetle (*Tomicus piniperda*) and the gypsy moth (*Lymantria dispar*), reducing their impact on forest ecosystems. Integrated pest management (IPM) emerged as a comprehensive framework for addressing forest insect problems, combining biological, chemical, cultural, and mechanical control methods. IPM emphasized the importance of monitoring pest populations, understanding ecological thresholds, and minimizing the use of pesticides to protect non-target organisms and ecosystem health. The adoption of IPM in forestry reflected a broader shift toward sustainable practices, aligning pest management with conservation goals and public health concerns.

The application of molecular biology and genetics in forest entomology has further advanced the field in recent decades. Researchers have used genetic markers to study pest population structures, trace the origins of invasive species, and develop genetically modified trees with enhanced resistance to insect damage. For example, the development of transgenic poplar trees resistant to the cottonwood leaf beetle (*Chrysomela scripta*) demonstrated the potential of genetic engineering in forestry. These innovations, while promising, have also raised ethical and ecological questions, highlighting the need for careful evaluation and regulation.

Another significant development in forest entomology is the use of remote sensing and geographic information systems (GIS) to monitor insect activity at landscape scales. Satellite imagery, aerial surveys, and drone technology have enabled the detection of defoliation, tree mortality, and pest movement patterns across large areas. These tools enhance the ability of forest managers to respond to outbreaks quickly and efficiently, reducing the risk of widespread damage. The integration of forest entomology with climate science has become increasingly important in the context of global environmental change. Rising temperatures, altered precipitation patterns, and increased frequency of extreme weather events have influenced insect behavior, population dynamics, and geographic ranges. For example, the mountain pine beetle (*Dendroctonus ponderosae*) has expanded its range into previously unsuitable areas, driven by warmer winters and stressed forests. Understanding these changes is critical for predicting future pest risks and developing adaptive management strategies.

1.4 The Growing Interconnection Between Pathology and Entomology

The Ecological Nexus: Pathogens, Insects, and Forest Dynamics

The interconnectedness of forest pathology and entomology reflects a broader ecological relationship within forest ecosystems, where pathogens and insects often act in concert to influence forest health and dynamics. This nexus is rooted in the fundamental roles of both groups: pathogens primarily impact forests through disease, altering tree physiology and ecosystem processes, while insects interact with forests both as pests and as key contributors to pollination, decomposition, and nutrient cycling. Together, these agents form a dynamic system of interactions that can either stabilize ecosystems or push them toward instability, depending on environmental conditions and forest resilience.

The relationship between pathogens and insects is complex and multifaceted. In many cases, insects serve as vectors for pathogens, spreading disease from one tree to another and amplifying the scale and severity of outbreaks. For instance, bark beetles, such as the European spruce bark beetle (*Ips typographus*), carry fungal pathogens like *Ophiostoma* species, which cause blue-stain disease. These fungi thrive in the tree's vascular tissues, disrupting water and nutrient transport, and ultimately leading to tree mortality. The symbiotic relationship between the beetle and the fungus exemplifies how insects and pathogens can co-evolve to exploit forest resources effectively.

Another example of this interdependence is the interaction between defoliating insects and secondary pathogens. Insects such as gypsy moths (Lymantria dispar) or forest tent caterpillars (Malacosoma disstria) can weaken trees by stripping their foliage, reducing their photosynthetic capacity and exposing them to abiotic stress. This weakened state often predisposes trees to infection by pathogens, such as root rot fungi (Armillaria spp.), which further reduce the tree's vitality and accelerate its decline. These cascading interactions demonstrate how the combined effects of insects and pathogens can have far-reaching consequences for forest structure and function. Pathogens can also indirectly influence insect populations by modifying tree physiology and chemistry. Trees infected with pathogens often produce secondary metabolites, such as phenolic compounds and terpenes, as a defensive response. While these compounds may deter some herbivorous insects, they can attract others that have adapted to exploit stressed or weakened trees. For example, bark beetles are particularly adept at colonizing trees with compromised defenses, using chemical signals emitted by stressed trees to locate suitable hosts. This feedback loop illustrates the tightly coupled interactions between forest pathology and entomology, where the activity of one directly shapes the dynamics of the other.

The broader ecological implications of these interactions are profound. Pathogen-insect dynamics can shape forest composition, succession, and biodiversity by selectively targeting certain tree species or age classes. For instance, widespread infestations of the mountain pine beetle (*Dendroctonus ponderosae*) in North America, exacerbated by fungal pathogens, have led to the large-scale mortality of lodgepole pine (*Pinus contorta*), altering forest landscapes and ecosystem processes. These changes affect not only timber production but also wildlife habitats, hydrological cycles, and carbon sequestration capacities. The ecological nexus between pathogens, insects, and forest dynamics is essential for predicting and managing forest health outcomes. This requires an integrative approach that considers the roles of environmental factors, such as climate change, in shaping these interactions. Rising temperatures, altered precipitation patterns, and increased frequency of extreme weather events are expected to exacerbate pathogen-insect interactions, creating new challenges for forest management and conservation.

Collaborative Approaches in Research and Management

The growing recognition of the interdependence between forest pathology and entomology has spurred the development of collaborative approaches in research and management. These approaches emphasize the need to integrate knowledge from both disciplines to address the complex challenges posed by forest diseases and insect infestations. By fostering interdisciplinary collaboration, researchers and practitioners can develop more effective strategies for diagnosing, monitoring, and mitigating forest health issues.

One of the key areas of collaboration is the development of integrated pest and disease management (IPDM) frameworks. IPDM integrates the principles of forest pathology and entomology with ecological theory to develop holistic management strategies that minimise environmental impacts while maintaining forest productivity. For example, IPDM approaches often include the use of biological control agents, such as predators, parasitoids, and antagonistic microbes, to regulate pest and pathogen populations. These methods are complemented by cultural practices, such as selective thinning, pruning, and sanitation logging, which reduce the availability of suitable hosts for both pathogens and insects. Collaborative research has also advanced the understanding of the genetic and molecular basis of pathogen-insect interactions. Genomic studies have revealed the mechanisms by which pathogens and insects overcome tree defenses, highlighting potential targets for intervention. For instance, the identification of genes involved in fungal virulence and beetle aggregation has led to the development of pheromone-based traps and resistant tree varieties. These innovations demonstrate the potential of interdisciplinary research to translate basic scientific knowledge into practical solutions for forest health management. Technology plays a central role in enabling collaboration between forest pathologists and entomologists. Remote sensing tools, such as satellite imagery and LiDAR, provide valuable data on forest health at landscape scales, allowing researchers to monitor disease and pest outbreaks in real-time. Geographic information systems (GIS) facilitate the integration of spatial data on pathogen and insect distributions, enabling the identification of hotspots and the prediction of outbreak dynamics. These tools not only enhance the efficiency of forest health monitoring but also foster communication and data sharing among researchers, forest managers, and policymakers.

Education and capacity-building initiatives further support collaborative approaches by training the next generation of forest health professionals in interdisciplinary methods. Workshops, field schools, and joint research programs provide opportunities for pathologists, entomologists, and ecologists to share expertise and develop integrated solutions. These initiatives are particularly important in regions with high biodiversity and limited resources, where the stakes of forest health management are especially high. Policy frameworks also play a crucial role in promoting collaboration between forest pathology and entomology. International agreements, such as the Convention on Biological Diversity (CBD) and the International Plant Protection Convention (IPPC), encourage member states to adopt integrated approaches to forest health management. National and regional policies that prioritize interdisciplinary research, funding, and stakeholder engagement create an enabling environment for collaboration. For example, the European Union's Forest Strategy emphasizes the importance of addressing forest health challenges through cross-disciplinary and cross-border cooperation.

Despite these advancements, challenges remain in achieving effective collaboration between forest pathology and entomology. These include differences in disciplinary priorities, limited funding for interdisciplinary research, and the complexity of translating scientific findings into actionable management practices. Addressing these challenges requires sustained investment in research, education, and infrastructure, as well as the establishment of platforms for dialogue and coordination among stakeholders. The growing interconnection between forest pathology and entomology underscores the need for integrated approaches to forest health science and management. By recognizing the ecological links between pathogens and insects and fostering collaboration across disciplines, researchers and practitioners can develop innovative solutions to protect forests from emerging threats. These efforts are essential for ensuring the resilience and sustainability of forest ecosystems in the face of global environmental change.

2. Pathogens and Insects as Key Agents in Forest Dynamics

2.1 The Dual Role of Pathogens: Regulators and Threats

Pathogens in Natural Ecosystem Balance

Pathogens, often perceived as destructive forces within forest ecosystems, play a far more nuanced role, functioning as both regulators of ecological processes and agents of change. These microorganisms—including fungi, bacteria, viruses, and nematodes—are integral components of forest dynamics, influencing tree health, species interactions, and nutrient cycling. In natural systems, pathogens act as regulators by selectively targeting specific trees or populations, thereby maintaining diversity, preventing dominance by a single species, and promoting forest succession. This regulatory role is vital for sustaining the balance of forest ecosystems, where competition for resources such as light, water, and nutrients can otherwise lead to homogenization and reduced biodiversity.

One of the primary ways pathogens contribute to ecological balance is through density-dependent effects, where the likelihood of disease increases with the density of susceptible hosts. This phenomenon, often referred to as the "Janzen-Connell hypothesis," suggests that pathogens limit the proliferation of dominant tree species by increasing mortality near parent trees, where pathogen load is highest. As a result, space and resources are freed for other species, fostering heterogeneity and enabling the coexistence of diverse plant communities. For instance, soil-borne pathogens like *Phytophthora spp.* and *Pythium spp.* are known to affect seedling survival in tropical forests, promoting the establishment of less common tree species and contributing to the forest's overall resilience.

Pathogens also play a crucial role in nutrient cycling and decomposition, particularly in old-growth and unmanaged forests. Fungal pathogens, for example, facilitate the breakdown of deadwood and organic matter, releasing nutrients back into the soil and enhancing its fertility. This process supports the growth of new vegetation and maintains the productivity of the forest ecosystem. Saprotrophic fungi, such as species of *Armillaria* and *Ganoderma*, exemplify this dual role, functioning as decomposers in some contexts and as pathogens in others, depending on host health and environmental conditions. In addition to their ecological functions, pathogens contribute to forest resilience by acting as selective pressures that drive evolutionary adaptation. Trees exposed to pathogens over generations develop resistance mechanisms, such as the production of secondary metabolites, lignification of tissues, and hypersensitive responses to infection. These adaptations not only enhance individual tree survival but also shape the genetic diversity of forest populations, ensuring their capacity to withstand future disturbances. However, the regulatory role of pathogens is highly context-dependent, influenced by factors such as forest type, climate, and human activities. In undisturbed forests, pathogens typically function as part of a self-regulating system, maintaining balance without causing widespread destruction. In contrast, anthropogenic influences, such as habitat fragmentation, pollution, and climate change, can disrupt these natural dynamics, exacerbating pathogen impacts and pushing ecosystems toward decline.

The Role of Pathogens in Forest Decline and Recovery

While pathogens are vital for ecosystem balance, their role as agents of forest decline becomes apparent when their natural regulatory functions are overwhelmed by external pressures. Under certain conditions, pathogens can cause significant damage, leading to tree mortality, loss of canopy cover, and disruptions in ecosystem processes. These conditions often arise when forests are stressed by factors such as drought, poor soil quality, or human-induced changes, which weaken tree defenses and increase susceptibility to infection.

Forest decline associated with pathogens is frequently a multifactorial process involving interactions between biotic and abiotic stressors. For example, root rot fungi such as *Heterobasidion annosum* thrive in stressed coniferous forests, causing root decay that compromises tree stability and water uptake. Similarly, *Phytophthora ramorum*, the pathogen responsible for sudden oak death, has devastated oak and tanoak populations in North America, with environmental factors such as high humidity and mild temperatures facilitating its spread. These outbreaks not only affect individual trees but also alter forest composition and structure, with cascading effects on wildlife habitats, hydrology, and carbon storage. Despite their destructive potential, pathogens also play a critical role in forest recovery by shaping successional pathways and facilitating regeneration. In the aftermath of pathogen-induced tree mortality, light and space become available for understory vegetation, pioneer species, and seedling establishment. This process contributes to the natural renewal of forests, albeit with a shift in species composition and structure. For instance, the decline of American chestnut (*Castanea dentata*) due to chestnut blight (*Cryphonectria parasitica*) in the early 20th century led to the proliferation of oak, hickory, and other hardwood species in eastern U.S. forests, illustrating how pathogen impacts can drive long-term ecological change.

The capacity for forests to recover from pathogen outbreaks depends on several factors, including the availability of resistant or tolerant species, the presence of genetic diversity, and the mitigation of external stressors. Reforestation efforts, such as the breeding and planting of disease-resistant tree varieties, play a pivotal role in facilitating recovery. Advances in genetic engineering and molecular biology have enabled the development of transgenic trees with enhanced resistance to pathogens, offering new possibilities for restoring forest health. For example, efforts to reintroduce blight-resistant American chestnut trees demonstrate the potential of combining traditional breeding techniques with modern biotechnology to address pathogen challenges.

Climate change adds another layer of complexity to the role of pathogens in forest decline and recovery. Rising temperatures, altered precipitation patterns, and increased frequency of extreme weather events can influence pathogen behavior, host susceptibility, and the interactions between them. Warmer climates, for instance, may expand the range of pathogens traditionally confined to tropical or subtropical regions, exposing temperate and boreal forests to new threats. Additionally, prolonged droughts and heatwaves can weaken tree defenses, creating conditions conducive to pathogen outbreaks. These changes highlight the need for adaptive management strategies that account for the evolving role of pathogens in forest dynamics.

2.2 Functional Importance of Insects in Forest Ecosystems

Pollination, Nutrient Cycling, and Other Ecological Roles

Insects are foundational to the structure and functioning of forest ecosystems, performing a multitude of ecological roles that sustain biodiversity, promote forest productivity, and ensure the stability of ecosystem services. Among their most critical contributions are pollination, nutrient cycling, seed dispersal, and decomposition. These processes collectively influence tree regeneration, forest composition, and the flow of energy and nutrients within ecosystems. The significance of insects extends beyond forests, as their activities also impact global ecological processes and the resilience of ecosystems to environmental change.

Pollination is perhaps the most widely recognized ecological service provided by insects in forests. A significant proportion of tree species, particularly those in tropical and subtropical forests, rely on insects for pollination. Bees, butterflies, beetles, and flies are among the most prominent insect pollinators, transferring pollen between flowers and facilitating sexual reproduction in plants. This process is essential for the production of seeds and fruits, which serve as food sources for a wide range of forest-dwelling organisms. For example, in tropical rainforests, species like the orchid bee (*Euglossa spp.*) play a critical role in pollinating orchids and other plants, ensuring their survival and propagation. Similarly, beetles are primary pollinators of many economically important tree species, such as the African oil palm (*Elaeis guineensis*).

Beyond pollination, insects contribute significantly to nutrient cycling through their roles as decomposers and consumers. Detritivorous insects, such as termites, wood-boring beetles, and various fly larvae, break down organic matter, including deadwood, leaf litter, and animal remains. This decomposition process releases essential nutrients back into the soil, enriching it and supporting the growth of new vegetation. Termites, for instance, are renowned for their ability to decompose cellulose, transforming coarse organic matter into forms that are more readily assimilated by soil microorganisms and plants. By accelerating the decomposition of organic material, these insects play a pivotal role in maintaining soil fertility and forest productivity.

Seed dispersal is another key function performed by certain insects, such as ants and beetles, which interact with seeds in ways that facilitate their germination and establishment. Many ants participate in a process known as myrmecochory, where they transport seeds to their nests, consume the nutrient-rich appendages (elaiosomes), and leave the seeds in nutrient-rich refuse piles. This interaction benefits both ants and plants, ensuring the spread of seeds to favorable germination sites while providing ants with a valuable food resource. Such interactions exemplify the mutualistic relationships that underpin forest ecosystems. In addition to these direct contributions, insects also serve as a critical food source for a wide range of forest species, including birds, mammals, reptiles, and amphibians. The abundance and diversity of insects support complex food webs, with cascading effects on higher trophic levels. For instance, caterpillars, which are rich in protein, are a primary food source for many forest birds during the breeding season, influencing avian reproductive success and population dynamics. The loss of insect diversity or abundance can therefore have far-reaching implications for forest biodiversity and ecosystem resilience.

Insect-Mediated Stress on Forest Health: Defoliators, Borers, and Suckers

While insects perform indispensable roles in forest ecosystems, they can also become significant stressors under certain conditions, particularly when their populations reach outbreak levels. In such cases, insects transition from being integral ecosystem components to agents of forest health decline, causing widespread damage to trees and altering ecosystem processes. Defoliators, borers, and sap-sucking insects are among the most impactful groups of forest pests, each exerting unique pressures on trees and forest ecosystems.

Defoliators, as the name suggests, feed on the leaves of trees, reducing their photosynthetic capacity and weakening their overall health. Prominent examples include gypsy moths (*Lymantria dispar*), forest tent caterpillars (*Malacosoma disstria*), and spruce budworms (*Choristoneura fumiferana*). During outbreaks, these insects can strip entire forest stands of their foliage, causing significant stress to trees and increasing their susceptibility to secondary pathogens and pests. Repeated defoliation over consecutive years often results in tree mortality, with cascading effects on forest structure, biodiversity, and ecosystem services. For instance, the severe defoliation caused by gypsy moth outbreaks in North American oak forests has led to substantial changes in forest composition, favoring less palatable or more resistant tree species.

Borers, including bark beetles and wood-boring insects, pose another major threat to forest health. These insects typically infest tree trunks, branches, or roots, disrupting vascular tissues and impeding the flow of water and nutrients. Bark beetles, such as the mountain pine beetle (*Dendroctonus ponderosae*), are particularly notorious for their ability to cause large-scale tree mortality. By boring into the phloem and cambium layers, these beetles create galleries that interfere with nutrient transport, eventually killing the tree. The damage caused by borers is often exacerbated by their association with fungal pathogens, which colonize the tree's vascular system and accelerate its decline. For example, blue-stain fungi, carried by bark beetles, contribute to wood discoloration and structural degradation, reducing the commercial value of timber and complicating forest recovery efforts.

Sap-sucking insects, such as aphids, scale insects, and whiteflies, feed on the phloem sap of trees, extracting nutrients and water while introducing stress. These insects are particularly problematic for young trees and seedlings, which have limited energy reserves to withstand prolonged feeding. In addition to directly weakening trees, sap-sucking insects can serve as vectors for plant pathogens, spreading viruses, bacteria, or fungi. For instance, adelgids, such as the hemlock woolly adelgid (*Adelges tsugae*), have caused extensive declines in hemlock forests in eastern North America by feeding on tree sap and facilitating secondary infections. The impacts of insect-mediated stress on forest health are often amplified by environmental and anthropogenic factors. Climate change, for example, has altered the behavior, distribution, and population dynamics of many insect species, leading to more frequent and severe outbreaks. Warmer temperatures have allowed pests such as the mountain pine beetle to expand into higher altitudes and latitudes, infesting previously unaffected forests. Similarly, habitat fragmentation and monoculture forestry have created conditions that favor pest outbreaks, as these practices reduce the natural diversity and complexity that buffer ecosystems against insect proliferation.

Addressing the dual role of insects in forest ecosystems requires a nuanced approach that balances their ecological contributions with the need to mitigate their negative impacts. Integrated pest management (IPM) strategies offer a promising framework for achieving this balance, combining biological control, cultural practices, and targeted interventions to manage insect populations sustainably. For example, the introduction of natural predators, such as parasitoid wasps, has been effective in controlling defoliator outbreaks, while pheromone traps have proven useful in monitoring and managing bark beetle infestations.

2.3 Synergistic Interactions Between Pathogens and Pests

Case Studies of Pathogen-Pest Interactions in Major Forests

The interactions between pathogens and pests in forest ecosystems are often synergistic, meaning their combined effects exceed the sum of their impacts. These interactions can dramatically influence forest health, structure, and composition, often resulting in widespread tree mortality, changes in species dynamics, and long-lasting ecological consequences. Case studies from forests around the world illustrate the profound effects of these interactions and highlight the complexity of managing forest health in the face of such challenges.

One of the most widely studied examples of pathogen-pest interaction is the relationship between bark beetles (*Dendroctonus spp.*) and blue-stain fungi (*Ophiostoma* species) in conifer forests of North America and Europe. Bark beetles bore into the phloem and cambium layers of trees, disrupting nutrient transport and creating galleries where they lay eggs. As they tunnel through the tree, they introduce spores of blue-stain fungi, which colonize the vascular tissues and rapidly spread within the tree. This dual assault overwhelms the tree's defenses, causing death within weeks or months. The mountain pine beetle (*Dendroctonus ponderosae*), a key player in this interaction, has devastated millions of hectares of lodgepole pine (*Pinus contorta*) forests in western North America, with outbreaks exacerbated by climate change and fire suppression policies. The symbiotic relationship between the beetles and the fungi highlights the evolutionary adaptation of pests and pathogens to exploit forest resources cooperatively.

A similar synergistic interaction is observed in oak forests of the eastern United States, where the gypsy moth (*Lymantria dispar*) and root rot fungi (*Armillaria spp.*) have contributed to widespread oak decline. Gypsy moth caterpillars are voracious defoliators, stripping trees of their foliage and severely reducing their photosynthetic capacity. The stress induced by repeated defoliation renders trees vulnerable to infection by *Armillaria* fungi, which invade root systems and cause decay. The combination of defoliation and root rot often leads to tree mortality, with cascading effects on forest structure and biodiversity. The oak decline syndrome exemplifies how pests and pathogens can exploit weakened trees, amplifying their individual impacts and accelerating forest degradation. In tropical forests, similar interactions between pathogens and pests have been documented, though they are often less studied due to the complexity and diversity of these ecosystems. One notable example is the interaction between leaf-cutting ants (*Atta spp.*) and fungal pathogens in Central and South American rainforests. Leaf-cutting ants harvest leaves to cultivate fungal gardens, which serve as their primary food source. However, these ants also create wounds on trees that act as entry points for opportunistic pathogens, such as *Phytophthora* and *Fusarium* species. These pathogens can cause localized infections that weaken trees, creating vulnerabilities for further pest or disease attacks. Such interactions underscore the interconnectedness of forest organisms and the importance of studying multiple agents of disturbance together.

Co-Infections and Multispecies Infestations

The phenomenon of co-infections and multispecies infestations further demonstrates the amplified impacts of pathogen-pest interactions on forest ecosystems. Co-infections occur when a tree is simultaneously or sequentially attacked by multiple pathogens, pests, or both, leading to compounded stress and a heightened likelihood of mortality. These complex interactions often arise in environments where ecological disturbances, such as climate change, logging, or habitat fragmentation, create conditions conducive to the proliferation of multiple agents of forest decline.

Co-infections involving pathogens are particularly prevalent in stressed forests, where weakened trees are unable to mount effective defenses against multiple attackers. For example, in eucalyptus plantations in Australia, trees infected with the fungal pathogen *Cylindrocladium scoparium* often become more susceptible to infestation by eucalyptus beetles (*Chrysomelidae*), which further damage foliage and exacerbate stress. The combined effects of fungal infection and insect herbivory result in reduced growth, increased mortality, and decreased timber yields, illustrating the economic and ecological implications of such interactions. Multispecies infestations involving multiple pests are another form of amplified impact that poses significant challenges for forest management. In the boreal forests of Canada, the interaction between spruce budworms (*Choristoneura fumiferana*) and secondary bark beetles has been identified as a key driver of forest decline. Spruce budworms defoliate spruce and fir trees, creating stress conditions that attract bark beetles, which infest and kill weakened trees. The synergistic effects of defoliation and bark beetle infestation accelerate tree mortality and create large-scale gaps in forest cover. These gaps alter microclimatic conditions, soil properties, and wildlife habitats, demonstrating the far-reaching consequences of multispecies interactions.

The impacts of co-infections and multispecies infestations are often amplified by environmental factors, particularly climate change. Warmer temperatures and prolonged droughts weaken tree defenses and expand the ranges of many pests and pathogens, increasing the likelihood of overlapping outbreaks. For instance, in California's mixed-conifer forests, drought-stressed trees have experienced simultaneous attacks by bark beetles, fungal pathogens like *Fusarium circinatum*, and parasitic mistletoes (*Arceuthobium spp.*). These compounded stressors have resulted in extensive tree mortality and raised concerns about the long-term viability of these forests under changing climatic conditions.

The ecological consequences of co-infections and multispecies infestations extend beyond individual tree mortality. These interactions can alter species composition, shift successional trajectories, and disrupt ecosystem services such as carbon sequestration and water regulation. For example, the loss of dominant tree species due to pathogen-pest interactions can create opportunities for invasive species to establish and spread, further complicating forest recovery. Additionally, the large amounts of deadwood generated by widespread tree mortality can increase fuel loads, raising the risk of wildfires and further disrupting forest ecosystems. Addressing the challenges posed by synergistic interactions between pathogens and pests requires an integrative approach that combines monitoring, modeling, and management. Advanced diagnostic tools, such as DNA-based assays and remote sensing technologies, can enhance the early detection of co-infections and multispecies infestations, enabling timely intervention. Predictive models that incorporate climatic, biological, and ecological data can help forecast outbreak dynamics and identify potential hotspots. These tools are particularly valuable for proactive forest management in the context of climate change and other global stressors.

Management strategies should focus on enhancing forest resilience by promoting diversity, reducing stress, and mitigating the spread of pathogens and pests. Silvicultural practices, such as selective thinning, mixed-species planting, and the removal of infected material, can reduce the density of susceptible hosts and disrupt the transmission pathways of pathogens and pests. Biological control agents, such as natural predators, parasitoids, and antagonistic microbes, can also be employed to regulate pest and pathogen populations sustainably. Additionally, fostering collaboration between forest pathologists, entomologists, and ecologists is essential for developing holistic solutions that address the multifaceted nature of pathogen-pest interactions.

2.4 Ecological Impacts of Pathogen and Pest Outbreaks

Disruptions in Forest Succession and Habitat Stability

Pathogen and pest outbreaks are among the most significant natural disturbances that disrupt forest succession and habitat stability, altering the long-term trajectory of forest ecosystems. These outbreaks often trigger a cascade of ecological changes, starting with the immediate mortality or decline of host trees and extending to the alteration of forest structure, species composition, and ecosystem processes.

The disruption caused by pathogens and pests can vary in scale and severity, but their cumulative impacts shape the ecological dynamics of forests in profound ways.

One of the primary disruptions caused by pathogen and pest outbreaks is the interruption of forest succession, the natural process by which forests develop and mature over time. For example, in temperate forests, defoliating pests like the gypsy moth (*Lymantria dispar*) or fungal pathogens such as *Cryphonectria parasitica* (the causal agent of chestnut blight) target dominant tree species, reducing their abundance and creating openings in the canopy. These gaps allow light to penetrate to the forest floor, enabling the establishment of shade-intolerant pioneer species that can outcompete slower-growing, shade-tolerant trees. This shift in species composition can delay or redirect successional pathways, leading to forests dominated by different tree species than those present before the outbreak. In the case of chestnut blight, the near-total loss of American chestnuts (*Castanea dentata*) in eastern North America permanently altered forest composition, with oaks, hickories, and maples filling the ecological void left by the chestnut.

The structural changes induced by outbreaks also destabilize habitats and disrupt the ecological functions of forests. Tree mortality caused by bark beetles, such as the mountain pine beetle (*Dendroctonus ponderosae*), results in the accumulation of standing deadwood and fallen logs, which change the physical structure of the forest. While these structural changes provide new habitats for certain species, such as cavity-nesting birds and decomposers, they also make forests more susceptible to secondary disturbances, such as wildfires and soil erosion. The loss of canopy cover further alters microclimatic conditions, increasing temperature fluctuations and reducing moisture levels on the forest floor. These changes affect seed germination, understory vegetation growth, and the survival of shade-dependent species, perpetuating a cycle of ecological instability.

In tropical forests, where biodiversity and ecological interactions are particularly complex, pathogen and pest outbreaks can have even more pronounced effects on forest succession. Pathogens like *Phytophthora* species or pests such as leaf-cutting ants (*Atta spp.*) can decimate specific tree populations, disrupting mutualistic relationships between trees and their associated fauna, such as pollinators and seed dispersers. The resulting decline in tree diversity not only alters the successional dynamics but also reduces the forest's capacity to recover from subsequent disturbances, such as storms or droughts. Additionally, the loss of keystone species due to outbreaks can trigger cascading effects throughout the ecosystem, affecting the survival and behavior of other organisms.

The destabilization of habitats caused by pathogen and pest outbreaks has far-reaching consequences for ecosystem services, including carbon sequestration, water regulation, and soil fertility. Forests undergoing rapid structural and compositional changes often experience declines in biomass and productivity, reducing their ability to capture and store carbon. Similarly, the loss of root systems in affected trees increases soil erosion and nutrient runoff, impacting water quality and downstream ecosystems. These disruptions underscore the interconnectedness of forest health and the broader landscape, highlighting the importance of understanding and managing pathogen and pest dynamics to maintain ecosystem stability.

The Cascading Effects on Wildlife and Plant Biodiversity

The ecological impacts of pathogen and pest outbreaks extend beyond the immediate effects on trees and forest structure to influence wildlife and plant biodiversity. The loss of host trees and the changes in habitat structure caused by outbreaks disrupt species interactions, food webs, and community dynamics, often with cascading effects that reverberate through the ecosystem. These changes can result in shifts in species distributions, population declines, and, in some cases, the local extinction of specialized organisms.

Wildlife species that depend on affected trees for food, shelter, or reproduction are particularly vulnerable to pathogen and pest outbreaks. For instance, the decline of hemlock trees (*Tsuga canadensis*) in eastern North America due to the hemlock woolly adelgid (*Adelges tsugae*) has significantly impacted wildlife communities that rely on hemlock-dominated forests. Bird species such as the black-throated green warbler (*Setophaga virens*), which prefers hemlock habitats, have experienced population declines as these forests transition to hardwood dominance. Similarly, aquatic ecosystems downstream from hemlock stands have been affected by changes in leaf litter inputs and stream shading, altering the habitat and food availability for fish and invertebrates.

The cascading effects of outbreaks also influence plant biodiversity by altering competition dynamics and resource availability. When dominant tree species are removed from the forest canopy, the increased light and space create opportunities for invasive plant species to establish and spread. These invasive species often outcompete native vegetation, reducing overall plant diversity and homogenizing the forest understory. For example, the loss of ash trees (*Fraxinus spp.*) in North America due to the emerald ash borer (*Agrilus planipennis*) has facilitated the invasion of non-native species such as buckthorn (*Rhamnus spp.*) and honeysuckle (*Lonicera spp.*), which thrive in disturbed habitats.

The interactions between wildlife and plant biodiversity are further complicated by the role of pathogens and pests in mediating species-specific effects. Certain tree species that are disproportionately affected by outbreaks serve as keystone species, meaning their loss has a disproportionate impact on ecosystem structure and function. For example, the decline of whitebark pine (*Pinus albicaulis*) in western North America due to white pine blister rust (*Cronartium ribicola*) and mountain pine beetles has affected grizzly bears (*Ursus arctos horribilis*), which rely on whitebark pine seeds as a critical food source. The loss of this keystone resource has forced grizzlies to shift their diets, increasing competition with other species and altering predator-prey dynamics. Pathogen and pest outbreaks also have indirect effects on wildlife through changes in ecosystem services. For example, the reduction in canopy cover caused by outbreaks increases exposure to predators for many ground-dwelling species, such as small mammals and amphibians. Additionally, the accumulation of deadwood and debris provides habitat for some species but can also increase the risk of wildfires, further altering habitat availability and quality. These compounded effects create complex feedback loops that influence the stability and resilience of wildlife populations. In the context of global environmental change, the impacts of pathogen and pest outbreaks on biodiversity are likely to become more pronounced. Climate change is altering the geographic ranges and population dynamics of both pests and their host species, creating novel interactions and exacerbating existing pressures on forests. The introduction of invasive species through global trade and travel further complicates these dynamics, as many pathogens and pests lack natural predators or competitors in their new environments. These trends underscore the need for proactive monitoring and management strategies that address the interconnected challenges of biodiversity loss, forest health, and climate resilience.

3.1 Emerging Patterns in Disease and Pest Outbreaks

Climate Change as a Driver of New Outbreaks

Climate change is reshaping the landscape of forest pathology and entomology, acting as a powerful driver of emerging disease and pest outbreaks. The increasing frequency of extreme weather events, rising temperatures, and altered precipitation patterns are creating conditions that favor the proliferation and spread of pathogens and pests in forest ecosystems. These changes are not only altering the geographic distribution of many species but also intensifying their impacts, posing unprecedented challenges to forest health and management.

Rising global temperatures have extended the range of pathogens and pests into regions once inhospitable. For instance, the mountain pine beetle (Dendroctonus ponderosae), once confined to lower elevations, has expanded its range into higher altitudes and latitudes due to warmer winters. This shift has resulted in large-scale infestations of lodgepole pine (Pinus contorta) and whitebark pine (Pinus albicaulis) forests in western North America, ecosystems that were historically protected by cold temperatures. Similarly, pathogens such as Phytophthora cinnamomi, a root rot agent, are expected to invade cooler regions as temperatures rise, threatening forests in temperate and boreal zones. The phenology, or timing of biological events, of pathogens and pests is also being altered by climate change, leading to mismatches in ecological interactions and compounding forest health problems. Many insect pests are completing additional generations within a single year, leading to higher population densities and more severe outbreaks. For example, the spruce budworm (*Choristoneura fumiferana*) in Canada has exhibited faster life cycles, resulting in greater defoliation of spruce and fir trees. Similarly, earlier spring warming has allowed pathogens like *Cryphonectria parasitica*, responsible for chestnut blight, to become active earlier in the growing season, increasing the duration of infection periods.

Changes in precipitation patterns are another critical factor driving new disease and pest outbreaks. Drought stress, exacerbated by climate change, weakens tree defenses and makes forests more susceptible to attack. For instance, prolonged droughts in California have intensified the impacts of bark beetles and fungal pathogens on pine and oak forests, resulting in widespread mortality. Conversely, increased rainfall and humidity in some regions create favorable conditions for foliar diseases, such as needle cast in conifer plantations and leaf spots in hardwood species. These shifts highlight the intricate connections between climate variability and the epidemiology of forest pathogens and pests.

The interaction between climate change and other environmental factors amplifies the complexity of managing forest health. For example, warmer temperatures can enhance the virulence of certain pathogens, increasing their ability to infect host trees. At the same time, climate-driven changes in host tree physiology, such as reduced resin production, further exacerbate susceptibility. These feedback loops underscore the need for adaptive management strategies that incorporate climate projections and prioritize the resilience of forest ecosystems.

The Role of Anthropogenic Activities in Spread and Intensification

Human activities have played a central role in the emergence and intensification of disease and pest outbreaks in forests, often acting in tandem with climate change to exacerbate their impacts. Globalization, land-use change, and forestry practices have created pathways for the introduction and spread of invasive species, disrupted natural ecological balances, and heightened the vulnerability of forests to pathogens and pests.

Global trade and travel have facilitated the movement of pathogens and pests across continents, introducing species into regions where they lack natural predators or competitors. For example, the emerald ash borer (*Agrilus planipennis*), native to Asia, was inadvertently introduced to North America through infested wood packaging materials. Since its discovery in Michigan in 2002, this beetle has decimated ash tree populations across the United States and Canada, causing billions of dollars in economic losses and permanently altering forest composition. Similarly, the fungal pathogen responsible for Dutch elm disease (*Ophiostoma ulmi*) was introduced to North America and Europe through timber imports, leading to widespread mortality of elm trees (*Ulmus spp.*). Urbanization and land-use change have further contributed to the spread and intensification of forest diseases and pests. The fragmentation of forests due to agricultural expansion, road construction, and urban development creates edge effects that expose trees to environmental stress and facilitate the movement of pathogens and pests. Fragmented landscapes also disrupt ecological interactions, such as predator-prey dynamics, that naturally regulate pest populations. For instance, the proliferation of scale insects in fragmented oak forests of California has been linked to reduced abundance of natural predators, such as parasitic wasps.

Intensive forestry practices, including monoculture plantations and short rotation cycles, have amplified the vulnerability of forests to disease and pest outbreaks. Monocultures, where a single tree species is planted over large areas, provide ideal conditions for the rapid spread of host-specific pathogens and pests. For example, plantations of radiata pine (*Pinus radiata*) in New Zealand have been heavily impacted by *Dothistroma* needle blight, a fungal disease that thrives in the dense, uniform stands of this species. Similarly, eucalyptus plantations in Brazil and India have faced repeated outbreaks of pests like the eucalyptus gall wasp (*Leptocybe invasa*), highlighting the risks associated with homogeneity in forest management.

The intensification of disease and pest outbreaks is also linked to the introduction of novel stressors, such as pollution and pesticide use. Air pollution, including ozone and sulfur dioxide, weakens tree defenses and predisposes them to infection. For instance, polluted environments have been associated with increased susceptibility of beech trees (*Fagus sylvatica*) to bark beetles and canker fungi. The indiscriminate use of pesticides, while aimed at controlling pests, can disrupt non-target organisms, including natural enemies of pests, creating ecological imbalances that worsen outbreak dynamics. Anthropogenic activities not only drive the spread of diseases and pests but also complicate their management by creating socio-economic and logistical challenges. The global nature of these problems requires coordinated efforts across jurisdictions, yet variations in regulatory frameworks, enforcement capacities, and economic priorities often hinder effective action. For example, while some regions have implemented strict phytosanitary measures to prevent the import of invasive species, others lack the infrastructure or political will to enforce such regulations, perpetuating the cycle of introductions and outbreaks. Emerging patterns in forest disease and pest outbreaks underscore the interconnectedness of ecological, climatic, and human factors in shaping forest health. Addressing these challenges requires a multi-faceted approach that combines scientific research, policy interventions, and public engagement. Efforts to mitigate the impacts of climate change, enhance biosecurity, and promote sustainable forestry practices are critical for reducing the risks associated with emerging pathogens and pests.

By understanding the drivers of these outbreaks and their ecological consequences, forest managers can develop proactive strategies to safeguard the health and resilience of global forest ecosystems.

3.2 Advances in Diagnostic Technologies and Surveillance Systems

Remote Sensing and AI in Forest Disease Detection

The rapid evolution of diagnostic technologies and surveillance systems has transformed forest health monitoring, enabling researchers and forest managers to detect and respond to disease and pest outbreaks with unprecedented speed and accuracy. Among the most impactful advancements are remote sensing technologies and artificial intelligence (AI), which have revolutionized the ability to monitor forest health at local, regional, and global scales. These innovations provide critical tools for early detection, outbreak prediction, and the assessment of long-term ecological impacts.

Remote sensing involves the use of satellite imagery, aerial photography, and drone-based systems to collect data on forest conditions without direct contact. This approach offers numerous advantages, including the ability to monitor vast and often inaccessible forest areas, detect subtle changes in vegetation health, and identify potential hotspots of disease and pest activity. High-resolution satellite imagery from platforms such as Landsat, Sentinel-2, and MODIS has been instrumental in detecting signs of stress, such as canopy discoloration, defoliation, and tree mortality, that are indicative of pathogen or pest presence. For example, satellite imagery has been used to map the spread of sudden oak death (*Phytophthora ramorum*) in California, revealing spatial patterns of disease progression and informing targeted management efforts.

Drones have added a new dimension to remote sensing by providing high-resolution, close-range observations of forest health. Equipped with advanced sensors, including multispectral and hyperspectral cameras, drones can capture detailed information about tree physiology, such as chlorophyll content, water stress, and leaf area index. These data allow for the precise identification of disease symptoms and the differentiation of biotic and abiotic stress factors. In the context of pest management, drones have been used to monitor bark beetle infestations in conifer forests, detecting early signs of attack through subtle changes in tree color and infrared reflectance.

Artificial intelligence (AI) has emerged as a powerful complement to remote sensing technologies, enhancing the capacity to analyze large and complex datasets generated by these systems. Machine learning algorithms, a subset of AI, are particularly effective in identifying patterns and anomalies that indicate disease or pest presence. For instance, deep learning models trained on satellite or drone imagery can distinguish healthy from diseased trees with high accuracy, even when visual differences are minimal. These models can also predict the spread of outbreaks by analyzing environmental variables, host tree distributions, and historical outbreak data. For example, AI-driven tools have been used to model the potential expansion of the mountain pine beetle (Dendroctonus ponderosae) under different climate scenarios, aiding in proactive management planning. The integration of remote sensing and AI has also enabled real-time monitoring and early warning systems for forest health. Networks of sensors deployed in forests can transmit continuous data on temperature, humidity, soil moisture, and other variables that influence pathogen and pest dynamics. These data are analyzed by AI algorithms to detect deviations from normal conditions that may signal an emerging outbreak. Such systems have been implemented in eucalyptus plantations to monitor fungal diseases, allowing for timely interventions that prevent large-scale losses. Despite their potential, the implementation of remote sensing and AI technologies faces challenges, including high costs, technical expertise requirements, and data accessibility. Collaboration among researchers, governments, and private sector stakeholders is essential to overcome these barriers and maximize the benefits of these tools. By providing actionable insights into forest health, remote sensing and AI are reshaping the way forests are managed in the face of mounting environmental and biological threats.

Molecular Tools for Rapid Pathogen and Pest Identification

The advent of molecular diagnostic tools has revolutionized the field of forest pathology and entomology, enabling the rapid and accurate identification of pathogens and pests at the genetic level. These tools address one of the critical challenges in forest health management: distinguishing between similar-looking species or strains that may have vastly different ecological impacts or management requirements. Molecular diagnostics has become an indispensable component of disease and pest surveillance, facilitating early detection, biosecurity measures, and targeted interventions.

One of the most widely used molecular techniques in forest health diagnostics is polymerase chain reaction (PCR), which amplifies specific DNA or RNA sequences to detect the presence of pathogens or pests. PCR-based assays are highly sensitive and can identify organisms even in low concentrations, making them ideal for early-stage detection. For example, PCR has been used to detect *Phytophthora cinnamomi*, a root rot pathogen, in soil and water samples before visible symptoms appear on host trees. This capability allows for the implementation of quarantine measures to prevent the pathogen's spread. Real-time PCR, also known as quantitative PCR (qPCR), has further enhanced diagnostic capabilities by quantifying pathogen or pest load in samples. This information is critical for assessing the severity of infections or infestations and determining appropriate management responses. For instance, qPCR has been employed to measure the fungal load of *Cryphonectria parasitica* in chestnut trees, providing insights into disease progression and resistance levels. Similarly, qPCR assays have been developed to detect the DNA of invasive pests, such as the emerald ash borer (*Agrilus planipennis*), in wood samples, aiding in regulatory inspections and biosecurity efforts.

Next-generation sequencing (NGS) represents a significant leap forward in molecular diagnostics, allowing for the simultaneous identification of multiple organisms in a single sample. NGS technologies, such as Illumina and Oxford Nanopore, can sequence entire genomes or transcriptomes, revealing detailed information about the genetic makeup, virulence factors, and evolutionary history of pathogens and pests. This approach has been particularly valuable in identifying novel or cryptic species that are difficult to detect using traditional methods. For example, NGS was instrumental in characterizing the genomes of *Ophiostoma ulmi* and *Ophiostoma novo-ulmi*, the fungal pathogens responsible for Dutch elm disease, shedding light on their mechanisms of pathogenicity and resistance to control measures.

Metagenomics, an application of NGS, has expanded the scope of molecular diagnostics to include entire microbial and insect communities. By analyzing environmental DNA (eDNA) from soil, water, or air samples, metagenomics provides a comprehensive snapshot of the biodiversity present in a given area. This technique has been used to monitor pathogen and pest diversity in forest ecosystems, identifying potential threats before they cause noticeable damage. For instance, eDNA analysis has detected invasive fungi in water bodies near forestry operations, enabling preventive actions to mitigate their spread.

CRISPR-based diagnostic tools are an emerging frontier in molecular diagnostics, offering rapid, portable, and cost-effective solutions for pathogen and pest detection. These tools leverage the precision of CRISPR-Cas systems to target specific DNA or RNA sequences, generating results within minutes. Early applications of CRISPR diagnostics in forestry include the detection of fungal pathogens in timber and the identification of pest species in field samples. The simplicity and scalability of CRISPR technologies hold promise for widespread adoption in forest health monitoring.

The integration of molecular diagnostics with other technologies, such as remote sensing and AI, is creating synergistic capabilities that enhance forest health surveillance. For example, molecular data can be used to validate remote sensing observations, while AI algorithms can analyze molecular datasets to predict outbreak dynamics. This multi-dimensional approach provides a robust framework for addressing complex forest health challenges in real-time. While molecular tools offer unparalleled precision and speed, their adoption faces challenges related to cost, technical expertise, and infrastructure. Capacity-building initiatives, such as training programs and the establishment of diagnostic laboratories, are essential to ensure the effective use of these technologies. Additionally, international collaboration is crucial for standardizing protocols, sharing data, and addressing transboundary threats.

3.3 The Role of International Collaboration in Forest Health Management

Cross-Border Cooperation to Address Invasive Species

Invasive species represent one of the most significant threats to forest health globally, capable of devastating ecosystems, reducing biodiversity, and inflicting severe economic losses. Addressing this challenge necessitates international collaboration, as the spread of invasive pathogens and pests often transcends national borders. Cross-border cooperation involves coordinated efforts in surveillance, prevention, control, and mitigation, leveraging the expertise and resources of multiple nations to combat shared threats.

Global trade and travel are primary drivers of the introduction and dissemination of invasive species. Pathogens and pests are frequently transported inadvertently through pathways such as timber shipments, nursery plants, and wood packaging materials. For example, the emerald ash borer (*Agrilus planipennis*), native to Asia, was introduced to North America via infested wood packaging. Since its arrival, it has caused widespread mortality of ash trees, with ecological and economic consequences that extend across the United States and Canada. Similarly, the fungus *Cryphonectria parasitica*, responsible for chestnut blight, spread from Asia to Europe and North America through infected nursery stock, leading to the near-extinction of the American chestnut (*Castanea dentata*).

Preventing the introduction of invasive species requires robust biosecurity measures that involve international coordination. The International Plant Protection Convention (IPPC), established under the Food and Agriculture Organization (FAO), plays a critical role in facilitating cross-border efforts to mitigate the risks of invasive species. The IPPC sets global standards for phytosanitary measures, such as the inspection and treatment of imported plants, timber, and packaging materials. For instance, the ISPM 15 standard requires wood packaging to be heat-treated or fumigated to eliminate pests before international shipment, reducing the likelihood of new introductions. Regional organizations also contribute to cross-border cooperation by addressing invasive species within specific geographical contexts.

The European and Mediterranean Plant Protection Organization (EPPO) collaborates with member states to monitor and manage invasive pests and pathogens in Europe. One of its notable initiatives is the creation of regional pest risk analyses, which identify potential threats and recommend control measures tailored to the region's ecological and economic conditions. Similarly, the North American Plant Protection Organization (NAPPO) fosters collaboration among the United States, Canada, and Mexico to harmonize biosecurity policies and facilitate information sharing on invasive species.

Surveillance and monitoring are key components of cross-border cooperation. Collaborative networks of researchers and forest managers use advanced technologies such as satellite imagery, remote sensing, and environmental DNA (eDNA) analysis to detect invasive species and monitor their spread. For example, the Global Environment Facility (GEF) has funded projects that employ satellite-based surveillance to track forest health across multiple countries, enabling early detection of invasive outbreaks. These efforts are complemented by field surveys and diagnostic tools that provide ground-level validation of remote observations.

Cross-border collaboration extends to the development and implementation of management strategies for invasive species. Biological control, which involves the introduction of natural predators, parasitoids, or pathogens to regulate invasive populations, often requires international coordination to ensure safety and efficacy. For instance, the release of parasitic wasps to control the emerald ash borer in North America was preceded by extensive research and testing in the beetle's native range in Asia. Similarly, collaborative research has identified fungal pathogens that can suppress the spread of invasive pests like the hemlock woolly adelgid (*Adelges tsugae*), demonstrating the value of international partnerships in developing innovative solutions. Despite these successes, cross-border cooperation faces challenges such as inconsistent enforcement of regulations, limited funding, and differing national priorities. Addressing these barriers requires strengthening international frameworks, increasing resource allocation, and fostering trust and communication among stakeholders. By recognizing invasive species as a global issue, nations can work together to protect forest ecosystems from the far-reaching impacts of these threats.

Global Frameworks and Agreements for Forest Protection

The complexity and scale of forest health challenges necessitate global frameworks and agreements that provide a unified approach to managing diseases, pests, and invasive species. These frameworks enable countries to align their policies, share knowledge, and mobilize resources to address common goals. They also promote sustainable forest management, biodiversity conservation, and climate resilience, recognizing the interconnectedness of forest health with global environmental and socio-economic systems.

One of the most influential global frameworks for forest protection is the Convention on Biological Diversity (CBD), which emphasizes the importance of conserving biodiversity and managing ecosystems sustainably. Under the CBD, member countries commit to protecting forests from the impacts of invasive species, pathogens, and pests. The Aichi Biodiversity Targets, established in 2010, include specific goals related to the prevention and management of invasive alien species, highlighting their role in biodiversity loss. The post-2020 Global Biodiversity Framework, currently under negotiation, is expected to strengthen commitments to forest health by integrating invasive species management with broader conservation and climate goals.

The United Nations Framework Convention on Climate Change (UNFCCC) and its associated initiatives, such as REDD+ (Reducing Emissions from Deforestation and Forest Degradation), also play a vital role in forest protection. REDD+ provides financial incentives for developing countries to conserve and sustainably manage their forests, with co-benefits for biodiversity and ecosystem services. By addressing the drivers of deforestation and forest degradation, REDD+ indirectly contributes to reducing the spread and impacts of pests and pathogens, which often thrive in disturbed and fragmented landscapes. Another key global agreement is the International Treaty on Plant Genetic Resources for Food and Agriculture (ITPGRFA), which promotes the conservation and sustainable use of genetic resources, including forest trees. The treaty facilitates the exchange of genetic material for research and breeding programs, enabling the development of disease-resistant and pest-tolerant tree varieties. For instance, collaborative efforts under the treaty have supported the reintroduction of blight-resistant American chestnut trees, demonstrating the potential of genetic resources in restoring forest health. In addition to these formal agreements, global organizations and networks provide platforms for information sharing, capacity building, and research collaboration. The International Union of Forest Research Organizations (IUFRO) is a leading body that brings together researchers from around the world to address forest health challenges. IUFRO's working groups on forest pathology and entomology facilitate the exchange of knowledge on emerging threats, diagnostic tools, and management strategies. Similarly, the Global Invasive Species Programme (GISP) guides prevention and management of invasive species, with a focus on capacity building in developing countries.

Technological advancements have enhanced the effectiveness of global frameworks and agreements by enabling real-time data sharing and collaborative decision-making. Platforms such as the Forest Invasive Species Network (FISN) and the Global Forest Watch (GFW) use satellite imagery, field data, and machine learning to provide actionable insights into forest health. These tools allow countries to monitor outbreaks, assess risks, and coordinate responses across borders, ensuring a more integrated approach to forest protection. Despite the progress made through global frameworks and agreements, significant challenges remain. Limited funding, political will, and enforcement mechanisms often hinder the implementation of international commitments. Additionally, the uneven capacity of countries to participate in global initiatives creates disparities in forest health management, particularly in regions with high biodiversity and limited resources. Addressing these challenges requires a renewed focus on equity, capacity building, and the integration of forest health into broader sustainable development agendas. The role of international collaboration in forest health management is indispensable for addressing the complex and interconnected challenges posed by invasive species, pathogens, and pests. By fostering cross-border cooperation and strengthening global frameworks, nations can protect forest ecosystems from emerging threats while promoting sustainability, biodiversity, and resilience. As forests face increasing pressures from climate change, trade, and land-use change, the importance of international collaboration will only continue to grow.

3.4 Future Directions in Forest Pathology and Entomology

Integration of Genomics and Big Data in Forest Health Sciences

The integration of genomics and big data analytics into forest pathology and entomology represents a transformative shift in the field, enabling unprecedented insights into the complex interactions between pathogens, pests, and forest ecosystems. Genomic technologies, such as whole-genome sequencing, transcriptomics, and metagenomics, provide detailed information about the genetic makeup, functional traits, and evolutionary dynamics of forest pathogens and pests. When combined with big data analytics, these technologies allow researchers to uncover patterns, predict outbreaks, and develop targeted interventions to protect forest health.

Genomics has already revolutionized our understanding of forest pathogens and pests by revealing their mechanisms of pathogenicity, host specificity, and adaptation to environmental changes. For example, the complete genome sequencing of *Ophiostoma ulmi* and *Ophiostoma novo-ulmi*, the fungal pathogens responsible for Dutch elm disease, has uncovered genes associated with virulence and resistance to fungicides. These insights have informed the development of disease-resistant elm varieties and new treatment strategies. Similarly, genomic studies of bark beetles, such as the mountain pine beetle (*Dendroctonus ponderosae*), have identified genes involved in pheromone production and detoxification, which are critical for understanding their behavior and resilience in changing climates. Metagenomics, which analyzes the collective genetic material of microbial communities, has expanded the scope of pathogen and pest research to include the broader ecological context. This approach allows scientists to study the microbiomes associated with forests, including beneficial and antagonistic microorganisms that influence pathogen and pest dynamics. For instance, metagenomic analyses have identified bacterial and fungal communities in soil that suppress the growth of root rot pathogens, offering potential avenues for biological control. Similarly, the study of endophytic fungi and bacteria within trees has revealed their roles in enhancing host resistance to diseases and insect attacks.

Big data analytics complements genomics by enabling the analysis of vast and complex datasets generated from genomic studies, remote sensing, and environmental monitoring. Machine learning algorithms and artificial intelligence (AI) tools are increasingly used to process and interpret these datasets, uncovering correlations and trends that would be difficult to detect through traditional methods. For example, AI-driven models can predict the spread of pathogens and pests by integrating genomic data with climate variables, host distributions, and landscape connectivity. These models provide actionable insights for forest managers, allowing them to prioritize surveillance and intervention efforts in high-risk areas. The integration of genomics and big data also facilitates the development of precision forestry, an approach that tailors management practices to the specific needs of individual forest stands. Genomic screening can identify tree species or genotypes that are most resistant to local pathogens and pests, guiding reforestation and afforestation efforts. Big data tools can then monitor the performance of these trees over time, optimizing management strategies to enhance forest resilience. For instance, precision forestry initiatives in Scandinavia have used genomic markers to select disease-resistant Norway spruce (*Picea abies*) for planting, reducing the impacts of root rot and needle cast diseases. Despite the promise of genomics and big data, their implementation in forest health sciences faces challenges, including high costs, technical complexity, and data integration issues. Establishing accessible genomic databases and standardized protocols for data collection and analysis is essential to maximizing the utility of these technologies. International collaboration and capacity-building efforts are also critical to ensure that countries with limited resources can participate in and benefit from these advancements. By addressing these challenges, the integration of genomics and big data has the potential to revolutionize forest pathology and entomology, enabling more effective and sustainable management of forest ecosystems.

Adaptive Management Strategies for Resilient Forest Ecosystems

The increasing frequency and severity of pathogen and pest outbreaks, exacerbated by climate change and anthropogenic pressures, underscore the need for adaptive management strategies that promote the resilience of forest ecosystems. Adaptive management is a dynamic approach that emphasizes flexibility, learning, and iterative decision-making to address the uncertainties and complexities of forest health challenges. By incorporating ecological, social, and economic factors, adaptive management aims to balance short-term needs with long-term sustainability, ensuring that forests can recover and thrive in the face of emerging threats.

A cornerstone of adaptive management is the concept of resilience, which refers to the capacity of forests to absorb disturbances, recover from damage, and adapt to changing conditions. Enhancing resilience involves maintaining or restoring the diversity, connectivity, and functionality of forest ecosystems, which serve as buffers against pathogen and pest outbreaks. For example, mixed-species planting is a widely recommended strategy for reducing the vulnerability of forests to host-specific pathogens and pests. Diverse forests provide a more complex and stable environment, limiting the spread and impact of outbreaks while supporting a broader range of ecosystem services. Another key component of adaptive management is the use of early warning systems and real-time monitoring to inform decision-making. Advances in remote sensing, molecular diagnostics, and AI-driven analytics have enabled the development of tools that detect and predict outbreaks at their earliest stages. These systems allow forest managers to implement targeted interventions, such as quarantine measures, biological control, or selective harvesting, before outbreaks escalate. For instance, early detection of *Phytophthora ramorum* in Californian forests has enabled the removal of infected trees and the treatment of adjacent areas, reducing the spread of sudden oak death.

Adaptive management also emphasizes the importance of stakeholder engagement and collaboration in developing and implementing forest health strategies. Involving local communities, indigenous peoples, and private landowners ensures that management practices are informed by diverse perspectives and address the needs of all stakeholders. Collaborative approaches, such as community-based monitoring programs, have been effective in detecting and responding to outbreaks in remote or resource-constrained areas. For example, participatory forest health initiatives in tropical regions have trained residents to identify symptoms of diseases and pests, contributing valuable data to national and international monitoring networks. Climate-smart forestry is an emerging framework that aligns with the principles of adaptive management, integrating climate change mitigation and adaptation into forest health strategies. This approach recognizes that managing pathogens and pests requires addressing the broader environmental and socio-economic drivers of forest decline. Climate-smart forestry practices include enhancing carbon storage through reforestation, reducing greenhouse gas emissions from forestry operations, and promoting the use of forest products that sequester carbon. These practices not only mitigate climate change but also create conditions that enhance forest resilience to pathogens and pests.

The implementation of adaptive management strategies is not without challenges, including uncertainties in predicting pathogen and pest dynamics, limited resources for monitoring and intervention, and conflicts among stakeholders with differing priorities. Overcoming these challenges requires investments in research, technology, and capacity building, as well as the development of policy frameworks that support adaptive approaches. Additionally, fostering a culture of experimentation and learning is essential to refining and improving management practices over time. Future directions in forest pathology and entomology must embrace the principles of adaptive management to address the rapidly evolving threats to forest health. By integrating genomics, big data, and innovative management strategies, forest health professionals can enhance the resilience and sustainability of forest ecosystems, ensuring their continued provision of ecological, economic, and cultural benefits in an uncertain and changing world.

Chapter - 2

Plant Pathogens in Forest Ecosystems

1. Introduction to Plant Pathogens in Forests

Plant pathogens in forests are a critical aspect of forest ecosystem dynamics, influencing biodiversity, nutrient cycling, and overall health of these complex environments. Their study is integral to understanding the intricate web of interactions that sustain forest ecosystems, which are among the most biodiverse and ecologically significant systems on Earth. Forests provide a vast array of ecological services, including carbon sequestration, habitat for countless species, and regulation of the water cycle. The presence of plant pathogens, however, introduces a level of complexity to these systems, shaping their structure and function in ways that are often subtle but sometimes result in devastating outbreaks. In order to comprehend and mitigate the impact of plant pathogens, it is essential to explore their roles, dynamics, and interactions within forest ecosystems.

Forest pathogens are important because they influence both ecological and economic aspects of forests. Diseases caused by pathogens can lead to significant declines in tree populations, thereby disrupting ecosystem services and causing economic losses in industries reliant on forest resources. Additionally, forest pathogens affect the biodiversity of associated flora and fauna, altering food webs and habitat availability. Forests are dynamic systems where the interaction of abiotic and biotic factors creates a constantly shifting equilibrium. Pathogens are a fundamental component of this balance, acting as both regulators of population densities and agents of change that drive evolutionary processes. Consequently, studying these pathogens provides insight into the resilience and vulnerability of forests, allowing for the development of informed strategies for their management and conservation.

The characteristics of forest ecosystems play a pivotal role in shaping pathogen dynamics. Forests are characterized by their structural complexity, diverse species composition, and variability in microclimatic conditions. These factors create a mosaic of niches where pathogens can establish and propagate. The density and diversity of host plants influence the prevalence and severity of pathogen outbreaks, as high biodiversity can either dilute pathogen impacts or provide opportunities for pathogen specialization. Microclimatic conditions such as temperature, humidity, and light availability further modulate the growth and survival of pathogens, affecting their ability to infect and cause disease. Moreover, the interconnectedness of forest systems, through root networks and mycorrhizal associations, facilitates the spread of pathogens across large areas, amplifying their impact.

DOI: 10.1201/9781042011063-2

The interactions between plants and pathogens in forests are highly complex and involve a delicate balance between infection and resistance. Pathogens employ various strategies to invade host plants, including the production of enzymes, toxins, and other virulence factors that overcome plant defenses. On the other hand, plants have evolved intricate defense mechanisms to recognize and combat pathogen attacks. These include physical barriers, such as thickened cell walls, and biochemical responses, such as the production of antimicrobial compounds and activation of signaling pathways. The outcome of these interactions depends on a multitude of factors, including the genetic makeup of both the pathogen and the host, environmental conditions, and the presence of other organisms within the ecosystem. These interactions not only determine the health of individual plants but also influence the structure and function of entire forest communities. Plant pathogens in forest ecosystems encompass a diverse array of organisms, including fungi, bacteria, viruses, viroids, and nematodes, each with distinct ecological roles and impacts. Among these, fungi are the most dominant and well-studied pathogens, responsible for a wide range of diseases such as root rots, cankers, and leaf blights. Their ability to decompose organic matter and thrive in nutrient-rich forest soils gives them a key role in disease dynamics. Bacteria, although less studied in forest systems compared to fungi, play significant roles in diseases such as bacterial wilt and blight. Their rapid multiplication and ability to produce biofilms enable them to colonize plant tissues effectively. Viruses and viroids are emerging as significant threats to forest health, with their ability to alter plant metabolism and reproduction causing long-term impacts on tree populations. Nematodes, microscopic worms that infect plant roots, further contribute to tree decline by impairing nutrient and water uptake.

The ecological roles of plant pathogens extend beyond their direct impacts on host plants. In natural forest ecosystems, pathogens regulate plant populations, preventing dominance by a single species and thereby promoting biodiversity. By selectively targeting certain species or individuals, pathogens facilitate ecological succession, enabling the establishment of new plant communities. They also play a crucial role in nutrient cycling by decomposing organic material and recycling nutrients back into the soil. This decomposition process, while detrimental to individual plants, supports the overall productivity and stability of the forest ecosystem. Furthermore, pathogens contribute to maintaining species balance by exerting pressure on dominant species, allowing for the coexistence of a diverse array of plants and associated organisms. These ecological roles highlight the dual nature of pathogens as both harmful agents and essential components of healthy ecosystems.

The interaction between pathogens and their hosts in forest ecosystems involves complex mechanisms of infection, disease development, and resistance. Pathogens use a variety of strategies to invade host tissues, including the secretion of enzymes that break down plant cell walls and the production of toxins that suppress plant defenses. Once inside the host, pathogens manipulate plant cellular processes to establish themselves and reproduce. In response, plants have evolved sophisticated defense mechanisms that include the recognition of pathogen-associated molecular patterns (PAMPs) and the activation of immune responses. These responses often involve producing reactive oxygen species, antimicrobial compounds, and resistance genes. Over time, the co-evolution of pathogens and their hosts has led to a dynamic arms race, with each side developing new strategies to outcompete the other. This co-evolutionary process shapes the genetic diversity of both pathogens and forest trees, influencing the resilience of forest ecosystems to disease.

The impact of plant pathogens on forest health can be observed through the symptoms of diseases, economic losses, and ecological disruptions they cause. Symptoms such as leaf spots, cankers, root rot, and wilting are indicative of pathogen activity and often lead to reduced growth and vigor of affected trees. Severe outbreaks can result in widespread tree mortality, altering forest structure and composition. Economically, these outbreaks affect industries dependent on timber, non-timber forest products, and tourism. Ecologically, the loss of tree species can have cascading effects on wildlife, soil stability, and water cycles. Historical case studies, such as the chestnut blight and Dutch elm disease epidemics, illustrate the profound and long-lasting impacts of plant pathogens on forest ecosystems. These examples underscore the importance of early detection and effective management to prevent similar outcomes in the future.

Environmental factors play a crucial role in determining the prevalence and severity of forest diseases. Climate change, with its associated shifts in temperature, precipitation, and extreme weather events, has emerged as a significant driver of forest disease dynamics. Warmer temperatures can expand the range of pathogens, while altered precipitation patterns can create conditions conducive to disease outbreaks. Soil conditions, including nutrient availability and pH, also influence pathogen activity and host susceptibility. The structure of the forest, such as canopy density and species composition, affects microclimatic conditions and pathogen spread. Anthropogenic activities, including logging, land-use changes, and the introduction of non-native species, further exacerbate the risk of pathogen outbreaks. These factors highlight the need for a holistic understanding of the environmental context in managing forest health.

Studying forest pathogens requires a multidisciplinary approach that combines traditional and modern techniques. Field surveys and morphological studies remain essential for identifying pathogens and assessing disease prevalence. Advances in molecular techniques, such as genomics, proteomics, and high-throughput sequencing, have revolutionized our ability to detect and characterize pathogens at a genetic level. These methods provide insights into pathogen diversity, virulence factors, and evolutionary history. Remote sensing and spatial analysis offer powerful tools for monitoring forest health over large areas, enabling the early detection of disease outbreaks and assessment of environmental risk factors. Together, these methods provide a comprehensive framework for understanding the dynamics of forest pathogens and informing management strategies.

The management and control of forest pathogens involve a combination of preventive and curative measures aimed at reducing disease incidence and impact. Prevention strategies, such as quarantine, the use of resistant tree species, and biosecurity measures, are essential for minimizing the introduction and spread of pathogens. Biological control methods, which involve the use of beneficial microbes and natural predators, offer sustainable alternatives to chemical treatments. Integrated forest disease management approaches combine these strategies with ecological and silvicultural practices to create resilient forest ecosystems. By addressing the underlying causes of disease and promoting ecosystem health, these approaches aim to mitigate the impact of pathogens while preserving the ecological and economic value of forests.

1.1 Importance of Understanding Forest Pathogens

The health of forests is critical to the well-being of life on Earth. Forests provide food, timber, medicine, and recreational opportunities, while also serving as carbon sinks that mitigate climate change. However, the presence of pathogens poses significant challenges to these benefits. Understanding plant pathogens in forests goes beyond simply identifying harmful organisms; it involves comprehending their role in ecosystem functioning, their potential to cause disease outbreaks, and the strategies needed to mitigate their effects. The importance of understanding forest pathogens can be discussed from multiple perspectives, including ecological, economic, and environmental dimensions.

1.1.1 Ecological Significance

Forest pathogens are an integral part of natural ecosystems. While often perceived as purely destructive, they also serve ecological functions by acting as agents of natural selection. For example, pathogens can regulate plant populations by targeting specific species or individuals that are genetically weaker or stressed, thereby promoting genetic diversity within forests. This selective pressure helps maintain a balance among plant species, prevents monocultures, and fosters biodiversity. However, when this natural balance is disrupted, due to climate change, invasive species, or human activities, pathogens can become aggressive and cause widespread damage. For instance, chestnut blight (*Cryphonectria parasitica*) and Dutch elm disease (*Ophiostoma ulmi*) have decimated native tree populations, leading to significant ecological imbalances in affected regions.

1.1.2 Economic Impacts

The economic importance of forests cannot be overstated. They provide raw materials for industries, generate revenue through ecotourism, and support local and global economies. Pathogens can have devastating effects on these economic benefits. Diseases that result in widespread tree mortality reduce timber yield, increase management costs, and disrupt industries dependent on forest resources. For example, *Phytophthora ramorum*, the pathogen responsible for sudden oak death, has not only altered ecosystems in North America but also imposed significant economic costs on forest management and affected industries reliant on oak timber. Additionally, the loss of forest health often leads to secondary outbreaks of pests, further compounding economic losses.

1.1.3 Environmental Implications

Forests play a central role in global carbon dynamics. Through photosynthesis, trees sequester vast amounts of carbon dioxide, helping to mitigate the effects of climate change. When pathogens attack forests, they increase tree mortality, releasing stored carbon back into the atmosphere and reducing the forest's ability to absorb future emissions. This creates a feedback loop: climate change exacerbates pathogen activity, and pathogen-induced forest decline accelerates climate change. Additionally, the loss of healthy forests affects water cycles, soil stability, and wildlife habitats, leading to cascading environmental impacts.

1.1.4 Threats from Invasive Pathogens

Globalization and the increased movement of goods and people have facilitated the spread of invasive pathogens to new regions. Native plant species often lack resistance to these introduced pathogens, leading to catastrophic disease outbreaks. The introduction of Hymenoscyphus fraxineus (ash dieback) to Europe is a prime example that has devastated ash populations across the continent. Understanding invasive pathogens and their pathways of introduction is essential for developing preventative measures, such as quarantine regulations and monitoring systems.

1.1.5 Monitoring and Early Detection

The early detection of forest pathogens is critical for mitigating their impact. Advances in molecular biology, remote sensing, and field diagnostics have provided powerful tools for monitoring pathogen populations and identifying emerging threats.

1.2 Characteristics of Forest Ecosystems That Affect Pathogen Dynamics

Forest ecosystems are unique environments with structural, biological, and environmental features that distinguish them from other plant systems, such as agriculture or horticulture. These features play a pivotal role in shaping the dynamics of plant pathogens, influencing their survival, reproduction, spread, and interaction with host species. The specific characteristics of forest ecosystems are essential for predicting pathogen behaviour and designing effective management strategies.

1.2.1 High Biodiversity and Host Complexity

Forests are characterized by high levels of biodiversity, with multiple tree species coexisting alongside shrubs, herbs, and other plant types. This diversity can influence pathogen dynamics in two contrasting ways. On one hand, biodiversity can dilute pathogen pressure by reducing the prevalence of any single host species, thereby limiting the pathogen's ability to spread. On the other hand, forests often contain closely related tree species that share similar susceptibilities, providing opportunities for pathogens with broad host ranges to thrive. For example, *Armillaria* root rot affects a wide variety of tree species, making it particularly problematic in mixed forests.

1.2.2 Vertical Stratification and Microclimatic Variation

Forests exhibit vertical stratification, with distinct layers such as the canopy, understory, and forest floor. Each layer creates unique microclimatic conditions that influence pathogen behavior. For example, fungal pathogens that rely on moisture often thrive in the shaded, humid conditions of the understory, while pathogens that require high light levels may target leaves in the upper canopy. Additionally, seasonal changes in temperature, humidity, and light intensity further affect pathogen activity, creating temporal variations in disease severity.

1.2.3 Soil Ecosystem Interactions

The soil in forest ecosystems is rich in organic matter, nutrients, and microbial communities, creating a dynamic environment for plant-pathogen interactions. Soilborne pathogens, such as *Phytophthora* species, can persist for years in the soil, infecting trees through their roots and causing widespread damage. The interaction between pathogens and other soil organisms is also significant; for example, mycorrhizal fungi may enhance tree resistance to certain pathogens by improving nutrient uptake and activating plant defense mechanisms. Conversely, some microbes may facilitate pathogen survival by creating favorable conditions for infection.

1.2.4 Longevity of Hosts and Disease Persistence

Trees in forests are long-lived organisms, often persisting for decades or centuries. This longevity influences pathogen dynamics in several ways. Chronic infections can persist in individual trees for many years, serving as reservoirs for pathogen populations. Additionally, long generation times mean that trees evolve resistance to pathogens at a slower pace compared to shorter-lived plants. This slow evolutionary response makes forests particularly vulnerable to rapidly adapting or introduced pathogens.

1.2.5 Limited Human Intervention

Unlike agricultural systems, which are intensively managed to minimize disease outbreaks, forest ecosystems are largely unmanaged. This allows pathogens to spread unchecked, particularly in remote or inaccessible areas. While this lack of intervention fosters natural ecological processes, it also means that pathogen outbreaks can reach epidemic levels before being detected or addressed.

1.2.6 Influence of Climate and Weather

Climate and weather conditions have a profound impact on pathogen activity in forests. Temperature, humidity, and precipitation are critical factors that determine the survival, reproduction, and dispersal of pathogens. For instance, warmer temperatures associated with climate change have been linked to increased activity of pathogens like *Dothistroma septosporum*, which causes red band needle blight in conifers. Similarly, prolonged drought conditions can weaken tree defenses, making them more susceptible to infections.

1.3 Interactions Between Plants and Pathogens

The relationship between plants and pathogens in forest ecosystems is dynamic and complex. Unlike agricultural systems, where pathogen-host interactions are often limited to a few crop species and controlled environments, forests present a more intricate web of interactions involving multiple host species, pathogens, and environmental factors.

1.3.1 Mechanisms of Pathogen Infection

Pathogens use various strategies to infect their hosts. Fungal pathogens often penetrate plant tissues through natural openings, such as stomata, or through wounds. Bacterial pathogens typically invade water-conducting tissues, leading to systemic infections, while viruses rely on insect vectors for transmission. Soilborne pathogens, such as nematodes and Phytophthora, attack roots, disrupting nutrient and water uptake. Once inside the host, pathogens deploy an array of enzymes, toxins, and effectors to suppress plant defenses and colonize tissues.

1.3.2 Host Defense Mechanisms

Plants have evolved a range of defense mechanisms to counteract pathogen attacks. These include physical barriers, such as thick bark and waxy cuticles, as well as chemical defenses, including the production of antimicrobial compounds. Additionally, plants can recognize specific pathogen-associated molecular patterns (PAMPs) and activate immune responses to limit infection. However, pathogens often evolve counterstrategies to evade these defenses, leading to a co-evolutionary arms race between hosts and pathogens.

1.3.3 Ecological Consequences of Plant-Pathogen Interactions

Plant-pathogen interactions have significant ecological consequences. Pathogen outbreaks can alter forest structure by selectively removing susceptible species, creating gaps in the canopy that allow light to reach the forest floor and promote succession. These changes can cascade through the ecosystem, affecting herbivores, pollinators, and decomposers. Moreover, severe outbreaks can disrupt nutrient cycling, water dynamics, and other ecological processes.

2. Types of Plant Pathogens in Forest Ecosystems

Plant pathogens are a critical component of forest ecosystems, acting as both natural regulators of ecological balance and agents of forest decline. These organisms interact with trees and other plants at microscopic and macroscopic levels, influencing the health, structure, and dynamics of forests. The diversity of plant pathogens in forest ecosystems reflects the immense variability in environmental conditions, host species, and ecological interactions present in these habitats. Pathogens can attack various parts of trees, including roots, stems, leaves, and reproductive organs, and their impacts range from localized infections to widespread epidemics capable of altering forest landscapes. Unlike agricultural systems, where plant-pathogen interactions often involve monocultures in controlled environments, forest ecosystems are characterized by their complexity, longevity of hosts, and lack of direct human management. This complexity adds unique challenges to the study, monitoring, and management of forest pathogens.

Forest pathogens are broadly classified into several groups based on their biological characteristics, modes of infection, and interactions with hosts. The dominant groups include fungi, which are by far the most prevalent and diverse; bacteria, which cause both localized and systemic infections; viruses and viroids, which are increasingly recognized as significant threats; and nematodes, which exert their influence primarily in the rhizosphere but can also infect above-ground parts of trees. Each of these groups plays a distinct role in forest disease dynamics, and their impacts are shaped by factors such as climate, forest management practices, and global trade. Understanding these pathogens requires a multidisciplinary approach that incorporates microbiology, plant pathology, ecology, and climate science.

2.1 Fungi

Fungi are the most significant group of plant pathogens in forest ecosystems, both in terms of their diversity and the scale of their impact. These organisms belong to a kingdom of eukaryotes characterized by their filamentous growth forms, spore production, and ability to degrade complex organic matter. Fungi are highly adaptable, thriving in diverse environments ranging from tropical rainforests to boreal woodlands. They infect all parts of trees, including roots, trunks, branches, leaves, and reproductive structures, causing a wide array of diseases that influence forest health and productivity. Fungi have evolved a variety of mechanisms to infect and exploit their hosts. They use specialized structures, such as appressoria, to penetrate plant tissues and secrete enzymes that degrade cell walls. Some fungi establish symbiotic relationships with their hosts, as seen in mycorrhizal associations, while others are strictly pathogenic, leading to severe damage or death. Fungi can reproduce sexually or asexually, and their spores are dispersed by wind, water, animals, or even human activity. This dispersal capability, combined with their adaptability, makes fungi formidable agents of forest disease.

Several major fungal diseases exemplify the impact of these pathogens on forests. Root rot diseases caused by *Armillaria* and *Phytophthora* are among the most destructive, affecting tree stability and nutrient uptake. *Armillaria* species, often referred to as "honey fungi," form extensive networks of rhizomorphs that spread through soil, infecting the roots of healthy trees. Similarly, *Phytophthora* species, such as *Phytophthora cinnamomi*, cause root rot in a wide range of tree species and are particularly devastating in regions with poorly drained soils. Canker diseases, such as chestnut blight caused by *Cryphonectria parasitica*, result in localized necrosis that can girdle and kill trees. Dutch elm disease, caused by *Ophiostoma* species and transmitted by bark beetles, has had catastrophic effects on elm populations in Europe and North America. Fungal pathogens also include leaf and needle blight pathogens, such as *Dothistroma septosporum* and *Mycosphaerella pini*, which reduce photosynthetic efficiency and weaken trees over time. These diseases are especially problematic in conifer plantations, where dense canopies create microclimates conducive to fungal growth. In addition to direct damage, fungi often act as facilitators for other pathogens and pests, creating synergistic effects that amplify their impact. For example, trees weakened by fungal infections are more susceptible to insect infestations, which can further accelerate their decline. The ecological role of fungal pathogens extends beyond their immediate impact on individual trees. By selectively targeting certain species, fungi can alter forest composition and biodiversity, influencing successional dynamics and habitat availability for other organisms. Fungi also play a critical role in nutrient cycling by breaking down organic matter and releasing nutrients into the soil. However, their adaptability to changing environmental conditions poses significant challenges for forest management. Climate change, in particular, is expected to exacerbate fungal diseases by altering temperature and moisture regimes, expanding the range of certain pathogens, and increasing the frequency of favourable conditions for disease outbreaks.

2.2 Bacteria and Their Role in Forest Diseases

Bacterial pathogens, though less diverse than fungi, are nonetheless important contributors to forest diseases. These microscopic, single-celled organisms infect a variety of tree species, causing localized and systemic infections that impair tree health and reduce forest productivity. Bacteria typically enter plants through natural openings, such as stomata and lenticels, or through wounds caused by mechanical damage or insect activity. Once inside the host, they multiply rapidly and produce toxins, enzymes, and other virulence factors that disrupt normal physiological processes.

The diseases caused by bacterial pathogens are varied and often highly visible. Bacterial blights, such as those caused by *Pseudomonas syringae* and *Xanthomonas* species, manifest as necrotic spots on leaves, stems, and fruits. These pathogens are particularly problematic in temperate forests, where wet and humid conditions favor their proliferation. Crown gall disease, caused by *Agrobacterium tumefaciens*, results in tumor-like growths on tree trunks and roots, which interfere with water and nutrient transport. Bacterial cankers, such as those caused by *Pseudomonas syringae pv. morsprunorum*, create sunken lesions on stems and branches, leading to dieback and increased susceptibility to secondary infections.

Bacterial wilt diseases, exemplified by *Ralstonia solanacearum*, are another significant concern in forests and agroforestry systems. These diseases block the xylem vessels of trees, causing wilting, yellowing, and eventual death. In some cases, bacteria can also interact synergistically with other pathogens, as seen in fire blight, where *Erwinia amylovora* infections are often exacerbated by fungal or viral co-infections. Despite their relatively small size and simple structure, bacterial pathogens have profound ecological and economic impacts. They contribute to forest decline, reduce timber quality, and disrupt ecological processes such as water cycling and soil stabilization. The management of bacterial diseases in forests is particularly challenging due to the rapid reproduction and adaptability of these pathogens. Antibiotics, while effective in agricultural systems, are rarely used in forests due to environmental concerns and the potential for resistance development. Instead, management strategies focus on preventive measures, such as the selection of resistant tree species, proper sanitation practices, and the control of insect vectors.

2.3 Viruses and Viroids

Viruses and viroids are among the most enigmatic groups of plant pathogens, largely due to their microscopic size and reliance on host cells for replication. These pathogens are composed of genetic material, either DNA or RNA, encased in a protein coat (in the case of viruses) or as naked RNA molecules (in the case of viroids). They are transmitted through vectors, such as insects, nematodes, and fungi, or through mechanical wounds and vegetative propagation. Despite their small size and minimal genetic content, viruses and viroids can cause significant damage to forest trees, particularly in the context of global trade and climate change. The diseases caused by viral pathogens often go undetected until they reach advanced stages. Symptoms include mosaic patterns on leaves, stunted growth, yellowing, and dieback, all of which reduce the photosynthetic capacity and overall vigor of infected trees. For example, *Tobacco mosaic virus* (TMV) and related viruses have been reported to affect various forest species, causing economic losses in timber and non-timber forest products. Viroids, such as *Coconut cadang-cadang viroid* (CCCVd), are even more insidious, leading to severe decline and mortality in economically important species like coconuts and oil palms. The management of viral diseases in forests is particularly challenging due to the lack of effective treatments. Control measures focus on preventing the introduction and spread of viruses through the use of virus-free planting material, quarantine regulations, and the management of vector populations. Advances in molecular diagnostics have improved the detection of viral pathogens, but the integration of these tools into forest management remains limited.

2.4 Nematodes and Their Impact on Tree Health

Nematodes are microscopic, worm-like organisms that inhabit soil and plant tissues. While many nematodes are beneficial, playing roles in nutrient cycling and soil health, certain species are pathogenic and cause significant damage to forest trees. Pathogenic nematodes typically feed on plant roots or invade vascular tissues, disrupting water and nutrient uptake. They can act as primary pathogens or facilitate secondary infections by fungi and bacteria, creating complex disease syndromes. The most notable nematode-associated disease in forests is pine wilt disease, caused by *Bursaphelenchus xylophilus*. This nematode is transmitted by beetle vectors and leads to rapid wilting and death of pine trees, particularly in Asia and Europe. Other nematodes, such as root-knot nematodes (*Meloidogyne* species) and lesion nematodes (*Pratylenchus* species), cause gall formation and root damage, weakening trees and reducing their ability to withstand environmental stressors. Nematode infestations have profound ecological and economic consequences. They reduce forest productivity, increase susceptibility to other pathogens and pests, and disrupt soil-plant interactions. Management strategies for nematodes are limited and often focus on preventive measures, such as the use of resistant tree species and biological control agents. However, the application of these methods in forest ecosystems is constrained by logistical and environmental challenges.

3. Ecological Roles of Plant Pathogens

Plant pathogens are often perceived solely as destructive agents that compromise forest health, reduce productivity, and disrupt ecological processes. However, this narrow perspective overlooks their essential and multifaceted ecological roles within forest ecosystems. In reality, plant pathogens act as natural components of these systems, participating in processes that shape biodiversity, regulate populations, and drive ecological succession. While their activities can have devastating consequences when ecosystems are stressed or imbalanced, plant pathogens also play critical roles in maintaining forest structure, nutrient cycling, and species dynamics. These roles are particularly evident in unmanaged, natural forest ecosystems, where pathogens act as agents of selective pressure, promoting diversity and ecological resilience.

3.1 Plant Pathogens as Natural Regulators in Forest Ecosystems

Plant pathogens serve as natural regulators in forest ecosystems by influencing plant population dynamics, controlling species dominance, and driving evolutionary processes. Unlike agricultural or managed systems, where diseases often emerge as outbreaks due to monocultures and human interventions, natural forest ecosystems operate within a framework of ecological balance. In this context, pathogens act as selective forces that help maintain the integrity and stability of these systems.

One of the primary ways pathogens regulate forest ecosystems is by limiting the dominance of specific tree species. This phenomenon, often referred to as the **Janzen-Connell hypothesis**, suggests that pathogens and herbivores disproportionately affect seedlings near parent trees, reducing their survival rates. This spatially dependent mortality creates opportunities for other species to establish, preventing any single species from dominating the ecosystem. For instance, pathogens targeting dominant canopy trees can create light gaps, allowing shade-intolerant species to regenerate and enhancing overall diversity. In addition to regulating species composition, plant pathogens also act as agents of natural selection. By selectively infecting individuals with specific vulnerabilities, pathogens drive the evolution of resistance traits in host populations. This evolutionary arms race between plants and pathogens fosters genetic diversity, which is essential for the long-term resilience of forest ecosystems. For example, trees in pathogen-rich environments often exhibit greater genetic variation in defense mechanisms, such as the production of secondary metabolites or structural barriers. Pathogens also play a role in controlling invasive species in forest ecosystems. While invasive plants often thrive due to a lack of natural enemies in their new habitats, pathogens can sometimes adapt to exploit these novel hosts. This "enemy release" hypothesis highlights the potential of native pathogens to limit the spread of invasive species and restore ecological balance. However, this dynamic is not universal and depends on the ecological compatibility between pathogens and introduced species. Although plant pathogens regulate ecosystems in natural ways, their regulatory role can be disrupted by anthropogenic factors, such as habitat fragmentation, pollution, and climate change. These disruptions often lead to the emergence of new disease dynamics, where pathogens become disproportionately destructive, leading to tree mortality and loss of forest cover. Thus, while pathogens act as natural regulators, their impacts must be carefully managed in the context of changing environmental conditions.

3.2 Contribution to Nutrient Cycling and Ecological Succession

Plant pathogens play a pivotal role in nutrient cycling by breaking down organic matter, facilitating the redistribution of nutrients within ecosystems, and influencing ecological succession. Their contributions to these processes underscore their importance as integral components of forest ecosystems, driving both short-term nutrient availability and long-term forest dynamics.

3.2.1 Decomposition and Nutrient Release

Many plant pathogens, particularly fungal species, act as decomposers, breaking down organic matter and releasing nutrients into the soil. When pathogens infect and kill trees or other plants, they accelerate the process of decomposition, converting complex organic compounds into simpler forms that can be readily absorbed by other organisms. For instance, pathogens that cause root rot or vascular wilt contribute to the breakdown of woody tissues, releasing carbon, nitrogen, phosphorus, and other essential nutrients back into the ecosystem. These nutrients are then taken up by neighboring plants, fueling growth and productivity. In addition to decomposing plant tissues, some pathogens interact with other soil organisms, such as saprophytic fungi and bacteria, to enhance nutrient cycling. These interactions create a dynamic soil environment where nutrients are constantly being recycled and redistributed. This process is particularly important in nutrient-poor forests, such as tropical rainforests or boreal forests, where nutrient availability limits plant growth.

3.2.2 Role in Ecological Succession

Pathogens also play a critical role in ecological succession, the process by which forest ecosystems change over time. By selectively targeting certain species, pathogens create openings in the forest canopy or alter competition dynamics, driving the progression of successional stages. For example, pathogens that kill mature canopy trees create light gaps that allow pioneer species to establish and grow. Over time, these pioneer species are replaced by shade-tolerant species, leading to the development of a more diverse and stable forest structure. Pathogen-mediated succession is particularly evident in post-disturbance scenarios, such as after wildfires, storms, or logging. In these contexts, pathogens often target weakened or stressed trees, accelerating their decline and creating conditions for new species to colonize. This process not only promotes species turnover but also contributes to the recovery and regeneration of forest ecosystems.

3.2.3 Interactions with Other Ecological Processes

The contributions of pathogens to nutrient cycling and succession are closely linked to other ecological processes, such as herbivory, predation, and symbiosis. For instance, pathogens and herbivores often act synergistically to decompose plant material, while symbiotic interactions between pathogens and soil microbes can enhance nutrient availability. These interactions highlight the interconnected nature of forest ecosystems and the central role of pathogens in maintaining their functionality.

3.3 Pathogens in Maintaining Biodiversity and Species Balance

Biodiversity is a cornerstone of forest ecosystem resilience and functionality, and pathogens play a crucial role in shaping and maintaining this diversity. By influencing species composition, population dynamics, and competitive interactions, pathogens contribute to the balance of species within forests. Their impact on biodiversity extends beyond plants, affecting the entire web of life within these ecosystems.

3.3.1 Reducing Dominance of Competitive Species

One of the key ways pathogens maintain biodiversity is by reducing the dominance of highly competitive species. In many forests, a small number of tree species tend to dominate due to their superior growth rates, resource acquisition strategies, or reproductive success. Pathogens that preferentially target these dominant species act as a check on their proliferation, preventing them from excluding other species. This process, known as **density-dependent regulation,** ensures that less competitive species have the opportunity to thrive, thereby enhancing overall diversity.

3.3.2 Promoting Coexistence of Species

Pathogens also promote the coexistence of multiple species by creating spatial and temporal variations in disease pressure. For example, pathogens may be more prevalent in areas with high host density, creating refuges for less abundant species. Similarly, seasonal fluctuations in pathogen activity can create windows of opportunity for different species to establish and reproduce. These dynamics help maintain a diverse array of species within forest ecosystems, contributing to their resilience and stability.

3.3.3 Cascading Effects on Ecosystem Diversity

The influence of plant pathogens on biodiversity extends beyond trees and plants to other organisms in the ecosystem. For instance, the loss of a dominant tree species due to pathogen activity can have cascading effects on herbivores, pollinators, decomposers, and predators. These effects create new niches and opportunities for species to adapt and evolve, further enhancing ecosystem diversity. At the same time, pathogens can indirectly shape the diversity of microbial communities by altering the availability of nutrients and substrates in the soil.

3.3.4 Challenges to Biodiversity Conservation

While pathogens play a positive role in maintaining biodiversity under natural conditions, their impacts can become detrimental when ecosystems are stressed or imbalanced. Factors such as habitat fragmentation, pollution, and climate change can exacerbate the effects of pathogens, leading to the decline or extinction of vulnerable species. For example, the introduction of invasive pathogens, such as *Cryphonectria parasitica* (chestnut blight) and *Hymenoscyphus fraxineus* (ash dieback), has significantly reduced the genetic and species diversity of affected forests. These examples underscore the need for careful monitoring and management of pathogen activity in forest ecosystems.

4. Pathogen-Host Interactions in Forest Ecosystems

The interactions between pathogens and their plant hosts are fundamental to understanding the dynamics of forest ecosystems. These interactions are intricate, multifaceted, and shaped by millions of years of evolution. Pathogen-host interactions not only determine the health and survival of individual trees but also influence forest composition, biodiversity, and ecological balance. Pathogens exploit specific mechanisms to infect plants, overcome their defenses, and establish disease. In response, plants have developed a wide array of defense strategies that range from physical barriers to highly specialized biochemical and molecular responses. The constant interplay between plant defenses and pathogen adaptations has resulted in a co-evolutionary arms race, where both sides continually adapt to gain an advantage.

4.1 Mechanisms of Pathogen Infection and Disease Development

Pathogens employ a variety of strategies to infect plants, establish themselves within host tissues, and cause disease. The infection process is often highly specialized, reflecting the intricate interactions that have evolved between pathogens and their hosts. These mechanisms vary widely among different types of pathogens, including fungi, bacteria, viruses, viroids, and nematodes, but certain general principles can be identified.

4.1.1 Modes of Entry

The first step in the infection process is the entry of the pathogen into the host plant. Pathogens gain access to plant tissues through natural openings, such as stomata, lenticels, or hydathodes, or through wounds caused by environmental damage, herbivory, or human activity. For example, bacterial pathogens like *Pseudomonas syringae* enter through stomata, while fungal pathogens such as *Magnaporthe oryzae* penetrate directly through the plant cuticle using specialized infection structures called appressoria. Nematodes invade plants by mechanically piercing the epidermis with their stylets, while viruses often rely on vectors, such as aphids or beetles, to breach plant defenses.

4.1.2 Colonization of Host Tissues

Once inside the host, pathogens must colonize plant tissues to establish infection. Fungal pathogens spread through the plant using hyphae, which grow intercellularly or intracellularly, depending on the pathogen's lifestyle. Hemibiotrophic fungi, such as *Phytophthora infestans*, initially grow in a biotrophic phase, extracting nutrients from living cells, before switching to a necrotrophic phase, where they kill host cells and feed on the dead tissue. In contrast, obligate biotrophs like *Puccinia graminis* maintain the host cells' viability to ensure their survival. Bacterial pathogens colonize intercellular spaces and vascular tissues, often producing toxins and enzymes that degrade cell walls and facilitate movement through plant tissues. For instance, *Ralstonia solanacearum*, a vascular wilt pathogen, invades xylem vessels, blocking water transport and causing wilting. Viruses replicate within plant cells, hijacking the host's machinery to produce viral particles, which then spread to adjacent cells through plasmodesmata. Nematodes establish feeding sites, such as giant cells or syncytia, within the roots, diverting nutrients for their growth and reproduction.

4.1.3 Suppression of Host Defenses

To successfully establish infection, pathogens must suppress or evade the host's immune responses. This is often achieved through the secretion of effector proteins, which interfere with plant defense signaling pathways. For example, fungal pathogens like *Blumeria graminis* produce effectors that suppress the production of reactive oxygen species (ROS), a key component of the plant's early defense response. Similarly, bacterial pathogens secrete type III effectors through specialized secretion systems, targeting host proteins to disable immune signaling.

4.1.4 Disease Development

The final stage of the infection process is disease development, which occurs when the pathogen successfully disrupts the plant's physiological processes. Symptoms of disease include chlorosis, necrosis, wilting, cankers, and galls, depending on the type of pathogen and the tissues affected. In forest ecosystems, disease development can have significant ecological consequences, such as increased tree mortality, reduced forest productivity, and changes in species composition.

4.2 Plant Defense Responses in Forest Species

Plants have evolved a sophisticated array of defense mechanisms to detect and respond to pathogen attacks. These defenses operate at multiple levels, from physical barriers that prevent pathogen entry to complex molecular signaling pathways that activate immune responses. Forest species, with their long lifespans and exposure to diverse pathogen pressures, exhibit particularly robust and varied defense strategies.

4.2.1 Physical and Chemical Barriers

The first line of defense against pathogens is the plant's physical and chemical barriers. The cuticle and cell wall form a protective shield that prevents pathogen entry. Some forest species have evolved thick bark, resin production, or waxy leaf surfaces that make it more difficult for pathogens to penetrate. For example, conifers produce oleoresins, which are rich in antimicrobial compounds and act as both a physical and chemical deterrent against pathogens. Chemical defenses include the production of secondary metabolites, such as alkaloids, terpenoids, and phenolics, which have antimicrobial properties. For instance, tannins in oak trees inhibit fungal growth, while phytoalexins, such as stilbenes in pines, are synthesized in response to infection and help limit pathogen spread.

4.2.2 Innate Immune Responses

Plants possess an innate immune system that allows them to recognize and respond to pathogen-associated molecular patterns (PAMPs), such as flagellin or chitin. This recognition triggers PAMP-triggered immunity (PTI), which includes the production of reactive oxygen species, the strengthening of cell walls through callose deposition, and the activation of defense-related genes. When pathogens suppress PTI through the action of effectors, plants activate a second layer of defense known as effector-triggered immunity (ETI). ETI is often stronger and more specific than PTI, leading to localized cell death (the hypersensitive response) that limits pathogen spread. For example, trees infected by rust fungi may develop necrotic lesions at the infection site as part of their ETI response.

4.2.3 Systemic Defense Mechanisms

In addition to localized responses, plants can activate systemic defense mechanisms that protect uninfected tissues. Systemic acquired resistance (SAR) is a long-lasting defense response that is induced after an initial infection. SAR involves the production of signaling molecules, such as salicylic acid, and the expression of pathogenesis-related (PR) proteins, which provide broad-spectrum resistance. Induced systemic resistance (ISR), mediated by beneficial microbes, is another form of systemic defense that enhances the plant's ability to resist subsequent pathogen attacks.

4.2.4 Trade-Offs and Resource Allocation

Defense responses are energetically costly, and plants must balance the allocation of resources between growth and defense. In forest ecosystems, this trade-off is particularly pronounced, as trees must invest in long-term survival strategies while responding to intermittent pathogen pressures. For instance, trees in nutrient-poor soils may prioritize growth over defense, making them more susceptible to pathogens.

4.3 Co-Evolution of Pathogens and Forest Trees

The interactions between pathogens and forest trees are shaped by a continuous process of co-evolution, where both parties adapt to gain an advantage. This evolutionary arms race has resulted in highly specialized relationships, with pathogens developing sophisticated strategies to overcome plant defenses and trees evolving diverse resistance mechanisms to counteract infection.

4.3.1 Evolution of Pathogen Virulence

Pathogens evolve virulence traits that enhance their ability to infect and exploit host plants. These traits include the production of enzymes that degrade plant cell walls, toxins that disrupt cellular functions, and effectors that suppress immune responses. For example, rust fungi have evolved complex life cycles that involve multiple host species, allowing them to persist and spread in diverse environments. Pathogen populations are highly dynamic, with genetic variation arising through mutation, recombination, and horizontal gene transfer. This variability enables pathogens to rapidly adapt to changes in host resistance or environmental conditions. In forest ecosystems, this adaptability is evident in the emergence of new pathogen strains that overcome resistance in previously resistant tree populations.

4.3.2 Evolution of Host Resistance

In response to pathogen pressure, forest trees evolve resistance mechanisms that limit infection and disease development. Resistance can be qualitative, involving specific gene-for-gene interactions, or quantitative, involving multiple genes that confer partial resistance. For example, the deployment of resistance (R) genes in conifers has been shown to limit infection by rust fungi and other pathogens. Forest tree populations exhibit high levels of genetic diversity, which is crucial for their ability to adapt to pathogen pressures. This diversity is maintained through natural selection, where individuals with resistant traits are more likely to survive and reproduce. Over time, this process shapes the genetic composition of tree populations, enhancing their resilience to pathogens.

4.3.3 Co-Evolutionary Dynamics

The co-evolutionary relationship between pathogens and forest trees is characterized by cycles of adaptation and counter-adaptation. When a pathogen evolves a new virulence trait, it may temporarily gain an advantage, leading to increased disease prevalence. However, this advantage is often short-lived, as host plants evolve new resistance mechanisms in response. This dynamic equilibrium ensures that neither the pathogen nor the host gains a permanent upper hand. Co-evolution is also influenced by environmental factors, such as climate change and habitat fragmentation, which can alter the balance of pathogen-host interactions. For example, warmer temperatures may accelerate pathogen evolution or disrupt the synchrony between host resistance and pathogen virulence, leading to increased disease outbreaks.

4.3.4 Implications for Forest Management

The co-evolutionary dynamics between pathogens and forest trees have important implications for forest management. Strategies that promote genetic diversity, such as the conservation of wild populations and the use of mixed-species plantations, can enhance the resilience of forests to pathogen pressures.

Additionally, breeding programs that incorporate knowledge of pathogen virulence mechanisms and host resistance traits can help develop tree varieties that are better equipped to withstand disease.

5. Impact of Plant Pathogens on Forest Health

Plant pathogens exert profound influences on forest health, with their effects ranging from localized infections to large-scale epidemics capable of transforming entire ecosystems. The impact of these pathogens is multifaceted, encompassing visible symptoms of disease, disruptions to ecological processes, and significant economic losses. In natural forests, pathogens are integral components of the ecosystem, shaping species composition and driving evolutionary dynamics. However, when environmental conditions or human activities disrupt this balance, pathogens can become devastating agents of decline. Forest health is particularly vulnerable in the face of increasing global challenges, such as climate change, habitat fragmentation, and the spread of invasive species, all of which exacerbate the impact of plant pathogens.

5.1 Symptoms of Major Diseases in Forest Ecosystems

The symptoms of plant pathogen infections in forest ecosystems are as diverse as the pathogens themselves, reflecting the wide variety of strategies pathogens use to infect hosts and disrupt their physiological functions. These symptoms often serve as critical indicators of forest health, providing insights into the underlying causes of tree decline. Pathogen-induced symptoms can manifest at the cellular, tissue, organ, or whole-plant level, with severity depending on the type of pathogen, the susceptibility of the host, and environmental conditions.

5.1.1 Root and Vascular Symptoms

Root infections, often caused by soilborne pathogens such as *Phytophthora* and *Armillaria*, result in root rot, which impairs water and nutrient uptake. Trees with root infections exhibit symptoms such as wilting, chlorosis (yellowing of leaves), and stunted growth. Severe infections can lead to tree instability and uprooting, particularly during storms. Vascular wilt diseases, caused by pathogens like *Ceratocystis* and *Fusarium*, block xylem vessels, disrupting water transport and causing rapid wilting, leaf drop, and dieback.

5.1.2 Cankers and Stem Symptoms

Canker diseases, such as those caused by *Cryphonectria parasitica* (chestnut blight), result in localized necrosis on stems and branches. These cankers girdle the tree, cutting off the flow of water and nutrients and often leading to death. Symptoms include sunken, discolored lesions that may ooze resin or sap, creating entry points for secondary infections.

5.1.3 Leaf and Needle Blights

Pathogens targeting foliage, such as *Dothistroma septosporum* and *Mycosphaerella pini*, cause leaf and needle blights characterized by discoloration, necrotic spots, and premature leaf drop. These symptoms reduce the photosynthetic capacity of trees, weakening them over time and increasing their susceptibility to other stresses.

5.1.4 Galls and Abnormal Growths

Gall-forming pathogens, including nematodes and bacteria like *Agrobacterium tumefaciens*, induce abnormal growths on roots, stems, and leaves. These galls disrupt nutrient flow and can serve as reservoirs for other pathogens and pests. Galls are often visible and can help diagnose specific diseases.

5.1.5 Dieback and Mortality

In severe cases, pathogen infections lead to dieback, characterized by the progressive death of branches and twigs, and eventual tree mortality. This is common in diseases like sudden oak death caused by *Phytophthora ramorum*, where extensive necrosis and cankers result in widespread tree death. Symptoms often overlap or are exacerbated by environmental stresses, such as drought, which complicates diagnosis. However, understanding the specific symptoms associated with major forest diseases is crucial for early detection and effective management.

5.2 Economic and Ecological Consequences of Pathogen Outbreaks

The consequences of plant pathogen outbreaks extend far beyond the infected trees, impacting both the economic value of forests and the ecological functions they provide. These impacts are often interlinked, with economic losses stemming from ecological disruptions and vice versa.

5.2.1 Economic Consequences

Forests are a major economic resource, providing timber, non-timber forest products, and ecosystem services that underpin livelihoods and industries. Pathogen outbreaks can significantly reduce the economic value of forests through tree mortality, loss of timber quality, and increased management costs. For example, Dutch elm disease (*Ophiostoma ulmi*), which has devastated elm populations across Europe and North America, caused substantial losses in the timber and ornamental tree industries. Similarly, sudden oak death has affected the nursery and forestry industries, with costly quarantines and removal efforts. The economic burden of pathogen outbreaks also includes indirect costs, such as reduced property values in urban areas, loss of recreational opportunities, and diminished ecosystem services like carbon sequestration and water regulation. In regions where communities depend on forests for their livelihoods, outbreaks can lead to economic instability and increased vulnerability to poverty.

5.2.2 Ecological Consequences

Pathogen outbreaks have profound ecological consequences, altering forest composition, structure, and functionality. By selectively targeting specific species, pathogens can shift competitive dynamics and reduce biodiversity. For instance, chestnut blight led to the near-extinction of the American chestnut (*Castanea dentata*), which once dominated eastern U.S. forests. This loss not only altered forest structure but also impacted wildlife species that depended on chestnuts for food and habitat. Pathogens also disrupt nutrient cycling and energy flows within ecosystems. Tree mortality resulting from pathogen outbreaks increases the availability of dead organic matter, fueling decomposition processes and altering soil nutrient dynamics. However, the loss of canopy cover can lead to soil erosion, reduced water retention, and changes in microclimate conditions, with cascading effects on understory vegetation and wildlife. The impact of pathogens on forest health is further amplified by their interactions with other stressors, such as climate change and invasive species. Warmer temperatures and altered precipitation patterns can increase pathogen activity and expand their geographic range, while invasive pathogens often encounter naive host populations, leading to severe outbreaks.

5.3 Case Studies of Significant Forest Disease Epidemics

Several case studies illustrate the devastating impact of pathogen outbreaks on forest ecosystems, highlighting the complexity and scale of these events. These examples underscore the need for proactive monitoring and management strategies to mitigate the effects of future outbreaks.

5.3.1 Chestnut Blight

One of the most infamous forest disease epidemics, chestnut blight, caused by *Cryphonectria parasitica*, was introduced to North America in the early 20th century. The pathogen, native to Asia, spread rapidly through eastern U.S. forests, infecting and killing billions of American chestnuts within a few decades. This epidemic not only eliminated a keystone species but also disrupted ecosystems that relied on chestnuts for food, shade, and soil stabilization. Efforts to restore the American chestnut have focused on breeding blight-resistant hybrids, but full ecological recovery remains elusive.

5.3.2 Dutch Elm Disease

Dutch elm disease, caused by *Ophiostoma ulmi* and later *Ophiostoma novo-ulmi*, has devastated elm populations in Europe and North America since the early 20th century. The disease, spread by bark beetles, blocks the vascular tissues of trees, causing wilting and death. Urban and rural landscapes have been profoundly altered, with elms replaced by less vulnerable species. Management efforts, including fungicides, sanitation, and resistant cultivars, have slowed the spread but have not eradicated the disease.

5.3.3 Sudden Oak Death

Sudden oak death, caused by *Phytophthora ramorum*, emerged in California and Oregon in the late 20th century, causing widespread mortality in tanoaks, coast live oaks, and other species. The pathogen infects a wide range of hosts, including ornamental plants, making it a significant threat to both natural ecosystems and the horticultural industry. The disease has led to changes in forest composition and increased fire risk due to the accumulation of dead biomass. Efforts to control sudden oak death have focused on monitoring, quarantines, and selective removal of infected trees.

5.3.4 Pine Wilt Disease

Pine wilt disease, caused by the pinewood nematode (*Bursaphelenchus xylophilus*), has caused severe damage to pine forests in Asia and Europe. The nematode, transmitted by beetle vectors, infects xylem tissues, disrupting water transport and causing rapid tree death. Pine wilt disease has had significant economic impacts on the forestry industry and has prompted intensive management efforts, including vector control, removal of infected trees, and the use of resistant pine varieties.

5.3.5 Ash Dieback

Ash dieback, caused by *Hymenoscyphus fraxineus*, has devastated ash populations across Europe since its introduction in the 1990s. The disease causes leaf necrosis, dieback, and eventual tree mortality, with young trees particularly vulnerable. The loss of ash trees has altered forest ecosystems, affecting species that depend on ash for food and habitat. Research efforts are focused on identifying resistant ash genotypes and developing strategies to mitigate the disease's impact.

6. Environmental Factors Influencing Pathogen Prevalence

The prevalence and impact of plant pathogens in forest ecosystems are not solely determined by the biology of the pathogens and their hosts; they are profoundly influenced by environmental factors that govern pathogen survival, reproduction, and dissemination. Environmental conditions create the context in which pathogens interact with their hosts, shaping disease dynamics at both local and global scales. Among these factors, climate change, soil conditions, forest structure, and anthropogenic activities play pivotal roles. As forests face increasing environmental pressures due to global warming, habitat fragmentation, pollution, and the movement of invasive species, understanding these influences has become more critical than ever. Each environmental factor operates on multiple levels, affecting not just individual pathogen-host interactions but also broader ecosystem processes that determine the vulnerability and resilience of forests. These influences are not static; they evolve and interact with one another, creating complex feedback loops that amplify or suppress disease outbreaks.

6.1 Role of Climate Change in Forest Disease Dynamics

Climate change is one of the most significant drivers of forest disease dynamics in the 21st century, altering the relationships between pathogens, hosts, and their environment. Rising global temperatures, shifting precipitation patterns, and increased frequency of extreme weather events are transforming the conditions that determine pathogen survival and activity. These changes are not uniform but vary across regions, species, and ecosystems, creating a complex and unpredictable landscape for forest diseases.

Higher temperatures have direct and indirect effects on plant pathogens. Many fungal pathogens, for instance, thrive in warmer conditions, which accelerate their growth, reproduction, and spore dispersal. Warmer winters reduce the mortality of overwintering pathogen populations, enabling earlier and more severe outbreaks in subsequent growing seasons. For example, pathogens such as *Dothistroma septosporum*, which causes red band needle blight in conifers, have expanded their geographic range and increased in severity due to warmer temperatures in northern regions. Similarly, bark beetle outbreaks, which facilitate the spread of fungal pathogens like *Ophiostoma* species (Dutch elm disease), have intensified under warming conditions, leading to widespread tree mortality in temperate forests.

Shifts in precipitation patterns further complicate forest disease dynamics. Increased rainfall and humidity create favorable conditions for many waterborne and foliar pathogens, such as *Phytophthora* species and rust fungi. Conversely, prolonged droughts weaken tree defenses, making them more susceptible to opportunistic pathogens. Drought-stressed trees, unable to produce sufficient defensive compounds or maintain adequate water transport, are particularly vulnerable to vascular wilt diseases and root rot pathogens. This interplay between drought stress and pathogen pressure has been observed in Mediterranean forests, where *Phytophthora cinnamomi* has caused extensive oak decline. Extreme weather events, including storms, floods, and heatwaves, exacerbate the impacts of pathogens by creating stress conditions that weaken tree resilience. Storms and high winds cause physical injuries to trees, providing entry points for pathogens, while flooding promotes the proliferation of soilborne pathogens by creating anaerobic conditions. Heatwaves, on the other hand, can disrupt the synchrony between host resistance mechanisms and pathogen activity, tipping the balance in favor of the pathogen.

Climate change also affects the phenology and geographic distribution of both pathogens and their hosts. Pathogens may shift their ranges to higher altitudes or latitudes as temperatures rise, encountering new hosts and potentially causing novel disease outbreaks. For example, *Hymenoscyphus fraxineus*, the causative agent of ash dieback, has expanded its range across Europe, with devastating effects on ash populations. Changes in phenology, such as earlier bud break or prolonged growing seasons, can alter the timing and severity of disease outbreaks by increasing the window of vulnerability for host plants. Moreover, the interaction between climate change and other environmental factors amplifies disease risks. For instance, warming temperatures combined with habitat fragmentation and pollution can reduce genetic diversity in forest species, weakening their collective resistance to pathogens. Similarly, invasive pathogens, introduced through global trade, may gain an advantage in climates that differ from those in their native range, leading to severe outbreaks in new regions. The cumulative impact of climate change on forest pathogens is evident in the increasing frequency and severity of disease outbreaks worldwide. Addressing these challenges requires integrating climate models with disease monitoring and management strategies, emphasizing the need for proactive approaches to forest conservation.

6.2 Influence of Soil Conditions and Forest Structure

Soil conditions and forest structure are foundational components of forest ecosystems that shape pathogen prevalence and disease dynamics. Soil provides the physical, chemical, and biological environment in which many pathogens live and interact with their hosts, while forest structure determines the spatial and temporal distribution of resources, microclimates, and species interactions. Together, these factors create the conditions that enable or suppress pathogen activity, influencing the vulnerability and resilience of forests to disease.

Soilborne pathogens, such as *Phytophthora* species, *Armillaria* root rot fungi, and nematodes like *Bursaphelenchus xylophilus*, are particularly sensitive to soil properties. Factors such as soil moisture, temperature, pH, nutrient availability, and organic matter content directly affect the survival, reproduction, and infectivity of these pathogens. For example, *Phytophthora cinnamomi*, a notorious root rot pathogen, thrives in waterlogged soils with poor drainage, causing severe damage in forests with compacted or poorly aerated soils. Conversely, well-drained soils with high organic matter content and active microbial communities can suppress soilborne pathogens by promoting competition and antagonism from beneficial microbes.

The biological composition of the soil microbiome plays a critical role in regulating pathogen activity. Beneficial organisms, including mycorrhizal fungi, rhizobacteria, and decomposers, contribute to soil health and plant defense by competing with pathogens for nutrients, producing antimicrobial compounds, and enhancing host resistance. Forest management practices that disrupt soil microbiomes, such as clear-cutting, intensive tillage, or chemical inputs, can reduce microbial diversity and resilience, increasing susceptibility to pathogens. Forest structure, encompassing factors such as tree density, species composition, canopy cover, and vertical stratification, further influences pathogen prevalence. Dense forests with closed canopies create humid microclimates that are ideal for the proliferation of foliar pathogens, such as rusts and powdery mildews. In contrast, open-canopy forests with greater airflow and sunlight penetration tend to have lower disease pressure but may be more vulnerable to abiotic stresses like drought. Mixed-species forests, which include a diversity of tree species with varying susceptibility to pathogens, are generally more resilient to disease outbreaks than monocultures, where a single pathogen can rapidly spread and devastate the entire stand.

Forest age and successional stage also affect pathogen dynamics. Young, rapidly growing forests may be more susceptible to certain pathogens due to their high nutrient demands and thinner bark, which provides less protection against infection. Mature forests, on the other hand, may harbor chronic infections or latent pathogens that become active under stress conditions. The spatial arrangement of trees, including the proximity of susceptible hosts and the presence of natural barriers such as understory vegetation, can further modulate disease spread and severity. Interactions between soil conditions and forest structure are critical in shaping disease dynamics. For example, nutrient-poor soils can limit tree growth and reduce canopy cover, altering microclimates and increasing vulnerability to foliar pathogens. Similarly, disturbances such as logging or fire can compact soils, reduce organic matter content, and disrupt forest structure, creating conditions favorable for opportunistic pathogens.

6.3 Anthropogenic Activities and Pathogen Introduction

Human activities have become one of the most significant factors influencing pathogen prevalence and disease dynamics in forest ecosystems. The scale and intensity of anthropogenic disturbances, including habitat fragmentation, pollution, global trade, and forest management practices, have created conditions that favor the emergence and spread of plant pathogens. These activities not only introduce pathogens to new environments but also weaken the natural defenses of forests, increasing their disease susceptibility.

One of the most direct ways humans affect pathogen prevalence is through introducing invasive species. Global trade and transportation have facilitated the movement of plant pathogens across continents, often with devastating consequences. Invasive pathogens, such as Cryphonectria parasitica (chestnut blight), Hymenoscyphus fraxineus (ash dieback), and Phytophthora ramorum (sudden oak death), have caused widespread damage in regions where native plants lack resistance. These introductions are often unintentional, occurring through contaminated soil, nursery stock, or wooden packaging materials. Once established, invasive pathogens are difficult to control, spreading rapidly and causing long-term ecological and economic damage.

Habitat fragmentation and deforestation are another major drivers of pathogen prevalence. Fragmentation reduces genetic diversity within tree populations, making them more vulnerable to pathogens. Small, isolated forest patches are less able to support healthy populations of beneficial organisms, such as predators and competitors of pathogens. Additionally, fragmented forests often experience edge effects, where altered microclimates and increased human activity create conditions conducive to pathogen outbreaks. Pollution, including heavy metals, pesticides, and nitrogen deposition, further exacerbates pathogen pressures. These pollutants can weaken tree defenses, disrupt soil microbial communities, and alter nutrient cycling, creating conditions that favor pathogen survival and infection. For example, acid rain has been linked to increased susceptibility of spruce and fir trees to root rot pathogens in temperate forests. Forest management practices, such as monoculture plantations, clear-cutting, and excessive thinning, can also contribute to pathogen prevalence. Monocultures provide a continuous host resource for pathogens, enabling rapid spread and high disease severity. Clear-cutting and other intensive practices disrupt soil structure, reduce microbial diversity, and create stress conditions that weaken tree resilience. Conversely, sustainable management practices, such as promoting mixed-species stands, maintaining soil health, and minimizing disturbance, can mitigate these risks.

Anthropogenic activities also interact with climate change to amplify pathogen pressures. For instance, warmer temperatures and increased trade create opportunities for pathogens to establish in regions previously unsuitable for their survival. Similarly, habitat fragmentation combined with extreme weather events can create stress conditions that trigger latent pathogen infections. The cumulative impact of anthropogenic activities on forest health is profound, highlighting the need for integrated approaches to pathogen management. Addressing these challenges requires global cooperation, robust biosecurity measures, and the adoption of sustainable forestry practices that balance economic, ecological, and social objectives.

7. Methods for Studying Forest Pathogens

Studying forest pathogens requires a multidisciplinary approach that combines traditional methods, advanced molecular technologies, and spatial analysis tools. The complexity of forest ecosystems, coupled with the diversity of pathogens and their interactions with hosts, necessitates robust methods for identifying pathogens, understanding their biology, and monitoring their spread. Effective study methods allow researchers to diagnose diseases, predict outbreaks, and develop management strategies that mitigate the impact of pathogens on forest health.

7.1 Traditional Approaches: Field Surveys and Morphological Studies

Traditional approaches to studying forest pathogens form the foundation of forest pathology and remain indispensable in understanding disease dynamics. These methods involve direct observation of diseased trees, laboratory-based morphological studies, and experimental techniques to identify pathogens and their impacts. Although these approaches are sometimes labor-intensive and limited in scale, they provide critical baseline data and are essential for establishing pathogen-host relationships.

7.1.1 Field Surveys and Disease Mapping

Field surveys are a primary method for studying forest pathogens in their natural environments. These surveys involve systematically assessing forested areas to identify diseased trees, record symptoms, and collect samples for further analysis. Field surveys provide critical information on the prevalence, distribution, and severity of diseases, which is essential for understanding the spatial dynamics of pathogen outbreaks. Researchers often use disease mapping to visualize the spread of pathogens over time and identify environmental factors that influence disease hotspots. For example, during the spread of sudden oak death caused by *Phytophthora ramorum*, field surveys were instrumental in identifying affected regions, hosts, and environmental conditions conducive to disease development.

7.1.2 Morphological Identification of Pathogens

Morphological studies involve examining the physical characteristics of pathogens, such as spores, hyphae, bacterial colonies, or nematodes, to identify species and understand their biology. Microscopy, staining techniques, and culture-based methods are commonly used to study the structure and growth of pathogens. For instance, rust fungi are identified based on the morphology of their urediniospores and teliospores, while nematodes are distinguished by their stylet structure and feeding sites. These methods are particularly useful for identifying fungal and nematode pathogens, which often exhibit distinctive morphological features.

7.1.3 Koch's Postulates and Pathogenicity Tests

To confirm the causal relationship between a pathogen and a disease, researchers apply Koch's postulates, a set of criteria that involve isolating the pathogen from a diseased plant, inoculating it into a healthy host, and observing disease symptoms. Pathogenicity tests, including artificial inoculations and controlled infection experiments, are used to study the host range, infection mechanisms, and virulence of pathogens. These tests are critical for understanding the conditions under which pathogens cause disease and for developing effective management strategies.

7.1.4 Limitations of Traditional Methods

While traditional approaches provide valuable insights, they have limitations. Morphological studies often require expertise and are time-consuming, and some pathogens exhibit overlapping or cryptic features that make identification challenging. Field surveys can be resource-intensive, especially in large or remote forest areas, and are sometimes limited in detecting pathogens at early stages of infection. These limitations highlight the need for integrating traditional methods with advanced technologies for a comprehensive understanding of forest pathogens.

7.2 Molecular Techniques: Genomics, Proteomics, and Diagnostics

The advent of molecular biology has revolutionized the study of forest pathogens, providing powerful tools to unravel their genetics, biochemistry, and interactions with hosts. Molecular techniques offer unparalleled precision and scalability, enabling researchers to identify pathogens, trace their origins, and study their mechanisms of infection. Genomics, proteomics, and molecular diagnostics have become indispensable in modern forest pathology.

7.2.1 Genomics and Transcriptomics

Genomics involves sequencing and analyzing the complete genetic material of pathogens to understand their biology, evolution, and virulence factors. Whole-genome sequencing has been applied to a wide range of forest pathogens, including *Phytophthora ramorum*, *Cryphonectria parasitica*, and *Hymenoscyphus fraxineus*. Genomic data provide insights into pathogen diversity, population structure, and adaptive traits that influence disease dynamics. Transcriptomics, which examines gene expression patterns during infection, reveals how pathogens respond to host defenses and environmental conditions. For instance, transcriptomic studies of *Armillaria* root rot fungi have identified genes involved in lignin degradation and host colonization.

7.2.2 Proteomics and Metabolomics

Proteomics, the study of the complete set of proteins produced by a pathogen, sheds light on the molecular mechanisms underlying infection and host manipulation. Proteomic analyses identify effector proteins secreted by pathogens to suppress host immune responses, such as those used by rust fungi and bacterial pathogens. Metabolomics complements proteomics by analyzing the chemical compounds produced during pathogen-host interactions, including toxins, enzymes, and secondary metabolites. These approaches are particularly useful for identifying targets for disease management and developing resistance strategies.

7.2.3 Molecular Diagnostics

Molecular diagnostic tools, such as polymerase chain reaction (PCR), quantitative PCR (qPCR), and loop-mediated isothermal amplification (LAMP), enable the rapid and accurate detection of pathogens in forest samples. These techniques amplify specific DNA or RNA sequences unique to the pathogen, allowing for early detection and monitoring. For example, qPCR assays have been developed to detect *Phytophthora* species in soil and water samples, facilitating the management of sudden oak death. High-throughput sequencing technologies, such as next-generation sequencing (NGS), provide comprehensive insights into microbial communities and identify novel or emerging pathogens.

7.2.4 Applications in Disease Management

Molecular techniques have practical applications in breeding resistant tree species, developing diagnostic kits, and tracing the origins of invasive pathogens. Genomic tools have been used to identify resistance genes in forest species, such as the resistance of American chestnut hybrids to chestnut blight. Molecular diagnostics are widely employed in quarantine programs to prevent the introduction of pathogens through international trade.

7.2.5 Limitations and Challenges

Despite their advantages, molecular techniques require specialized equipment, technical expertise, and significant financial investment. The vast amount of data generated by genomics and proteomics also necessitates robust bioinformatics tools for analysis and interpretation. Additionally, molecular diagnostics may not always distinguish between viable and non-viable pathogens, which can complicate disease management.

7.3 Remote Sensing and Spatial Analysis in Pathogen Monitoring

Advances in remote sensing and spatial analysis have opened new avenues for studying forest pathogens at landscape and regional scales. These technologies provide tools for monitoring disease outbreaks, predicting pathogen spread, and assessing environmental factors that influence disease dynamics. Remote sensing and spatial analysis are particularly valuable for studying pathogens in large, inaccessible, or remote forest areas.

7.3.1 Satellite and Aerial Remote Sensing

Remote sensing technologies, including satellite imagery and aerial surveys, allow researchers to monitor forest health over large areas. High-resolution multispectral and hyperspectral imaging can detect changes in canopy reflectance associated with disease symptoms, such as chlorosis, defoliation, and dieback. For instance, hyperspectral data have been used to identify areas affected by sudden oak death, enabling targeted management efforts. Thermal imaging, which detects temperature changes in infected trees, is another tool for identifying stressed or diseased trees.

7.3.2 Unmanned Aerial Vehicles (Drones)

Unmanned aerial vehicles (UAVs) or drones equipped with cameras and sensors provide high-resolution data for studying forest pathogens. Drones offer flexibility and precision, allowing researchers to monitor individual trees or specific forest areas. For example, drones have been used to map the spread of *Hymenoscyphus fraxineus* in ash forests and to assess the severity of bark beetle outbreaks, which facilitate fungal pathogen infections.

7.3.3 Spatial Analysis and Geographic Information Systems (GIS)

Spatial analysis tools, including GIS, enable researchers to integrate field data, remote sensing imagery, and environmental variables to model disease dynamics. GIS can identify spatial patterns of disease prevalence, assess the influence of environmental factors, and predict the spread of pathogens. For example, spatial models have been used to predict the distribution of *Phytophthora cinnamomi* based on soil conditions, climate, and host distribution. These models help prioritize areas for monitoring and intervention.

7.3.4 Disease Prediction and Early Warning Systems

Remote sensing and spatial analysis are critical for developing disease prediction and early warning systems. By combining environmental data, pathogen biology, and host susceptibility, researchers can identify areas at risk of disease outbreaks and implement preventive measures. For example, early warning systems for pine wilt disease use climatic data and vector activity to predict outbreaks and guide management efforts.

7.3.5 Limitations and Challenges

While remote sensing and spatial analysis provide powerful tools, they have limitations. Satellite imagery may lack the resolution needed to detect early-stage infections, and aerial surveys can be affected by weather conditions or canopy density. Integrating remote sensing data with ground-based observations and molecular diagnostics is essential for accurate pathogen monitoring and management.

8. Management and Control of Forest Pathogens

The management and control of forest pathogens represent a critical challenge in modern forestry, where the increasing frequency and severity of disease outbreaks threaten the health and sustainability of global forest ecosystems. Plant pathogens are inherently dynamic, adapting to changing environments, evolving alongside their hosts, and spreading through both natural and anthropogenic pathways. These characteristics make forest pathogen management an intricate and multifaceted endeavor. Unlike managed agricultural systems, forests are complex and largely uncontrolled ecosystems, where pathogens interact with a diverse array of hosts, environmental variables, and competing organisms. The global movement of goods and people, coupled with the accelerating effects of climate change, has amplified the risks associated with forest pathogens, creating an urgent need for comprehensive, adaptable, and sustainable management strategies. Three major pillars underpin the control of forest pathogens: prevention strategies that focus on stopping the introduction and spread of pathogens, biological control methods that harness the natural enemies of pathogens to mitigate their impacts, and integrated forest disease management approaches that combine multiple strategies to address the complexities of forest ecosystems. These approaches emphasize the importance of proactive measures, ecological balance, and collaboration across disciplines and stakeholders. While the challenges are significant, advances in technology, scientific understanding, and global cooperation offer new opportunities for effective pathogen management.

8.1 Prevention Strategies: Quarantine, Resistant Species, and Biosecurity

Prevention remains the cornerstone of forest pathogen management, offering the most effective and economically viable means of protecting forest ecosystems. Once a pathogen is introduced and established, eradication is often impossible, and control measures become costly, disruptive, and less effective. Preventive strategies focus on reducing the risk of pathogen introduction, limiting their spread within and between ecosystems, and enhancing the resilience of forests to withstand potential infections. Key components of prevention include quarantine measures, the development and deployment of resistant tree species, and the implementation of comprehensive biosecurity protocols.

Quarantine measures form the first line of defense against the introduction of invasive pathogens, which are often transported across borders through trade in plants, seeds, wood products, and soil. International agreements and national regulations, such as those outlined by the International Plant Protection Convention (IPPC) and regional plant health organizations, establish protocols for the inspection, certification, and treatment of materials to minimize the risk of pathogen introduction. For instance, the movement of nursery stock, which has been implicated in the spread of pathogens like *Phytophthora ramorum* (sudden oak death), is tightly regulated in many regions to prevent further outbreaks. Quarantine measures also include the establishment of pest-free zones, buffer areas around infected regions, and restrictions on the import of high-risk species. The success of quarantine relies heavily on enforcement, surveillance, and public compliance, as even a single breach can lead to widespread consequences. The introduction of *Cryphonectria parasitica* into North America, which caused the devastating chestnut blight epidemic, underscores the catastrophic impacts of inadequate quarantine practices.

The use of resistant tree species is another critical preventive measure, providing a sustainable and long-term solution to managing forest pathogens. Resistance can be achieved through natural selection, where populations of trees evolve defenses against local pathogens, or through deliberate breeding programs that enhance resistance traits. Advances in molecular biology, including genomic sequencing and CRISPR-based gene editing, have revolutionized the development of resistant species by enabling the identification and manipulation of resistance genes. For example, efforts to restore the American chestnut (*Castanea dentata*) have focused on breeding hybrids with resistance to chestnut blight, offering a pathway for reintroducing this iconic species into eastern U.S. forests. However, the deployment of resistant species must be carefully managed to maintain genetic diversity and prevent the emergence of more virulent pathogen strains, which can result from selective pressures imposed by widespread resistance.

Biosecurity encompasses a broad range of practices designed to prevent the introduction and spread of pathogens within and between forested areas. These practices include early detection systems, sanitation measures, public awareness campaigns, and the development of contingency plans for disease outbreaks. Monitoring programs that use molecular diagnostics, remote sensing, and field surveys are essential for detecting pathogens at early stages of infection, allowing for rapid response and containment. Sanitation measures, such as the removal and destruction of infected material, decontamination of equipment, and restrictions on the movement of soil and plant debris, help to limit the spread of pathogens within forests. Public awareness and engagement are also critical, as human activities, including recreational use of forests, are significant pathways for pathogen spread.

For example, the spread of *Hymenoscyphus fraxineus* (ash dieback) in Europe has been linked to the movement of contaminated plant material and soil, highlighting the importance of public education and stakeholder involvement in biosecurity efforts.

8.2 Biological Control: Beneficial Microbes and Natural Predators

Biological control represents a sustainable and ecologically friendly approach to managing forest pathogens by leveraging the natural relationships between pathogens, their hosts, and other organisms within the ecosystem. This method involves the use of beneficial microbes, such as antagonistic fungi and bacteria, and natural predators or parasites that suppress pathogen populations through competition, predation, or other interactions. Unlike chemical treatments, which can have negative environmental impacts and limited long-term efficacy, biological control methods are integrated into the natural dynamics of the forest, promoting resilience and reducing dependency on external inputs.

Beneficial microbes play a central role in biological control, particularly in the rhizosphere, where complex interactions between pathogens, plants, and microbial communities determine the health of forest soils. Mycorrhizal fungi, which form mutualistic associations with tree roots, enhance nutrient and water uptake while producing bioactive compounds that inhibit pathogen growth. For example, ectomycorrhizal fungi have been shown to suppress the activity of *Phytophthora* species by competing for root colonization sites and altering soil chemistry. Similarly, rhizobacteria such as *Pseudomonas fluorescens* and *Bacillus subtilis* produce antibiotics, siderophores, and enzymes that disrupt pathogen metabolism and enhance plant immune responses. The application of these microbes as biocontrol agents has been explored in various contexts, from preventing root rot in conifers to reducing the severity of foliar diseases in hardwoods.

Natural predators and parasites also contribute to the biological control of forest pathogens. Hyperparasites, such as *Trichoderma* species, target pathogenic fungi by colonizing their structures, disrupting their growth, and outcompeting them for resources. These fungi are widely used in commercial biocontrol products to manage root and stem diseases, including those caused by *Armillaria* and *Fusarium* species. In addition to fungal hyperparasites, insect predators can indirectly reduce pathogen impacts by controlling vector populations. For instance, predatory beetles that feed on bark beetles help limit the spread of fungal pathogens like *Ophiostoma* species (Dutch elm disease). While these methods show significant promise, their effectiveness often depends on environmental conditions, such as soil moisture, temperature, and forest structure, which can influence the behavior and survival of biocontrol agents.

Despite their advantages, biological control methods face several challenges. The establishment and persistence of biocontrol agents in natural forest ecosystems are not always guaranteed, as they must compete with native microbial communities and adapt to variable environmental conditions. Additionally, the introduction of non-native biocontrol agents carries ecological risks, including potential disruption of existing microbial balances and unintended effects on non-target organisms. These risks necessitate rigorous testing, monitoring, and regulatory oversight to ensure the safety and efficacy of biocontrol programs.

8.3 Integrated Forest Disease Management Approaches

Integrated forest disease management (IFDM) provides a holistic framework for controlling forest pathogens by combining multiple strategies that address the complexities of forest ecosystems. This approach recognizes that no single method is sufficient to manage the diverse and dynamic interactions between pathogens, hosts, and the environment. Instead, IFDM integrates prevention, biological control, cultural practices, and, where necessary, chemical treatments, tailoring interventions to the specific pathogens, forest conditions, and management objectives. Central to IFDM is the establishment of robust monitoring and early detection systems, which provide the foundation for effective decision-making. By integrating traditional field surveys with advanced technologies such as remote sensing, molecular diagnostics, and geographic information systems (GIS), IFDM enables real-time assessment of pathogen prevalence, distribution, and progression. For example, hyperspectral imaging and drone-based surveys have been used to detect early symptoms of sudden oak death in California forests, allowing for targeted interventions that minimize the spread of the disease. Early detection also facilitates the implementation of quarantine measures, reducing the risk of pathogen introduction into unaffected areas.

Forest management practices play a critical role in IFDM, as they influence the physical, biological, and environmental factors that affect pathogen dynamics. Thinning dense stands improves airflow and light penetration, reducing humidity levels that favor the growth of fungal pathogens. Pruning infected branches and removing diseased trees helps to limit pathogen reservoirs and reduce the risk of secondary infections. Mixed-species plantations, which include a diversity of tree species with varying susceptibility to pathogens, are particularly effective in enhancing forest resilience by disrupting the continuous host availability required for pathogen proliferation. For instance, interplanting resistant tree species with susceptible ones has been shown to mitigate the impact of pine wilt disease caused by *Bursaphelenchus xylophilus*.

While IFDM prioritizes sustainable and ecological approaches, chemical treatments are sometimes necessary to manage severe outbreaks or protect high-value forests. Fungicides, bactericides, and nematicides can provide short-term relief in situations where other methods are insufficient. However, their use must be carefully regulated to minimize environmental impacts, prevent the development of pathogen resistance, and ensure compatibility with other management strategies. Advances in precision application technologies, such as drone-based spraying, have improved the efficiency and specificity of chemical treatments, reducing their ecological footprint. Adaptive management is a defining feature of IFDM, emphasizing the need for continuous evaluation and refinement of strategies based on monitoring data, research findings, and changing environmental conditions. This flexibility allows forest managers to respond to emerging threats, such as new pathogen strains or climate-induced shifts in disease dynamics, ensuring the long-term health and sustainability of forest ecosystems. By integrating scientific knowledge, technological innovations, and collaborative efforts among stakeholders, IFDM offers a comprehensive and effective approach to managing forest pathogens in an era of unprecedented environmental change.

Conclusion

The study of plant pathogens in forest ecosystems is fundamental to understanding the dynamic interplay between biotic and abiotic factors that shape these complex environments. Plant pathogens, encompassing fungi, bacteria, viruses, viroids, and nematodes, play multifaceted roles that extend beyond disease causation to influence biodiversity, nutrient cycling, ecological succession, and species balance. While their impacts can be destructive, often manifesting in the form of severe disease outbreaks and widespread tree mortality, pathogens also contribute to the maintenance of ecological stability and resilience within forests.

One of the most significant reasons for studying plant pathogens in forests lies in their potential to disrupt ecological and economic systems. Diseases caused by pathogens affect forest structure, composition, and the delivery of vital ecosystem services, such as carbon sequestration, habitat provision, and water regulation. These disruptions are not confined to isolated events but often have cascading effects on global ecological processes and human economies reliant on forests. Consequently, understanding the dynamics of forest pathogens is not only an ecological imperative but also a socio-economic necessity.

Forest ecosystems exhibit characteristics that both facilitate and mitigate the spread and impact of pathogens. The structural complexity, species diversity, and microclimatic variability within forests create a dynamic backdrop where pathogens can establish and propagate. These same features, however, can also serve as buffers against widespread disease outbreaks by diluting pathogen-host interactions. The interplay between pathogen prevalence, host density, and environmental factors highlights the importance of maintaining forest biodiversity as a natural defense mechanism against pathogens. The interactions between pathogens and their host plants are defined by a co-evolutionary arms race, where each entity continuously adapts to the other's strategies for infection or defense. Pathogens deploy a variety of mechanisms, such as enzyme production, toxin secretion, and effector proteins, to overcome plant defenses. Conversely, plants exhibit a sophisticated array of defense responses, including structural barriers, immune signaling pathways, and antimicrobial compounds. The evolutionary pressure exerted by these interactions has shaped the genetic diversity and resilience of forest species, underscoring the intricate balance that governs pathogen-host dynamics.

Pathogens' ecological roles extend beyond direct plant infection. As natural regulators, they contribute to maintaining biodiversity by selectively suppressing dominant species and enabling ecological succession. Their role in nutrient cycling, through the decomposition of organic matter, supports the productivity and stability of forest ecosystems. These ecological benefits, however, do not diminish the need for vigilance against their potential to cause significant harm, particularly in the context of anthropogenic activities and environmental changes.

The impact of plant pathogens is amplified by environmental factors such as climate change, altered soil conditions, and human-mediated activities. Climate change, in particular, is emerging as a critical driver of forest disease dynamics, influencing pathogen survival, reproduction, and geographical range. Rising temperatures, altered precipitation patterns, and increased frequency of extreme weather events create conditions that favor the proliferation and spread of pathogens. Additionally, human activities, including logging, land-use changes, and global trade, exacerbate the risk of introducing invasive pathogens, posing a significant threat to native forest ecosystems. To mitigate the impact of plant pathogens, a multifaceted approach is required, integrating traditional and modern methods of study and management. Advances in molecular biology, genomics, and remote sensing have revolutionized the study of pathogens, enabling precise identification, monitoring, and characterization. These tools, combined with traditional field-based observations, provide a comprehensive understanding of pathogen dynamics and their environmental interactions. Management strategies, ranging from preventive measures such as quarantine and biosecurity to curative methods like biological control and silvicultural practices, offer pathways to mitigate disease impact while maintaining forest health and productivity. The future of forest pathogen management lies in adopting integrated and adaptive strategies that account for the complex and evolving nature of forest ecosystems. This requires a holistic approach that combines scientific research, ecological knowledge, and sustainable management practices. Promoting forest resilience through biodiversity conservation, fostering natural defense mechanisms, and addressing the root causes of disease outbreaks are essential components of such strategies.

Chapter - 3

Classification of Forest Diseases

1. Introduction to Forest Disease Classification

The classification of forest diseases is a cornerstone of forest pathology, providing a systematic framework to understand, diagnose, and manage the diverse ailments that affect forest ecosystems worldwide. Forest diseases encompass a wide array of biotic and abiotic factors that influence tree health, productivity, and ecological balance. Classifying these diseases helps organize the overwhelming diversity of disease manifestations into meaningful categories that can guide both scientific inquiry and practical management strategies. By categorizing diseases based on causal agents, affected plant parts, ecological dynamics, modes of spread, and other criteria, researchers and forest managers can develop tailored approaches to mitigate their impacts and safeguard forest health. Classification systems also facilitate communication among scientists, policymakers, and stakeholders, ensuring that knowledge about forest diseases is effectively shared and applied.

Classifying forest diseases is highly important, as forests are integral to global ecological stability and human well-being. Forest ecosystems regulate the Earth's climate, conserve biodiversity, and provide vital resources such as timber, non-timber products, and clean water. Diseases that affect forests, whether caused by fungi, bacteria, viruses, nematodes, or abiotic factors, can have devastating consequences on these functions. Proper classification enables the identification of disease patterns, aids in monitoring disease progression, and supports the implementation of targeted management interventions. Moreover, classification systems serve as a foundation for research into the biology and ecology of pathogens, host-pathogen interactions, and the environmental factors that drive disease dynamics. Understanding these elements is crucial not only for mitigating existing threats but also for anticipating and addressing emerging challenges in a changing world.

The historical development of forest disease classification reflects the evolution of forest pathology as a scientific discipline. Early classification systems were primarily descriptive, based on observable symptoms such as leaf spots, cankers, or root rots. These approaches laid the groundwork for understanding disease manifestations but were limited in their ability to explain underlying causes. As knowledge advanced, classification systems began to incorporate causal agents, such as fungi and bacteria, marking a shift toward cause-based frameworks. The advent of molecular biology and genomics in the 20th and 21st centuries further revolutionized disease classification, enabling

DOI: 10.1201/9781042011063-3

the identification of pathogens at genetic and biochemical levels. Modern approaches now integrate molecular data with ecological, physiological, and epidemiological criteria, reflecting the complexity of forest diseases and their interactions with environmental and human-mediated factors.

Classifying forest diseases based on causal agents provides insight into the diversity and biology of pathogens that affect trees. Among these, fungal pathogens are the most prominent and include a wide range of species responsible for diseases such as rusts, mildews, and root rots. Fungi often thrive in the moist, nutrient-rich environments of forest soils and play dual roles as both pathogens and decomposers. Bacterial pathogens, while less dominant in forest ecosystems, can cause significant diseases such as blights and wilts. Their mechanisms of infection often involve the production of toxins and biofilms, which allow them to invade and colonize plant tissues effectively. Viral and viroid diseases, though historically understudied in forests, are emerging as significant threats due to their ability to interfere with plant metabolism and reproduction. Nematodes, as soilborne pathogens, contribute to tree decline by damaging roots and impeding nutrient and water uptake. Abiotic factors, such as extreme weather, pollution, and soil nutrient imbalances, also play a critical role in forest diseases, often exacerbating the effects of biotic pathogens or acting as primary stressors.

Another important classification criterion is the part of the host plant affected. Diseases targeting different plant parts can have varied ecological and physiological impacts. Root diseases, for instance, disrupt water and nutrient uptake, leading to tree instability and susceptibility to windthrow. Stem and branch diseases, including cankers and galls, compromise the structural integrity of trees, making them vulnerable to breakage and secondary infections. Leaf and needle diseases affect photosynthesis and energy production, resulting in reduced growth and vigor. Diseases of flowers, fruits, and seeds, though less immediately visible, have long-term consequences for forest regeneration and genetic diversity by impairing reproductive processes. Classifying diseases based on the affected plant parts helps in diagnosing and addressing specific vulnerabilities within forest ecosystems.

The ecological context of forest diseases offers another dimension for classification. Endemic diseases are those that persist at relatively stable levels within forest ecosystems, often playing a role in maintaining ecological balance by regulating tree populations and promoting biodiversity. Epidemic diseases, by contrast, occur as outbreaks that can decimate large areas of forest, often triggered by specific environmental or anthropogenic factors. Invasive diseases, caused by non-native pathogens, pose a growing threat to forest health, as they disrupt established ecological relationships and often outcompete native species. Understanding the ecological dynamics of these disease categories is essential for predicting their behavior and mitigating their impacts.

The mode of pathogen spread also serves as a basis for classification. Soilborne diseases, transmitted through root contact or contaminated soil, often involve pathogens that persist in the soil for long periods, making them particularly challenging to manage. Airborne diseases, spread via spores, wind, or insect vectors, can travel vast distances and cause rapid outbreaks. Waterborne diseases, transmitted through rain, rivers, or irrigation, are influenced by hydrological conditions and play a significant role in riparian and wetland forests. Vector-transmitted diseases rely on insects, nematodes, or other organisms for dispersal, highlighting the interconnectedness of forest ecosystems. Classifying diseases by their modes of spread offers insights into their epidemiology and guides strategies for containment and prevention.

Temporal patterns in disease progression form another key classification criterion. Acute diseases are characterized by a rapid onset and short duration, often resulting in sudden tree mortality. Chronic diseases, on the other hand, develop slowly and persist over time, gradually weakening trees and reducing forest productivity. Latent diseases remain dormant within host plants until activated by environmental stressors or other triggers, posing hidden threats to forest health. Recognizing these temporal patterns aids in anticipating disease impacts and implementing timely interventions. The impact of diseases on host plants and ecosystems offers yet another perspective for classification. Primary diseases directly affect healthy trees, often caused by highly virulent pathogens. Secondary diseases, by contrast, occur in trees already weakened by stress or other factors, such as drought or pollution. Beyond their effects on individual trees, forest diseases can alter ecosystem services, such as carbon storage, water filtration, and habitat provision. They can also influence biodiversity by selectively targeting specific species or groups, thereby reshaping forest composition and function. Geographic distribution provides additional insights into disease classification. Some diseases are region-specific, influenced by local climatic, soil, and ecological conditions. Others have a global distribution, often facilitated by international trade, climate change, and human activity. Understanding the geographic patterns of disease occurrence is critical for developing region-specific management strategies and addressing global challenges.

Emerging approaches to classification reflect the integration of molecular, ecological, and technological advancements. Hybrid systems combine traditional criteria, such as symptoms and causal agents, with molecular data, enabling more precise and comprehensive disease categorization. The advent of omics technologies, including genomics, proteomics, and metabolomics, has transformed our ability to identify and understand pathogens at a molecular level. These advancements not only refine classification systems but also enhance our ability to predict and mitigate the impacts of forest diseases. By linking classification to disease management, these modern approaches provide a foundation for developing resilient forests capable of withstanding current and future challenges.

1.1 Importance of Classifying Forest Diseases

The classification of forest diseases is fundamental to understanding their biology, dynamics, and impacts on forest ecosystems. With thousands of plant pathogens affecting forests worldwide, ranging from fungi and bacteria to viruses and nematodes, an organized system of classification enables researchers, foresters, and policymakers to approach these threats systematically. The importance of classification extends across ecological, economic, and scientific dimensions.

1.1.1 Facilitating Diagnosis and Identification

Accurate diagnosis is the first step in managing forest diseases, and classification provides a reference framework for identifying pathogens and understanding their behavior. By grouping diseases based on shared characteristics, such as symptoms, host specificity, or causal agents, classification systems help standardize diagnostic protocols, enabling foresters to identify diseases more efficiently. For example, knowing whether a pathogen is soilborne or airborne guides the selection of diagnostic tools and monitoring methods.

1.1.2 Enhancing Disease Management

Classifying forest diseases is essential for designing targeted management strategies. Different classes of diseases require different approaches; soilborne diseases may be managed through improved drainage and sanitation, while foliar pathogens may necessitate fungicidal sprays or pruning. Classification also aids in prioritizing management efforts, focusing resources on diseases with the greatest ecological or economic impact. For instance, epidemic diseases caused by invasive pathogens, such as sudden oak death or chestnut blight, demand more immediate and intensive interventions compared to endemic diseases with localized impacts.

1.1.3 Advancing Research and Knowledge

A robust classification system fosters scientific understanding by organizing diseases into categories that reflect their biological and ecological traits. This organization facilitates comparative studies, enabling researchers to investigate patterns and relationships across different disease types. For example, studying similarities among rust fungi or vascular wilt pathogens can reveal common infection mechanisms, host adaptations, or evolutionary trends. Classification also supports the development of predictive models for disease emergence, spread, and impact under changing environmental conditions.

1.1.4 Supporting Forest Conservation and Restoration

In the context of forest conservation and restoration, classification informs the selection of tree species and management practices that minimize disease risks. For example, understanding which diseases are prevalent in a particular region or ecosystem allows conservationists to choose resistant species or design mixed-species plantations that reduce the spread of pathogens. Classification also guides restoration efforts in post-disease landscapes, ensuring that replanted forests are more resilient to future outbreaks.

1.1.5 Facilitating Communication and Policy Development

Classification systems provide a common language for communicating about forest diseases across disciplines, regions, and stakeholder groups. This standardization is critical for developing and implementing policies related to biosecurity, quarantine, and disease management. For example, international agreements on the trade of forest products, such as the International Plant Protection Convention (IPPC), rely on standardized disease classification to assess risks and enforce regulations.

1.2 Historical Perspectives on Forest Disease Classification

The classification of forest diseases has evolved over centuries, shaped by advances in plant pathology, microbiology, and ecology. Early systems were often descriptive and symptom-based, focusing on visible signs of disease, while modern approaches incorporate molecular, genetic, and ecological criteria to capture the complexity of pathogen-host-environment interactions.

1.2.1 Early Symptom-Based Classification

In the pre-modern era, forest diseases were classified primarily based on visible symptoms, such as wilting, leaf discoloration, cankers, or dieback. This approach, while practical for field observation, had significant limitations, as many diseases produce overlapping or nonspecific symptoms. For example, chlorosis (leaf yellowing) can result from fungal, bacterial, or abiotic factors, making it insufficient for accurate classification. Nevertheless, symptom-based classification provided an initial framework for distinguishing between major disease types, such as foliar, vascular, and root diseases.

1.2.2 Pathogen-Centric Classification

The discovery of plant pathogens in the 19th and early 20th centuries revolutionized the classification of forest diseases. Researchers such as Anton de Bary, considered the father of modern plant pathology, demonstrated the causal role of microorganisms in plant diseases. This led to the development of pathogen-centric classification systems, which grouped diseases based on their causal agents, such as fungi, bacteria, or viruses. For instance, diseases caused by rust fungi (*Puccinia* spp.) or oomycetes (*Phytophthora* spp.) were categorized separately, reflecting their distinct biology and life cycles. This shift from symptom-based to pathogen-based classification marked a major milestone in forest pathology.

1.2.3 Ecological and Epidemiological Approaches

As ecological and epidemiological perspectives gained prominence in the mid-20th century, forest disease classification began to consider the broader context of pathogen-host-environment interactions. Diseases were categorized based on their ecological roles (e.g., endemic vs. epidemic diseases) or modes of spread (e.g., soilborne, airborne, or vector-transmitted diseases). This approach recognized that the impacts of forest diseases are shaped not only by the pathogens themselves but also by environmental conditions, host susceptibility, and forest management practices.

1.2.4 Integration of Molecular and Genetic Criteria

The advent of molecular biology and genetics in the late 20th century transformed the classification of forest diseases by enabling the identification and characterization of pathogens at the DNA and RNA levels. Molecular techniques, such as PCR, genomic sequencing, and phylogenetic analysis, allowed researchers to classify pathogens based on genetic relationships, shedding light on their evolutionary history and mechanisms of virulence. For example, molecular studies revealed the genetic diversity within *Phytophthora* species, leading to the identification of new species and a deeper understanding of their pathogenicity. This molecular perspective has also facilitated the study of previously unculturable or cryptic pathogens, expanding the scope of forest disease classification.

1.3 Modern Approaches to Classification

Modern classification systems for forest diseases integrate multiple criteria, reflecting the complexity of pathogen biology and the diverse ways in which diseases affect forests. These approaches combine traditional symptom- and pathogen-based frameworks with molecular, ecological, and functional perspectives, creating a comprehensive and dynamic system.

1.3.1 Multidimensional Criteria

Modern classification recognizes that no single criterion is sufficient to capture the diversity of forest diseases. Instead, diseases are classified using a combination of factors, including:

- **Causal agents:** Grouping diseases by their pathogens (e.g., fungi, bacteria, viruses, nematodes).
- **Host plant parts affected:** Categorizing diseases based on the tissues they infect (e.g., roots, stems, leaves, or reproductive structures).

» **Ecological roles:** Distinguishing between endemic diseases that regulate forest dynamics and epidemic diseases that cause widespread damage.

» **Modes of spread:** Classifying diseases based on their transmission pathways (e.g., soilborne, airborne, or vector-borne).

» **Temporal patterns:** Differentiating between acute diseases with rapid onset and chronic diseases with long-term effects.

This multidimensional approach provides a holistic view of forest diseases, facilitating both scientific understanding and practical management.

1.3.2 Role of Omics Technologies

Advances in omics technologies, such as genomics, transcriptomics, proteomics, and metabolomics, have enhanced the precision and depth of forest disease classification. These technologies enable researchers to identify genetic markers, study gene expression patterns, and analyze pathogen-host interactions at the molecular level. For example, genomic studies of *Armillaria* species have revealed genetic adaptations associated with root rot pathogenicity, while proteomic analyses of rust fungi have identified effector proteins that suppress host defenses.

1.3.3 Implications for Forest Management

Modern classification systems are not only tools for scientific research but also practical instruments for forest management. By linking classification criteria to disease impacts and management options, these systems provide actionable insights for foresters and policymakers. For example, understanding that a disease is soilborne and chronic informs decisions about soil management, crop rotation, and resistant species. Similarly, recognizing that a disease is vector-transmitted and epidemic guides efforts to monitor and control insect vectors.

1.3.4 Dynamic and Adaptive Frameworks

Modern classification systems are dynamic, evolving with new discoveries and changing environmental conditions. As climate change, globalization, and urbanization introduce new challenges for forest health, classification systems must adapt to incorporate emerging pathogens, hybrid disease types, and novel ecological interactions. This adaptability ensures that classification remains relevant and effective in addressing the complex and rapidly changing landscape of forest pathology.

2. Classification Based on Causal Agents

Forest diseases are caused by a variety of agents, including living organisms such as fungi, bacteria, viruses, and nematodes, as well as non-living, abiotic factors such as climate extremes, soil conditions, and pollution. Classification based on causal agents provides a structured approach to understanding the origins, mechanisms, and impacts of forest diseases. Each type of causal agent operates through distinct biological or environmental processes, influencing tree health in unique ways.

2.1 Fungal Diseases: Characteristics and Examples

Fungi are the most diverse and significant group of pathogens causing forest diseases. They infect all parts of trees, including roots, stems, branches, leaves, and reproductive structures, resulting in a wide range of symptoms and ecological impacts. Fungal pathogens thrive in diverse environmental conditions, making them ubiquitous in forest ecosystems.

Characteristics of Fungal Diseases

Fungal pathogens exhibit diverse life cycles and infection strategies. Many fungi reproduce through spores, which are dispersed by wind, water, insects, or animals. Once they land on a susceptible host, they germinate and penetrate plant tissues through natural openings (stomata or lenticels) or wounds. Some fungi, such as *Phytophthora* species, form swimming zoospores that infect roots in waterlogged soils, while others, like rust fungi, rely on airborne spores to infect leaves and stems. Fungi are classified into several groups based on their morphology and reproduction, including Ascomycetes, Basidiomycetes, and Oomycetes. For example, rusts (*Puccinia* spp.), smuts (*Ustilago* spp.), and blights (*Dothistroma septosporum*) are well-known Basidiomycetes that cause devastating forest diseases.

Examples of Fungal Diseases in Forests

» **Root Rot Diseases:** *Armillaria* species cause widespread root rot, destabilizing trees and leading to mortality. This pathogen spreads through rhizomorphs and infects a wide range of tree species.

» **Canker Diseases:** *Cryphonectria parasitica* causes chestnut blight, characterized by sunken cankers on stems and branches, eventually killing the host.

» **Leaf and Needle Blights:** *Dothistroma septosporum*, responsible for red band needle blight, infects coniferous trees, reducing photosynthetic capacity and productivity.

» **Vascular Wilt Diseases:** *Ceratocystis* and *Fusarium* species block xylem vessels, causing water stress, wilting, and tree death.

Ecological and Economic Impacts

Fungal diseases can alter forest composition by selectively targeting specific species, leading to changes in biodiversity and ecosystem functioning. Economically, they reduce timber quality, increase management costs, and disrupt industries reliant on forest products.

2.2 Bacterial Diseases: Mechanisms and Notable Pathogens

Bacteria are less diverse than fungi but still play a significant role in forest diseases. They are unicellular microorganisms that infect trees through wounds or natural openings, such as stomata. Once inside the host, bacterial pathogens multiply rapidly, producing toxins and enzymes that disrupt normal physiological functions.

Mechanisms of Bacterial Infection

Bacterial pathogens colonize intercellular spaces or vascular tissues, interfering with water and nutrient transport. Many bacteria produce extracellular polysaccharides that block xylem vessels, causing wilting and dieback. Some bacterial pathogens secrete enzymes that degrade cell walls, facilitating their spread within host tissues.

Notable Bacterial Pathogens and Diseases

- **Crown Gall Disease:** Caused by *Agrobacterium tumefaciens*, this disease results in tumor-like growths on tree trunks and roots, disrupting nutrient flow.
- **Bacterial Wilt:** *Ralstonia solanacearum* infects xylem vessels, leading to systemic wilting and eventual death.
- **Bacterial Cankers:** *Pseudomonas syringae* causes necrotic lesions on stems and branches, leading to dieback and increased susceptibility to secondary infections.
- **Bacterial Blights:** Pathogens like *Xanthomonas* species cause leaf spots and blights, reducing the photosynthetic efficiency of infected trees.

Challenges in Management

Bacterial diseases are difficult to manage due to their rapid reproduction and ability to form biofilms that protect them from environmental stress and chemical treatments. Sanitation, quarantine, and resistant species are critical tools for managing bacterial diseases in forests.

2.3 Viral and Viroid Diseases: Emerging Challenges in Forests

Viruses and viroids are submicroscopic pathogens that are becoming increasingly recognized for their role in forest health. Viruses consist of genetic material (DNA or RNA) encased in a protein coat, while viroids are simpler entities composed solely of RNA. These pathogens rely on host cells for replication and are often transmitted by insect vectors, nematodes, or mechanical wounds.

Characteristics of Viral and Viroid Diseases

Viral infections are systemic, affecting multiple parts of the plant. Symptoms include mosaic patterns on leaves, stunted growth, yellowing, and reduced vigor. Viroids cause similar symptoms but often result in more severe decline due to their ability to interfere with cellular metabolism.

Examples of Viral and Viroid Diseases

- **Mosaic Diseases:** Caused by viruses like *Tobacco mosaic virus* (TMV), these diseases reduce photosynthetic capacity and growth.
- **Citrus Tristeza Virus:** Although primarily associated with fruit trees, this virus can also impact forest species in mixed agroforestry systems.
- **Coconut Cadang-Cadang Viroid (CCCVd):** This viroid causes severe decline in palm species, with significant implications for tropical forests.

Emerging Threats and Challenges

The global movement of plants has facilitated the spread of viral pathogens to new regions, where they can infect naive host populations. The subtle and often asymptomatic nature of viral infections complicates detection and management, underscoring the need for advanced diagnostic tools.

2.4 Nematode-Induced Diseases: Role of Soilborne Pathogens

Nematodes are microscopic, worm-like organisms that inhabit soil and plant tissues. While many nematodes are beneficial, certain species are pathogenic and cause significant damage to forest trees. Pathogenic nematodes feed on plant roots or invade vascular tissues, disrupting water and nutrient uptake.

Characteristics and Mechanisms of Nematode Pathogens

Nematodes use specialized structures, such as stylets, to pierce plant cells and extract nutrients. They may act as primary pathogens or facilitate secondary infections by creating entry points for fungi and bacteria. For example, *Bursaphelenchus xylophilus* (pinewood nematode) is transmitted by beetles and causes rapid wilting in pine trees.

Examples of Nematode Diseases

- **Pine Wilt Disease:** Caused by *Bursaphelenchus xylophilus*, this disease leads to wilting, discoloration, and death of pines.
- **Root-Knot Disease:** *Meloidogyne* species induce galls on roots, reducing water and nutrient absorption.
- **Lesion Nematode Damage:** *Pratylenchus* species create lesions on roots, weakening trees and predisposing them to secondary infections.

Management Challenges

Nematode control in forest ecosystems is challenging due to their microscopic size and soilborne nature. Management strategies include the use of resistant species, biological control agents, and improved soil health practices.

2.5 Diseases Caused by Abiotic Factors: Climate, Pollution, and Soil Conditions

In addition to biotic pathogens, abiotic factors such as climate extremes, pollution, and soil degradation can cause forest diseases. These factors often weaken trees, making them more susceptible to infections by biotic pathogens.

Climate-Related Diseases

Extreme temperatures, drought, and excessive rainfall can stress trees, leading to symptoms such as leaf scorch, dieback, and mortality. For example, drought-induced water stress predisposes trees to vascular wilt diseases, while flooding promotes root rot by creating anaerobic conditions.

Pollution-Induced Diseases

Air pollution, including ozone, sulfur dioxide, and heavy metals, damages foliage and disrupts physiological processes. Acid rain, for instance, leaches nutrients from the soil and weakens trees, making them vulnerable to pathogens like *Phytophthora* species.

Soil-Related Disorders

Poor soil conditions, such as compaction, nutrient deficiencies, or salinity, can cause root dysfunction and stunted growth. For example, compacted soils reduce oxygen availability, leading to root asphyxiation and increased susceptibility to pathogens.

Management of Abiotic Diseases

Addressing abiotic factors requires a combination of ecological restoration, soil amendment, and adaptive management practices. Enhancing soil health, reducing pollution, and mitigating climate impacts are essential for building forest resilience.

3. Classification Based on Host Plant Parts Affected

The classification of forest diseases based on the specific parts of the host plant they affect provides valuable insights into how pathogens and environmental factors disrupt the growth, development, and functionality of trees. Each disease type targets distinct plant structures, leading to a range of physiological, structural, and ecological impacts. Forest diseases categorized in this manner include those affecting roots, stems, branches, leaves, needles, flowers, fruits, and seeds. These parts not only serve specific roles in the life cycle of the tree but are also critical to the overall health and stability of forest ecosystems. Diseases targeting these parts can lead to declines in forest productivity, biodiversity, and ecological balance.

3.1 Root Diseases

Root diseases are among the most destructive types of forest diseases, often causing significant damage to tree health and stability by impairing the tree's ability to absorb water and nutrients from the soil. Roots form the foundation of a tree, anchoring it in the soil and acting as the primary interface for nutrient and moisture exchange. Pathogens that target roots disrupt this critical function, leading to a cascade of physiological failures throughout the plant. Root diseases are typically caused by soilborne pathogens such as fungi, oomycetes, nematodes, and occasionally bacteria.

One of the most common and impactful root diseases is *Armillaria* root rot, caused by fungi in the genus *Armillaria*. These pathogens infect a wide range of tree species, spreading through underground rhizomorphs and causing significant decay of root tissues. Symptoms of root diseases include reduced canopy growth, wilting, chlorosis (yellowing of leaves), and eventual dieback. Infected trees may also exhibit poor structural stability, leading to uprooting during storms or other disturbances. Another prominent example is root rot caused by *Phytophthora cinnamomi*, a waterborne pathogen that thrives in poorly drained soils. This disease has devastated oak forests in Mediterranean regions, where the combination of waterlogged soils and drought stress exacerbates tree susceptibility. Root diseases often progress invisibly underground, making early detection and management challenging. The ecological impacts of root diseases extend beyond individual trees, influencing soil microbial communities and altering nutrient cycling processes. For instance, the death of root systems increases the volume of decaying organic matter, which in turn can create conditions favorable for other pathogens and pests. Management strategies for root diseases typically involve improving soil drainage, removing infected stumps and roots, and promoting the growth of resistant tree species. However, in natural forests, the control of root diseases is often limited by logistical and ecological constraints.

3.2 Stem and Branch Diseases

Diseases that target stems and branches are particularly significant because they compromise the structural integrity and mechanical strength of trees, making them vulnerable to breakage, dieback, and secondary infections. These diseases can result from fungal, bacterial, or environmental factors, and their effects are often visible in the form of cankers, galls, cracks, and necrotic lesions. Stem and branch diseases are frequently exacerbated by environmental stressors, such as drought or frost, which weaken the tree's natural defenses.

Canker diseases, such as those caused by *Cryphonectria parasitica* (chestnut blight) and *Nectria* species, are among the most destructive stem diseases. These pathogens infect the cambium layer of the tree, causing localized necrosis that disrupts the flow of water and nutrients. Over time, cankers can girdle the tree, cutting off vascular transport and leading to dieback and death. Symptoms include sunken or swollen areas on the bark, oozing sap or resin, and discoloration around the affected regions. In addition to pathogens, abiotic factors such as frost and mechanical injury can create entry points for canker-causing organisms, further complicating management efforts. Another major category of stem diseases involves heart rot fungi, such as *Ganoderma* and *Phellinus* species, which decay the inner wood tissues of trees. These pathogens primarily infect older or weakened trees, hollowing out the core and significantly reducing their strength. Although heart rot does not always kill the host, it increases the risk of breakage, posing hazards in urban areas or recreational forests. Similarly, gall-forming pathogens like *Agrobacterium tumefaciens* cause abnormal growths on stems and branches, interfering with nutrient flow and reducing tree vigor. The ecological impacts of stem and branch diseases include altered forest structure and habitat availability. Dead and decaying branches create niches for wildlife, including insects, birds, and fungi, but excessive damage can lead to large-scale tree mortality, affecting carbon storage and nutrient cycling. Management of these diseases often involves pruning infected branches, applying protective treatments to wounds, and selecting resistant tree varieties for reforestation projects.

3.3 Leaf and Needle Diseases

Diseases affecting leaves and needles are significant because these organs are the primary sites of photosynthesis, the process that sustains tree growth and productivity. Pathogens that attack foliage can cause discoloration, necrosis, defoliation, and premature leaf drop, reducing the tree's ability to capture sunlight and produce energy. Foliar diseases are caused by fungi, bacteria, and occasionally viruses, and they tend to spread rapidly in humid conditions or densely planted forests. Fungal pathogens are the most common culprits behind leaf and needle diseases. For example, *Dothistroma septosporum*, which causes red band needle blight in conifers, is a major concern in pine plantations worldwide. The disease is characterized by red-brown lesions on needles, which eventually lead to defoliation and reduced growth. Similarly, *Rhytisma acerinum*, the causative agent of tar spot on maple leaves, produces black, tar-like spots that, while often not fatal, reduce the aesthetic and economic value of ornamental trees.

Bacterial foliar diseases, though less common than fungal ones, can also have significant impacts. *Xanthomonas* species, for instance, cause bacterial leaf blight, which manifests as water-soaked lesions that coalesce into large necrotic areas. Viral infections, such as those caused by mosaic viruses, result in mottled patterns on leaves, stunted growth, and reduced vigor. The ecological consequences of leaf and needle diseases include shifts in species competition, as heavily defoliated trees lose their competitive advantage to healthier neighbors. In severe cases, widespread foliar diseases can lead to cascading effects on entire ecosystems, including changes in understory vegetation and altered food availability for herbivores. Management practices for foliar diseases include thinning to improve air circulation, applying fungicides or bactericides in high-value areas, and breeding resistant tree species.

3.4 Flower, Fruit, and Seed Diseases: Consequences for Reproduction

Diseases that target flowers, fruits, and seeds have critical implications for the reproductive success and regeneration of tree populations. These diseases can reduce seed viability, impair pollination, and disrupt the natural cycles of forest renewal. Pathogens affecting reproductive structures often have specialized infection mechanisms, targeting flowers during bloom or seeds during maturation. One notable example is blossom blight caused by *Monilinia* species, which infect flowers and developing fruits, leading to their premature death. This disease is particularly devastating in mixed-use forests that include fruit-bearing trees, as it reduces both ecological and economic yields. Similarly, rust fungi such as *Cronartium ribicola* cause white pine blister rust, infecting cones and seeds of five-needle pines, thereby impairing their reproductive capacity. Seed-borne diseases, caused by fungi like *Fusarium* species, are a major concern in forestry nurseries and seed banks. These pathogens infect seeds during storage or germination, reducing their viability and leading to poor seedling establishment. The spread of seed-borne pathogens through trade and transport further complicates their management, posing risks to native forest species in new regions. The impact of these diseases extends to wildlife, as reduced fruit and seed production affects the availability of food resources for birds, mammals, and insects. Additionally, the loss of reproductive capacity in key tree species can alter forest composition over time, favoring less affected species and potentially reducing biodiversity. Management strategies focus on improving seed storage conditions, applying fungicidal treatments, and implementing quarantine measures to prevent the spread of seed-borne pathogens.

4. Classification Based on Disease Ecology

The ecological classification of forest diseases focuses on their behavior, dynamics, and interactions within forest ecosystems. This approach examines not only the biological characteristics of pathogens but also their roles in shaping forest structure, composition, and resilience. Forest diseases can be categorized into endemic, epidemic, and invasive types based on their ecological roles and the patterns they exhibit within ecosystems. Endemic diseases are natural components of forest systems, often maintaining ecological balance and contributing to biodiversity. Epidemic diseases, in contrast, represent large-scale outbreaks driven by specific triggers, such as environmental stress or human activity, often causing extensive damage. Invasive diseases involve non-native pathogens that disrupt ecosystems, leading to significant ecological, economic, and social consequences.

4.1 Endemic Diseases: Role in Maintaining Ecological Balance

Endemic diseases are those that exist naturally within forest ecosystems, typically in a state of equilibrium with their host populations and the environment. These diseases are often caused by native pathogens that have co-evolved with their hosts over long periods, leading to a balance where the pathogens persist without causing widespread damage. Endemic diseases play a critical role in maintaining ecological balance by regulating host populations, promoting genetic diversity, and facilitating ecosystem processes such as nutrient cycling.

Ecological Roles of Endemic Diseases

Endemic diseases act as natural regulators of plant populations, preventing any single species from dominating the ecosystem. This phenomenon, often described by the Janzen-Connell hypothesis, suggests that pathogens disproportionately affect seedlings growing near parent trees, reducing their survival and promoting species diversity. For example, root pathogens such as *Armillaria* species, which are widespread in temperate forests, selectively target stressed or weakened trees, creating opportunities for other species to establish and grow. This selective mortality contributes to forest succession and the maintenance of biodiversity. Another important role of endemic diseases is their contribution to nutrient cycling. By infecting and killing certain trees, these pathogens accelerate the decomposition of organic matter, releasing nutrients into the soil and supporting the growth of other plants. For instance, foliar pathogens that cause leaf blight or needle cast increase litterfall and decomposition rates, enriching the soil with organic matter.

Challenges of Endemic Diseases

While endemic diseases generally operate within the natural dynamics of forests, their impacts can become more severe under certain conditions. Environmental changes, such as prolonged drought, pollution, or habitat fragmentation, can weaken host defenses and increase the prevalence of endemic pathogens, leading to localized dieback or mortality. Additionally, human activities, such as logging and urbanization, can disrupt the balance between pathogens and their hosts, causing endemic diseases to behave more aggressively.

Management Implications

The management of endemic diseases focuses on maintaining ecological balance rather than eradication. Strategies include preserving genetic diversity within tree populations, promoting mixed-species forests, and minimizing environmental stressors that exacerbate disease impacts. Understanding the ecological roles of these diseases allows forest managers to work with, rather than against, natural processes, enhancing forest resilience and sustainability.

4.2 Epidemic Diseases

Epidemic diseases are characterized by sudden and widespread outbreaks that cause significant damage to forest ecosystems. These diseases often occur when environmental conditions or human activities create favorable circumstances for pathogen proliferation, overwhelming the natural defenses of host populations. Epidemic diseases can have devastating ecological, economic, and social consequences, transforming forest landscapes and altering their functionality.

Causes of Epidemic Diseases

Several factors contribute to the emergence of epidemic diseases in forests. Environmental stress, such as drought, flooding, or extreme temperatures, weakens trees and makes them more susceptible to pathogens. For example, drought stress has been linked to outbreaks of vascular wilt diseases caused by *Ceratocystis* species, which block water transport in xylem vessels and lead to rapid wilting and mortality. Human activities are another major driver of epidemics. Monoculture plantations, which lack the genetic and species diversity needed to buffer against pathogens, are particularly vulnerable.

For instance, the spread of *Dothistroma septosporum*, the causative agent of red band needle blight, has been exacerbated by the extensive planting of susceptible pine species in commercial forests. Logging, road construction, and other disturbances can also create entry points for pathogens and facilitate their spread. Epidemic diseases often involve pathogens with high reproductive rates and efficient dispersal mechanisms. Airborne fungal spores, waterborne zoospores, and vector-transmitted bacteria and viruses can spread rapidly under favorable conditions, leading to large-scale infections. For example, the rapid spread of Dutch elm disease, caused by *Ophiostoma* species and transmitted by bark beetles, has led to the near elimination of elm populations in many regions.

Consequences of Epidemic Diseases

The impacts of epidemic diseases are far-reaching, affecting forest composition, productivity, and ecological services. Widespread tree mortality reduces carbon sequestration, disrupts habitat availability, and alters hydrological cycles. The loss of dominant tree species can lead to cascading effects on understory vegetation, wildlife, and microbial communities. For instance, the chestnut blight epidemic caused by *Cryphonectria parasitica* fundamentally transformed eastern U.S. forests, replacing American chestnuts with less ecologically significant species. Economically, epidemic diseases result in significant losses for the forestry industry, including reduced timber yields, increased management costs, and decreased property values. Socially, they impact communities that depend on forests for livelihoods, recreation, and cultural identity.

Management Approaches

Managing epidemic diseases requires a combination of prevention, early detection, and rapid response. Monitoring systems that integrate field surveys, molecular diagnostics, and remote sensing are critical for detecting outbreaks at an early stage. Quarantine measures, such as restricting the movement of infected plant material, are essential for containing the spread of pathogens. In affected areas, sanitation measures, including the removal and destruction of infected trees, can help reduce pathogen loads. Promoting genetic diversity and mixed-species planting can also enhance forest resilience to future outbreaks.

4.3 Invasive Diseases: Impact of Non-Native Pathogens on Native Forests

Invasive diseases involve non-native pathogens that are introduced into new regions, where they encounter host populations with little or no resistance. These diseases often cause disproportionate damage, disrupting forest ecosystems and leading to significant ecological and economic losses. Invasive diseases are a growing concern in an era of globalization, where the movement of plants, wood products, and soil has increased the likelihood of pathogen introductions.

Characteristics of Invasive Diseases

Non-native pathogens often exhibit high virulence in their introduced ranges, as they encounter naive host populations that lack evolved defenses. The absence of natural enemies, such as predators, competitors, or co-evolved pathogens, further amplifies their impacts. For example, *Hymenoscyphus fraxineus*, the causative agent of ash dieback, was introduced from Asia to Europe, where it has caused widespread mortality of European ash trees. Invasive diseases frequently spread rapidly due to their efficient dispersal mechanisms and the connectivity of modern trade and transportation networks.

For instance, *Phytophthora ramorum*, the pathogen responsible for sudden oak death, was introduced to North America through infected nursery stock and has since spread to natural forests, killing millions of oak and tanoak trees.

Ecological and Economic Impacts

The invasive pathogens can lead to the collapse of keystone species, triggering profound ecological changes. The loss of dominant tree species reduces habitat availability, alters competitive dynamics among plants, and affects the ecosystem services forests provide, such as water regulation and carbon storage. Invasive diseases can also lead to the homogenization of forest ecosystems, as native species are replaced by less diverse or invasive alternatives. Economically, invasive diseases impose significant costs related to quarantine enforcement, eradication efforts, and lost timber revenues. They also disrupt international trade, as affected regions may face restrictions on the export of forest products. The spread of *Bursaphelenchus xylophilus* (pine wilt nematode) from North America to Asia and Europe has caused significant economic losses in the forestry industry.

Management Challenges and Strategies

Managing invasive diseases is particularly challenging due to their rapid spread and high impact. Prevention is the most effective strategy, involving stringent biosecurity measures, such as inspection and certification of plant materials, treatment of wood products, and the establishment of pest-free zones. For established invasions, eradication is often difficult or impossible, requiring long-term management to mitigate impacts. Efforts include breeding resistant tree species, applying biological control agents, and restoring affected ecosystems with alternative species.

5. Classification Based on Mode of Pathogen Spread

The classification of forest diseases based on the mode of pathogen spread provides crucial insights into the dynamics of disease transmission and helps guide effective management strategies. Pathogens employ diverse mechanisms to disperse across forest ecosystems, exploiting soil, air, water, and living organisms as vectors. These pathways not only determine the speed and extent of disease outbreaks but also influence the effectiveness of intervention measures. The modes of pathogen spread are particularly important in natural forests, where controlling disease transmission can be challenging due to the scale and complexity of the environment.

5.1 Soilborne Diseases: Mechanisms of Infection and Dispersal

Soilborne diseases are caused by pathogens that reside and propagate in the soil, often infecting trees through their root systems. These diseases are among the most persistent and difficult to manage, as the soil provides a stable environment that protects pathogens from external stresses. Soilborne pathogens include fungi, oomycetes, nematodes, and certain bacteria, which can remain viable in the soil for extended periods, either as spores, survival structures, or active populations.

Mechanisms of Infection

Soilborne pathogens typically infect trees by penetrating root tissues or invading wounds in the root system. For example, *Phytophthora cinnamomi*, a notorious soilborne pathogen, produces motile zoospores that swim through waterlogged soils and infect roots at points of active growth. Similarly, *Armillaria* species spread through rhizomorphs, which grow underground and invade healthy roots. Nematodes like *Meloidogyne* species use specialized stylets to pierce root cells, creating feeding sites that disrupt water and nutrient uptake.

Mechanisms of Dispersal

Pathogens in the soil are dispersed through natural and anthropogenic processes. Water is a primary driver of soilborne pathogen movement, as rainfall, irrigation, or flooding can carry spores and other propagules to new locations. Human activities, such as the movement of contaminated soil, tools, or nursery stock, significantly contribute to the spread of soilborne pathogens. For instance, the global spread of *Phytophthora ramorum* has been linked to the trade of infected plants and soil.

Examples of Soilborne Diseases

» **Root Rot Diseases:** Caused by pathogens like *Phytophthora* and *Armillaria*, these diseases disrupt root function, leading to wilting, dieback, and mortality.

» **Damping-Off:** Fungal pathogens such as *Fusarium* and *Rhizoctonia* infect seedlings in nurseries, causing decay and poor establishment.

» **Nematode Diseases:** Root-knot nematodes (*Meloidogyne* spp.) form galls on roots, impairing water and nutrient uptake.

Management Challenges and Strategies

Managing soilborne diseases is challenging due to the persistence and wide distribution of pathogens in the soil. Strategies include improving soil drainage, applying biological control agents, using resistant tree species, and implementing sanitation measures to prevent the introduction of contaminated soil or plant material.

5.2 Airborne Diseases: Role of Spores, Wind, and Vectors

Airborne diseases are caused by pathogens that spread through the air, primarily as spores, conidia, or other propagules. These diseases often affect the above-ground parts of trees, such as leaves, stems, and branches, and are responsible for some of the most widespread and destructive forest disease outbreaks. Airborne pathogens include fungi, bacteria, and viruses, many of which are highly adapted for long-distance dispersal.

Mechanisms of Infection

Airborne pathogens typically infect trees by entering through natural openings, such as stomata, or through wounds on leaves, stems, or branches. Once established, these pathogens can spread rapidly within the host, causing symptoms such as leaf spots, blights, cankers, and dieback. For instance, rust fungi (*Puccinia* spp.) infect leaves and stems, forming pustules that release large quantities of spores into the air.

Role of Wind and Environmental Factors

Wind is a primary driver of airborne pathogen dispersal, carrying spores over long distances and facilitating the colonization of new hosts. Environmental factors such as humidity, temperature, and rainfall influence spore germination and infection. For example, high humidity promotes the germination of *Botrytis cinerea* spores, leading to severe outbreaks of gray mold in susceptible forests.

Examples of Airborne Diseases

- **Rust Diseases:** Caused by fungi like *Cronartium ribicola* (white pine blister rust), these diseases produce airborne spores that infect pine needles and stems.
- **Needle Blight:** *Dothistroma septosporum* spreads through windborne spores, causing defoliation and reduced growth in pine plantations.
- **Powdery Mildew:** Fungi such as *Erysiphe* species form whitish coatings on leaves, reducing photosynthetic efficiency.

Management Challenges and Strategies

The rapid spread of airborne pathogens poses significant challenges for disease control. Management strategies include thinning to improve airflow, applying fungicides in high-value areas, and planting resistant tree species. Monitoring airborne pathogens using spore traps and predictive models helps in detecting and mitigating outbreaks.

5.3 Waterborne Diseases: Spread Through Rain, Rivers, and Irrigation

Waterborne diseases are caused by pathogens that spread through water, including rain splash, rivers, streams, and irrigation systems. These diseases primarily affect roots and lower stems but can also infect foliage in wet conditions. Waterborne pathogens include fungi, oomycetes, and bacteria, many of which are highly virulent in water-saturated environments.

Mechanisms of Infection

Waterborne pathogens infect trees by entering root systems or splashing onto lower stems and leaves. Oomycetes like *Phytophthora* species produce motile zoospores that actively seek out root tips in waterlogged soils. Rain splash facilitates the dispersal of spores from infected leaves to nearby healthy plants, creating secondary infections.

Examples of Waterborne Diseases

- **Phytophthora Root Rot:** *Phytophthora cinnamomi* spreads through saturated soils, causing severe root decay and mortality.
- **Leaf Spot Diseases:** Fungal pathogens such as *Mycosphaerella* species spread through rain splash, creating lesions on foliage.
- **Bacterial Wetwood:** Caused by waterborne bacteria, this disease leads to the oozing of liquid from infected trunks.

Management Challenges and Strategies

Managing waterborne diseases requires addressing both pathogen sources and environmental conditions. Improving drainage, reducing irrigation frequency, and avoiding overwatering are key strategies. In natural forests, efforts focus on planting species that tolerate wet soils and monitoring waterways for pathogen presence.

5.4 Vector-Transmitted Diseases: Role of Insects, Nematodes, and Other Carriers

Vector-transmitted diseases involve pathogens that rely on living organisms, such as insects, nematodes, or birds, for their dispersal. These diseases are often highly specific, with pathogens and vectors co-evolving to optimize transmission. Vector-transmitted pathogens include bacteria, viruses, fungi, and nematodes, many of which cause severe damage to forest ecosystems.

Mechanisms of Transmission

Vectors play an active role in transmitting pathogens by wounding host plants and introducing infectious agents. For example, bark beetles transmit fungi like *Ophiostoma* species, which cause Dutch elm disease, by boring into the vascular tissues of trees. Similarly, nematodes like *Bursaphelenchus xylophilus* (pinewood nematode) rely on insect vectors for long-distance dispersal.

Examples of Vector-Transmitted Diseases

» **Dutch Elm Disease:** Transmitted by bark beetles, this fungal disease has decimated elm populations worldwide.

» **Pine Wilt Disease:** The pinewood nematode, vectored by longhorn beetles, causes rapid wilting and death in pines.

» **Viral Diseases:** Aphids and whiteflies transmit plant viruses, causing mosaic patterns and growth deformities.

Management Challenges and Strategies

Controlling vector-transmitted diseases requires addressing both the pathogen and its vector. Strategies include applying biological or chemical controls to reduce vector populations, breeding resistant tree species, and removing infected trees to disrupt the transmission cycle. Monitoring vector activity and implementing quarantine measures are critical for preventing the spread of these diseases.

6. Classification Based on Temporal Patterns

The classification of forest diseases based on temporal patterns provides a way to understand how pathogens interact with hosts over time. Temporal patterns describe the duration, progression, and activation of diseases, highlighting the diversity of infection dynamics in forest ecosystems. This classification is critical for predicting the course of disease outbreaks, assessing their impacts, and designing appropriate management strategies. Forest diseases can be categorized into three major temporal types: acute diseases, which are characterized by rapid onset and short duration; chronic diseases, which persist over extended periods and lead to gradual decline; and latent diseases, which remain dormant until activated by environmental stress or changes in host susceptibility. These temporal patterns underscore the complex interplay between pathogens, hosts, and environmental conditions.

6.1 Acute Diseases

Acute diseases are those that manifest quickly, causing noticeable symptoms and damage within a short time frame. These diseases are often driven by highly virulent pathogens that reproduce and spread rapidly under favorable conditions. Acute diseases typically target specific parts of the host, such as leaves, needles, or vascular tissues, and their impacts can range from localized infections to widespread mortality. The intensity of acute diseases makes them particularly disruptive to forest ecosystems, especially when they occur as outbreaks.

Characteristics of Acute Diseases

Acute diseases are marked by their rapid progression and visible symptoms, which may include wilting, chlorosis, necrosis, defoliation, and dieback. Pathogens causing acute diseases often have short life cycles, enabling them to produce multiple generations within a single growing season. For example, rust fungi such as *Puccinia graminis* produce spores that spread rapidly, infecting large areas within weeks. Similarly, vascular wilt pathogens like *Ceratocystis* species can block xylem vessels, causing trees to wilt and die within days.

Environmental factors, such as high humidity, temperature fluctuations, or sudden stress events, often trigger the onset of acute diseases. In monoculture plantations, where genetic uniformity provides little resistance, acute diseases can spread unchecked, resulting in severe economic losses and ecological disruptions.

Examples of Acute Diseases

- **Sudden Oak Death:** Caused by *Phytophthora ramorum*, this disease spreads rapidly in wet conditions, leading to extensive dieback in oaks and tanoaks.
- **Red Band Needle Blight:** *Dothistroma septosporum* infects conifer needles, causing rapid defoliation and reduced growth.
- **Fire Blight:** A bacterial disease caused by *Erwinia amylovora*, which spreads quickly during warm, humid conditions, affecting fruit and ornamental trees.

Impacts and Management

The rapid onset of acute diseases often leaves little time for intervention, requiring immediate action to prevent further spread. Management strategies include monitoring environmental conditions, applying fungicides or bactericides, and removing infected material. In forests, thinning and improving air circulation can help reduce the conditions that favor acute disease outbreaks. Early detection systems, such as spore traps and remote sensing, are critical for identifying and mitigating acute diseases before they escalate.

6.2 Chronic Diseases: Long-Term Effects and Persistence

Chronic diseases develop slowly, persisting over months or years, and often lead to a gradual decline in tree health. These diseases are typically caused by pathogens that establish long-term relationships with their hosts, either as latent infections that progress slowly or as continuous infections that exert a low but consistent pathogenic load. Chronic diseases are particularly challenging to manage because their impacts accumulate over time, often becoming apparent only after significant damage has occurred.

Characteristics of Chronic Diseases

Chronic diseases are characterized by subtle, progressive symptoms such as stunted growth, gradual dieback, reduced vigor, and decreased reproductive output. Pathogens associated with chronic diseases often infect perennial tissues, such as roots, stems, or vascular systems, creating a persistent reservoir of infection. For example, *Armillaria* root rot fungi establish extensive underground networks that slowly degrade root systems, weakening trees over many years. Unlike acute diseases, chronic diseases are less dependent on environmental triggers for their progression. Instead, they exploit host weaknesses, such as age, stress, or injury, to maintain a steady infection rate. This slow progression allows pathogens to spread undetected, particularly in natural forests where monitoring is limited.

Examples of Chronic Diseases

» **Armillaria Root Rot:** Caused by *Armillaria* species, this disease leads to slow root decay, increasing tree vulnerability to windthrow and secondary infections.

» **Heart Rot:** Fungal pathogens like *Ganoderma* and *Phellinus* species infect the inner wood of trees, gradually reducing structural integrity over decades.

» **Ash Dieback:** *Hymenoscyphus fraxineus* slowly spreads through ash populations, causing chronic dieback and mortality.

Impacts and Management

The long-term nature of chronic diseases makes them particularly harmful to forest ecosystems. Over time, they reduce forest productivity, alter species composition, and create opportunities for secondary pathogens and pests. Management strategies for chronic diseases focus on maintaining forest health and resilience. These include promoting genetic diversity, reducing stressors such as drought or nutrient deficiencies, and implementing sanitation measures to remove infected trees. In managed forests, breeding programs that develop resistant tree varieties play a crucial role in mitigating the impacts of chronic diseases.

6.3 Latent Diseases

Latent diseases involve pathogens that remain dormant or asymptomatic within their hosts until triggered by specific environmental conditions or stress factors. These diseases are particularly insidious because they can persist undetected for years, only becoming active when hosts are weakened or exposed to favorable conditions for pathogen proliferation. Latent diseases highlight the complexity of pathogen-host interactions, as they depend on a delicate balance between the host's defenses and the pathogen's ability to exploit vulnerabilities.

Characteristics of Latent Diseases

Latent pathogens typically infect their hosts during periods of high susceptibility, such as early development or after injury, but do not immediately cause symptoms. Instead, they remain quiescent within plant tissues, awaiting favorable conditions for activation. Environmental stressors such as drought, frost, pollution, or mechanical damage often trigger the activation of latent infections. For example, *Cytospora* cankers in poplars and willows become active after drought stress, causing dieback and reduced vigor. Latent diseases are commonly associated with endophytic pathogens, which exist harmlessly within plant tissues under normal conditions but shift to a pathogenic state when the host is compromised. These pathogens exploit reduced host defenses, rapidly colonizing tissues and causing visible symptoms.

Examples of Latent Diseases

» **Cytospora Canker:** Fungi in the genus *Cytospora* remain latent in tree bark until stress events activate their pathogenicity.

» **Latent Vascular Wilt:** Pathogens like *Verticillium* species infect hosts as seedlings and remain dormant until maturity, when they cause wilting and dieback.

» **Post-Fire Diseases:** Certain fungal pathogens activate after forest fires, infecting weakened trees and accelerating mortality.

Impacts and Management

Latent diseases pose significant challenges for detection and management, as they often go unnoticed until they cause extensive damage. The unpredictability of activation events makes it difficult to implement timely interventions. Management strategies focus on reducing stressors that trigger latent infections, such as improving soil health, managing water availability, and minimizing mechanical injuries. Regular monitoring of forest health, combined with molecular diagnostics, can help detect latent pathogens before they become active. In some cases, proactive treatments, such as protective fungicides or biological control agents, can be applied to high-risk areas to prevent activation.

7. Classification Based on Impact on Host and Ecosystem

The classification of forest diseases based on their impact on the host and ecosystem highlights the ecological and functional roles of these diseases. Forest pathogens influence not only individual trees but also broader ecosystem processes, including biodiversity, nutrient cycling, and ecosystem services. The effects of these diseases can be categorized into primary diseases, which directly affect healthy trees; secondary diseases, which exploit stressed or weakened hosts; and broader impacts on biodiversity and ecosystem functionality. Understanding these impacts is essential for developing holistic management strategies that address the interconnectedness of forest health and ecological balance.

7.1 Primary Diseases

Primary diseases are caused by aggressive pathogens that infect and damage healthy trees. These pathogens often have specialized mechanisms to overcome host defenses, enabling them to establish infections even in robust and unstressed individuals. The direct effects of primary diseases can include tissue necrosis, vascular blockage, defoliation, and structural damage, often leading to significant reductions in growth, vigor, and reproductive success.

Characteristics of Primary Diseases

Primary diseases typically involve highly virulent pathogens with efficient infection and dissemination strategies. These diseases can affect various parts of the tree, including roots, stems, leaves, and reproductive structures, and their symptoms are often severe and rapid in progression. For example, vascular wilt diseases caused by fungi like *Ceratocystis* species block water-conducting tissues, causing sudden wilting and mortality in otherwise healthy trees. Primary diseases are often facilitated by environmental conditions that favor pathogen activity, such as high humidity or temperature extremes. For instance, *Phytophthora ramorum*, the causative agent of sudden oak death, thrives in moist, temperate climates, infecting healthy oaks and tanoaks with devastating results.

Examples of Primary Diseases

» **Chestnut Blight:** Caused by *Cryphonectria parasitica*, this fungal disease infects healthy American chestnuts, causing cankers that girdle stems and kill the tree.

» **Dutch Elm Disease:** A vascular wilt disease caused by *Ophiostoma* species, which spreads through bark beetles and rapidly kills healthy elm trees.

» **Sudden Oak Death:** Caused by *Phytophthora ramorum*, this disease infects the vascular tissues of oaks, leading to rapid decline and mortality.

Impacts and Management

The direct effects of primary diseases can lead to significant losses in forest productivity, timber quality, and ecological functionality. The removal of healthy trees from the canopy disrupts habitat structure and alters species interactions, potentially triggering cascading ecological effects. Management of primary diseases involves early detection, removal of infected trees, and the use of resistant species. In high-value forests, fungicides and biological control agents may be applied to reduce disease severity.

7.2 Secondary Diseases

Secondary diseases occur when pathogens exploit trees that are already stressed or weakened by environmental factors, pests, or mechanical damage. Unlike primary diseases, which directly target healthy hosts, secondary diseases rely on pre-existing vulnerabilities to establish infections. These diseases often act as opportunistic agents, further compromising the health of affected trees and accelerating their decline.

Characteristics of Secondary Diseases

Secondary pathogens are typically less virulent than primary pathogens, requiring a weakened host to overcome natural defenses. Stressors such as drought, nutrient deficiency, pollution, and physical injury create entry points and favorable conditions for these pathogens. For example, *Armillaria* root rot fungi often infect trees weakened by drought or competition, spreading through the root systems and causing structural decay. Secondary diseases are also commonly associated with post-disturbance environments, such as areas affected by storms, fires, or human activities. In these contexts, pathogens capitalize on the reduced resilience of trees to establish infections and spread.

Examples of Secondary Diseases

» **Armillaria Root Rot:** Caused by *Armillaria* species, this disease infects stressed trees, leading to root decay and increased susceptibility to windthrow.

» **Cytospora Canker:** Fungi in the genus *Cytospora* infect trees weakened by drought or frost, causing dieback and reduced vigor.

» **Heart Rot Diseases:** Fungal pathogens like *Ganoderma* species infect older or injured trees, degrading structural integrity over time.

Impacts and Management

Secondary diseases exacerbate the effects of environmental stressors, contributing to widespread tree mortality and forest decline. The cumulative impact of these diseases can change forest composition, favoring resilient species but reducing overall biodiversity. Management strategies focus on reducing stressors that predispose trees to infection, such as improving soil health, optimizing water availability, and minimizing mechanical damage. In some cases, sanitation measures, such as the removal of infected material, are necessary to prevent the spread of secondary pathogens.

7.3 Impact on Biodiversity and Ecosystem Services

Forest diseases extend their impacts beyond individual trees, influencing the biodiversity and ecological processes that sustain forest ecosystems. The loss or decline of tree species due to disease can disrupt trophic interactions, alter competitive dynamics, and affect the provision of ecosystem services such as carbon storage, water regulation, and habitat provision. These broader impacts underscore the interconnectedness of forest health and ecosystem functionality.

Effects on Biodiversity

Pathogens that selectively target specific tree species can significantly alter forest composition and structure. For example, the near-complete loss of American chestnuts due to chestnut blight (*Cryphonectria parasitica*) transformed eastern U.S. forests, reducing habitat complexity and changing the competitive balance among tree species. Similarly, the decline of European ash (*Fraxinus excelsior*) due to ash dieback (*Hymenoscyphus fraxineus*) has led to the loss of associated species, including lichens, fungi, and invertebrates that depend on ash trees. Forest diseases can also create opportunities for invasive species to establish and spread, further reducing native biodiversity. The loss of dominant tree species often leads to the homogenization of forest ecosystems, diminishing their resilience to future stressors.

Impacts on Ecosystem Services

The decline or mortality of trees caused by forest diseases affects a wide range of ecosystem services. For example:

» **Carbon Sequestration:** Widespread tree mortality reduces the capacity of forests to absorb and store carbon, contributing to increased atmospheric CO_2 levels.

» **Water Regulation:** Root diseases and deforestation caused by pathogens can disrupt water cycles, leading to increased runoff, erosion, and altered hydrological patterns.

» **Habitat Provision:** Dead and dying trees change the availability and quality of habitats for wildlife, affecting species that rely on specific tree types for nesting, foraging, or shelter.

Examples of Ecosystem Impacts

» **Sudden Oak Death:** The decline of oak species in California forests has reduced habitat availability for acorn-dependent wildlife, such as birds and mammals.

- **Ash Dieback:** The loss of ash trees has disrupted riparian ecosystems, altering shade, nutrient input, and water temperature regulation.
- **Pine Wilt Disease:** Caused by *Bursaphelenchus xylophilus*, this disease reduces the economic and ecological value of pine forests, particularly in Asia and Europe.

Management Implications

Addressing the broader impacts of forest diseases requires integrated approaches that consider ecological, economic, and social dimensions. Conservation strategies, such as preserving genetic diversity and promoting mixed-species forests, enhance ecosystem resilience to disease. Restoration efforts in affected areas focus on replanting with resistant species and rebuilding ecological functions. Additionally, monitoring and research are essential for understanding disease dynamics and mitigating their long-term effects on biodiversity and ecosystem services.

8. Classification Based on Geographic Distribution

Classifying forest diseases based on geographic distribution gives insights into pathogen distribution, environmental dependencies, and ecosystem impacts. Geographic distribution often reflects the adaptability of pathogens to specific climatic conditions, host availability, and modes of dispersal. Diseases may be regional, confined to particular climatic zones or forest types, or global, with pathogens capable of infecting hosts across continents due to natural adaptability or human-mediated spread.

8.1 Regional Diseases: Diseases Specific to Certain Climates or Zones

Regional diseases are those that occur in specific geographic areas, often due to the unique climatic, edaphic (soil-related), or ecological conditions that define these regions. These diseases tend to be closely associated with the forest types and tree species native to these areas, and their occurrence is often limited by environmental factors such as temperature, humidity, and soil properties. Regional diseases play a significant role in shaping the composition and dynamics of local forest ecosystems.

Characteristics of Regional Diseases

Regional diseases typically arise from native pathogens that have co-evolved with local tree species, leading to a balance where the disease is endemic and causes minimal disruption under normal conditions. However, changes in environmental factors, such as drought or increased precipitation, can trigger outbreaks by altering the balance between host resistance and pathogen virulence. For example, certain foliar pathogens are more prevalent in humid, tropical regions, where continuous leaf wetness facilitates spore germination and infection.

Climatic and Ecological Dependencies

Regional diseases are heavily influenced by climatic zones. In temperate forests, fungal diseases such as powdery mildew (*Erysiphe* spp.) and needle cast (*Lophodermium* spp.) are common, thriving in moist and cool conditions. In contrast, tropical forests are more susceptible to foliar pathogens like *Pestalotiopsis* species, which flourish in consistently high temperatures and humidity. Soilborne pathogens, such as *Phytophthora cinnamomi*, exhibit regional specificity in areas with poorly drained soils, particularly in Mediterranean-type ecosystems.

Examples of Regional Diseases

- » **Red Band Needle Blight:** *Dothistroma septosporum* predominantly affects conifers in temperate and boreal regions with high humidity.
- » **Cypress Canker:** Caused by *Seiridium cardinale*, this disease is prevalent in arid or Mediterranean climates, particularly affecting cypress species.
- » **Sheath Blight:** Common in tropical regions, this disease is caused by *Rhizoctonia solani*, affecting palms and other monocots in forested areas.

Impacts and Management

Regional diseases influence local forest composition, often acting as natural regulators of plant populations. While they typically operate within ecological equilibrium, environmental changes such as global warming and altered precipitation patterns may expand their range, threatening previously unaffected areas. Management strategies for regional diseases focus on monitoring local environmental conditions, promoting genetic diversity, and reducing anthropogenic stressors that increase host susceptibility. Early detection through regional surveillance programs is critical for addressing outbreaks effectively.

8.2 Global Diseases: Pathogens with Worldwide Distribution and Impact

Global diseases are caused by pathogens with broad adaptability, capable of infecting tree species across continents. These diseases are often associated with invasive pathogens introduced through human activities, such as trade, travel, and the movement of plant materials. Unlike regional diseases, global diseases tend to have devastating impacts on forest ecosystems due to the lack of co-evolution between pathogens and their hosts, leading to widespread susceptibility.

Characteristics of Global Diseases

Global pathogens exhibit a combination of traits that facilitate their spread and adaptability, including high virulence, wide host ranges, and efficient dispersal mechanisms. Many global diseases are caused by fungi, bacteria, or oomycetes that produce resilient propagules, such as spores or cysts, capable of surviving harsh conditions during transport. For example, rust fungi such as *Puccinia graminis* produce windborne spores that can travel long distances, infecting crops and forest trees on multiple continents. Human-mediated dispersal plays a critical role in the globalization of forest pathogens. The introduction of invasive pathogens through contaminated nursery stock, wood products, or soil has led to the emergence of significant global diseases. For instance, *Phytophthora ramorum*, the pathogen responsible for sudden oak death, has spread from its native range to North America and Europe, causing widespread mortality in oak and tanoak forests.

Examples of Global Diseases

- » **Chestnut Blight:** Caused by *Cryphonectria parasitica*, this disease originated in Asia but has spread to Europe and North America, nearly eradicating American chestnut populations.
- » **Dutch Elm Disease:** Fungal pathogens *Ophiostoma ulmi* and *Ophiostoma novo-ulmi*, transmitted by bark beetles, have caused global declines in elm populations.

- **Pine Wilt Disease:** The pinewood nematode (*Bursaphelenchus xylophilus*), vectored by beetles, has spread from North America to Asia and Europe, threatening pine forests worldwide.
- **Sudden Oak Death:** *Phytophthora ramorum* infects a wide range of hosts, spreading globally through trade in infected plants and soil.

Impacts on Ecosystems and Economies

The effects of global diseases are often severe and widespread, leading to significant ecological and economic losses. Ecologically, global diseases disrupt forest composition and function by decimating keystone or dominant species. For example, the loss of elms to Dutch elm disease has reduced habitat availability for associated wildlife and altered competitive dynamics among tree species. Economically, these diseases impose high costs on forestry industries, conservation efforts, and urban landscaping. The trade restrictions and quarantine measures imposed to prevent further spread can also affect international markets.

Challenges in Management

Managing global diseases poses significant challenges due to their rapid spread and the scale of their impacts. Prevention through strict biosecurity measures is the most effective strategy, focusing on quarantine protocols, inspection of plant materials, and sanitation of equipment. For established invasions, eradication is often impractical, requiring long-term containment and mitigation efforts. Research into resistant tree species and biological control agents is critical for reducing the impacts of global diseases. International collaboration, such as information sharing and coordinated response efforts, is essential for addressing the transboundary nature of these threats.

9. Hybrid and Emerging Classifications

The dynamic nature of forest ecosystems and the complexity of pathogen-host-environment interactions demand new approaches to classifying forest diseases. Hybrid and emerging classifications combine traditional and modern methods, integrating molecular, ecological, and technological advances to provide a more comprehensive framework. These classifications are essential for addressing contemporary challenges, such as emerging pathogens, climate change, and the globalization of disease spread. Hybrid systems not only refine our understanding of existing diseases but also enable the identification and categorization of novel pathogens.

9.1 Integration of Molecular and Ecological Criteria

The integration of molecular and ecological criteria in disease classification represents a significant advancement in forest pathology. While traditional classifications were based on symptoms, morphology, and host specificity, modern approaches incorporate molecular data to uncover genetic relationships, evolutionary histories, and the ecological roles of pathogens. Molecular techniques, such as DNA sequencing and phylogenetic analysis, allow for the precise identification of pathogens, even those that are morphologically indistinguishable or unculturable. By combining molecular data with ecological insights, researchers can classify diseases based on both their genetic makeup and their interactions within forest ecosystems.

For example, molecular tools have revealed cryptic species within previously defined pathogen groups, such as *Phytophthora* species, many of which were indistinguishable using traditional methods. These insights have enabled the identification of species-specific traits, such as host range, virulence factors, and environmental preferences, leading to more accurate classifications. Similarly, ecological criteria, such as modes of transmission, host specificity, and environmental dependencies, provide context for understanding how pathogens influence and are influenced by forest ecosystems. Integrating these criteria allows for the development of multidimensional classifications that reflect both the genetic and ecological complexity of forest diseases. This hybrid approach also facilitates the classification of pathogens with wide-ranging ecological roles. For instance, endophytic fungi, which exist harmlessly within plant tissues under normal conditions, may transition to pathogenic behavior under environmental stress. Understanding such transitions requires both molecular identification and ecological analysis, enabling a more nuanced classification system that captures the dual roles of these organisms. Furthermore, hybrid classifications help to bridge the gap between basic research and applied forest management by linking genetic data to practical outcomes, such as predicting disease outbreaks or identifying resistant tree species.

9.2 Role of Omics Technologies in Disease Classification

Omics technologies, including genomics, transcriptomics, proteomics, and metabolomics, have transformed forest disease classification by providing comprehensive insights into pathogen biology, host-pathogen interactions, and the molecular basis of disease. These technologies allow researchers to analyze the entire genetic or biochemical landscape of pathogens, uncovering novel traits and relationships that were previously inaccessible.

Genomics, the study of the complete genetic material of organisms, has been instrumental in resolving phylogenetic relationships and identifying genetic markers for pathogen identification. Whole-genome sequencing of pathogens such as *Hymenoscyphus fraxineus* (ash dieback) and *Cryphonectria parasitica* (chestnut blight) has provided critical information about their origins, evolutionary trajectories, and mechanisms of virulence. Genomic data also enable the identification of genes associated with host adaptation, fungicide resistance, and pathogenicity, offering new criteria for classification.

Transcriptomics, which examines gene expression profiles, sheds light on how pathogens respond to environmental conditions and host defenses. For instance, transcriptomic studies of rust fungi have revealed effector proteins that suppress host immune responses, highlighting molecular strategies that underpin disease development. Such data are invaluable for classifying pathogens based on their functional traits, rather than solely on morphological or ecological characteristics.

Proteomics and **metabolomics** provide complementary insights by analyzing the proteins and metabolites produced during infection. These approaches reveal the biochemical pathways involved in disease progression, enabling the identification of biomarkers for rapid disease detection and monitoring. For example, proteomic analyses of *Armillaria* root rot have identified enzymes responsible for lignin degradation, a key process in host colonization. Metabolomic studies of oomycetes like *Phytophthora* species have uncovered secondary metabolites that play roles in host-pathogen interactions, further enriching classification frameworks. Omics technologies also facilitate the study of microbial communities associated with forest diseases. High-throughput sequencing techniques, such as metagenomics, allow researchers to analyze the composition and dynamics of microbiomes in infected forests, identifying interactions between pathogens, beneficial microbes, and environmental factors. This holistic perspective supports the development of systems-based classifications that account for the multifaceted nature of forest diseases.

9.3 Implications for Disease Management and Research

The integration of hybrid and emerging classifications has profound implications for both the management of forest diseases and the advancement of research. By providing more precise and nuanced categorizations, these systems enhance our ability to predict, detect, and respond to disease outbreaks. The linkage of molecular data with ecological and functional traits enables the development of targeted strategies for disease control, ranging from early diagnostics to the breeding of resistant tree species. In disease management, hybrid classifications improve the accuracy and efficiency of pathogen detection and monitoring. Molecular diagnostic tools, such as PCR and next-generation sequencing, allow for the rapid identification of pathogens based on genetic markers, even at low infection levels. These tools are particularly valuable for detecting emerging pathogens or monitoring the spread of invasive species. For example, the identification of specific *Phytophthora* species in nursery stock has enabled quarantine measures to prevent the introduction of these pathogens into natural forests. Hybrid classifications also inform the development of integrated disease management strategies by linking pathogen traits to environmental conditions and host susceptibility. For instance, understanding the molecular basis of fungicide resistance in *Dothistroma septosporum* has guided the selection of effective chemical treatments in pine plantations. Similarly, the identification of resistance genes in tree species through genomic studies has accelerated the breeding of resistant varieties, providing long-term solutions for disease management.

Conclusion

The classification of forest diseases is a cornerstone of forest pathology, providing an organized framework for understanding, diagnosing, and managing the diverse factors that affect forest health. It serves as a bridge between scientific research and practical management, enabling forest pathologists, ecologists, and managers to address complex disease dynamics systematically. By categorizing diseases based on causal agents, affected plant parts, ecological roles, modes of spread, temporal patterns, impacts on hosts and ecosystems, and geographic distribution, classification systems offer a multidimensional perspective that is essential for both theoretical understanding and applied forest management.

The importance of classifying forest diseases is underscored by the critical ecological and economic roles that forests play. Forest ecosystems are among the most biodiverse environments on Earth, providing essential services such as carbon sequestration, biodiversity conservation, and the regulation of water cycles. Diseases, whether caused by biotic agents like fungi, bacteria, viruses, nematodes, or by abiotic factors such as climate and pollution, can severely compromise these functions. Effective classification not only facilitates the identification and diagnosis of diseases but also aids in predicting disease outbreaks, understanding their underlying causes, and designing targeted intervention strategies. This systematic approach ensures that limited resources are utilized efficiently and that management actions are tailored to specific threats.

Historically, the classification of forest diseases has evolved in parallel with advancements in forest pathology and related scientific disciplines. Early systems were largely descriptive, focusing on visible symptoms such as leaf discoloration, cankers, or root decay. While these methods provided an essential foundation, they were limited in their ability to explain the underlying causes or ecological roles of diseases. The introduction of etiology-based classification, which considers the causal agents of diseases, marked a significant shift toward a more comprehensive understanding of forest pathology. With the advent of molecular biology and omics technologies, classification systems have become increasingly sophisticated, integrating genetic, biochemical, and ecological data. These modern approaches allow for precise identification of pathogens and their interactions with host plants and environments, reflecting the complexity of forest disease dynamics.

Classifying forest diseases based on causal agents reveals the vast diversity of pathogens and their mechanisms of action. Fungi, for instance, are among the most common and destructive forest pathogens, responsible for diseases like root rots, cankers, and foliar blights. Their ability to thrive in forest soils and decompose organic matter underscores their dual role as both pathogens and ecosystem regulators. Bacterial diseases, though less prevalent, can cause severe damage, particularly through mechanisms like toxin production and biofilm formation. Viral and viroid diseases, historically underappreciated in forest ecosystems, are emerging as significant threats, especially in the context of climate change and global trade. Nematodes, as soilborne pathogens, further complicate forest disease dynamics by impairing root function and contributing to tree decline. Abiotic factors, including extreme weather events, pollution, and soil nutrient imbalances, act either as primary stressors or as exacerbating conditions for biotic diseases, highlighting the interconnectedness of these factors.

The classification of diseases based on host plant parts affected provides critical insights into their physiological and ecological impacts. Root diseases, for example, compromise water and nutrient uptake, leading to tree instability and increased susceptibility to windthrow. Stem and branch diseases, such as cankers and galls, undermine structural integrity and create entry points for secondary pathogens. Leaf and needle diseases directly affect photosynthesis, reducing energy production and growth. Diseases targeting flowers, fruits, and seeds, though less immediately visible, have long-term implications for forest regeneration and genetic diversity. This aspect of classification emphasizes the need to address specific vulnerabilities within forest ecosystems, enabling more precise diagnosis and intervention.

Ecological classifications offer an understanding of the broader roles that diseases play within forest ecosystems. Endemic diseases, for instance, often serve as natural regulators of plant populations, contributing to biodiversity and ecological balance. Epidemic diseases, on the other hand, can cause widespread tree mortality and disrupt ecosystem functions, often triggered by environmental changes or anthropogenic activities. Invasive diseases, caused by non-native pathogens, pose unique challenges by disrupting established ecological relationships and often outcompeting native species. Recognizing these ecological roles is essential for managing forests in a way that balances conservation and disease mitigation.

The mode of pathogen spread provides another critical dimension for classification. Soilborne diseases, which persist in the soil and infect through root contact, are particularly challenging to manage due to their longevity and resilience. Airborne diseases, spread via spores or vectors, can cause rapid and widespread outbreaks, often influenced by climatic conditions. Waterborne diseases, transmitted through rain or irrigation, are significant in riparian and wetland forests, while vector-transmitted diseases underscore the interconnectedness of forest ecosystems, as they rely on organisms such as insects or nematodes for dispersal. Understanding these mechanisms is vital for designing containment strategies and predicting disease trajectories. Temporal classifications further enhance our understanding of forest diseases by considering their progression and duration. Acute diseases, characterized by rapid onset and short duration, often result in immediate and visible impacts, such as sudden tree mortality. Chronic diseases, by contrast, develop slowly and persist over time, gradually weakening trees and reducing forest productivity. Latent diseases, which remain dormant until triggered by environmental stressors, pose hidden threats that can suddenly escalate under changing conditions. Recognizing these patterns is crucial for implementing timely and effective management interventions.

The impact of diseases on host plants and ecosystems offers yet another lens for classification. Primary diseases affect healthy trees directly, often caused by highly virulent pathogens, while secondary diseases occur in trees already weakened by stressors such as drought or pollution. Beyond their direct effects, forest diseases influence ecosystem services and biodiversity, reshaping forest composition and function. For instance, the selective pressure exerted by certain pathogens can lead to shifts in species dominance, affecting habitat availability and ecosystem stability.

Geographic distribution provides an additional perspective on disease classification, distinguishing between regional diseases influenced by specific climatic and ecological conditions and global diseases facilitated by international trade and human activity. These geographic patterns is critical for developing region-specific management strategies and addressing global challenges, particularly in the context of climate change and globalization. Emerging hybrid classification systems reflect the integration of traditional and modern approaches, incorporating molecular, ecological, and technological criteria. These systems provide a holistic view of forest diseases, bridging the gap between foundational knowledge and cutting-edge research. The application of omics technologies, such as genomics and proteomics, has revolutionized our ability to identify pathogens and understand their interactions with hosts and environments. By linking classification to disease management, these approaches enable the development of innovative strategies to mitigate the impacts of forest diseases while preserving ecosystem health and resilience.

Chapter - 4

Common Diseases in Forest Trees

Introduction

Root, stem, leaf, and vascular diseases of forest trees represent a significant area of concern in forest pathology, reflecting the intricate interactions between biotic pathogens, abiotic stressors, and host plants. These diseases impact various parts of the tree, from roots to flowers, with each category presenting unique challenges and requiring tailored strategies for diagnosis, monitoring, and management. Forest trees, as long-lived organisms, are particularly vulnerable to chronic and acute diseases that can compromise their health and longevity, disrupting ecosystem services, biodiversity, and the economic value of forests. The diverse range of diseases, caused by fungi, bacteria, nematodes, and environmental stressors, underscores the need for a comprehensive understanding of their biology, ecology, and management.

Root diseases form the foundation of tree health concerns, given that roots are essential for water and nutrient uptake and provide physical anchorage. Among the most significant of these diseases is Armillaria root rot, caused by fungi of the genus Armillaria. This disease is often referred to as the "forest pathogen par excellence" due to its widespread occurrence, ability to persist as mycelial networks in soil, and impact on both coniferous and deciduous trees. Symptoms include reduced growth, chlorosis, and eventual mortality. Its ecological role, however, extends beyond being a mere pathogen; it participates in nutrient cycling by decomposing deadwood, highlighting its dual nature. Similarly, Phytophthora root rot, caused by various Phytophthora species, is a silent killer in forest ecosystems. Often exacerbated by poorly drained soils, this disease disrupts the vascular connection between roots and shoots, causing water stress and dieback. Root-knot nematodes (*Meloidogyne species*) and Fusarium root rot, on the other hand, highlight the challenges posed by belowground pathogens that interact with abiotic stressors like drought and soil compaction, compounding their impact on tree health.

Stem and branch diseases, often visually conspicuous, have profound implications for forest structure and productivity. Chestnut blight, caused by *Cryphonectria parasitica*, remains one of the most catastrophic examples of stem disease, responsible for decimating American chestnut populations in the 20th century. Its global impact is a testament to the devastation that invasive pathogens can cause. Canker diseases, characterized by localized necrosis on stems and branches, lead to girdling

DOI: 10.1201/9781042011063-4

and dieback, weakening the tree's structural integrity. Heart rot diseases, caused by decay fungi such as Ganoderma and Trametes species, exploit older trees, decomposing heartwood and creating hollow cavities. While often viewed as detrimental, these diseases contribute to forest ecology by providing habitats for wildlife. Pitch canker, primarily affecting conifers, is a rising threat due to its association with climate variability and global trade, which facilitate the spread of its fungal pathogen, Fusarium circinatum.

Leaf and needle diseases directly impair photosynthesis, the cornerstone of tree growth and productivity. Red band needle blight, caused by *Dothistroma* species, manifests as red lesions on conifer needles, leading to defoliation and reduced growth. Powdery mildew, a ubiquitous fungal disease, affects a broad range of tree species, causing powdery white patches on foliage and stunted growth. Tar spot disease, characterized by black tar-like spots on leaves, is caused by *Rhytisma* species and has both aesthetic and ecological impacts. Needle cast diseases, caused by fungi such as Lophodermium and Rhizosphaera species, are particularly detrimental to conifers, leading to premature needle shedding and tree stress. The study and management of these diseases are critical for maintaining the productivity and aesthetics of forested landscapes.

Diseases of flowers, fruits, and seeds, though often overlooked, have significant implications for forest regeneration and genetic diversity. Blossom blights caused by Monilinia and related pathogens affect reproductive structures, leading to poor seed set and reduced reproductive success. Seed-borne pathogens, such as Fusarium, threaten nursery stock and young plantations, serving as primary inoculum sources for forest-wide outbreaks. Rust diseases, exemplified by *Cronartium* species, cause cone infections, which directly impact seed viability and regeneration efforts, highlighting the importance of protecting reproductive structures in forest management strategies.

Vascular diseases, often systemic and highly destructive, pose some of the most severe challenges to forest health. Dutch elm disease, caused by *Ophiostoma* species and spread by elm bark beetles, is a catastrophic vascular wilt that has decimated elm populations across Europe and North America. Oak wilt, caused by *Bretziella fagacearum*, spreads through root grafts and insect vectors, leading to rapid wilt and mortality in susceptible oak species. Verticillium wilt, caused by Verticillium species, represents a persistent challenge for deciduous forests, affecting a broad range of hosts and often exacerbated by environmental stressors. These diseases underscore the importance of understanding pathogen-host interactions at the vascular level, where disruptions have systemic consequences. Diseases caused by invasive pathogens represent a growing threat to forest ecosystems globally. Sudden oak death, caused by *Phytophthora ramorum*, has devastated oak and tanoak populations in parts of the United States and Europe, disrupting forest structure and associated biodiversity. Ash dieback, caused by *Hymenoscyphus fraxineus*, has rapidly spread across Europe, threatening the survival of ash species and the ecosystems dependent on them. Pine wilt disease, caused by the pinewood nematode (*Bursaphelenchus xylophilus*), highlights the role of invasive nematodes in altering forest dynamics, with its impact amplified by global trade and climate change. These examples underscore the urgent need for biosecurity measures and proactive management to curb the spread of invasive forest pathogens.

Abiotic stressors such as frost, drought, soil compaction, and pollution contribute significantly to forest diseases, either as primary stressors or by predisposing trees to biotic pathogens. Frost and drought-associated cankers result from environmental extremes that weaken bark tissues, creating entry points for secondary infections. Soil compaction disrupts root function by limiting oxygen and nutrient availability, often leading to root dysfunction and decline. Pollution-induced foliar disorders, caused by atmospheric pollutants such as sulfur dioxide and ozone, manifest as chlorosis, necrosis, and reduced growth, highlighting the interplay between human activities and forest health.

Emerging and cryptic diseases pose unique challenges in forest pathology, often linked to climate change and global trade. Climate change, through its influence on temperature, precipitation, and extreme weather events, creates favorable conditions for the emergence of novel pathogens or the resurgence of dormant ones. Cryptic fungal pathogens, which remain undetected until environmental conditions trigger disease, complicate early detection and management efforts. Novel diseases in urban and plantation forests, influenced by human activities and environmental stressors, highlight the need for proactive research and monitoring to address these emerging threats. Integrated disease management in forests combines traditional and modern approaches to mitigate the impacts of forest diseases. Monitoring systems, including field surveys and remote sensing technologies, provide critical data for disease detection and progression. Molecular tools, such as PCR-based diagnostics and genomic analyses, enable early detection and precise identification of pathogens. Combining silvicultural practices, such as selective thinning and resistant species planting, with biocontrol measures, such as beneficial microbes and natural predators, represents a holistic approach to forest disease management. These strategies ensure the sustainability of forest ecosystems while mitigating the economic and ecological impacts of diseases.

1. Root Diseases

Root diseases are among the most significant challenges to forest health, as they directly impact the foundational structures of trees, compromising their ability to absorb water and nutrients. These diseases are caused by a variety of pathogens, including fungi, oomycetes, and nematodes, which infect root systems and create conditions that often lead to tree mortality. Root diseases not only weaken individual trees but also disrupt the ecological balance of forest ecosystems, reducing biodiversity and altering nutrient cycling processes.

1.1 Armillaria Root Rot

Armillaria root rot, caused by fungi in the genus Armillaria, is one of the most widespread and destructive root diseases in forest ecosystems. These pathogens are versatile and infect a wide range of tree species across temperate and tropical forests. Armillaria species are facultative parasites, capable of surviving as saprophytes on dead organic matter while also infecting living hosts under favorable conditions.

The biology of Armillaria is characterized by its production of rhizomorphs, dark, root-like fungal structures that spread through the soil, seeking out and infecting tree roots. Once inside the host, the fungus produces enzymes that degrade lignin and cellulose, leading to the decay of root tissues. Symptoms of Armillaria root rot include reduced canopy density, chlorosis (yellowing of leaves), dieback, and increased susceptibility to windthrow due to weakened root systems. White mycelial mats often form beneath the bark of infected roots, providing a diagnostic feature of the disease. Ecologically, Armillaria root rot plays a dual role. While it contributes to the natural process of decomposition and nutrient cycling, its pathogenic phase can lead to significant tree mortality, particularly in stressed or densely planted forests. The spread of the disease is exacerbated by environmental factors such as drought, soil compaction, and poor forest management practices. In commercial forests, Armillaria root rot results in substantial economic losses by reducing timber quality and productivity. Management strategies for Armillaria root rot include removing infected stumps and roots to reduce inoculum levels, improving soil drainage, and promoting mixed-species planting to enhance forest resilience. Biological control agents, such as antagonistic fungi like Trichoderma, have also shown promise in suppressing Armillaria infections.

1.2 Phytophthora Root Rot

Phytophthora root rot, caused by various Phytophthora species, is a devastating disease that affects a wide range of forest tree species. Phytophthora is an oomycete, a group of water molds that thrive in moist soils and environments. The disease is often referred to as a "silent killer" because infections typically begin unnoticed, with symptoms manifesting only after significant damage has occurred. The life cycle of Phytophthora involves motile zoospores that swim through waterlogged soils, seeking out root tips and wounds for infection. Once inside the host, the pathogen disrupts water and nutrient transport by causing necrosis of root tissues and blocking vascular pathways. Symptoms of Phytophthora root rot include wilting, stunted growth, collar rot, and in severe cases, tree mortality.

Phytophthora cinnamomi is among the most notorious species, causing extensive damage in Mediterranean forests, Australian eucalyptus woodlands, and temperate regions worldwide. This pathogen has been implicated in the decline of oak, chestnut, and avocado trees, as well as native plant communities. The ecological impact of Phytophthora root rot extends beyond individual trees, as the loss of key species can alter forest composition and reduce biodiversity. Managing Phytophthora root rot is challenging due to the pathogen's persistence in soil and its broad host range. Preventative measures, such as improving soil drainage, avoiding overwatering, and implementing strict quarantine protocols, are critical to reducing the risk of infection. Fungicides can provide temporary control but are not sustainable solutions for natural forests. Advances in molecular diagnostics have improved early detection of Phytophthora infections, enabling more effective management strategies.

1.3 Root-knot Nematodes

Root-knot nematodes, primarily from the genus Meloidogyne, are microscopic, soil-dwelling organisms that cause significant damage to the root systems of forest trees. These nematodes are endoparasites, meaning they invade root tissues and establish feeding sites, disrupting normal root function and reducing tree vigor. Root-knot nematodes are particularly problematic in nurseries, where they infect seedlings and reduce their ability to establish in forest environments. The infection process begins when nematode juveniles penetrate root tissues, migrating to the vascular cylinder and inducing the formation of galls or knots. These galls interfere with water and nutrient uptake, leading to symptoms such as wilting, stunted growth, and chlorosis. Infected roots are also more susceptible to secondary infections by fungi and bacteria, further exacerbating the damage. Root-knot nematodes are highly adaptable and thrive in various soil types, making them a persistent threat in both natural and managed forests. They are often spread through contaminated soil, water, or planting materials, highlighting the importance of sanitation and quarantine measures. The ecological impact of root-knot nematodes includes reduced forest productivity and altered species dynamics, as infected trees are unable to compete effectively for resources. Management strategies focus on improving soil health, using resistant tree varieties, and applying biological control agents, such as predatory nematodes or fungi. Crop rotation and soil amendments, such as organic compost, can also reduce nematode populations in nurseries and reforestation sites.

1.4 Fusarium Root Rot

Fusarium root rot, caused by species in the genus Fusarium, is a fungal disease that primarily affects seedlings and young trees, leading to poor establishment and growth. The disease is closely associated with abiotic stressors such as drought, nutrient deficiency, and soil compaction, which weaken host defenses and create conditions conducive to infection.

Fusarium species are opportunistic pathogens that colonize root tissues through wounds or natural openings. Once established, they produce toxins and enzymes that degrade plant cells, leading to root necrosis and reduced water uptake. Symptoms of Fusarium root rot include stunted growth, wilting, and discoloration of root tissues. In advanced cases, the disease can progress to vascular wilt, causing systemic infection and mortality. The interaction between Fusarium root rot and abiotic stressors highlights the importance of environmental conditions in disease development. For example, prolonged drought weakens root systems, making them more susceptible to infection. Similarly, compacted soils reduce oxygen availability, creating anaerobic conditions that favour fungal growth. Management of Fusarium root rot involves reducing environmental stressors and improving soil conditions. This includes practices such as deep tillage to alleviate compaction, proper irrigation to prevent water stress, and the use of organic amendments to enhance soil microbial diversity. Biological control agents, such as beneficial fungi and bacteria, can also suppress Fusarium populations and promote root health. Early detection through soil testing and root examinations is critical for preventing the spread of this disease in nurseries and reforestation projects.

2. Stem and Branch Diseases

Stem and branch diseases are a significant category of forest diseases that primarily affect the structural integrity and functionality of trees. These diseases, caused by fungi, bacteria, and other pathogens, often lead to cankers, decay, dieback, and in severe cases, complete tree mortality. The pathogens responsible for these diseases target the vital vascular tissues within the stem or disrupt the cambium, reducing the tree's ability to transport water and nutrients. Over time, these infections can weaken the tree's structural stability, making it susceptible to breakage, windthrow, and secondary infections. Stem and branch diseases not only affect individual trees but also have broader ecological and economic impacts, altering forest composition, reducing timber quality, and disrupting ecosystem services such as carbon sequestration.

2.1 Chestnut Blight

Chestnut blight, caused by the fungal pathogen *Cryphonectria parasitica,* is one of the most catastrophic stem diseases in forest history, particularly in North America and parts of Europe. Originally native to Asia, where it exists in relative balance with its host species, the pathogen was introduced to North America in the late 19th century, where it devastated populations of the American chestnut (Castanea dentata), a keystone species in eastern forests. The disease is also a significant concern for European chestnuts (*Castanea sativa*), although the damage has been somewhat mitigated by breeding programs and natural resistance in some populations. The life cycle of Cryphonectria parasitica begins when fungal spores, dispersed by wind, rain, or insects, land on the bark of a susceptible tree. The fungus enters through wounds or natural openings and colonizes the cambium and phloem layers beneath the bark. As the infection progresses, it forms cankers, sunken, necrotic lesions that girdle the stem or branch. This girdling disrupts the flow of nutrients and water, effectively killing the tissues above the canker. The visible symptoms include orange stromata (spore-producing structures) on the bark, dieback of branches, and eventual tree death. Unlike many pathogens that primarily affect weakened or stressed hosts, *Cryphonectria parasitica* can infect and kill healthy chestnut trees.

The ecological impact of chestnut blight has been profound. Before the blight, the American chestnut was one of the dominant canopy trees in eastern North America, providing food and habitat for countless wildlife species and serving as a valuable source of timber and nuts. The near-total elimination of mature chestnuts from these forests has drastically altered ecosystem dynamics, leading to the proliferation of less ecologically significant species like oaks and maples. Economically, the loss of the American chestnut devastated industries reliant on its durable, rot-resistant wood. Efforts to manage chestnut blight include breeding programs aimed at developing resistant hybrids by crossing American chestnuts with resistant Asian species. Biocontrol methods, such as the use of hypovirulent strains of the fungus (infected with a virus that reduces pathogenicity), have shown promise in limiting the spread of the disease in Europe. However, the large-scale restoration of chestnut populations remains a significant challenge, requiring extensive collaboration between researchers, conservationists, and forest managers.

2.2 Canker Diseases

Canker diseases are a diverse group of infections caused by fungal and bacterial pathogens that attack the cambium and phloem tissues of trees, leading to the formation of necrotic lesions. These cankers often disrupt the vascular system, causing localized dieback and, in severe cases, the death of the entire tree. Canker diseases affect a wide range of forest species and are prevalent in both natural and managed ecosystems. Common pathogens associated with canker diseases include *Nectria, Cytospora*, and Seiridium species, each with distinct infection mechanisms and ecological impacts. The formation of cankers begins when pathogens enter the tree through wounds caused by environmental factors, such as frost cracks, wind damage, or pruning cuts. Once inside, the pathogen colonizes the cambium and produces toxins or enzymes that kill surrounding tissues, resulting in sunken, discolored lesions. Over time, these lesions expand, girdling branches or stems and causing dieback. Secondary infections by other pathogens or pests often exacerbate the damage, leading to a decline in tree vigor.

Canker diseases are particularly problematic in stressed forests, where factors such as drought, pollution, or overstocking reduce tree resilience. For example, Cytospora canker, caused by fungi in the genus *Cytospora*, is a common problem in aspen, poplar, and spruce trees. The disease typically affects weakened or stressed trees, leading to dieback and the production of orange or black pycnidia (spore-producing structures) on the bark. Similarly, Seiridium canker, which affects cypress trees, is exacerbated by water stress and poor soil conditions, resulting in extensive branch dieback and reduced aesthetic and economic value. Ecologically, canker diseases play a complex role. While they can reduce forest productivity and timber quality, they also contribute to natural processes such as decomposition and nutrient cycling. Dead and dying trees create habitat for a variety of organisms, including insects, birds, and fungi, which depend on decaying wood. However, severe outbreaks can disrupt forest structure and reduce biodiversity. Managing canker diseases involves a combination of preventive and remedial measures. Maintaining tree health through proper silvicultural practices, such as thinning and pruning during dry periods, can reduce susceptibility. The removal and destruction of infected branches or trees help to limit pathogen spread. In high-value forests or urban landscapes, fungicidal treatments may be used, although they are not practical for large-scale applications in natural forests.

2.3 Heart Rot

Heart rot refers to a group of diseases caused by fungi that decay the central, non-living (heartwood) tissues of trees. These diseases are common in mature or over-mature trees and are a significant concern in both natural forests and commercial plantations. Heart rot fungi include species such as Ganoderma, Phellinus, and Trametes, which colonize trees through wounds or natural openings and degrade lignin and cellulose, the structural components of wood. The process of heart rot begins when fungal spores enter a tree through injuries caused by storms, fire, insect activity, or human interventions such as logging. Once inside, the fungi produce enzymes that break down the structural components of the heartwood, leading to the formation of cavities, hollow trunks, and weakened stems. Symptoms of heart rot are often subtle, making early detection difficult. Visible signs may include fungal fruiting bodies (conks) on the trunk or branches, reduced crown density, and structural instability. Ecologically, heart rot plays an essential role in forest dynamics. By decomposing wood, heart rot fungi contribute to nutrient cycling, returning vital elements to the soil and supporting the growth of new vegetation. Hollowed-out trees created by heart rot provide habitat for a wide range of wildlife, including cavity-nesting birds, bats, and small mammals. However, in commercial forestry, heart rot represents a significant problem, reducing timber quality and economic value. Trees with advanced decay are prone to windthrow, posing safety risks in urban and recreational areas. Management of heart rot focuses on minimizing wounding and stress to trees, as well as promoting species that are less susceptible to decay. In commercial forests, careful logging practices and the application of wound protectants can reduce the risk of infection. Removing and replacing severely affected trees may be necessary in urban areas to prevent hazards. Biological control agents, such as antagonistic fungi, are being explored as potential tools for managing heart rot in natural ecosystems.

2.4 Pitch Canker

Pitch canker, caused by the fungal pathogen Fusarium circinatum, is a significant disease affecting coniferous trees, particularly pines. First identified in the southeastern United States, pitch canker has since spread to other parts of the world, including Europe, Asia, and South Africa, posing a growing threat to both natural forests and commercial plantations. The disease primarily affects the vascular tissues of branches, stems, and cones, leading to resinous lesions, dieback, and reduced growth.

The life cycle of Fusarium circinatum involves the production of airborne spores that infect trees through wounds created by insects, pruning, or mechanical damage. Once inside the tree, the fungus colonizes the vascular tissues, disrupting the flow of water and nutrients. The infection triggers the production of resin, which accumulates around the lesions, forming the characteristic "pitch" associated with the disease. Over time, affected branches and stems die back, and the tree's overall vigor declines. Pitch canker is often exacerbated by insect activity, particularly bark beetles, which create entry points for the fungus and facilitate its spread. This synergistic relationship between pathogens and pests highlights the complexity of managing pitch canker in forest ecosystems. The disease is particularly destructive in monoculture plantations, where the lack of genetic diversity increases the vulnerability of tree populations.

The ecological and economic impacts of pitch canker are significant. In natural forests, the loss of mature pines reduces habitat availability for wildlife and alters forest composition. In commercial plantations, the disease reduces timber yields and quality, increasing production costs. The export of infected seeds and seedlings has also contributed to the global spread of the pathogen, raising concerns about biosecurity measures in international trade. Managing pitch canker requires an integrated approach that combines prevention, monitoring, and remediation. Preventive measures include avoiding mechanical injuries to trees, controlling insect vectors, and implementing quarantine protocols to prevent the introduction of infected materials. Breeding programs aimed at developing resistant pine varieties offer a long-term solution, while fungicides and biological control agents provide temporary relief in high-value areas.

3. Leaf and Needle Diseases

Leaf and needle diseases represent a major threat to forest ecosystems, significantly impacting tree growth, photosynthetic capacity, and overall forest productivity. These diseases are caused primarily by fungal pathogens, but bacterial and viral agents can also contribute. Leaf and needle diseases often exhibit high rates of spread due to favorable environmental conditions, such as high humidity and mild temperatures, and they can cause widespread defoliation, leading to long-term declines in tree health and survival.

3.1 Red Band Needle Blight: Symptoms, Spread, and Control

Red band needle blight, caused by the fungal pathogen *Dothistroma septosporum* (and less commonly *Dothistroma pini*), is one of the most damaging needle diseases affecting conifers worldwide. This disease is particularly prevalent in monoculture plantations of pines, where its ability to spread rapidly can lead to significant reductions in growth, timber quality, and forest health. The symptoms of red band needle blight are distinctive and easily recognizable. The disease begins with small, dark-green to yellow spots on infected needles, which eventually expand and coalesce into red bands encircling the needle. The tips of infected needles turn brown while the base remains green, creating a two-tone effect. Severe infections result in premature needle drop, leaving trees with sparse crowns and reducing their photosynthetic capacity. In young trees, the disease can cause complete defoliation, stunting growth, and, in extreme cases, lead to mortality. The spread of red band needle blight is facilitated by the production of conidia, asexual spores that are dispersed by rain splash or wind. The fungus thrives in warm, humid conditions, making plantations in temperate and subtropical climates particularly susceptible. The disease is exacerbated by poor forest management practices, such as high planting densities and inadequate thinning, which create microclimates favorable for pathogen proliferation.

Controlling red band needle blight requires an integrated approach. Silvicultural practices, such as reducing planting density and pruning infected branches, help improve air circulation and reduce humidity within stands. Fungicides containing copper compounds are effective in protecting high-value trees, particularly in nurseries and young plantations. Breeding programs aimed at developing resistant pine species or hybrids are crucial for long-term management, as genetic resistance offers the most sustainable solution. Regular monitoring and early detection through surveys and molecular diagnostics are essential for managing outbreaks effectively.

3.2 Powdery Mildew

Powdery mildew is a common foliar disease caused by a group of fungal pathogens in the order *Erysiphales*, including species such as Erysiphe alphitoides, which is particularly problematic for oak trees. This disease affects a wide range of broadleaf and coniferous tree species, often reducing photosynthetic efficiency and overall tree vigor. Powdery mildew is especially prevalent in regions with mild temperatures and moderate humidity, where it can cause recurring infections that weaken trees over time. The disease is characterized by a white, powdery coating on the surface of leaves, stems, and, in some cases, flowers or fruits. This coating consists of fungal mycelia and conidia, which are produced in large quantities and easily dispersed by wind. Infected leaves may curl, discolor, and drop prematurely, reducing the tree's ability to capture sunlight and produce energy. Young seedlings and saplings are particularly vulnerable, as repeated infections can stunt growth and increase susceptibility to other pathogens or environmental stressors.

Powdery mildew pathogens are obligate parasites, meaning they require living host tissues to complete their life cycle. They germinate on the surface of leaves and penetrate the epidermis to extract nutrients, causing localized damage to host cells. Unlike many other foliar pathogens, powdery mildew does not require free water for spore germination, allowing it to thrive in relatively dry environments. Management of powdery mildew involves a combination of cultural, chemical, and biological strategies. Increasing air circulation through thinning and pruning can reduce the microclimatic conditions that favor the disease. Fungicides, such as sulfur-based or systemic treatments, are effective in controlling severe outbreaks, particularly in nurseries and urban landscapes. Biological control agents, including certain species of Bacillus and Trichoderma, have shown potential in suppressing powdery mildew infections. In forest systems, maintaining genetic diversity and planting resistant tree species are critical for reducing the long-term impacts of the disease.

3.3 Tar Spot Disease

Tar spot disease, caused by fungi in the genus *Rhytisma*, is a visually striking foliar disease that affects deciduous trees, particularly maples (Acer spp.). Although tar spot is not typically fatal to trees, it can significantly reduce aesthetic and economic value, especially in ornamental and urban forests. The disease is widely distributed in temperate regions and thrives in areas with prolonged periods of leaf wetness.

The primary symptom of tar spot is the development of black, tar-like spots on infected leaves, which give the disease its name. These spots begin as small, yellowish lesions that darken and enlarge as the fungus matures. In severe cases, the spots can cover large portions of the leaf surface, leading to premature leaf drop and reduced photosynthetic efficiency. The disease primarily affects older leaves, and while its impact on tree health is usually minimal, repeated defoliation can weaken trees over time. The life cycle of *Rhytisma* involves the production of ascospores within the tar-like spots, which are released in the spring to infect new leaves. The fungus overwinters in fallen leaves, making sanitation an important component of disease management. Environmental factors, such as high humidity and cool temperatures during leaf emergence, favor infection and spore dispersal. Controlling tar spot disease in forest settings is often unnecessary, as its effects on tree health are generally limited. However, in urban and ornamental landscapes, where aesthetic value is a concern, removing and destroying fallen leaves can significantly reduce inoculum levels and prevent new infections. Fungicides may be applied in severe cases, but their use is typically restricted to high-value trees or small-scale plantings. Maintaining tree health through proper nutrition and irrigation can also reduce the severity of tar spot infections.

3.4 Needle Cast Diseases: Fungal Pathogens in Conifers

Needle cast diseases, caused by a variety of fungal pathogens, are among the most significant threats to coniferous forests worldwide. These diseases result in premature needle drop, reducing the photosynthetic capacity of trees and leading to growth losses, reduced timber quality, and, in severe cases, mortality. Common pathogens associated with needle cast include Lophodermium, Rhizosphaera, and Mycosphaerella species. The symptoms of needle cast diseases vary depending on the pathogen but typically include discoloration of needles, often starting as yellow or brown spots that expand to cover the entire needle. Infected needles are shed prematurely, leaving trees with sparse crowns. The disease often progresses slowly, with symptoms becoming more pronounced over several years as new infections compound the effects of previous ones.

Needle-cast fungi produce spores that are dispersed by wind or rain splash. Infection occurs during wet conditions, with spores germinating on needle surfaces and penetrating through stomata. The fungi colonize the needle tissue, eventually forming fruiting bodies that release new spores. Dense, overcrowded stands and poor air circulation exacerbate the disease by creating humid microclimates that favor pathogen development. Management of needle cast diseases involves a combination of cultural and chemical strategies. Silvicultural practices, such as thinning and pruning, can improve air circulation and reduce humidity within stands, limiting the spread of the disease. Fungicides, applied as preventative treatments during periods of high risk, are effective in protecting valuable trees, particularly in nurseries and young plantations. Breeding programs aimed at developing resistant conifer species or hybrids are essential for long-term control. Regular monitoring and early detection are critical for managing needle cast diseases, as early intervention can prevent severe outbreaks.

4. Flower, Fruit, and Seed Diseases

Diseases affecting flowers, fruits, and seeds are critical not only for individual tree health but also for the reproductive success and regeneration of forest ecosystems. These diseases often disrupt processes such as pollination, seed development, and germination, significantly impacting forest sustainability and biodiversity. Pathogens targeting these reproductive structures include fungi, bacteria, and other microorganisms that exploit flowers, fruits, and seeds as nutrient sources or propagation platforms. Flower, fruit, and seed diseases are particularly significant in nurseries, where they compromise seedling production, and in forests, where they hinder natural regeneration.

4.1 Blossom Blights: Monilinia and Related Pathogens

Blossom blights, caused by pathogens such as Monilinia, Botrytis, and Sclerotinia species, are diseases that attack the flowers of various tree species, often leading to the failure of fruit and seed development. These blights are particularly destructive in temperate forests and agroforestry systems, where the timing of their infections coincides with the flowering period. While primarily affecting crops, Monilinia species, such as *Monilinia fructigena* and *Monilinia laxa*, have significant impacts on forest ecosystems as well, particularly when they infect wild or semi-wild fruiting trees. Blossom blight pathogens typically infect flowers during periods of high humidity and moderate temperatures, conditions that promote the germination of fungal spores. Infection begins when spores land on flower surfaces, where they germinate and penetrate tissues through natural openings or micro-wounds. Infected flowers turn brown or black, shrivel, and often fail to produce fruit. The disease may spread to adjacent flowers, fruits, or even twigs, forming cankers that further reduce reproductive output.

The ecological impacts of blossom blights are profound, particularly in forests where fruiting trees serve as a critical food source for wildlife. By reducing flower viability and subsequent fruiting, these pathogens disrupt the availability of resources for birds, mammals, and insects that depend on fruits and seeds. The long-term consequences of repeated infections include reduced regeneration potential and changes in forest composition, as species with higher susceptibility are gradually outcompeted. Managing blossom blights involves a combination of preventive and curative strategies. In nurseries and high-value forests, fungicides are commonly applied during flowering periods to protect flowers from infection. Improving air circulation through pruning or thinning can help reduce humidity levels, limiting the spread of the disease. In forest settings, biological control agents, such as antagonistic fungi or bacteria, are increasingly being explored to suppress Monilinia and related pathogens without disrupting ecosystem balance.

4.2 Seed-borne Pathogens

Seed-borne pathogens, such as those in the genus Fusarium, pose significant challenges in forestry nurseries and seed banks, where they affect seed viability, germination, and seedling health. These pathogens often infect seeds during maturation on the tree or through contamination during storage or sowing. Once infected, seeds serve as reservoirs for pathogens, introducing diseases into nurseries or forest regeneration sites.

Fusarium species, particularly *Fusarium oxysporum* and *Fusarium solani*, are among the most problematic seed-borne pathogens in forest ecosystems. These fungi infect seeds by penetrating the seed coat or colonizing the interior tissues. Infected seeds may fail to germinate or produce weak seedlings that are highly susceptible to secondary infections. Symptoms in germinating seeds or seedlings include damping-off, root rot, and discoloration of the stem and foliage. The impact of seed-borne pathogens extends beyond individual seed or seedling mortality. Infected seeds introduce pathogens into the soil, where they persist and infect subsequent plantings, creating a cycle of disease that is difficult to break. In natural forests, seed-borne infections reduce regeneration success, particularly for species with low seed production or high susceptibility. Managing seed-borne pathogens requires a combination of preventative measures and targeted treatments. Pre-sowing treatments, such as seed cleaning, fungicide application, or hot water treatments, are effective in reducing pathogen loads. In nurseries, maintaining strict sanitation protocols, including the use of pathogen-free soil and equipment, helps minimize the risk of infection. Biological control agents, such as beneficial fungi or bacteria, are increasingly being used to suppress Fusarium infections in seeds and seedlings. Additionally, molecular diagnostic tools enable the early detection of infected seeds, allowing for the removal of contaminated lots before sowing.

4.3 Rust Diseases: Cronartium Species and Cone Infections

Rust diseases, caused by fungi in the order *Pucciniales,* are among the most visually striking and ecologically significant diseases affecting cones, seeds, and other reproductive structures of forest trees. Cronartium species, in particular, are responsible for several important rust diseases, including white pine blister rust (*Cronartium ribicola*) and fusiform rust (*Cronartium quercuum*). These diseases not only reduce the reproductive success of affected trees but also impact forest structure and productivity by causing damage to stems and branches.

Rust pathogens have complex life cycles, often requiring alternate hosts to complete their development. For example, *Cronartium ribicola*, which infects white pines, alternates between pines and Ribes species (currants and gooseberries). The disease begins when fungal spores infect needles and spread to stems, forming cankers that disrupt vascular tissues. In cones, infections result in deformed or aborted seeds, reducing the reproductive potential of the tree. The visible symptoms of rust infections include bright orange pustules or spore masses on affected tissues, which facilitate the spread of the disease through wind dispersal.

The ecological impact of rust diseases is multifaceted. Infected trees often suffer from reduced growth and vigor, making them more susceptible to other stressors such as drought or insect infestations. The loss of viable seeds affects natural regeneration processes, particularly in monoculture plantations or forests with low species diversity. Furthermore, the dependence of rust fungi on alternate hosts creates intricate ecological interactions, influencing the dynamics of plant communities at both local and landscape scales. Managing rust diseases is challenging due to the complexity of their life cycles and the difficulty of controlling infections in natural forests. Quarantine measures, such as restricting the movement of infected plant material, are critical for preventing the introduction of rust pathogens to new areas. In plantations, resistant tree varieties offer a long-term solution, although breeding programs must account for the potential evolution of pathogen virulence. Removing alternate hosts, such as Ribes species in the case of white pine blister rust, has been used as a control measure, although it is often controversial due to its ecological consequences. Fungicide applications are generally impractical for large-scale forests but may be used in nurseries or high-value conservation areas to protect vulnerable trees.

5. Vascular Diseases

Vascular diseases are among the most devastating threats to forest health, targeting the vascular system of trees and disrupting the transport of water, nutrients, and photosynthates. These diseases are caused primarily by fungal pathogens and occasionally by bacterial species, which invade the xylem or phloem tissues, leading to systemic dysfunction, wilting, and often mortality. Vascular diseases are particularly destructive because their effects extend beyond individual trees, influencing forest structure, composition, and ecosystem functions such as carbon sequestration and water regulation.

5.1 Dutch Elm Disease

Dutch elm disease (DED), caused by the fungal pathogens *Ophiostoma ulmi* and *Ophiostoma novo-ulmi,* is one of the most catastrophic vascular wilt diseases in forest history. First identified in the Netherlands in the early 20th century, the disease spread rapidly across Europe and North America, decimating populations of elm trees (Ulmus spp.). The introduction of the more aggressive Ophiostoma novo-ulmi in the mid-20th century further accelerated the decline, with some regions experiencing near-total losses of mature elms.

Dutch elm disease is primarily spread by bark beetles, such as the European elm bark beetle (*Scolytus multistriatus*) and the native elm bark beetle (*Hylurgopinus rufipes).* These beetles carry fungal spores as they burrow into tree bark, introducing the pathogen into the vascular system. Once inside, the fungus colonizes the xylem, producing toxins and occluding the vessels with tyloses and fungal mycelium. This blockage disrupts water transport, leading to wilting, yellowing, and browning of leaves, typically starting at the crown and progressing downward. Infected trees often die within one to two years, though some may succumb more rapidly under favorable conditions for the pathogen. The ecological impact of Dutch elm disease has been profound. Elms, which once dominated urban landscapes and riparian forests, provided critical habitat for wildlife, shade for urban areas, and aesthetic and cultural value. Their loss has significantly altered forest composition and urban greenery, with many regions transitioning to less susceptible tree species such as maples and oaks. The economic impact has also been severe, with municipalities spending millions on tree removal, replacement, and disease control.

Managing Dutch elm disease involves a combination of prevention, early detection, and intervention. Sanitation measures, such as removing and destroying infected trees, are essential for reducing beetle populations and pathogen spread. Systemic fungicides, such as propiconazole, can protect high-value trees but require repeated applications, making them impractical for large-scale use. Breeding programs have focused on developing resistant elm cultivars, such as Ulmus americana hybrids, offering hope for long-term restoration. Additionally, integrated pest management strategies targeting bark beetles, including pheromone traps and biological control agents, have been employed to reduce vector activity.

5.2 Oak Wilt

Oak wilt, caused by the fungal pathogen *Bretziella fagacearum*, is a lethal vascular disease affecting oak trees (*Quercus spp.)* in temperate forests, particularly in North America. The disease has had a devastating impact on oak populations, with red oaks (*Quercus rubra* group) being particularly susceptible. First identified in the early 20th century, oak wilt continues to spread, threatening both natural forests and urban landscapes.

The pathogen infects trees through root grafts or wounds created by insects, storms, or human activity. Once inside, *Bretziella fagacearum* colonizes the xylem, producing fungal mycelium and spores that block water-conducting vessels. This disruption of water transport causes leaves to wilt, discolor, and fall prematurely. In red oaks, symptoms progress rapidly, with trees often dying within weeks to months of infection. White oaks (Quercus alba group) are more resistant and typically exhibit slower disease progression, with localized infections that can persist for years without causing mortality. Oak wilt spreads through two primary mechanisms: root grafts between adjacent trees and spore dispersal by sap-feeding beetles, such as nitidulids, which are attracted to fungal spore mats on infected trees. Root graft transmission is the dominant mode of spread in densely planted stands, while beetle-mediated spread facilitates long-distance transmission.

The ecological impact of oak wilt is substantial, as oaks are keystone species in many temperate forests. Their loss reduces habitat availability for wildlife, including species that depend on acorns as a food source. The decline of oaks also alters forest composition, often favoring less desirable species such as invasive shrubs or less ecologically valuable trees. Economically, oak wilt affects the timber industry and increases management costs for forest owners and municipalities. Managing oak wilt requires proactive measures to prevent infection and limit spread. Severing root grafts using trenching or vibratory ploughs is an effective method for isolating infected trees. Infected trees should be removed and destroyed promptly to eliminate sources of fungal spores. Systemic fungicides, such as propiconazole, can protect high-value trees when applied preventatively. Public education campaigns emphasizing proper tree care, such as avoiding pruning during peak beetle activity periods, are essential for reducing human-mediated spread.

5.3 Verticillium Wilt

Verticillium wilt, caused by soilborne fungi in the genus Verticillium (primarily Verticillium dahliae and Verticillium albo-atrum), is a significant vascular disease affecting a wide range of deciduous tree species, including maples, ashes, and elms. Unlike Dutch elm disease and oak wilt, which rely on insect vectors or root grafts, Verticillium wilt is transmitted through the soil, making its management particularly challenging.

The pathogen infects trees by colonizing their root systems, entering through wounds or natural openings. Once inside, Verticillium fungi move upward through the xylem, producing toxins and occluding vessels with fungal mycelium and tyloses. This disruption of water transport results in wilting, chlorosis, and necrosis of leaves, often affecting only one side or branch of the tree initially. Over time, infections can spread, causing dieback, stunted growth, and, in severe cases, tree mortality. One of the distinguishing features of Verticillium wilt is the presence of dark streaks or discoloration in the xylem, which can be observed in cross-sections of infected branches. These streaks serve as a diagnostic feature for identifying the disease. However, symptoms can be intermittent, with trees recovering temporarily before relapsing during periods of stress.

The ecological impact of Verticillium wilt is more localized than that of Dutch elm disease or oak wilt, but it can still significantly affect forest dynamics by reducing the health and vigor of infected trees. In urban and ornamental settings, Verticillium wilt poses a major threat to landscape trees, leading to economic losses and aesthetic damage. Managing Verticillium wilt is complicated by the pathogen's persistence in soil, where it can survive as microsclerotia for years, even in the absence of a host. Preventive measures include avoiding planting susceptible species in areas with a history of the disease and improving soil health to enhance tree resilience. Biological control agents, such as mycorrhizal fungi, have shown promise in suppressing Verticillium populations and promoting tree health. Infected trees should be pruned to remove affected branches, although this is often only a temporary solution. Developing resistant cultivars and diversifying tree species in forests and urban landscapes are critical for reducing the long-term impact of Verticillium wilt.

6. Diseases Caused by Invasive Pathogens

Invasive pathogens represent one of the most significant threats to forest ecosystems worldwide, causing devastating losses to biodiversity, altering forest composition, and disrupting ecosystem functions. Unlike endemic pathogens, which co-evolve with their hosts in a state of relative balance, invasive pathogens are introduced to new regions where their hosts often lack natural resistance. This leads to widespread mortality, forest degradation, and cascading ecological impacts. Diseases caused by invasive pathogens are further exacerbated by globalization, climate change, and human activities, which facilitate the movement of pathogens across natural barriers.

6.1 Sudden Oak Death: *Phytophthora ramorum* and Forest Health

Sudden oak death, caused by the oomycete Phytophthora ramorum, is a devastating disease that has led to the widespread decline of oaks (*Quercus spp.*) and tanoaks (*Notholithocarpus densiflorus*) in North America and Europe. First identified in the mid-1990s in California, this invasive pathogen has since spread to several countries, primarily through infected nursery stock and contaminated soil. *Phytophthora ramorum* also infects a wide range of other plant species, including ornamental shrubs like rhododendrons and camellias, which act as reservoirs for the pathogen. The disease begins when *P. ramorum* spores, dispersed by rain splash or wind-driven water, infect leaves, stems, or bark. In susceptible trees like tanoaks, the pathogen colonizes the cambium and phloem, forming cankers that girdle the tree and disrupt vascular function. Symptoms of sudden oak death include dark, bleeding lesions on the bark, wilting leaves, dieback, and eventual mortality. On foliar hosts, the disease manifests as leaf spots and shoot blight, which facilitate the production and spread of spores.

The ecological impacts of sudden oak death are profound. Oaks and tanoaks are keystone species in many ecosystems, providing habitat, food, and structural diversity. Their loss reduces biodiversity and alters forest dynamics, often leading to the proliferation of less ecologically valuable species. Additionally, dead and dying trees increase fuel loads, raising the risk of severe wildfires. Managing sudden oak death is challenging due to the pathogen's broad host range and persistence in the environment. Quarantine measures, such as restricting the movement of infected plant material, are critical for preventing further spread. Fungicides like phosphonates can protect high-value trees, but their effectiveness is limited to early-stage infections. In natural forests, sanitation practices, including the removal and destruction of infected trees, help reduce pathogen loads. Long-term management strategies focus on breeding resistant tree species and developing integrated pest and disease monitoring systems.

6.2 Ash Dieback: *Hymenoscyphus fraxineus* and Its European Spread

Ash dieback, caused by the fungal pathogen *Hymenoscyphus fraxineus*, is one of the most destructive invasive diseases to affect European forests in recent decades. Native to East Asia, where it exists in balance with its host species, *H. fraxineus* was introduced to Europe in the early 1990s, likely through infected nursery stock. Since then, the disease has spread rapidly, causing widespread mortality in European ash (*Fraxinus excelsior*) and other *Fraxinus* species.

The disease begins with airborne ascospores released from fungal fruiting bodies on infected leaf litter. These spores infect leaves and spread to the shoots, bark, and branches, leading to symptoms such as wilting, blackened leaf stalks, and necrotic lesions. In advanced stages, the pathogen invades the vascular tissues, disrupting water transport and causing dieback of branches and stems. Trees weakened by ash dieback are also more susceptible to secondary pathogens and pests, such as Armillaria root rot and the emerald ash borer (*Agrilus planipennis*). The ecological impact of ash dieback is immense, as ash trees are foundational species in many temperate forests, riparian zones, and hedgerows. Their loss affects numerous dependent species, including lichens, fungi, and invertebrates, leading to declines in biodiversity. The economic impact is also significant, as ash is a valuable timber species used for furniture, flooring, and tools.

Managing ash dieback is complex, as no effective cure exists. Efforts focus on slowing the spread of the disease and mitigating its impacts. Quarantine measures and restrictions on the movement of ash material are essential for preventing further introductions. Infected trees are often removed to reduce spore loads, although this is not always feasible in natural forests. Breeding programs aimed at identifying and propagating resistant ash genotypes offer hope for long-term recovery. Additionally, research into biological control agents, such as endophytic fungi that inhibit *H. fraxineus*, is ongoing.

6.3 Pine Wilt Disease

Pine wilt disease, caused by the pinewood nematode (*Bursaphelenchus xylophilus*), is a severe invasive disease affecting pine forests worldwide. Native to North America, where it primarily affects stressed or weakened trees, the nematode was introduced to Asia and Europe, where it has caused significant damage to native pine species. The disease is particularly devastating in Japan, China, Korea, and Portugal, where susceptible pine species have experienced widespread mortality. The life cycle of the pinewood nematode involves transmission by longhorn beetles of the genus Monochamus. These beetles act as vectors, carrying the nematodes to healthy trees during feeding or egg-laying. Once inside the tree, the nematodes migrate to the vascular tissues, where they feed on plant cells and reproduce. Their activity disrupts water transport, leading to rapid wilting, browning of needles, and tree death, often within a few months of infection.

The ecological impact of pine wilt disease is profound, particularly in regions where pines are the dominant species. The loss of pines affects soil stabilization, watershed protection, and biodiversity, as many species depend on pine forests for habitat. In commercial plantations, the disease reduces timber production and increases management costs, posing significant economic challenges. Managing pine wilt disease requires an integrated approach targeting both the nematode and its vector. Preventative measures include monitoring and controlling Monochamus beetle populations through pheromone traps, insecticides, and biological control agents. Quarantine measures, such as restricting the movement of infected wood, are essential for preventing the spread of the disease to new areas. Infected trees should be removed and destroyed to eliminate nematode reservoirs. Research into resistant pine species and nematicidal treatments is ongoing, offering potential solutions for mitigating the long-term impacts of pine wilt disease.

7. Diseases Related to Abiotic Stressors

Abiotic stressors are non-living environmental factors that adversely affect tree health, rendering forests more susceptible to diseases and other ecological challenges. Diseases associated with abiotic stressors often arise as indirect consequences of prolonged or extreme stress conditions such as frost, drought, soil compaction, or pollution. These factors weaken tree defenses, disrupt physiological processes, and create opportunities for secondary infections by opportunistic pathogens. Abiotic stress-related diseases are particularly concerning in the context of climate change, which amplifies the frequency and severity of stress events.

7.1 Frost and Drought-Associated Cankers

Frost and drought are two of the most significant abiotic stressors affecting forest health, often resulting in the formation of cankers on tree stems and branches. These stressors weaken trees by disrupting normal growth processes, creating entry points for opportunistic pathogens, and exacerbating existing infections. Frost-associated cankers typically occur when rapid temperature fluctuations cause freezing and thawing cycles, damaging cambial tissues. This damage creates cracks or lesions in the bark, which serve as entry points for fungal pathogens such as *Nectria* or *Cytospora* species. Symptoms of frost-associated cankers include sunken, discolored lesions on the bark, exudation of sap, and dieback of branches. Trees in high-altitude or temperate regions are particularly vulnerable, especially during late frosts that coincide with active growth phases.

Drought-associated cankers, on the other hand, arise when prolonged water stress reduces the vigor of trees, making them susceptible to infections by canker-causing pathogens. During drought conditions, reduced water availability impairs the tree's ability to transport nutrients and maintain its defenses. Pathogens such as *Botryosphaeria* and *Diplodia* species exploit this vulnerability, infecting weakened stems and branches. Symptoms include cracking bark, resin flow, and branch dieback, which can progress to tree mortality under severe conditions.

The ecological and economic impacts of frost and drought-associated cankers are significant. Repeated infections can lead to the decline of forest stands, altering species composition and reducing timber yields. In urban and peri-urban forests, affected trees pose safety risks due to branch breakage and structural instability. Managing frost and drought-associated cankers involves a combination of prevention and mitigation strategies. For frost-prone areas, selecting frost-tolerant species and timing pruning to avoid vulnerable periods can help reduce damage. In drought-affected regions, improving soil moisture retention through mulching, irrigation, and maintaining a diverse understory can enhance tree resilience. Silvicultural practices that promote tree vigor, such as thinning to reduce competition, also play a crucial role in minimizing the impacts of these stressors.

7.2 Soil Compaction and Root Dysfunction

Soil compaction, caused by heavy machinery, foot traffic, or livestock grazing, is a common abiotic stressor in forests and urban landscapes. Compacted soils have reduced pore space, limiting the availability of oxygen and water to tree roots. Over time, this leads to root dysfunction, impaired nutrient uptake, and a decline in overall tree health. Soil compaction is particularly problematic in reforestation sites, managed forests, and recreational areas, where frequent human activity exacerbates its effects. The symptoms of soil compaction-related root dysfunction include stunted growth, chlorosis (yellowing of leaves), wilting, and dieback of branches. Severely compacted soils can also predispose trees to root rot diseases caused by pathogens like Phytophthora and Armillaria, as anaerobic conditions favor the proliferation of these organisms. Trees in compacted soils are also more susceptible to drought stress, as their compromised root systems are unable to access deep soil moisture.

Ecologically, soil compaction reduces forest productivity and alters species composition, favoring shallow-rooted or stress-tolerant species over deep-rooted, slow-growing trees. Compacted soils also hinder the establishment of seedlings, reducing natural regeneration rates and altering successional dynamics. Managing soil compaction and root dysfunction requires a combination of preventive and remedial measures. Preventive strategies include minimizing the use of heavy machinery, establishing designated paths to reduce foot traffic, and using protective mats or barriers in construction zones. For compacted soils, mechanical aeration (e.g., subsoiling) and the incorporation of organic amendments can improve soil structure and increase aeration and water infiltration. Mulching and planting ground cover species can further enhance soil health, promoting a more favorable environment for root growth.

7.3 Pollution-Induced Foliar Disorders

Airborne pollutants, such as ozone, sulfur dioxide, nitrogen oxides, and particulate matter, have detrimental effects on forest health, leading to pollution-induced foliar disorders. These disorders manifest as visible damage to leaves and needles, reduced photosynthetic efficiency, and long-term declines in growth and productivity. Urban and peri-urban forests are particularly vulnerable, as they are often located near industrial areas or major transportation corridors with high levels of air pollution. Ozone-induced foliar damage is one of the most common pollution-related disorders in forests. Ozone enters leaves through stomata, where it reacts with cellular components to produce reactive oxygen species (ROS). This oxidative stress damages cell membranes, chloroplasts, and proteins, leading to visible symptoms such as stippling, chlorosis, necrosis, and premature leaf drop. Sensitive species, such as pines and poplars, are particularly affected, with severe cases resulting in reduced growth and increased susceptibility to secondary pathogens. Sulfur dioxide and nitrogen oxides contribute to acid deposition, which damages foliage and alters soil chemistry. Acid rain leaches essential nutrients from the soil and releases toxic metals such as aluminum, impairing root function and reducing tree vigor. Foliar symptoms include yellowing, browning, and curling of leaves, often accompanied by dieback of shoots and branches. In conifers, needle browning and tip dieback are common indicators of pollution stress.

Pollution-induced foliar disorders have far-reaching ecological consequences. Reduced forest productivity and altered species composition affect carbon sequestration, hydrological cycles, and habitat availability. In addition, stressed trees are more prone to insect infestations and diseases, further compounding the impacts of pollution. Managing pollution-induced foliar disorders requires addressing both the source of pollution and the affected trees. Policy interventions, such as stricter emission standards and the establishment of buffer zones around industrial areas, are critical for reducing pollutant levels. In forests, maintaining tree vigor through proper nutrition and irrigation can mitigate the effects of pollution. Planting pollution-tolerant species, such as birch and spruce, in high-risk areas can also help reduce the vulnerability of forests to air pollution. Research into the physiological and molecular mechanisms of pollution tolerance continues to inform breeding programs aimed at developing more resilient tree species.

8. Emerging and Cryptic Diseases

Emerging and cryptic diseases are a growing concern in forest ecosystems, driven by complex interactions between environmental changes, human activities, and pathogen evolution. Emerging diseases are often associated with the introduction of novel pathogens or the sudden rise of previously dormant or low-impact pathogens, frequently triggered by climate change, globalization, and shifts in forest management practices. Cryptic diseases, on the other hand, are caused by pathogens that remain undetected or misdiagnosed due to their subtle symptoms, complex life cycles, or co-occurrence with other stressors. These diseases present significant challenges for forest health monitoring, management, and research.

8.1 Climate Change and Emerging Forest Pathogens

Climate change is one of the most significant drivers of emerging forest diseases, altering the distribution, virulence, and host range of pathogens. Rising temperatures, shifting precipitation patterns, and extreme weather events create favorable conditions for the proliferation and spread of many forest pathogens. These changes also stress trees, reducing their natural defenses and making them more susceptible to infection. One of the most notable impacts of climate change is the poleward and upward expansion of pathogens into regions where they were previously absent. For example, *Dothistroma septosporum*, the causative agent of red band needle blight, has expanded its range into cooler, higher-altitude regions due to warming temperatures. Similarly, the range of Phytophthora cinnamomi, a soilborne pathogen responsible for root rot, has increased in temperate forests, particularly in areas with wetter winters and drier summers. Climate change also promotes the emergence of pathogens previously suppressed by environmental conditions. Warmer winters have reduced the mortality of overwintering fungal spores and insect vectors, increasing the risk of outbreaks. For example, milder winters have been linked to the proliferation of bark beetles, which act as vectors for fungal pathogens like *Ophiostoma* species, the agents of Dutch elm disease.

The ecological impacts of emerging pathogens under climate change are profound. Increased pathogen activity leads to higher mortality rates, reduced forest productivity, and altered species composition. Keystone and foundation species, such as ash (Fraxinus spp.) and beech (Fagus spp.), are particularly vulnerable, with their decline affecting entire ecosystems. Additionally, the introduction of invasive pathogens into new regions through global trade and travel is exacerbated by climate change, creating opportunities for novel host-pathogen interactions. Managing the impacts of climate-driven emerging pathogens requires a proactive and integrated approach. Monitoring systems that combine field surveys with remote sensing and molecular diagnostics are critical for early detection. Silvicultural practices that promote genetic diversity and resilience, such as mixed-species planting, can help mitigate the impacts of emerging diseases. Additionally, international cooperation and stricter biosecurity measures are essential to prevent the introduction and spread of invasive pathogens.

8.2 Detection Challenges in Cryptic Fungal Pathogens

Cryptic fungal pathogens present unique challenges to forest health management due to their subtle symptoms, complex life cycles, and interactions with other biotic and abiotic factors. These pathogens often remain undetected until they cause significant damage, and even then, their role may be obscured by co-occurring stressors such as drought, pollution, or insect infestations. Advances in molecular and genomic technologies have begun to shed light on these elusive pathogens, but many remain poorly understood.

One of the primary challenges in detecting cryptic fungal pathogens is their ability to exist in asymptomatic or latent states within host tissues. For example, endophytic fungi, which colonize healthy plant tissues without causing visible symptoms, can transition to pathogenic behavior under stress conditions. Botryosphaeria species, for instance, are common endophytes in many tree species but can cause cankers and dieback during drought or other stress events. Another challenge is the identification of cryptic species within fungal complexes. Morphologically similar pathogens often differ significantly in their virulence, host range, and environmental requirements, complicating disease diagnosis and management. For example, the *Colletotrichum* genus includes numerous cryptic species that cause anthracnose diseases, but precise identification requires molecular tools such as DNA barcoding or whole-genome sequencing.

The role of cryptic pathogens in disease complexes further complicates detection and diagnosis. Many forest diseases result from synergistic interactions between pathogens, pests, and environmental stressors, making it difficult to attribute symptoms to a single cause. For instance, Armillaria root rot often occurs in conjunction with drought stress or secondary infections, obscuring the primary pathogen's role. Addressing these detection challenges requires a combination of advanced technologies and ecological insights. High-throughput sequencing and metagenomic approaches allow researchers to identify cryptic pathogens and their interactions within host microbiomes. Remote sensing technologies, such as hyperspectral imaging, can detect subtle changes in canopy health associated with latent infections. Additionally, long-term monitoring and field experiments are essential for understanding the ecological dynamics of cryptic fungal pathogens.

8.3 Novel Diseases in Urban and Plantation Forests

Urban and plantation forests are particularly vulnerable to the emergence of novel diseases due to their unique environmental conditions and management practices. In urban forests, high levels of stress from pollution, soil compaction, and human activity weaken tree defenses, creating opportunities for pathogens to establish and spread. Plantation forests, with their high-density, monoculture planting, provide ideal conditions for the rapid proliferation of diseases.

One of the key drivers of novel diseases in these settings is the introduction of non-native pathogens through global trade. For example, the invasive fungus Fusarium circinatum, which causes pitch canker in pines, has been introduced to several plantation regions worldwide, resulting in significant losses in timber production. Similarly, the introduction of Phytophthora ramorum, the agent of sudden oak death, into ornamental nurseries has facilitated its spread into natural forests. In urban forests, novel diseases often arise from interactions between introduced pathogens and stressed hosts. Xylella fastidiosa, a bacterial pathogen causing leaf scorch in several tree species, has become a major concern in Mediterranean urban landscapes, where drought and heat exacerbate its impacts. In plantations, pathogens such as Raffaelea species, associated with ambrosia beetles, have caused emerging diseases by exploiting the uniformity and density of host trees.

The impacts of novel diseases in urban and plantation forests extend beyond individual tree mortality. In urban areas, tree losses reduce ecosystem services such as air purification, temperature regulation, and stormwater management, while also diminishing aesthetic and recreational value. In plantations, outbreaks of novel diseases reduce timber yields, increase management costs, and disrupt global supply chains. Managing novel diseases in these contexts requires tailored strategies that address the specific challenges of urban and plantation environments. In urban forests, integrated pest and disease management (IPDM) practices that combine monitoring, sanitation, and targeted treatments can reduce pathogen spread. In plantations, diversifying tree species, implementing rotational planting, and breeding resistant varieties are critical for reducing vulnerability to novel pathogens. Public awareness campaigns and stricter biosecurity measures are also essential for preventing the introduction and establishment of invasive pathogens.

9. Integrated Disease Management in Forests

Integrated Disease Management (IDM) in forests represents a holistic approach to addressing the challenges posed by forest pathogens. It combines monitoring, early diagnosis, silvicultural practices, biological controls, and other management strategies to create a sustainable framework for forest health. Unlike traditional methods that often focus on controlling specific pathogens in isolation, IDM emphasizes the interplay between environmental conditions, host trees, and pathogen dynamics. It aims to mitigate disease impacts while maintaining ecosystem balance and productivity.

9.1 Monitoring Systems for Common Forest Diseases

Effective monitoring systems are the backbone of integrated disease management in forests. Timely and accurate detection of disease outbreaks is critical for preventing widespread damage and implementing targeted interventions. Monitoring systems involve a combination of field surveys, remote sensing, and data analysis to track the occurrence, severity, and spread of forest diseases over time. Field surveys remain a foundational component of disease monitoring. Trained personnel systematically assess tree health by observing symptoms such as leaf discoloration, dieback, cankers, or root decay. Ground-based sampling allows for the collection of infected tissues, which can then be analyzed in laboratories to confirm the presence of pathogens. However, field surveys are labor-intensive and often limited by accessibility, especially in large or remote forested areas.

Remote sensing technologies have revolutionized disease monitoring by providing large-scale, real-time data on forest health. Satellite imagery, drones, and hyperspectral imaging are increasingly used to detect subtle changes in canopy color, density, or reflectance that indicate the onset of disease. For example, spectral data can identify chlorophyll degradation or water stress in diseased trees before symptoms become visible to the naked eye. Thermal imaging is another powerful tool, capable of detecting heat patterns associated with pathogen activity or stressed trees. Integrated data platforms and geographic information systems (GIS) enhance monitoring efforts by compiling spatial and temporal data on disease outbreaks. These systems enable researchers and forest managers to visualize disease hotspots, predict future spread using climate and ecological models, and prioritize areas for intervention. Publicly accessible databases, such as pest and pathogen distribution maps, facilitate collaboration between countries and regions, improving global disease preparedness. To optimize the effectiveness of monitoring systems, it is essential to train personnel, invest in cutting-edge technologies, and promote collaborations between forestry agencies, researchers, and local communities. Continuous advancements in remote sensing and artificial intelligence are expected to further improve disease monitoring capabilities in the coming years.

9.2 Molecular Tools for Early Disease Diagnosis

Early and accurate diagnosis is crucial for mitigating the impacts of forest diseases. Molecular tools have emerged as indispensable components of integrated disease management, enabling the rapid and precise identification of pathogens at early stages of infection or even before symptoms appear. These tools are particularly valuable for detecting cryptic or emerging pathogens that are difficult to diagnose using traditional methods.

Polymerase chain reaction (PCR) is one of the most widely used molecular techniques for pathogen detection. PCR amplifies specific DNA or RNA sequences, allowing for the identification of pathogens with high sensitivity and specificity. Variants such as quantitative PCR (qPCR) and reverse transcription PCR (RT-PCR) enable real-time monitoring of pathogen load, providing insights into disease progression. For example, PCR-based assays have been successfully employed to detect Phytophthora ramorum in nursery stock and natural forests, helping to curb the spread of sudden oak death. Next-generation sequencing (NGS) has revolutionized forest pathology by providing a comprehensive view of the microbial communities associated with diseased trees. Metagenomic approaches allow researchers to identify multiple pathogens, including previously unknown species, from a single sample. This technology has been instrumental in uncovering the complex interactions between pathogens, hosts, and environmental factors, facilitating the development of targeted management strategies.

Other molecular tools, such as loop-mediated isothermal amplification (LAMP) and CRISPR-based diagnostics, are gaining popularity due to their simplicity, speed, and cost-effectiveness. LAMP, for instance, has been used to detect soilborne pathogens like Verticillium dahliae directly in field conditions without the need for extensive laboratory equipment. CRISPR-based diagnostics offer a promising future for portable, field-deployable systems that can rapidly detect forest pathogens with minimal technical expertise. Despite their advantages, the widespread adoption of molecular tools faces challenges, including high costs, the need for specialized training, and the lack of infrastructure in some regions. Addressing these barriers will require investments in capacity building, the development of user-friendly diagnostic kits, and collaboration between research institutions and forest management agencies.

9.3 Combining Silvicultural Practices with Biocontrol Measures

Silvicultural practices and biological control measures are key components of integrated disease management, providing sustainable alternatives to chemical interventions. Silvicultural practices focus on enhancing tree health and forest resilience through site preparation, species selection, thinning, pruning, and other management techniques. Biological control measures involve the use of beneficial organisms, such as fungi, bacteria, or insects, to suppress pathogens and promote forest health.

Silvicultural practices play a pivotal role in reducing disease susceptibility. Mixed-species planting, for example, increases biodiversity and reduces the likelihood of monoculture-related disease outbreaks. Tree species with varying resistance levels act as physical and biological barriers, slowing the spread of pathogens. Thinning overcrowded stands improves air circulation and light penetration, creating unfavorable conditions for pathogens like rust fungi or needle blight. Similarly, pruning infected branches can prevent the spread of pathogens within individual trees and across forest stands. Biological control measures complement silvicultural practices by directly targeting pathogens or their vectors. For instance, antagonistic fungi like Trichoderma species are used to suppress soilborne pathogens such as Phytophthora and Armillaria. These fungi colonize the root zone, outcompeting pathogenic fungi and enhancing tree root health. Similarly, entomopathogenic fungi like Beauveria bassiana are used to control insect vectors that spread diseases, such as bark beetles associated with Dutch elm disease. The integration of silvicultural and biocontrol approaches offers synergistic benefits. For example, maintaining healthy soils through organic amendments not only improves tree vigor but also supports populations of beneficial microbes that suppress pathogens. Similarly, mixed-species planting can create microhabitats for natural predators of insect vectors, enhancing biocontrol effectiveness. While these methods are environmentally friendly and sustainable, their implementation requires careful planning and monitoring. The effectiveness of biocontrol agents can be influenced by environmental conditions, pathogen populations, and forest management practices. Silvicultural interventions must also be tailored to specific forest types and disease pressures to maximize their benefits. Ongoing research into the interactions between silvicultural practices, biocontrol agents, and forest pathogens is critical for refining integrated management strategies.

Conclusion

The study of forest diseases, encompassing a wide range of pathogens and stressors affecting various parts of trees, highlights the complexity and interconnectedness of forest ecosystems. Root, stem, leaf, flower, and vascular diseases each present unique challenges and have distinct impacts on tree health, forest productivity, and ecosystem stability. Root diseases, such as Armillaria root rot and Phytophthora root rot, demonstrate the vulnerability of belowground systems that are crucial for water and nutrient uptake, while stem and branch diseases like chestnut blight and heart rot emphasize the structural and ecological consequences of these infections. Leaf and needle diseases, including red band needle blight and powdery mildew, disrupt photosynthesis and energy production, while diseases of flowers, fruits, and seeds, such as rust infections and seed-borne pathogens, threaten forest regeneration and genetic diversity.

Vascular diseases, such as Dutch elm disease and oak wilt, underscore the systemic and often catastrophic effects of pathogens that interfere with a tree's vascular system, leading to widespread mortality and ecological disruption. Invasive pathogens, including Phytophthora ramorum (Sudden Oak Death) and Hymenoscyphus fraxineus (Ash Dieback), highlight the global threat posed by non-native pathogens, which disrupt native ecosystems and spread rapidly due to global trade and climate change. Abiotic stressors, such as frost, drought, soil compaction, and pollution, act as both primary causes of tree decline and factors that exacerbate the effects of biotic pathogens. Emerging diseases linked to climate change and human activity further complicate the management of forest health, as novel pathogens and cryptic diseases challenge traditional monitoring and diagnostic methods.

The diversity of forest diseases, their causal agents, and their modes of impact underscore the need for a comprehensive and integrated approach to their study and management. Disease classification provides a critical framework for understanding the biology, ecology, and epidemiology of pathogens, enabling targeted interventions and proactive strategies. Advancements in molecular tools, remote sensing, and early detection systems have greatly enhanced our ability to identify pathogens, monitor their spread, and predict potential outbreaks. These innovations, combined with traditional silvicultural practices and biological control measures, form the foundation of integrated disease management systems aimed at reducing disease impacts while preserving forest ecosystems. The ecological and economic significance of forest diseases cannot be overstated. Beyond their immediate effects on tree health, these diseases alter forest composition, disrupt biodiversity, and compromise ecosystem services, such as carbon storage, water regulation, and habitat provision. Understanding the multifaceted impacts of forest diseases, from localized tree decline to global outbreaks driven by invasive pathogens, is essential for safeguarding forest health in the face of climate change, habitat loss, and increasing anthropogenic pressures.

Chapter - 5

Nursery Diseases and Parasitic Plant Diseases

Introduction

The study of nursery diseases and their management forms a foundational component of forest pathology, with direct implications for forest regeneration, biodiversity conservation, and reforestation programs worldwide. Nurseries play a crucial role in producing healthy seedlings, which serve as the foundation for forest establishment and restoration efforts. However, these environments are highly susceptible to a wide range of pathogens, owing to the dense planting of seedlings, high humidity, and often repetitive use of soil and irrigation systems. Diseases affecting nursery stock include damping-off, root rots, foliar infections, stem cankers, bacterial blights, and viral infections, each of which poses specific challenges to seedling growth and survival. Compounding these threats is the role of environmental stress, which creates conditions conducive to pathogen outbreaks, thereby amplifying the impacts of these diseases on nursery productivity and the success of reforestation efforts.

Among the most significant nursery diseases is damping-off, caused by soilborne pathogens such as *Pythium* and *Rhizoctonia*. This disease affects germinating seeds and young seedlings, resulting in poor establishment and high mortality rates. Root rot diseases, notably those caused by *Phytophthora* and *Fusarium*, further threaten seedling health by impairing water and nutrient uptake, causing wilting, chlorosis, and eventual death. Foliar diseases, including powdery mildew and leaf spots, affect photosynthesis and energy production, weakening seedlings and making them more susceptible to other stressors. Stem cankers and collar rot, often caused by fungal pathogens, girdle the base of seedlings, leading to structural instability and eventual collapse. Bacterial blights, caused by species such as *Pseudomonas* and *Xanthomonas*, are another significant challenge, characterized by water-soaked lesions, wilting, and necrosis. Viral diseases, though less common in nurseries, can severely impact plant vigor and quality, as viruses are often transmitted through infected seeds, soil, or vectors, making them difficult to control.

Environmental stressors, including temperature extremes, poor drainage, and high humidity, exacerbate nursery pathogen outbreaks. These stressors weaken seedling defenses, creating favorable conditions for pathogens to establish and proliferate. The interactions between abiotic factors and pathogens highlight the importance of maintaining optimal environmental conditions in nursery settings to minimize disease risks. Furthermore, the combination of diseases and stressors can lead

DOI: 10.1201/9781042011063-5

to compounded effects, resulting in significant economic losses and setbacks in reforestation and afforestation programs. Parasitic plants represent another layer of complexity in forest pathology, particularly their impact on young trees and nursery stock. Mistletoes, as hemiparasitic plants, extract water and nutrients from their hosts, reducing growth rates and sometimes leading to tree mortality. Dodder (*Cuscuta* species) is another parasitic plant that primarily affects young and nursery-grown trees by forming dense tangles that compete for resources and spread pathogens. Broomrapes (*Orobanche* species) and witchweeds (*Striga* species), both obligate root parasites, further highlight the challenges posed by parasitic plants in forestry. These parasites not only drain their hosts of essential resources but also alter forest composition and succession by selectively targeting specific species. The interactions between parasitic plants and fungal pathogens create synergistic effects that amplify damage, necessitating a comprehensive understanding of their combined impacts on forest ecosystems.

Parasitic plants also play complex roles in forest ecology, sometimes altering successional trajectories by influencing species competition and nutrient cycling. For example, heavy infestations of mistletoes or broomrapes can weaken dominant tree species, creating opportunities for understory growth and biodiversity changes. This dual role of parasitic plants, as both disruptors and ecological influencers, underscores the need to study their interactions with forest ecosystems in depth. Integrated management of nursery diseases and parasitic plant threats is essential for maintaining forest health and achieving successful regeneration efforts. Sanitation practices, such as the removal of infected plant material and disinfection of tools and containers, are fundamental preventive measures. Early detection techniques, including molecular diagnostics and remote sensing tools, enable timely intervention, reducing the spread and impact of pathogens and parasites. Biological and cultural control methods, such as the use of beneficial microbes, crop rotation, and optimal irrigation practices, offer sustainable alternatives to chemical treatments. The selection of resistant rootstocks and species further strengthens the resilience of nursery plants to both pathogens and parasitic plants. Integrated pest and disease management (IPDM) approaches, which combine these strategies, provide a holistic framework for mitigating disease risks while ensuring the sustainability of nursery operations.

The role of biotechnology in disease management is rapidly expanding, with advances in genomics, transcriptomics, and proteomics enabling the development of disease-resistant nursery varieties. Molecular tools, such as CRISPR-based gene editing and marker-assisted selection, offer unprecedented opportunities to enhance resistance traits in forest species. Additionally, policies and regulations governing the movement of plant materials and the use of biocontrol agents are critical for preventing the introduction and spread of pathogens and parasites. The implementation of strict biosecurity measures, coupled with public awareness campaigns, can further safeguard forest nurseries from emerging threats. Nursery diseases have far-reaching implications for forest regeneration and ecosystem sustainability. High seedling mortality rates caused by pathogens limit the success of reforestation efforts, delaying the recovery of degraded landscapes and reducing carbon sequestration potential. The loss of biodiversity in nurseries, driven by pathogen- and parasite-induced mortality, impacts the genetic diversity and ecological resilience of forests. Furthermore, the economic costs associated with nursery diseases, including lost productivity, increased management expenses, and reduced timber yields, underscore the importance of proactive disease management in supporting reforestation programs and sustainable forestry practices.

Climate change adds another layer of complexity to nursery pathogen dynamics, as rising temperatures, shifting precipitation patterns, and extreme weather events create conditions favorable to pathogen proliferation and spread. Warming temperatures accelerate the life cycles of many pathogens, while changing rainfall patterns influence the distribution and severity of waterborne and soilborne diseases. Adaptation strategies, such as the development of climate-resilient nursery practices and the use of drought-tolerant species, are essential for mitigating the impacts of climate change on nursery diseases. Advances in nursery disease research and technological innovations are paving the way for more effective management strategies. Genomics and molecular tools have revolutionized the identification and study of pathogens, enabling precise diagnostics. Artificial intelligence and machine learning applications are enhancing disease monitoring and forecasting, allowing for early detection and targeted interventions. The development of disease-resistant nursery varieties through traditional breeding and biotechnological approaches offers long-term solutions to recurring pathogen challenges.

1. Nursery Diseases

Nursery diseases are a major challenge to forestry operations, as they target young, vulnerable seedlings and saplings, often compromising their growth, establishment, and survival. These diseases are caused by a variety of pathogens, including fungi, oomycetes, bacteria, and viruses, which exploit the ideal conditions provided by nursery environments, such as moisture, close plant spacing, and high nutrient availability. Such diseases can reduce the quality and quantity of nursery stock, delay reforestation projects, and increase the cost of seedling production. Additionally, infected nursery plants can serve as vectors, introducing pathogens into previously uninfected forests.

1.1 Damping-Off

Damping-off is one of the most common and devastating nursery diseases, particularly affecting germinating seeds and young seedlings. Caused by a complex of pathogens, including species of *Pythium*, *Phytophthora*, *Fusarium*, and *Rhizoctonia*, damping-off is favored by the warm, moist conditions prevalent in nurseries. This disease manifests in two distinct phases: pre-emergence damping-off, where seeds rot before germination, and post-emergence damping-off, where seedlings collapse shortly after sprouting. Both forms severely impact seedling establishment and the overall success of nursery operations. Pre-emergence damping-off occurs when fungal pathogens invade the seed coat or germinating embryo, causing soft rot and preventing seedling emergence. In contrast, post-emergence damping-off is characterized by the infection of the seedling stem at or just below the soil surface. This results in a characteristic water-soaked lesion, which quickly girdles the stem, leading to seedling collapse. Affected seedlings often appear wilted, with yellowing cotyledons or leaves, and may display stunted growth before ultimately dying. The rapid progression of damping-off makes it particularly destructive, as entire seedbeds can be wiped out within a matter of days.

The conditions in nurseries make them particularly susceptible to damping-off. High humidity, poor soil drainage, overcrowding, and overwatering create an environment that encourages pathogen growth and spore production. Infected seeds or soil can serve as reservoirs for these pathogens, perpetuating the cycle of infection. The dense spacing of seedlings allows the disease to spread rapidly from plant to plant, compounding its effects. Managing damping-off requires an integrated approach that combines cultural, chemical, and biological methods. Proper nursery hygiene is essential, including the use of sterilized soil, clean tools, and pathogen-free seeds. Improving drainage and aeration in seedbeds can help reduce soil moisture levels, making conditions less favorable for pathogens. Fungicidal seed treatments and soil drenches, such as those containing captan or metalaxyl, provide effective chemical control, particularly in high-risk areas. Biological control agents, such as *Trichoderma* and *Gliocladium* species, have shown promise in suppressing damping-off pathogens through competition and antagonism. Additionally, maintaining optimal growing conditions, such as moderate soil moisture and appropriate spacing can reduce the stress on seedlings, enhancing their resistance to infection.

1.2 Root Rot Diseases in Nurseries

Root rot diseases caused by *Phytophthora* and *Fusarium* species are among the most significant threats to nursery plants, affecting the root systems of seedlings and saplings. These diseases lead to poor root development, reduced water and nutrient uptake, and increased susceptibility to environmental stressors and secondary pathogens. Root rot diseases can cause widespread losses in nursery operations, not only reducing the quantity of viable stock but also compromising the quality and vigor of seedlings destined for planting. *Phytophthora* root rot is caused by several species of the oomycete genus *Phytophthora*, with *P. cinnamomi* being particularly notorious. This pathogen thrives in waterlogged soils, producing motile zoospores that swim through soil water to infect root tips. Symptoms of *Phytophthora* root rot include root discoloration, reduced root mass, and the development of necrotic lesions. Above-ground symptoms may include stunted growth, chlorosis, wilting, and dieback, particularly during periods of water stress. In severe cases, entire seedlings may collapse and die.

Fusarium root rot, caused by fungi in the genus *Fusarium*, often affects seedlings under stress, such as those experiencing drought, nutrient deficiencies, or poor soil conditions. This disease is characterized by the invasion of roots through wounds or natural openings, leading to decay and the production of brown to black lesions. *Fusarium* also produces toxins that impair root function, further exacerbating the damage. Affected seedlings may exhibit similar above-ground symptoms to those caused by *Phytophthora*, including wilting, yellowing, and reduced vigor. The epidemiology of root rot diseases is strongly influenced by nursery conditions. Overwatering, poor drainage, and soil compaction create the anaerobic conditions favored by *Phytophthora*, while excessive fertilization and high temperatures can promote *Fusarium* growth. Contaminated soil, tools, and irrigation water are common sources of these pathogens, allowing them to persist and spread within nursery environments. Managing root rot diseases requires a multifaceted approach. Improving drainage and avoiding overwatering is critical for reducing the risk of *Phytophthora* infections, while minimizing stress through proper nutrition and irrigation can help seedlings resist *Fusarium*. Fungicides, such as mefenoxam and fosetyl-Al, are effective against *Phytophthora*, while treatments like thiophanate-methyl are used to combat *Fusarium*. Biological control agents, including mycorrhizal fungi and antagonistic bacteria like *Pseudomonas* species, can enhance root health and suppress pathogen activity. Regular monitoring and early intervention are essential, as symptoms often become apparent only after significant damage has occurred.

1.3 Foliar Diseases in Nursery Stock: Powdery Mildew and Leaf Spots

Foliar diseases, such as powdery mildew and leaf spots, are common in nursery stock and can significantly affect the health and appearance of seedlings. While these diseases are rarely fatal, they can reduce photosynthetic capacity, delay growth, and make plants more vulnerable to other stressors.

Foliar diseases are caused by a range of fungal and bacterial pathogens, which thrive in the warm, humid conditions typical of nursery environments. Powdery mildew, caused by fungi in the family Erysiphaceae, is one of the most prevalent foliar diseases in nurseries. It affects a wide range of plant species, covering leaves, stems, and other aerial parts with a characteristic white, powdery coating. This coating consists of fungal mycelia and conidia, which are easily dispersed by air currents. Infected leaves may become distorted, discolored, and prone to premature senescence, reducing the vigor of affected seedlings. Powdery mildew is particularly problematic in crowded seedbeds and during periods of high humidity, although the disease does not require free water for infection.

Leaf spot diseases, caused by pathogens such as *Cercospora*, *Colletotrichum*, and *Alternaria* species, are another significant concern in nurseries. These diseases produce characteristic lesions on leaves, ranging from small, circular spots to larger, irregular patches. Lesions are often surrounded by chlorotic halos and may coalesce, leading to extensive defoliation. In severe cases, leaf spot infections can reduce the photosynthetic area of seedlings, stunting their growth and affecting their quality for outplanting. Managing foliar diseases requires a combination of cultural, chemical, and biological strategies. Reducing humidity through improved ventilation, spacing, and pruning can help limit the spread of pathogens. Fungicides, such as sulfur and copper-based products, are effective against powdery mildew, while broader-spectrum fungicides can control leaf spots. Biological control agents, including *Bacillus subtilis* and *Trichoderma* species, have shown promise in suppressing foliar pathogens while promoting plant health. Regular monitoring and prompt removal of infected foliage are also critical for preventing the buildup of inoculum in nursery environments.

1.4 Stem Cankers and Collar Rot in Seedlings

Stem cankers and collar rot are significant nursery diseases that can severely damage seedlings and young plants, often resulting in substantial mortality if not managed effectively. These diseases are caused primarily by fungal pathogens such as *Botryosphaeria*, *Cytospora*, *Fusarium*, and *Phytophthora* species, which invade the stem and collar regions of seedlings. Environmental conditions that promote high humidity, excessive moisture, or mechanical damage to plant tissues increase the likelihood of infections. Stem cankers typically begin as localized lesions on the stem that expand over time. The bark around the infected area may appear sunken, discolored, or cracked, with visible resin or sap exudation in conifers. As the infection progresses, the cambium beneath the bark dies, leading to girdling of the stem, wilting of leaves, and eventual dieback of the seedling. Cankers are often initiated at wounds caused by pruning, frost cracks, or insect activity, providing an entry point for pathogens.

Collar rot primarily affects the basal portion of the stem near the soil line. This disease is commonly associated with *Phytophthora* species, which thrive in waterlogged soils and infect seedlings through the root collar. Symptoms include dark, water-soaked lesions at the soil line, wilting, stunted growth, and chlorosis. In advanced stages, the seedling's structural integrity is compromised, leading to lodging or collapse. Management of stem cankers and collar rot involves an integrated approach. Preventive measures include avoiding mechanical injuries to stems, maintaining proper drainage, and using fungicides such as mefenoxam or copper-based products as soil drenches. Biological control agents, such as *Trichoderma* and *Bacillus* species, can help suppress pathogens and protect seedlings. Pruning should be conducted during dry weather to minimize the risk of infection, and infected seedlings should be removed and destroyed promptly to prevent the spread of disease.

1.5 Bacterial Blights in Nursery Seedlings: Pathogenesis and Control

Bacterial blights are a significant category of nursery diseases caused by pathogenic bacteria such as *Pseudomonas syringae* and *Xanthomonas* species. These pathogens affect a wide range of seedlings, causing foliar lesions, stem infections, and even systemic disease in severe cases. The rapid spread of bacterial blights in humid nursery conditions can lead to widespread losses in seedling production. Bacterial blights often begin with water-soaked spots on leaves, which enlarge and turn brown or black over time. These lesions may be surrounded by yellow halos, a characteristic symptom of bacterial infections. When stems are infected, the bacteria can cause girdling lesions that disrupt water and nutrient flow, leading to wilting and dieback. In some cases, bacterial pathogens enter through natural openings like stomata or through wounds caused by pruning, hail, or insect feeding. The epidemiology of bacterial blights is heavily influenced by environmental conditions. High humidity, excessive irrigation, and overcrowded seedbeds create ideal conditions for bacterial multiplication and dispersal. Splashing water from rain or overhead irrigation spreads bacterial cells, while contaminated tools, soil, or plant material serve as reservoirs for infection. Managing bacterial blights requires a combination of cultural, chemical, and biological approaches. Proper sanitation is critical, including the use of sterilized tools, pathogen-free seeds, and clean irrigation water. Reducing humidity through improved ventilation and spacing can limit bacterial proliferation. Copper-based bactericides are commonly used to control bacterial blights, although their effectiveness may be limited under heavy disease pressure. Biological control agents, such as *Bacillus subtilis* and *Pseudomonas fluorescens*, have shown promise in suppressing bacterial pathogens while promoting seedling health.

1.6 Viral Diseases in Nursery Plants: Transmission and Impacts

Viral diseases in nursery plants, although less common than fungal or bacterial diseases, can cause significant damage by stunting growth, distorting leaves, and reducing the vigor of seedlings. Viruses are obligate pathogens, meaning they require living host cells to replicate. Some of the most prevalent viruses affecting nursery plants include Tobacco Mosaic Virus (TMV), Cucumber Mosaic Virus (CMV), and Tomato Spotted Wilt Virus (TSWV), which can infect a wide range of forest tree species. Symptoms of viral infections vary depending on the virus and host, but often include chlorotic or necrotic spots on leaves, vein banding, mosaic patterns, and malformations of leaves, stems, or roots. Infected seedlings may show reduced growth and vigor, making them less suitable for planting and more vulnerable to environmental stressors. Viral diseases are primarily spread through vectors, such as aphids, whiteflies, and thrips, which transmit viruses while feeding on plant sap. Mechanical transmission through contaminated tools, hands, or soil is also common in nursery settings. Once a plant is infected, the virus can spread systemically, often persisting throughout the plant's life. Managing viral diseases in nurseries is challenging because there are no direct chemical treatments for viruses. Prevention is the most effective strategy, including strict sanitation measures, the use of certified virus-free seeds, and the control of insect vectors through integrated pest management (IPM). Resistant plant varieties, when available, provide an additional line of defense. Eliminating infected plants from nurseries is crucial to prevent the spread of viruses to healthy seedlings and neighboring forest ecosystems.

1.7 Impact of Environmental Stress on Nursery Pathogen Outbreaks

Environmental stress is a critical factor that influences the outbreak and severity of nursery diseases. Stressors such as drought, temperature extremes, poor soil conditions, and nutrient imbalances weaken seedlings, making them more susceptible to infection by opportunistic pathogens. Stress also alters plant physiology, reducing the production of secondary metabolites and other defense compounds that play a key role in resisting infection. Drought stress, for example, reduces water availability to roots, impairing nutrient uptake and causing physiological stress that predisposes seedlings to root rot diseases caused by *Phytophthora* and *Fusarium*. Similarly, waterlogged soils caused by over-irrigation or poor drainage create anaerobic conditions that favor pathogens like *Pythium* and *Phytophthora*, leading to damping-off or collar rot. Temperature extremes, such as frost or heat waves, also increase disease susceptibility. Frost can cause physical damage to seedling tissues, providing entry points for canker-causing pathogens like *Cytospora* and *Botryosphaeria*. High temperatures, on the other hand, can disrupt plant metabolism and exacerbate foliar diseases like powdery mildew and leaf spots.

Nutrient imbalances, whether due to deficiencies or over-fertilization, further compound the problem. Excessive nitrogen, for instance, promotes lush, succulent growth that is more vulnerable to foliar pathogens like *Alternaria* and *Cercospora*. Conversely, deficiencies in essential nutrients such as potassium and calcium weaken plant cell walls, making seedlings more prone to pathogen invasion. Managing the impact of environmental stress on nursery pathogen outbreaks requires a proactive and holistic approach. Proper site selection, soil preparation, and irrigation management are essential for creating optimal growing conditions. Adjusting nutrient applications to match seedling growth requirements can reduce stress and improve overall resilience. Monitoring environmental conditions and implementing stress-reducing measures, such as shading during heatwaves or windbreaks to reduce frost exposure, can further enhance seedling health. Integrating these practices with pathogen monitoring and targeted interventions forms the cornerstone of effective nursery disease management.

2. Parasitic Plant Diseases

Parasitic plant diseases are a significant concern in forest ecosystems, impacting both natural and managed forests by directly siphoning resources from their host trees. These parasitic plants have evolved specialized structures to invade and extract nutrients and water from their hosts, often resulting in stunted growth, reduced productivity, and increased susceptibility to other stressors. While parasitic plants play a natural role in some ecosystems by influencing forest composition and dynamics, their proliferation, especially in stressed or disturbed environments, can lead to significant ecological and economic losses.

2.1 Mistletoes

Mistletoes are hemiparasitic plants from families such as *Loranthaceae* and *Viscaceae*. They have a dual ability to photosynthesize while extracting water and nutrients from their host trees using specialized structures called haustoria. These parasites are widely distributed across temperate, tropical, and subtropical regions, infecting a variety of tree species, including oaks, pines, and acacias. Mistletoe infestations can significantly weaken trees, reducing their growth, seed production, and structural stability. The biology of mistletoes is intricately linked to their dispersal mechanisms. Birds are the primary vectors, feeding on mistletoe berries and excreting seeds onto host branches. Once a seed lands on a suitable host, it germinates and produces a haustorium that penetrates the bark and establishes a connection with the host's xylem. Mistletoes extract water and nutrients from the host, leading to symptoms such as branch dieback, reduced foliage density, and decreased vigor. Severe infestations can kill branches or entire trees, particularly in stressed or drought-prone environments.

Ecologically, mistletoes play a complex role. While they can reduce the health and productivity of their hosts, they also provide food and habitat for various bird and insect species, contributing to biodiversity. However, in managed forests or urban landscapes, their negative impacts often outweigh their ecological benefits, leading to substantial economic losses. Managing mistletoe infestations requires an integrated approach. Mechanical removal of mistletoe clumps, though labor-intensive, is effective in reducing immediate damage, particularly in high-value trees. Pruning infected branches can limit the spread of the parasite but may compromise the structural integrity of the tree. Biological control methods, such as introducing natural predators or pathogens specific to mistletoes, have shown potential in certain regions. Long-term management strategies include planting resistant tree species or diversifying forests to reduce the spread and impact of mistletoe infestations.

2.2 Dodder

Dodder (*Cuscuta* spp.) is a holoparasitic plant that poses a significant threat to young and nursery trees by entirely relying on its hosts for survival. Unlike mistletoes, dodder lacks chlorophyll and is incapable of photosynthesis, extracting all its nutrients and water from the host through haustoria. Dodder is particularly problematic in nurseries, where its rapid growth and aggressive parasitism can severely impact seedlings and saplings. The life cycle of dodder begins with seed germination in the soil. The young seedling quickly seeks out a host plant using chemical cues and attaches itself by wrapping around the stem. Haustoria penetrate the host's vascular tissues, establishing a parasitic connection. As dodder grows, it forms a dense, tangled mass that smothers the host, reducing its access to light and further draining its resources. Symptoms of dodder infestation include stunted growth, yellowing leaves, and wilting, with young seedlings being particularly vulnerable to mortality.

Dodder spreads primarily through seeds, which can be transported by wind, water, animals, or contaminated nursery equipment. Its ability to parasitize a wide range of hosts makes it a significant threat in diverse nursery and forest settings. Infestations can quickly escalate in densely planted areas, making early detection and control crucial. Managing dodder infestations involves rigorous sanitation and monitoring. Removing and destroying infested plants is essential to prevent the spread of seeds. In nurseries, using sterile soil and ensuring that seeds and planting materials are free of contamination can minimize the risk of introduction. Herbicides, such as glyphosate, may be effective in controlling dodder, but their application must be carefully managed to avoid damage to non-target plants. Mulching and shading the soil can also suppress dodder germination by blocking sunlight, which is critical for its initial growth.

2.3 Broomrapes: Orobanche Species and Their Hosts

Broomrapes (*Orobanche* spp.) are obligate holoparasitic plants that infect the roots of a wide range of herbaceous and woody hosts, including many economically and ecologically important forest species. These parasitic plants lack chlorophyll and rely entirely on their hosts for water, nutrients, and carbon. Broomrape infestations can severely impact the growth and productivity of infected plants, particularly in nutrient-poor soils where competition for resources is already high. The life cycle of broomrape begins with seed germination triggered by chemical signals released by host roots. The emerging seedling produces a haustorium that attaches to the host root, penetrating the vascular tissues and establishing a parasitic connection. As the broomrape plant grows, it emerges above ground, producing small, tubular flowers that generate thousands of tiny seeds capable of remaining dormant in the soil for years.

The impacts of broomrape infestations are significant, particularly in stressed or degraded environments. Infected plants exhibit symptoms such as stunted growth, wilting, and reduced reproductive output, with severe infestations leading to host mortality. Broomrapes also exacerbate competition for water and nutrients, reducing the resilience of forest ecosystems to drought and other stressors. Managing broomrape infestations is challenging due to the persistence of their seeds in the soil and their wide host range. Preventive measures, such as using broomrape-free planting materials and rotating non-host crops, are essential in nurseries and reforestation programs. Soil solarization and fumigation can reduce seed banks in heavily infested areas, although these methods may not be feasible on a large scale. Biological control agents, such as certain fungi and bacteria that attack broomrape seeds or seedlings, offer a promising avenue for sustainable management.

2.4 Witchweed (*Striga* spp.): Parasitic Challenges in Forest Ecosystems

Witchweed (*Striga* spp.) is a hemiparasitic plant that poses a significant threat to forest ecosystems and agroforestry systems, particularly in tropical and subtropical regions. Like mistletoes, witchweed is capable of photosynthesis but relies on its host plants for water and nutrients, severely reducing their growth and productivity. While *Striga* species are best known for their impact on cereal crops, they also parasitize a variety of forest trees, particularly in disturbed or degraded areas. The life cycle of witchweed begins with the germination of tiny seeds in the soil. Germination is triggered by chemical signals released by host roots, prompting the witchweed seedling to produce a haustorium that attaches to the host. As the parasite grows, it emerges above ground, producing vibrant flowers and seeds that disperse widely through wind, water, and animal activity. A single *Striga* plant can produce tens of thousands of seeds, allowing infestations to escalate rapidly.

Witchweed inflicts significant damage on its hosts by extracting water and nutrients, leading to symptoms such as chlorosis, wilting, stunted growth, and reduced reproductive output. In forest ecosystems, *Striga* infestations can hinder natural regeneration and reduce the growth of young trees, particularly in degraded soils where competition for resources is already intense. Managing *Striga* infestations requires an integrated approach. Preventive measures, such as using clean planting material and avoiding the introduction of contaminated soil, are critical. Soil amendments, such as organic matter or biochar, can improve soil fertility and suppress *Striga* germination. Biological control agents, including fungal pathogens and insect herbivores specific to *Striga*, are being explored as environmentally friendly management options. Long-term strategies should focus on restoring soil health and enhancing the resilience of forest ecosystems through diverse planting schemes and sustainable land management practices.

2.5 Interactions Between Parasitic Plants and Fungal Pathogens

The interactions between parasitic plants and fungal pathogens are complex and can significantly influence forest dynamics. Parasitic plants, such as mistletoes, broomrapes, dodder, and witchweed, can create stress in host trees and disrupt their physiology, making them more susceptible to secondary infections by fungal pathogens. Conversely, fungal pathogens can weaken or alter the health of host plants, indirectly affecting the establishment and proliferation of parasitic plants. These interactions can amplify the impacts of both parasitism and fungal infections, leading to a cascade of negative effects on forest ecosystems. One of the primary ways in which parasitic plants interact with fungal pathogens is by creating entry points for infection. For example, mistletoe haustoria penetrate the vascular tissues of their host, often causing localized necrosis or wounds. These sites can serve as entry points for opportunistic pathogens like *Botryosphaeria*, *Fusarium*, or *Cytospora*, which further compromise the host's vascular system. Similarly, dodder infestations can weaken the stems of young trees, allowing fungal pathogens to invade and exacerbate the damage.

In some cases, fungal pathogens may alter the success of parasitic plants by affecting host vitality. For instance, root rot fungi like *Armillaria* or *Phytophthora* can reduce the water and nutrient flow available to a parasitic plant, potentially limiting its growth. On the other hand, fungal pathogens may also produce stress-related signals in the host plant, such as increased ethylene production or altered secondary metabolites, which may either promote or inhibit parasitic plant attachment. The combined impact of parasitic plants and fungal pathogens is particularly severe in degraded or stressed forests, where environmental factors like drought or soil compaction amplify the vulnerability of hosts. These interactions often lead to higher rates of host mortality, reduced regeneration, and shifts in forest composition. Understanding these relationships is critical for developing integrated management strategies that address both parasitism and fungal infections. For example, targeting fungal pathogens with biological control agents or fungicides may reduce their synergistic effects with parasitic plants, thereby limiting overall forest damage.

2.6 Role of Parasitic Plants in Altering Forest Succession

Parasitic plants play a significant role in shaping forest succession, influencing species composition, community dynamics, and ecosystem processes. By selectively targeting certain host species, parasitic plants can alter competitive relationships among tree species, creating opportunities for less dominant species to establish and thrive. These effects can cascade through the ecosystem, affecting not only plant communities but also the animals, fungi, and microbes that depend on them. One of the primary ways parasitic plants influence succession is by reducing the growth and reproduction of their hosts. For example, mistletoe infestations can weaken dominant canopy species, such as oaks or pines, leading to increased light availability on the forest floor. This change in light conditions can promote the growth of shade-intolerant understory plants or pioneer species, accelerating successional shifts. Similarly, parasitic plants like broomrapes can suppress fast-growing pioneer species, allowing more shade-tolerant or stress-tolerant species to gain a foothold. Parasitic plants also influence nutrient cycling and soil dynamics, further altering successional trajectories. For example, mistletoe infestations increase the deposition of nutrient-rich litter on the forest floor, as their leaves and berries often have higher nutrient contents than those of their hosts. This localized enrichment of the soil can create microhabitats favorable for certain plant species, fungi, and microbes, shaping the direction of forest succession.

In some cases, parasitic plants may delay succession by increasing host mortality or reducing seed production, particularly in forests recovering from disturbance. For instance, high densities of *Striga* or dodder in post-disturbance environments can suppress the establishment of key tree species, slowing the transition from grassland or shrubland to mature forest. Conversely, in nutrient-poor or degraded environments, parasitic plants may promote succession by targeting less stress-tolerant species, creating opportunities for hardier species to establish and stabilize the ecosystem. The role of parasitic plants in forest succession is context-dependent and influenced by factors such as host availability, environmental conditions, and the presence of other stressors like herbivory or pathogens.

2.7 Mechanisms of Host-Parasitic Plant Resistance and Adaptation

The interactions between hosts and parasitic plants are shaped by a dynamic evolutionary arms race, with hosts developing resistance mechanisms to fend off parasitic plants and parasites evolving strategies to overcome these defenses. Understanding these mechanisms provides valuable insights into forest health and offers opportunities for developing sustainable management practices. Host resistance to parasitic plants involves a variety of structural, chemical, and physiological defenses. Structural barriers, such as thick bark or lignified tissues, can prevent parasitic plants like mistletoes or dodder from successfully attaching to or penetrating the host. For example, some pine species produce resinous exudates at mistletoe infection sites, limiting the establishment of haustoria. Similarly, certain herbaceous hosts produce physical thickenings around broomrape attachment points, restricting the parasite's access to vascular tissues.

Many plants produce secondary metabolites, such as phenolic compounds, alkaloids, or flavonoids, that inhibit the germination, growth, or attachment of parasitic plants. For instance, hosts targeted by *Striga* may release allelopathic compounds into the soil, disrupting the germination of witchweed seeds. In some cases, hosts also produce reactive oxygen species (ROS) or antimicrobial compounds at the site of haustorial attachment, killing parasitic plant tissues and preventing further invasion. Physiological resistance mechanisms include the ability to regulate nutrient and water flow to parasitic plants. Some hosts can compartmentalize infections by altering xylem flow patterns, effectively "starving" the parasite. Others may induce localized cell death (hypersensitive response) at the site of parasitic attachment, preventing the spread of the infection.

Parasitic plants, in turn, have evolved adaptations to overcome host defenses. For example, mistletoes produce enzymes that degrade host cell walls, allowing their haustoria to penetrate resistant tissues. Dodder plants can adjust their attachment sites to avoid thick or lignified stems, while *Striga* and broomrape species release germination stimulants that mimic host root exudates, ensuring successful attachment. Understanding these resistance and adaptation mechanisms has practical implications for managing parasitic plant diseases. Breeding programs that focus on enhancing host resistance traits, such as thick bark or the production of inhibitory compounds, can reduce the impacts of parasitic plants in nurseries and forests. Similarly, strategies that exploit the vulnerabilities of parasitic plants, such as disrupting their chemical signaling pathways, offer promising avenues for control. By leveraging these insights, forest managers can develop integrated approaches that balance the ecological roles of parasitic plants with the need to maintain forest health and productivity.

3. Integrated Management of Nursery and Parasitic Plant Diseases

Integrated management of nursery and parasitic plant diseases is a holistic approach that combines preventive, cultural, biological, and monitoring strategies to mitigate the impact of these diseases effectively. Nurseries are critical environments for seedling production, and diseases in these controlled spaces can lead to significant losses in productivity, financial resources, and forest regeneration potential. Similarly, parasitic plants can severely reduce tree vigor, growth, and reproduction, complicating efforts to maintain healthy forests. Integrated management provides a framework for addressing these challenges comprehensively, balancing short-term control measures with long-term resilience strategies.

3.1 Sanitation Practices in Nurseries

Sanitation is one of the most fundamental practices in preventing and managing diseases in nurseries. By maintaining a clean growing environment, the risk of pathogen proliferation and parasitic plant establishment can be significantly reduced. Pathogens and parasitic plants often exploit poor sanitation conditions, such as contaminated soil, tools, and irrigation water, to spread rapidly in nursery systems. Key sanitation practices in nurseries include the sterilization of soil, containers, and equipment. Soil sterilization, achieved through methods such as steam treatment or solarization, effectively eliminates soilborne pathogens like *Phytophthora*, *Pythium*, and *Fusarium*. Similarly, disinfecting tools, trays, and surfaces with solutions containing hydrogen peroxide or sodium hypochlorite reduces the likelihood of spreading pathogens or parasitic plant seeds during routine nursery operations. Another crucial aspect of sanitation is the proper disposal of infected plant material. Seedlings showing symptoms of damping-off, root rot, or foliar diseases should be promptly removed and destroyed to prevent pathogen reservoirs from persisting in the nursery. Composting infected material is not advisable unless high temperatures are consistently maintained to kill pathogens and parasitic seeds. Instead, burning or deep burial is often recommended.

Water management is also critical to nursery sanitation. Contaminated irrigation water can serve as a conduit for spreading pathogens like *Phytophthora* and parasitic plant seeds. Using clean water sources, installing filtration systems, and avoiding overhead irrigation in favor of drip systems can minimize the risk of disease spread. Preventing the introduction of pathogens and parasitic plants through planting material is another key sanitation measure. Using certified pathogen-free seeds and seedlings ensures that the nursery remains free of harmful organisms. Regular inspections of incoming and existing stock help detect infections early, allowing for rapid intervention.

3.2 Early Detection Techniques for Parasitic Plants

Early detection is a cornerstone of integrated management for parasitic plant diseases. Identifying infestations in their initial stages prevents widespread damage and allows for more effective control measures. Traditional detection methods, such as visual inspections, are often limited by the cryptic nature of some parasitic plants and the delayed onset of symptoms. Advances in diagnostic technologies have significantly improved the ability to detect parasitic plants and their seeds before they cause major damage. Visual scouting remains an essential first step in early detection. Regular inspections of nursery stock and forest stands can help identify early signs of parasitism, such as the attachment of dodder stems, haustorial penetration by mistletoes, or localized chlorosis associated with *Striga* infestations. However, these methods require trained personnel who can differentiate parasitic plants from other growths or symptoms caused by environmental stress or pathogens. Molecular diagnostic tools have revolutionized the early detection of parasitic plants. Techniques such as polymerase chain reaction (PCR) and quantitative PCR (qPCR) allow for the detection of parasitic plant DNA in host tissues or soil samples, even before visible symptoms appear. These tools are highly sensitive and can differentiate between closely related species, making them invaluable for monitoring infestations in nurseries and forests. For example, qPCR assays have been developed to detect *Cuscuta* (dodder) and *Orobanche* (broomrape) species in contaminated soil or seed lots.

Remote sensing and imaging technologies offer non-invasive methods for detecting parasitic plants in larger areas. Hyperspectral imaging, which captures subtle variations in light reflectance, can identify stress patterns in host plants associated with parasitic infestations. Similarly, drone-based surveys equipped with thermal or multispectral cameras provide a cost-effective means of monitoring parasitic plant distribution in forests and plantations. Developing comprehensive monitoring protocols that combine traditional scouting, molecular diagnostics, and remote sensing ensures timely detection and intervention. Early detection programs should also include public awareness campaigns and training for nursery staff and forest managers to recognize parasitic plant infestations and report them promptly.

3.3 Biological and Cultural Control Methods

Biological and cultural control methods form the backbone of sustainable management strategies for nursery and parasitic plant diseases. Unlike chemical control measures, which can have negative environmental impacts and lead to resistance in pathogens, these methods work with natural ecological processes to suppress diseases and maintain healthy nursery and forest systems.

Biological Control Methods

Biological control involves the use of living organisms, such as fungi, bacteria, insects, or nematodes, to manage pathogens and parasitic plants. These organisms can act through direct antagonism, competition, or parasitism, reducing the survival and proliferation of harmful species. For nursery diseases, beneficial microbes such as *Trichoderma* and *Bacillus subtilis* are widely used to suppress soilborne pathogens like *Fusarium*, *Pythium*, and *Phytophthora*. These microbes colonize the rhizosphere, outcompeting pathogens for nutrients and space while promoting seedling health. Mycorrhizal fungi also play a vital role in protecting seedlings from root rot diseases by enhancing nutrient uptake and stimulating host defenses. Biological control of parasitic plants focuses on introducing natural enemies that target specific parasitic species. For example, fungal pathogens such as *Alternaria* and *Fusarium* species have been explored as biocontrol agents against *Striga* and broomrapes. Insect herbivores, such as the dodder moth (*Cuscuta epilobii*), feed on dodder vines, reducing their biomass and spread. However, the success of biological control depends on careful selection and evaluation of agents to ensure they do not harm non-target species or disrupt native ecosystems.

Cultural Control Methods

Cultural control methods aim to create unfavorable conditions for pathogens and parasitic plants while enhancing the resilience of host plants. These methods include crop rotation, mixed-species planting, soil management, and altering planting times. In nurseries, rotating crops with non-host species reduces the buildup of parasitic plant seeds and pathogen inoculum in the soil. For example, alternating susceptible nursery crops with cereals or legumes can disrupt the life cycle of *Striga* and broomrapes. Similarly, maintaining adequate spacing between plants improves air circulation, reducing the humidity levels that favor diseases like damping-off and powdery mildew.

Mixed-species planting in forests and nurseries helps dilute the effects of parasitic plants by providing alternative hosts or non-host barriers that slow the spread of infestation. For instance, planting resistant or less susceptible tree species alongside vulnerable ones can reduce the impact of mistletoe or dodder infestations. Soil management practices, such as the application of organic amendments and mulches, improve soil structure and fertility, promoting healthy root systems and reducing the germination of parasitic plant seeds. For example, biochar and compost suppress *Striga* by altering the soil microbiome and reducing the availability of host root exudates that trigger germination. Timing planting operations to avoid peak periods of parasitic plant germination or pathogen activity is another effective cultural strategy. Adjusting irrigation schedules to prevent overwatering and maintaining optimal soil moisture levels also limits the conditions that favor root rot pathogens and parasitic plant attachment.

3.4 Role of Resistant Rootstocks and Species Selection

The selection of resistant rootstocks and tree species is a cornerstone of integrated management for nursery and parasitic plant diseases. By planting resistant or tolerant varieties, nurseries and forests can reduce the prevalence of disease and minimize economic and ecological losses. Resistant rootstocks serve as a critical line of defense, especially against soilborne pathogens and parasitic plants, by providing a robust physiological barrier to infection while maintaining compatibility with desired scion varieties. Resistant rootstocks are particularly effective against root rot diseases caused by *Phytophthora*, *Fusarium*, and other pathogens. For instance, rootstocks of oak and citrus trees that exhibit resistance to *Phytophthora* species are widely used in commercial and reforestation nurseries. Similarly, resistant pine rootstocks are essential in mitigating the impact of pitch canker and nematode infestations. These rootstocks limit the spread of infection to above-ground tissues, preserving the overall health and productivity of the tree. Species selection also plays a critical role in managing parasitic plant infestations. Selecting tree species that are less susceptible to mistletoes, broomrapes, or *Striga* can limit the establishment and spread of these parasites. For example, fast-growing species like eucalyptus or acacia may outcompete parasitic plants, while certain resistant pines and hardwoods can suppress the growth of attached parasites through physiological and chemical defenses. In addition to reducing disease pressure, species and rootstock selection enhances biodiversity, improving the resilience of forests and nurseries to environmental stressors and emerging pathogens. Mixed-species planting schemes can create barriers that limit the spread of pathogens and parasitic plants, providing long-term stability to the ecosystem. Developing and utilizing resistant rootstocks and species requires robust research and breeding programs. Advances in genetic engineering and marker-assisted selection have accelerated the identification and propagation of resistant genotypes, enabling nurseries to provide high-quality planting materials. Regular testing and evaluation of rootstocks under local environmental conditions are essential to ensure their effectiveness against specific pathogens and parasites.

3.5 Integrated Pest and Disease Management in Forest Nurseries

Integrated Pest and Disease Management (IPDM) in forest nurseries is a comprehensive strategy that combines multiple approaches to minimize the impact of pests and diseases while reducing reliance on chemical controls. This holistic approach focuses on prevention, monitoring, and targeted intervention, ensuring sustainable and cost-effective nursery management. The foundation of IPDM lies in prevention. Maintaining nursery hygiene, using certified disease-free seeds and planting materials, and sterilizing soil and tools are critical preventive measures. Cultural practices, such as proper spacing, irrigation management, and crop rotation, reduce the environmental conditions favorable for pests and pathogens. For example, alternating susceptible crops with non-host species can disrupt the life cycle of parasitic plants like broomrapes and reduce pathogen inoculum levels in the soil.

Monitoring is another key component of IPDM. Regular scouting for signs of pest and disease activity, combined with advanced diagnostic tools like molecular assays and remote sensing, enables early detection and timely intervention. Threshold levels for action are established to prevent unnecessary interventions, ensuring that control measures are applied only when pest or pathogen populations exceed acceptable limits. Biological control agents play a central role in IPDM, targeting specific pests and pathogens without harming non-target organisms. For example, beneficial fungi like *Trichoderma* and *Gliocladium* suppress root rot pathogens, while bacterial species such as *Bacillus subtilis* inhibit foliar diseases. Natural enemies of parasitic plants, such as insects that feed on dodder or broomrape, are also important biological control tools. When chemical controls are necessary, IPDM emphasizes the use of targeted, environmentally friendly products. Systemic fungicides and bactericides, applied judiciously, can protect seedlings without disrupting beneficial soil microbiota. The use of resistant rootstocks and planting materials further enhances the effectiveness of IPDM by reducing the susceptibility of nursery stock to pathogens and parasites.

3.6 Biotechnological Tools for Disease Resistance

Biotechnology has emerged as a powerful tool in the fight against nursery and parasitic plant diseases, offering novel approaches for developing disease-resistant varieties and enhancing early detection capabilities. By leveraging genetic engineering, molecular biology, and omics technologies, researchers are creating innovative solutions to combat pathogens and parasitic plants more effectively. One of the most promising applications of biotechnology is the development of genetically modified (GM) tree species and rootstocks with enhanced resistance to diseases and parasites. For instance, transgenic trees engineered to express antifungal or antimicrobial proteins have shown significant resistance to root rot pathogens like *Phytophthora* and *Fusarium*. Similarly, GM trees with altered metabolic pathways can resist parasitic plants by disrupting their haustorial attachment or nutrient uptake mechanisms. Marker-assisted selection (MAS) is another valuable biotechnological tool, enabling researchers to identify and propagate resistant genotypes more efficiently. By linking specific genetic markers to disease-resistance traits, breeders can accelerate the development of resistant varieties without the need for extensive field testing. This approach has been successfully applied in breeding programs for pines, oaks, and other commercially important forest species.

Omics technologies, including genomics, transcriptomics, and proteomics, provide a deeper understanding of the interactions between hosts, pathogens, and parasitic plants. These insights are critical for identifying resistance genes, unraveling the molecular mechanisms of infection, and designing targeted interventions. For example, transcriptomic studies have revealed how certain tree species activate defense pathways in response to mistletoe or dodder attachment, guiding the development of resistant varieties. CRISPR-Cas9 gene-editing technology represents a cutting-edge tool for precise genetic modifications in forest species. By targeting specific genes associated with susceptibility or resistance, researchers can create trees with enhanced tolerance to pathogens and parasitic plants. For instance, CRISPR has been used to knock out susceptibility genes in crops, and similar approaches are being explored for forest species.

3.7 Policies and Regulations for Preventing Pathogen Spread

Effective policies and regulations are essential for preventing the spread of pathogens and parasitic plants, particularly in the context of global trade and increasing environmental stressors. International, national, and local frameworks play a critical role in safeguarding nurseries and forests from invasive species and disease outbreaks.

Quarantine regulations are a cornerstone of disease prevention, restricting the movement of potentially infected plant materials, soil, and equipment across borders. Agencies such as the International Plant Protection Convention (IPPC) establish guidelines for inspecting and certifying nursery stock to ensure that it is free of harmful organisms. Phytosanitary measures, including fumigation, heat treatment, and disinfection, are commonly required for imported materials to prevent the introduction of pathogens and parasitic seeds. National policies often include surveillance programs to detect and manage disease outbreaks. Early warning systems, supported by remote sensing and molecular diagnostics, enable authorities to respond quickly to emerging threats. For example, monitoring programs for invasive pathogens like *Phytophthora ramorum* have been implemented in many countries to prevent the spread of sudden oak death.

Integrated management plans, supported by governmental and non-governmental organizations, emphasize collaboration between researchers, forest managers, and policymakers. These plans often include funding for research, capacity building, and public awareness campaigns to educate stakeholders about best practices in nursery management and disease prevention. Enforcing compliance with regulations is critical for their effectiveness. Violations, such as the illegal movement of infected materials, can undermine efforts to control disease spread. Penalties, combined with incentives for adopting safe practices, encourage adherence to phytosanitary standards. In addition to preventing the spread of pathogens, policies should promote the adoption of sustainable management practices. Subsidies or technical support for nurseries implementing IPDM, using resistant varieties, or adopting biotechnological innovations can enhance the overall health and productivity of forests. International cooperation, facilitated through treaties and agreements, is also vital for addressing cross-border challenges posed by invasive species and climate change.

4. Impact of Nursery Diseases on Forest Regeneration

Nursery diseases significantly affect forest regeneration by disrupting the supply of healthy seedlings essential for reforestation and restoration programs. These diseases, caused by fungal, bacterial, viral, and parasitic pathogens, impact the quantity and quality of seedlings, leading to cascading effects on forest ecosystems, biodiversity, and the socio-economic frameworks tied to forestry. The damage extends beyond the nursery environment, as diseased seedlings planted in natural or managed forests can carry pathogens into previously unaffected areas, compounding ecological and economic challenges. This section explores three critical aspects of the impact of nursery diseases: seedling mortality as a barrier to reforestation, the effects on biodiversity and forest composition, and the economic implications for reforestation programs.

4.1 Seedling Mortality

Seedling mortality is one of the most direct and devastating consequences of nursery diseases, posing a significant barrier to successful reforestation and afforestation efforts. High mortality rates caused by diseases such as damping-off, root rot, and foliar blights can deplete nursery stock and limit the availability of seedlings for planting. The problem is particularly severe in large-scale reforestation projects, where achieving the desired planting density depends on the production of a high volume of healthy seedlings.

Damping-off, caused by pathogens such as *Pythium*, *Phytophthora*, and *Fusarium*, is a leading cause of seedling mortality in nurseries. This disease affects germinating seeds and young seedlings, causing pre-emergence seed decay or post-emergence stem collapse. Infected seedlings are unable to establish, resulting in bare patches in seedbeds and uneven stands in reforested areas. Similarly, root rot diseases caused by *Phytophthora cinnamomi* or *Armillaria* lead to the gradual decline and death of seedlings, particularly under conditions of poor drainage or over-irrigation.

High mortality rates in nurseries necessitate overproduction of seedlings to meet planting targets, increasing operational costs and resource use. Additionally, the loss of seedlings during early growth stages creates bottlenecks in the reforestation pipeline, delaying timelines for planting and compromising the success of restoration programs. In degraded landscapes or regions affected by deforestation, these delays can have long-term ecological impacts, including soil erosion, reduced water retention, and habitat loss. Addressing seedling mortality requires proactive measures to mitigate disease risks in nurseries. These include adopting integrated disease management practices, such as soil sterilization, biological controls, and resistant rootstocks, as well as optimizing environmental conditions to reduce pathogen proliferation. By minimizing seedling losses, nursery operators can ensure a steady supply of high-quality planting material, supporting large-scale reforestation initiatives.

4.2 Effects on Biodiversity and Forest Composition

Nursery diseases indirectly impact biodiversity and forest composition by influencing the survival and establishment of seedlings in natural and managed forests. Diseased or weak seedlings are less likely to survive the harsh conditions of the field, leading to altered species dynamics and competitive interactions. Additionally, the introduction of pathogens from nurseries into forest ecosystems can disrupt existing plant communities, with far-reaching consequences for associated fauna and microbial networks. The uneven survival of seedlings due to nursery diseases often results in a skewed species composition in reforested areas. For example, diseases like root rot or collar rot may disproportionately affect specific species, allowing others to dominate. This can lead to the loss of diversity in mixed-species plantations or the homogenization of forest stands, reducing ecosystem resilience to pests, diseases, and climate variability. In natural forests, the reduced establishment of key species due to nursery-borne diseases can disrupt successional processes, altering forest structure and function.

Pathogens introduced through diseased nursery stock can also have cascading ecological effects. Soilborne pathogens like *Phytophthora ramorum*, responsible for sudden oak death, not only affect host trees but also alter the understory vegetation, nutrient cycling, and hydrological processes of affected forests. Parasitic plants inadvertently transported with nursery seedlings can spread rapidly in the field, competing with native flora and exacerbating the stress on regenerating forests. Maintaining biodiversity in reforestation programs requires careful selection of species and the production of disease-free seedlings. Mixed-species planting can mitigate the risks associated with species-specific pathogens, promoting ecological balance and resilience. Regular monitoring of nursery and planting sites for emerging diseases is also critical for protecting forest ecosystems from pathogen introduction and proliferation.

4.3 Economic Implications for Reforestation Programs

The economic impacts of nursery diseases are significant, affecting the cost, efficiency, and success of reforestation programs. Disease outbreaks in nurseries result in direct financial losses due to the destruction of seedlings, increased production costs, and delays in planting schedules. Indirect costs, such as reduced timber yields, habitat restoration delays, and management expenses for controlling diseases introduced into forests, further amplify the economic burden. The production of nursery seedlings is a resource-intensive process involving investments in infrastructure, labor, fertilizers, irrigation, and pest management. Disease outbreaks, such as damping-off or powdery mildew, reduce seedling survival rates, forcing nursery operators to increase production to meet demand. This overproduction leads to higher input costs and may strain available resources, particularly in large-scale reforestation initiatives. Delays caused by nursery diseases also have financial repercussions. Postponing planting schedules can result in missed planting seasons, leading to additional costs for maintaining unused seedlings or re-establishing seedbeds. In government-funded or donor-supported programs, such delays can undermine confidence and reduce funding opportunities for future projects.

Diseased seedlings planted in reforestation sites can create long-term economic challenges. For example, the introduction of pathogens like *Fusarium* or *Phytophthora* into forest soils reduces the productivity of timber plantations and increases the cost of managing disease outbreaks in the field. In restoration projects aimed at protecting watersheds or combating desertification, the failure of seedlings to establish due to nursery diseases compromises ecosystem services, such as water regulation and carbon sequestration, which have economic and social value. Addressing the economic impacts of nursery diseases requires investment in preventive measures and efficient disease management practices. Policies that incentivize the production of disease-free seedlings, provide financial support for adopting advanced technologies, and promote training for nursery staff can reduce the costs associated with disease outbreaks. Additionally, integrating disease risk assessments into reforestation planning ensures that resources are allocated effectively, maximizing the economic and ecological benefits of forest regeneration programs.

5. Climate Change and Nursery Pathogen Dynamics

Climate change is profoundly altering the dynamics of pathogens in nursery systems, reshaping how diseases emerge, spread, and impact seedling production. Rising global temperatures, shifting precipitation patterns, and the increasing frequency of extreme weather events create conditions that favor the proliferation of many pathogens, while simultaneously challenging the resilience of nursery practices. Nurseries are especially vulnerable to these changes, as they serve as concentrated hubs of seedling production, where optimal growing conditions often inadvertently provide ideal environments for disease development. The interaction between climate change and pathogen dynamics in nurseries represents a critical concern for forestry operations, particularly in the context of reforestation and afforestation programs aimed at mitigating climate impacts.

5.1 Influence of Warming Temperatures on Pathogen Proliferation

Rising temperatures are one of the most direct consequences of climate change, significantly influencing the behavior, distribution, and virulence of nursery pathogens. Many fungal, bacterial, and viral pathogens thrive in warmer conditions, experiencing accelerated growth rates, enhanced reproduction, and increased activity. For instance, pathogens like *Phytophthora cinnamomi* and *Fusarium oxysporum*, which cause devastating root rot diseases, are particularly responsive to warmer temperatures, leading to more frequent and severe outbreaks in nurseries. Higher temperatures also affect the life cycles of pathogens by shortening the time required for spore germination and increasing the efficiency of infection processes. For example, fungal spores of *Pythium* species, which are responsible for damping-off in seedlings, germinate more rapidly in warm, moist soils, leading to faster disease progression. Similarly, bacterial pathogens like *Xanthomonas* species exhibit higher virulence in warmer environments, causing more aggressive foliar diseases in nursery stock. In addition to promoting pathogen activity, rising temperatures facilitate the expansion of pathogens into regions where they were previously limited by climatic conditions. For instance, pathogens such as *Botryosphaeria* species, which cause stem cankers, are increasingly observed in temperate regions, where milder winters and extended growing seasons create favorable conditions for their establishment. This geographic expansion poses new challenges for nurseries unprepared to manage unfamiliar pathogens. The indirect effects of warming temperatures on host plants further exacerbate pathogen impacts. Heat stress weakens seedling defenses, making them more susceptible to infections. For instance, higher temperatures can impair the production of defensive compounds like phenolics and alkaloids, which play critical roles in deterring pathogens. Heat stress also reduces seedling vigor, limiting their ability to recover from infections.

Managing the effects of warming temperatures on pathogen proliferation requires a proactive approach. Nursery managers must invest in monitoring systems to track the emergence and spread of temperature-responsive pathogens, ensuring early detection and intervention. Research into heat-tolerant tree species and resistant rootstocks is also essential, providing nursery operators with planting material better adapted to changing climatic conditions. Incorporating shading structures or temperature-regulating technologies into nursery designs can mitigate heat stress on seedlings, reducing their vulnerability to pathogens.

5.2 Changing Precipitation Patterns and Their Effects on Disease Spread

Climate change is also altering precipitation patterns, with significant implications for the spread and severity of nursery diseases. Increased rainfall in some regions creates waterlogged conditions that favor the proliferation of waterborne pathogens, while prolonged droughts in others exacerbate stress-related diseases. These shifting patterns complicate disease management in nurseries, where water availability and quality are critical factors influencing pathogen dynamics. Excessive rainfall and waterlogging promote the activity of soilborne pathogens like *Phytophthora* and *Pythium*, which rely on water to disseminate their motile zoospores. Waterlogged soils reduce oxygen availability, impairing root function and creating ideal conditions for root rot diseases to take hold. For example, *Phytophthora cinnamomi* thrives in saturated soils, rapidly infecting seedlings and causing significant mortality in nursery beds.

Heavy rains also facilitate the spread of foliar pathogens by splashing spores onto adjacent plants. Diseases like anthracnose, caused by *Colletotrichum* species, and bacterial leaf spots, caused by *Pseudomonas syringae*, are particularly prone to rapid proliferation during periods of frequent rainfall. Overhead irrigation, often necessary in nurseries, exacerbates this issue by mimicking the dispersal effects of natural precipitation. Conversely, reduced rainfall and prolonged droughts increase the prevalence of stress-induced diseases in nursery stock. Drought conditions weaken seedlings, making them more susceptible to opportunistic pathogens like *Fusarium* and *Botryosphaeria*. Drought stress also disrupts the soil microbiome, reducing populations of beneficial microbes that suppress pathogens and support seedling health. The variability in precipitation patterns further complicates disease management, as nurseries must adapt to both extremes of excess water and water scarcity. Efficient water management systems are essential for mitigating these challenges. Drip irrigation, for example, delivers water directly to the root zone, reducing the risk of foliar pathogen spread while conserving water during drought periods. Improving soil drainage and aeration can prevent waterlogging, limiting the activity of soilborne pathogens. Additionally, mulching and organic amendments can enhance soil water retention during dry periods, protecting seedlings from drought stress.

5.3 Adaptation Strategies for Climate-Resilient Nursery Practices

Adapting nursery practices to the challenges posed by climate change is essential for ensuring the production of healthy seedlings capable of supporting reforestation and afforestation efforts. Climate-resilient practices must address the direct and indirect impacts of warming temperatures, changing precipitation patterns, and other climatic stressors on pathogen dynamics and seedling health.

1. Climate-Responsive Nursery Design

Designing nurseries to mitigate the impacts of extreme temperatures and precipitation is a key adaptation strategy. Shading structures, such as shade nets or retractable roofs, can regulate temperature and reduce heat stress on seedlings, minimizing their susceptibility to pathogens. Elevated nursery beds and well-draining substrates prevent waterlogging during heavy rains, while water catchment systems can store excess rainfall for use during dry periods.

2. Diversification of Planting Material

Promoting genetic diversity in nursery stock is critical for enhancing resilience to climate change. Mixed-species planting reduces the risk of monoculture-related disease outbreaks, while selecting tree species and rootstocks with proven tolerance to heat, drought, or specific pathogens provides long-term stability. Research into climate-resilient genotypes, supported by molecular tools like marker-assisted selection, is essential for developing resistant planting material.

3. Integrated Disease Monitoring and Management

Advanced monitoring systems combining traditional scouting with molecular diagnostics and remote sensing can provide early warnings of disease outbreaks. For instance, qPCR assays can detect pathogen DNA in soil or irrigation water, enabling nursery managers to take preventive measures before visible symptoms appear. Remote sensing technologies, such as hyperspectral imaging, can identify stress-related changes in seedlings that may indicate the onset of disease. Integrated disease management (IDM) strategies must also incorporate biological controls, such as beneficial microbes and mycorrhizal fungi, to enhance seedling defenses against pathogens. These organisms can outcompete or suppress pathogens, reducing the need for chemical interventions. For example, *Trichoderma* species have shown promise in controlling root rot pathogens, while *Bacillus subtilis* can inhibit foliar diseases.

4. Water Management Innovations

Efficient water use is critical for addressing the dual challenges of water scarcity and excess precipitation. Drip irrigation systems, moisture sensors, and automated watering schedules can optimize water delivery, reducing pathogen proliferation while conserving resources. Treating irrigation water with ultraviolet light or filtration systems minimizes the risk of introducing waterborne pathogens like *Phytophthora*.

5. Education and Policy Support

Building capacity among nursery operators through training programs and knowledge-sharing initiatives is essential for implementing climate-resilient practices. Policymakers must also support nurseries by providing financial incentives for adopting sustainable technologies, investing in research on climate-pathogen interactions, and developing regulatory frameworks that promote best practices in disease management.

6. Advances in Nursery Disease Research and Innovations

The field of nursery disease research is rapidly evolving, with technological advancements and innovative approaches transforming how pathogens are identified, monitored, and managed. These innovations are critical for addressing the growing challenges posed by emerging pathogens, climate change, and increasing demand for disease-free nursery stock to support global reforestation and afforestation efforts. Genomics, artificial intelligence, and the development of disease-resistant varieties are at the forefront of these advancements, enabling nursery operators to enhance disease management while maintaining sustainability.

6.1 Genomics and Molecular Tools for Disease Identification

Genomic and molecular technologies have revolutionized the identification and characterization of nursery pathogens, offering unprecedented precision and speed in diagnosing diseases. Traditional methods of pathogen identification, such as microscopy and culture-based techniques, are often time-consuming and require expertise in pathogen morphology. Molecular tools, on the other hand, provide rapid, accurate, and scalable solutions for detecting pathogens at various stages of their life cycles. One of the most widely used molecular tools is the **polymerase chain reaction (PCR)**, which amplifies specific DNA sequences of pathogens. PCR-based methods allow for the detection of pathogens like *Phytophthora*, *Fusarium*, and *Pythium* directly from soil, water, or plant tissues, even when symptoms are not yet visible. Advances in quantitative PCR (qPCR) enable researchers to not only detect pathogens but also quantify their load, providing insights into disease severity and progression.

Next-generation sequencing (NGS) has further expanded the capabilities of pathogen diagnostics. Metagenomics, a subset of NGS, allows for the simultaneous identification of multiple pathogens within a sample, uncovering complex microbial interactions that may influence disease outcomes. For instance, NGS has been instrumental in studying the microbiomes of nursery soils, revealing the presence of beneficial microbes that suppress pathogens or enhance seedling resilience.

Emerging technologies, such as **CRISPR-based diagnostics**, are introducing even more efficient and field-deployable methods for pathogen detection. These systems use CRISPR-Cas proteins to identify specific DNA or RNA sequences associated with pathogens, providing highly specific results within minutes. The portability of CRISPR diagnostic kits makes them ideal for real-time disease monitoring in nurseries. The integration of molecular tools into nursery operations has significant implications for disease management. Early and accurate pathogen detection enables nursery managers to implement targeted interventions, reducing reliance on broad-spectrum chemical controls. Genomic studies of pathogens provide insights into their virulence factors, life cycles, and environmental preferences, guiding the development of more effective management strategies.

6.2 Use of Artificial Intelligence in Disease Monitoring

Artificial intelligence (AI) is emerging as a transformative tool in nursery disease monitoring, offering the ability to analyze large datasets, recognize patterns, and make predictive decisions with remarkable accuracy. AI-powered systems are particularly valuable for identifying diseases at early stages, tracking pathogen dynamics, and optimizing nursery management practices.

Image recognition technology is one of the most impactful applications of AI in disease monitoring. Machine learning algorithms trained on extensive datasets of diseased and healthy plants can accurately identify visual symptoms of infections, such as leaf spots, wilting, or discoloration. These systems can differentiate between similar symptoms caused by different pathogens or abiotic stressors, providing actionable insights to nursery operators. Drones equipped with high-resolution cameras and AI software are increasingly used to survey large nursery operations, identifying disease hotspots with speed and precision. AI is also being applied to **analyze environmental data** and predict disease outbreaks. Pathogen activity is often influenced by factors such as temperature, humidity, and soil moisture, which can be monitored in real-time using Internet of Things (IoT) sensors. AI models integrate this data with historical disease records to predict the likelihood of outbreaks, enabling nursery managers to take preventive measures. For example, AI-driven systems have been developed to forecast outbreaks of damping-off diseases caused by *Pythium* and *Phytophthora*, helping nurseries optimize irrigation and soil management practices.

Another promising application of AI is in **genomic data analysis**. AI algorithms can process the vast amounts of data generated by next-generation sequencing to identify novel pathogens, track genetic variations in known pathogens, and predict their virulence or resistance to treatments. This capability accelerates research on emerging nursery diseases and informs the development of tailored control strategies. The adoption of AI in nursery disease management also facilitates precision agriculture, reducing the overuse of fungicides and other chemical controls. By targeting interventions based on AI-driven predictions, nurseries can lower costs, minimize environmental impacts, and improve the health of their seedlings. However, the successful implementation of AI systems requires investment in infrastructure, training for nursery personnel, and collaboration between technologists, plant pathologists, and forestry experts.

6.3 Development of Disease-Resistant Nursery Varieties

The development of disease-resistant nursery varieties is a cornerstone of sustainable disease management, reducing the need for chemical interventions and enhancing the resilience of reforestation programs. Advances in genetics, breeding technologies, and biotechnological tools have significantly accelerated the process of developing resistant tree species and rootstocks tailored to specific nursery pathogens.

Conventional breeding programs remain a foundational approach to developing disease-resistant varieties. These programs involve selecting parent trees with desirable resistance traits, crossbreeding them, and testing the resulting progeny for disease tolerance. While effective, traditional breeding is time-intensive, requiring multiple generations to achieve stable resistance. However, modern techniques like **marker-assisted selection (MAS)** have streamlined the process by identifying genetic markers linked to resistance traits. MAS enables breeders to select resistant individuals at early stages, reducing the time and resources required for field trials.

Genetic engineering and gene editing have introduced revolutionary possibilities for developing resistant varieties. Transgenic trees engineered to express antifungal or antimicrobial proteins have demonstrated increased resistance to pathogens like *Phytophthora* and *Fusarium*. For example, genes encoding chitinase enzymes, which break down fungal cell walls, have been successfully introduced into several tree species, enhancing their ability to combat root rot diseases. The advent of **CRISPR-Cas9 technology** has further advanced the development of disease-resistant varieties by enabling precise edits to plant genomes. Unlike traditional genetic engineering, which often involves inserting foreign genes, CRISPR allows for targeted modifications to existing genes, such as enhancing the expression of resistance-related pathways or knocking out susceptibility genes. For instance, CRISPR has been used to modify wheat and rice for resistance to fungal pathogens, and similar applications are being explored for forest tree species.

Another promising innovation is the use of **RNA interference (RNAi)** to enhance disease resistance. RNAi involves silencing specific genes in pathogens or their hosts that are critical for infection. This technology has shown potential in controlling viral diseases and parasitic plants in nursery systems, providing a highly specific and environmentally friendly alternative to chemical treatments. Field validation is a critical step in the development of resistant varieties, ensuring that they perform effectively under diverse environmental conditions and pathogen pressures. Collaboration between research institutions, nurseries, and forest management agencies is essential for scaling up the production and deployment of resistant stock.

Conclusion

The management of nursery diseases and parasitic plants is central to ensuring the successful establishment and regeneration of forest ecosystems. Forest nurseries are critical for producing healthy seedlings that form the foundation for reforestation and afforestation efforts worldwide. However, the high-density planting, repetitive use of soil, and conducive environmental conditions in nurseries create ideal conditions for the proliferation of pathogens and parasitic plants. Diseases like damping-off, root rots, foliar infections, stem cankers, and bacterial blights significantly threaten seedling survival, while parasitic plants such as mistletoes, dodder, and broomrapes compound the challenges by competing for essential resources and introducing secondary stressors. The effective management of these issues is essential for sustaining global forestry efforts, biodiversity conservation, and the provision of critical ecosystem services.

Nursery diseases, including those caused by fungi, bacteria, and viruses, disrupt seedling development and reduce productivity. Damping-off, for instance, destroys germinating seeds and young seedlings, leading to severe establishment losses. Root rots, caused by pathogens like *Phytophthora* and *Fusarium*, compromise the ability of seedlings to absorb water and nutrients, leading to stunted growth and high mortality rates. Foliar diseases such as powdery mildew and leaf spots impair photosynthesis, weakening seedlings and predisposing them to other infections. Stem cankers and bacterial blights further exacerbate these threats, while viral infections silently undermine plant vigor and quality. The combined effects of these diseases often translate into significant setbacks for forest nurseries, with economic losses that ripple through reforestation and conservation programs.

Parasitic plants, including mistletoes, dodder, broomrapes, and witchweed, introduce additional challenges by siphoning water and nutrients from host plants. These parasitic species not only weaken their hosts but also alter forest dynamics by selectively targeting specific trees, which can lead to shifts in species composition and succession patterns. The interactions between parasitic plants and fungal pathogens amplify the damage, creating a complex web of biotic stresses that require integrated management approaches. At the same time, parasitic plants play a dual role in forest ecology, influencing biodiversity and nutrient cycling. This highlights the need for a balanced approach to their management that considers both their ecological roles and their detrimental impacts.

Integrated disease management offers a comprehensive solution for addressing nursery diseases and parasitic plant infestations. Practices such as sanitation, early detection, and biological control form the foundation of effective management strategies. The use of resistant rootstocks and careful species selection enhances the resilience of nursery plants, while integrated pest and disease management (IPDM) combines biological, cultural, and technological tools for holistic protection. Advances in biotechnology, including the use of molecular tools and gene editing technologies, have opened new avenues for developing disease-resistant nursery varieties. These innovations, coupled with policies and regulations to prevent the spread of pathogens and parasitic plants, are critical for maintaining healthy nurseries and ensuring successful forest regeneration. The broader implications of nursery diseases extend beyond individual seedlings to affect forest health, biodiversity, and economic stability. High seedling mortality rates reduce the efficiency of reforestation programs and delay the recovery of degraded landscapes. Pathogen outbreaks also affect the genetic diversity of forests by selectively eliminating susceptible species, potentially reducing the ecological resilience of forest ecosystems. Moreover, the economic costs of managing nursery diseases, including losses in seedling production, increased management expenses, and impacts on long-term timber yields, underscore the need for proactive disease prevention and control. Climate change further complicates the dynamics of nursery pathogens and parasitic plants, as warming temperatures and shifting precipitation patterns create conditions that favor disease proliferation and pathogen spread. These changes demand adaptive strategies, such as climate-resilient nursery practices, improved monitoring systems, and the development of drought- and disease-resistant species. By addressing these challenges proactively, forest nurseries can enhance their capacity to withstand environmental pressures and support sustainable forestry practices.

Chapter - 6

Principles of Tree Disease Control

1. Introduction to Tree Disease Control

Tree disease control is a cornerstone of forest management, encompassing a range of strategies and practices aimed at preserving the health, productivity, and ecological stability of forest ecosystems. Forests, as one of the planet's most vital natural resources, provide essential ecosystem services, including carbon sequestration, water regulation, biodiversity conservation, and timber production. However, the pervasive threat of tree diseases, driven by complex interactions between pathogens, hosts, and environmental conditions, poses significant challenges to these ecosystems. Effective disease control strategies are crucial not only for mitigating the direct impacts of tree diseases but also for maintaining the long-term resilience and sustainability of forests. This requires a holistic understanding of the underlying biological, ecological, and environmental dynamics that influence disease outbreaks and progression. The importance of tree disease control in forest ecosystems cannot be overstated, as diseases have profound impacts on forest health and functionality. Pathogens such as fungi, bacteria, viruses, and nematodes, along with abiotic stressors like pollution and climate variability, contribute to tree morbidity and mortality. These diseases affect individual trees and have cascading effects on forest structure, biodiversity, and ecosystem services. By weakening or eliminating certain species, diseases can alter competitive dynamics, disrupt ecological balance, and reduce the capacity of forests to provide critical services. In managed forests, disease outbreaks translate into significant economic losses, including reduced timber yields, higher management costs, and the loss of genetic and ecological diversity. Thus, effective disease control is essential for both ecological and economic sustainability.

Understanding pathogen-host-environment interactions forms the basis for devising effective disease control strategies. Forest diseases emerge and spread as a result of intricate interactions between pathogens and their host trees, mediated by environmental factors such as temperature, humidity, soil conditions, and human activities. Pathogens employ a range of mechanisms to infect hosts, including enzymatic degradation of tissues, production of toxins, and evasion of host immune responses. Hosts, in turn, have evolved complex defense mechanisms, such as physical barriers, chemical defenses, and immune signaling pathways. The environment plays a dual role, either facilitating or suppressing pathogen activity and host susceptibility. For instance, prolonged wet conditions may promote fungal spore germination, while nutrient deficiencies can weaken tree defenses. Recognizing these interactions is critical for predicting disease dynamics and developing targeted control measures. The historical

DOI: 10.1201/9781042011063-6

evolution of tree disease control practices reflects the progression of forestry science and technology. Early approaches relied heavily on observation and manual interventions, such as removing infected trees or applying rudimentary treatments. As scientific knowledge expanded, particularly with advancements in microbiology and plant pathology, more sophisticated strategies emerged, including chemical control, breeding for disease resistance, and the implementation of quarantine measures. In recent decades, the integration of molecular biology, biotechnology, and remote sensing has transformed disease control, enabling early detection, precise diagnostics, and innovative management techniques. This historical trajectory underscores the importance of adapting control practices to evolving threats and technological advancements.

Preventive measures are a fundamental aspect of tree disease control, aiming to minimize the introduction and establishment of pathogens. Sanitation and hygiene practices, such as removing infected plant material and disinfecting tools, reduce the risk of pathogen spread within forests and nurseries. The use of resistant tree varieties and rootstocks is another critical strategy, leveraging genetic diversity to enhance resilience against specific pathogens. Quarantine and biosecurity measures, including the regulation of plant material movement and monitoring of trade pathways, are vital for preventing the introduction of invasive pathogens. Additionally, climate-adapted silvicultural practices, such as selecting species suited to local environmental conditions and implementing water management strategies, help mitigate the impacts of abiotic stressors and reduce disease vulnerability. Early detection and monitoring are key to effective disease control, as they enable timely interventions to prevent outbreaks. Field surveys and symptom identification remain foundational techniques for detecting diseases in their early stages. However, advances in molecular diagnostics, including polymerase chain reaction (PCR) and next-generation sequencing (NGS), have revolutionized pathogen identification, allowing for rapid and precise detection of even cryptic pathogens. Remote sensing and geographic information systems (GIS) further enhance disease surveillance, providing spatial and temporal data on disease distribution and progression across large forested areas. These tools collectively enable proactive disease management and resource allocation. Cultural practices play a significant role in reducing pathogen prevalence and enhancing tree health. Site preparation and soil health management, including practices such as soil amendment and organic matter addition, improve tree vigor and reduce susceptibility to soilborne pathogens. Mixed-species planting mitigates disease spread by disrupting host-pathogen cycles, while proper timing and spacing of planting reduce pathogen load by improving airflow and minimizing pathogen transmission. Effective nutrient and water management further strengthens tree defenses, ensuring robust growth and resilience against pathogens.

Chemical control methods, including fungicides and bactericides, remain integral to tree disease management, particularly in high-value forests and nurseries. Fungicides, available in various formulations, target a wide range of fungal pathogens and are applied preventively or curatively depending on the disease lifecycle. Bactericides, though less commonly used, are effective against specific bacterial infections. However, the reliance on chemical treatments raises concerns about resistance development in pathogens, environmental pollution, and non-target effects on beneficial organisms. These challenges necessitate judicious use and the integration of chemical methods with other control strategies. Biological control approaches offer environmentally sustainable alternatives to chemical treatments, leveraging natural mechanisms to suppress pathogen activity. Beneficial microorganisms, including certain fungi, bacteria, and viruses, act as antagonists to pathogens, either directly inhibiting their growth or enhancing host resistance. Mycorrhizal fungi, which form symbiotic associations with tree roots, improve nutrient uptake and confer resistance to soilborne pathogens. Natural predators and competitors, such as nematodes and hyperparasitic fungi, further contribute to biological control efforts. These approaches are particularly valuable in maintaining ecosystem health and reducing dependency on chemical inputs.

Integrated Disease Management (IDM) represents a holistic approach to tree disease control, combining preventive, cultural, chemical, and biological strategies. By addressing multiple aspects of disease dynamics, IDM maximizes effectiveness while minimizing environmental impacts. Community participation in IDM programs is crucial, as collective efforts ensure comprehensive implementation and monitoring. Adaptive management, which involves continuously refining strategies based on emerging threats and new knowledge, is particularly important in the context of climate change and global trade, which exacerbate disease risks. Biotechnology has emerged as a powerful tool in tree disease control, offering innovative solutions to longstanding challenges. Genetically modified trees with enhanced resistance to specific pathogens are being developed using advanced breeding techniques and genetic engineering. CRISPR and other gene-editing technologies allow for precise modifications in tree genomes, targeting susceptibility genes or enhancing immune responses. RNA interference (RNAi) provides a novel method for suppressing specific pathogens at the molecular level. These biotechnological advancements hold promise for creating resilient forests capable of withstanding the pressures of emerging diseases and changing environments.

1.1 Importance of Tree Disease Control in Forest Ecosystems

The importance of tree disease control in forest ecosystems cannot be overstated, as the impacts of unmanaged diseases ripple through every layer of forest functionality. Healthy trees are fundamental to forest ecosystems, serving as primary producers that convert sunlight into energy, providing habitat for countless species, and stabilizing soil and water cycles. When diseases compromise tree health, the entire ecosystem suffers. For example, pathogens like *Phytophthora ramorum*, which causes sudden oak death, have decimated oak populations in North America, resulting in habitat loss for species reliant on these trees and altering forest dynamics in ways that may take decades or even centuries to recover. Beyond ecological implications, tree diseases have profound economic consequences. Commercial forestry operations depend on healthy trees to supply timber, pulp, and other forest products. Pathogens such as *Fusarium circinatum*, which causes pitch canker in pines, can devastate plantations, leading to substantial financial losses. In urban settings, diseases like Dutch elm disease (*Ophiostoma ulmi*) require costly tree removal and replacement efforts, disrupting city planning and diminishing urban green spaces. Moreover, forests play an indispensable role in regulating global carbon cycles. Widespread tree mortality due to pathogens contributes to carbon release and diminishes the capacity of forests to act as carbon sinks, exacerbating climate change. The importance of tree disease control is further underscored by the growing threats posed by globalization and climate change. Global trade has facilitated the movement of invasive pathogens across continents, often introducing diseases to regions where native trees lack resistance. For instance, the introduction of *Hymenoscyphus fraxineus* into Europe has led to the widespread decline of ash trees, with devastating ecological and cultural consequences. Simultaneously, climate change is altering pathogen dynamics, creating conditions conducive to the emergence of new diseases or the resurgence of previously controlled ones. Addressing these challenges requires robust disease control measures that integrate prevention, monitoring, and adaptive management.

1.2 Pathogen-Host-Environment Interactions

The dynamics of tree diseases are governed by the intricate interactions between pathogens, host trees, and their surrounding environment. This triadic relationship, often represented as the "disease triangle," underscores the complexity of disease development and the necessity for a comprehensive approach to control. Pathogens, as the causative agents of diseases, vary widely in their biology, life cycles, and modes of infection. Fungal pathogens such as *Armillaria* species, which cause root rot, infect host trees through mycelial networks, spreading underground and killing roots. Bacterial pathogens like *Pseudomonas syringae* exploit wounds or natural openings in tree bark to invade vascular tissues, leading to systemic infections. Viruses and viroids, while less common in forest ecosystems, pose significant threats by hijacking cellular processes to propagate, often causing stunted growth or deformities.

The susceptibility of host trees to pathogens depends on their genetic makeup, physiological condition, and ecological context. Trees that are under stress due to factors such as drought, nutrient deficiencies, or mechanical damage are more vulnerable to infections. For example, heat stress can weaken a tree's defense mechanisms, making it more susceptible to canker-causing pathogens like *Botryosphaeria*. Furthermore, the genetic diversity of forest stands plays a critical role in determining their resilience to disease outbreaks. Monocultures, often established in commercial forestry, are particularly susceptible to pathogen spread due to the lack of genetic variation, which limits natural resistance mechanisms. Environmental factors, including climate, soil conditions, and forest management practices, serve as key drivers in the development and spread of tree diseases. Temperature and humidity directly influence pathogen activity, with many fungi and bacteria thriving in warm, moist conditions. Soil properties, such as pH, texture, and organic matter content, affect the proliferation of soilborne pathogens and the health of tree root systems. Human activities, such as logging, road construction, and fire suppression, can disrupt ecological balance and create opportunities for pathogens to establish and spread. For example, soil compaction from heavy machinery reduces aeration and drainage, promoting the growth of root rot pathogens like *Phytophthora cinnamomi*. Understanding the interplay between pathogens, hosts, and the environment is essential for devising effective control strategies. Disease management approaches must consider not only the biology of pathogens and the vulnerabilities of hosts but also the environmental factors that contribute to disease outbreaks. This holistic perspective enables forest managers to anticipate risks, implement preventive measures, and respond to emerging threats with targeted interventions.

1.3 Historical Perspectives and Evolution of Disease Control Practices

The control of tree diseases has a long history, shaped by advances in science, changes in forestry practices, and growing awareness of ecological sustainability. Early approaches to disease control were often reactive, focusing on the removal of visibly infected trees to contain outbreaks. For instance, during the 20th-century outbreak of chestnut blight (*Cryphonectria parasitica*) in North America, efforts were largely centered on felling infected trees to prevent the spread of the disease. While this strategy reduced immediate pathogen proliferation, it was insufficient to save the American chestnut (*Castanea dentata*), which was nearly eradicated from its native range. The development of chemical controls marked a significant milestone in tree disease management. Fungicides, bactericides, and other chemical treatments provided effective tools for controlling specific pathogens, particularly in nurseries and urban environments. For example, copper-based fungicides have been used to manage foliar diseases like anthracnose, while systemic fungicides like phosphonates have shown efficacy against root rot pathogens. However, the reliance on chemical controls has raised concerns about environmental toxicity, non-target effects, and the emergence of resistant pathogen strains. The late 20th century saw a paradigm shift toward integrated disease management (IDM), emphasizing a combination of cultural, biological, and chemical approaches tailored to specific contexts. Advances in biological control, such as the use of antagonistic fungi like *Trichoderma* and beneficial bacteria like *Bacillus subtilis*, have provided sustainable alternatives to chemical treatments. Simultaneously, the adoption of cultural practices, such as crop rotation, mixed-species planting, and soil amendments, has enhanced the resilience of forest ecosystems to disease.

Recent decades have witnessed the emergence of cutting-edge technologies that have transformed tree disease control. Molecular diagnostics, such as polymerase chain reaction (PCR) and next-generation sequencing (NGS), enable early detection of pathogens at unprecedented levels of accuracy. These tools have been instrumental in identifying the spread of invasive pathogens, such as *Phytophthora ramorum*, across continents. Biotechnological advancements, including the development of genetically modified (GM) trees and CRISPR-based gene editing, hold promise for creating disease-resistant varieties capable of withstanding both biotic and abiotic stresses. The evolution of tree disease control practices reflects a growing recognition of the interconnectedness of ecological, economic, and social factors in forestry. Contemporary approaches prioritize sustainability, integrating scientific innovation with traditional knowledge and community involvement. As forests face increasing pressures from climate change, invasive species, and human activities, the principles of tree disease control must continue to evolve, guided by a commitment to preserving the health and diversity of these vital ecosystems.

2. Preventive Measures

Preventive measures form the foundation of effective tree disease management, aiming to reduce the incidence and impact of diseases before they become widespread. Unlike reactive strategies, which address diseases after their manifestation, preventive approaches emphasize minimizing the conditions and opportunities for pathogen proliferation. This proactive stance involves maintaining hygiene in forests and nurseries, utilizing disease-resistant varieties and rootstocks, enforcing quarantine and biosecurity measures, and adapting silvicultural practices to evolving climatic conditions.

2.1 Sanitation and Hygiene in Forests and Nurseries

Sanitation and hygiene are critical components of disease prevention, particularly in nurseries and managed forests, where dense planting and intensive management create ideal conditions for pathogen growth and spread. Pathogens often thrive in organic debris, infected plant material, and contaminated soil or water. Proper sanitation practices disrupt their life cycles, limiting the sources of inoculum and reducing the risk of infection.

In nurseries, maintaining clean growing environments is paramount. This begins with the sterilization of soil, containers, and tools to eliminate soilborne pathogens like *Phytophthora*, *Pythium*, and *Fusarium*. Soil sterilization can be achieved through steam treatment, solarization, or chemical fumigation. Tools and equipment should be disinfected with agents such as sodium hypochlorite or hydrogen peroxide after each use to prevent cross-contamination between plants. Regular removal of infected or decaying plant material is another essential practice. Diseased seedlings, fallen leaves, and pruned branches should be promptly collected and destroyed, as they often serve as reservoirs for pathogens and parasitic plant seeds. Composting infected material should be avoided unless high temperatures can be consistently maintained to kill pathogens. Burning or deep burial is often the most effective disposal method. In forests, sanitation efforts focus on minimizing pathogen spread through infected trees, stumps, and logging debris. For example, trees affected by root rot pathogens like *Armillaria* should be removed entirely, including their roots, to prevent the spread of mycelial networks. Debris from felled trees should be cleared or treated to reduce the survival of pests and pathogens. Water management is also critical to hygiene in nurseries and forests. Contaminated irrigation water is a common vector for pathogens like *Phytophthora*, which spread via motile zoospores. Using clean water sources, installing filtration systems, and avoiding practices like overhead irrigation can mitigate this risk. Drip irrigation systems, which deliver water directly to the root zone, further reduce the potential for foliar infections.

2.2 Use of Resistant Varieties and Rootstocks

The development and deployment of resistant tree varieties and rootstocks is a cornerstone of preventive disease management. By selecting species or cultivars with inherent resistance to specific pathogens, forest managers and nursery operators can significantly reduce disease incidence and improve the long-term health and productivity of forests. Resistant varieties are particularly effective in combating diseases with predictable and well-understood pathogen-host relationships. For example, resistant rootstocks have been developed for species susceptible to *Phytophthora* root rot, such as oaks, pines, and citrus trees. These rootstocks provide a robust defense against soilborne pathogens while maintaining compatibility with desired scions.

In addition to root rot diseases, resistant varieties are instrumental in managing foliar and vascular diseases. For instance, elm cultivars bred for resistance to Dutch elm disease have been successfully introduced into urban landscapes, helping to restore populations of this iconic tree species. Similarly, ash trees with genetic resistance to ash dieback (*Hymenoscyphus fraxineus*) are being developed and tested in response to the widespread decline of ash populations in Europe. The effectiveness of resistant varieties depends on their genetic diversity and adaptability to local conditions. Monocultures, even of resistant species, remain vulnerable to pathogen evolution and environmental stressors. Therefore, promoting genetic diversity within and among species is essential for enhancing forest resilience. Advances in molecular biology and genomics have accelerated the development of resistant varieties. Marker-assisted selection (MAS) allows researchers to identify and propagate resistance traits more efficiently, while CRISPR and other gene-editing technologies enable precise modifications to enhance resistance. These tools offer promising avenues for addressing emerging diseases and adapting trees to changing climatic conditions.

2.3 Quarantine and Biosecurity to Prevent Pathogen Spread

Quarantine and biosecurity measures are vital for preventing the introduction and spread of pathogens across regions and countries. In an increasingly globalized world, the movement of plant materials, soil, and forestry products has facilitated the spread of invasive pathogens, often with devastating consequences for native ecosystems. Quarantine regulations and biosecurity protocols provide a first line of defense against these threats. At the international level, organizations such as the International Plant Protection Convention (IPPC) establish guidelines for phytosanitary measures. These include inspections, certifications, and treatments for plant materials intended for export or import. For example, the use of heat treatments or fumigation ensures that nursery stock and timber products are free of harmful organisms like *Phytophthora ramorum* or *Bursaphelenchus xylophilus* (pine wilt nematode).

National and regional quarantine measures often target specific pathogens or high-risk areas. For instance, quarantine zones are established around areas affected by sudden oak death to restrict the movement of potentially infected plants and soil. Similarly, biosecurity measures at ports of entry include the inspection and treatment of imported plant material to prevent the introduction of invasive pathogens. Local biosecurity practices play a crucial role in nurseries and forest management. These include ensuring that seeds and seedlings are sourced from certified pathogen-free suppliers, sterilizing vehicles and equipment used in forest operations, and restricting access to infected areas. Public awareness campaigns and stakeholder training are essential for fostering compliance with biosecurity measures and promoting collective responsibility for forest health.

2.4 Climate-Adapted Silvicultural Practices

Adapting silvicultural practices to changing climatic conditions is an essential preventive measure for managing tree diseases. Climate change is altering pathogen dynamics, creating conditions that favor the emergence and spread of diseases. For example, rising temperatures accelerate the life cycles of many pathogens, while changes in precipitation patterns influence the prevalence of root rot and foliar diseases. Silvicultural practices must evolve to address these challenges and enhance the resilience of forests. Mixed-species planting is a key strategy for mitigating disease risks. Diverse forests are less susceptible to disease outbreaks than monocultures, as pathogens often specialize in infecting specific hosts. By including a variety of species with different resistance traits, forest managers can reduce the spread of pathogens and promote ecological stability. For example, interplanting resistant tree species with susceptible ones can act as a buffer, slowing the progression of diseases like sudden oak death or ash dieback.

Site preparation and soil management are critical components of climate-adapted silviculture. Maintaining healthy soil through organic amendments, mulching, and reduced compaction enhances root health and suppresses soilborne pathogens. Practices such as contour planting and terracing can improve drainage, reducing the risk of waterlogging and root rot in areas with heavy rainfall. Adjusting planting schedules to align with favorable climatic conditions can minimize stress on seedlings and reduce their vulnerability to diseases. For example, planting during cooler seasons may help seedlings establish before pathogens become active. Similarly, thinning overcrowded stands improves air circulation and light penetration, reducing humidity levels that favor foliar pathogens like powdery mildew or rust. Climate-adapted silviculture also involves the selection of tree species and provenances suited to future climatic conditions. Predictive models and climate scenarios can guide these decisions, ensuring that planted species thrive under anticipated environmental changes. Integrating these practices with broader forest management plans supports the long-term health and productivity of forests.

3. Early Detection and Monitoring

Early detection and monitoring are critical components of tree disease management, enabling forest managers and nursery operators to identify and address disease outbreaks before they escalate into widespread epidemics. Proactive monitoring reduces the reliance on reactive measures, which are often more expensive and less effective, and minimizes ecological and economic losses. Advances in field surveys, molecular diagnostics, and remote sensing technologies have significantly enhanced the ability to detect pathogens early, providing actionable insights into disease dynamics.

3.1 Field Surveys and Disease Symptom Identification

Field surveys remain one of the most fundamental and widely used methods for detecting and monitoring tree diseases. These surveys involve systematic observations of trees in nurseries, plantations, or natural forests to identify visual symptoms of infections. Despite their reliance on human expertise, field surveys provide invaluable data on disease prevalence, distribution, and progression, forming the foundation for disease management strategies. Disease symptoms observed during field surveys vary widely depending on the pathogen and the affected tree species. Common symptoms include wilting, chlorosis, necrotic lesions, cankers, leaf spots, and abnormal growth patterns. For example, root rot caused by *Phytophthora cinnamomi* often presents as wilting and dieback, while foliar diseases like powdery mildew appear as white fungal growth on leaves. Vascular diseases such as Dutch elm disease (*Ophiostoma ulmi*) are characterized by streaking in the sapwood and crown dieback.

Experienced surveyors play a critical role in identifying diseases during field inspections. They rely on their knowledge of tree physiology, pathogen life cycles, and environmental conditions to differentiate between biotic and abiotic stressors. However, accurate identification can be challenging, particularly when symptoms overlap between diseases or are masked by environmental factors. For example, drought stress and fungal infections may both cause wilting, requiring further investigation to confirm the cause. Standardized protocols enhance the effectiveness of field surveys by ensuring consistency in data collection. These protocols often include the use of survey grids, transects, and sampling guidelines to cover representative areas within a forest or nursery. Surveyors may also collect samples of infected tissue, soil, or water for laboratory analysis to confirm the presence of pathogens. Although field surveys provide valuable insights, they are labor-intensive and time-consuming, particularly for large or remote areas. Integrating field surveys with other monitoring tools, such as molecular diagnostics and remote sensing, can overcome these limitations, providing a more comprehensive understanding of disease dynamics.

3.2 Molecular Diagnostics and Genomic Tools

Molecular diagnostics and genomic tools have revolutionized the detection and identification of tree pathogens, offering unparalleled precision and efficiency. These technologies enable the identification of pathogens at the molecular level, often before visible symptoms develop, providing critical time for intervention. Molecular tools are especially valuable for detecting cryptic pathogens, mixed infections, and emerging diseases that may be difficult to diagnose using traditional methods. One of the most widely used molecular tools is the **polymerase chain reaction (PCR)**, which amplifies specific DNA sequences of pathogens. PCR-based assays are highly sensitive and can detect pathogens such as *Phytophthora*, *Fusarium*, and *Pseudomonas* directly from soil, water, or plant tissues. Quantitative

PCR (qPCR) enhances this capability by quantifying pathogen load, providing insights into disease severity and progression.

Next-generation sequencing (NGS) represents a significant advancement in molecular diagnostics, allowing for the simultaneous identification of multiple pathogens in a single sample. NGS-based metagenomics analyzes the entire microbial community within a sample, uncovering complex interactions between pathogens, hosts, and environmental factors. This approach has been instrumental in identifying new and emerging pathogens in forest ecosystems. Emerging technologies like **CRISPR-based diagnostics** offer even greater potential for rapid and field-deployable pathogen detection. CRISPR systems use guide RNAs to target specific DNA or RNA sequences, enabling precise identification of pathogens within minutes. These portable systems are particularly useful for on-site diagnostics in nurseries or remote forest areas. The integration of genomic tools into disease monitoring provides several advantages, including early detection, species-specific identification, and insights into pathogen evolution. For example, genomic studies of *Hymenoscyphus fraxineus*, the causative agent of ash dieback, have revealed genetic factors contributing to its virulence, guiding the development of resistant tree varieties. Despite their benefits, molecular diagnostics require specialized equipment, expertise, and resources, which may limit their accessibility in certain regions. Efforts to develop user-friendly diagnostic kits and expand training for forestry professionals are essential for maximizing the utility of these tools.

3.3 Role of Remote Sensing and GIS in Disease Surveillance

Remote sensing and geographic information systems (GIS) have emerged as powerful tools for disease surveillance, offering large-scale and real-time insights into forest health. These technologies provide spatial and temporal data that complement field surveys and molecular diagnostics, enabling the detection and monitoring of diseases across vast and inaccessible areas.

Remote sensing involves the use of satellites, drones, or aircraft to capture imagery of forested landscapes. Advanced sensors, such as hyperspectral and multispectral cameras, detect changes in tree physiology that may indicate disease. For instance, hyperspectral imaging can identify subtle variations in leaf reflectance caused by chlorophyll degradation, an early sign of infection. Thermal imaging detects heat patterns associated with stressed or diseased trees, providing another layer of diagnostic information. Drones equipped with remote sensing technology are increasingly used for targeted surveillance in nurseries and small forest plots. These unmanned aerial vehicles (UAVs) offer high-resolution imagery and flexibility, allowing for frequent monitoring and rapid assessment of disease outbreaks. For example, drones have been successfully employed to track the spread of sudden oak death in California and monitor foliar diseases in commercial plantations.

GIS enhances the utility of remote sensing by integrating spatial data with disease monitoring records, environmental variables, and predictive models. GIS-based platforms enable forest managers to map disease hotspots, track pathogen spread, and identify areas at high risk of infection. By combining historical data with real-time observations, GIS tools provide valuable insights into disease dynamics and inform targeted interventions. Predictive modeling, facilitated by GIS, plays a critical role in proactive disease management. By incorporating climate data, host distribution, and pathogen behavior, these models forecast the likelihood of disease outbreaks under different scenarios. For example, models predicting the spread of *Phytophthora ramorum* have guided quarantine measures and management plans in affected regions. The adoption of remote sensing and GIS for disease surveillance offers numerous benefits, including scalability, efficiency, and integration with other monitoring approaches. However, these technologies require substantial investment in equipment, software, and training. Collaborative efforts between research institutions, government agencies, and forestry organizations are essential to expand their accessibility and implementation.

4. Cultural Practices for Disease Control

Cultural practices form an essential component of sustainable tree disease management, addressing the environmental and biological factors that influence pathogen dynamics. These practices focus on creating unfavorable conditions for pathogen proliferation while enhancing the resilience of host trees. Chemical or biological interventions, cultural practices are preventive in nature, often relying on modifications to site preparation, planting techniques, and resource management. They are integral to integrated disease management (IDM), complementing other strategies to ensure the long-term health and productivity of forests.

4.1 Site Preparation and Soil Health Management

Site preparation and soil health management are foundational to controlling tree diseases, particularly those caused by soilborne pathogens. Healthy soil not only supports vigorous tree growth but also fosters a diverse microbiome that suppresses pathogens through natural biological processes. Conversely, degraded or compacted soils create conditions that favor root rot, damping-off, and other diseases.

Site preparation begins with a thorough assessment of soil conditions, including texture, pH, organic matter content, and drainage capacity. Poorly drained soils are particularly prone to harboring pathogens like *Phytophthora* and *Pythium*, which thrive in waterlogged environments. Improving drainage through techniques such as contouring, terracing, or the installation of drainage systems can significantly reduce the risk of disease. In areas with heavy clay soils, amending the soil with organic matter or sand improves aeration and water infiltration.

Soil health management focuses on enhancing the physical, chemical, and biological properties of the soil. Organic amendments, such as compost, biochar, and green manures, improve soil structure and nutrient availability while promoting the activity of beneficial microbes. These microbes, including mycorrhizal fungi and antagonistic bacteria, play a crucial role in suppressing pathogens and enhancing root health. For example, mycorrhizal associations protect tree roots from root rot pathogens by outcompeting them for space and nutrients. Soil solarization, a technique that uses clear plastic to trap solar heat, is another effective method for reducing pathogen loads in the soil. This approach is particularly useful in nurseries or reforestation sites where soilborne diseases are a persistent challenge. Soil fumigation, while effective, is less desirable due to its environmental impact and potential harm to non-target organisms.

4.2 Mixed-Species Planting for Disease Mitigation

Mixed-species planting is a powerful cultural practice for reducing disease incidence in forests and plantations. Monocultures, commonly used in commercial forestry, are highly susceptible to disease outbreaks because they provide uniform hosts for specialized pathogens. In contrast, mixed-species stands increase biodiversity, creating barriers to pathogen spread and reducing the overall vulnerability of the forest. The benefits of mixed-species planting extend beyond disease control. Diverse forests support a range of ecological processes, including nutrient cycling, pollination, and pest regulation, which contribute to overall forest health. By introducing species with varying resistance traits, mixed planting disrupts the life cycles of pathogens that rely on specific hosts. For example, interplanting resistant tree species with susceptible ones can slow the progression of diseases like sudden oak death or ash dieback. The effectiveness of mixed-species planting depends on careful species selection and spatial arrangement. Compatible species should be chosen based on their ecological requirements, growth habits, and resistance to local pathogens. Proper spacing between species is also critical to minimize competition and ensure adequate light, water, and nutrient availability. Mixed-species planting is particularly beneficial in the context of climate change, which is altering pathogen dynamics and increasing the likelihood of emerging diseases. Diverse forests are more adaptable to these changes, providing a buffer against the unpredictable impacts of shifting environmental conditions.

4.3 Timing and Spacing of Planting to Reduce Pathogen Load

The timing and spacing of planting play a crucial role in managing tree diseases by influencing the environmental conditions that pathogens require to thrive. Proper timing ensures that seedlings establish during periods of favorable weather, reducing stress and increasing their resistance to infections. Meanwhile, optimal spacing minimizes competition and improves airflow, reducing humidity levels that favor pathogen development.

Planting timing should align with the life cycles of pathogens and the growth requirements of tree species. For example, planting during cooler months can reduce the risk of foliar diseases like rusts and powdery mildew, which are more prevalent during warm and humid conditions. Similarly, planting during the early rainy season ensures adequate soil moisture for root establishment while avoiding the peak activity of soilborne pathogens.

Spacing is equally important for disease control. Dense planting creates microclimates with high humidity and limited airflow, ideal conditions for the proliferation of foliar pathogens such as anthracnose and needle blight. Proper spacing between trees improves ventilation, allowing leaves to dry quickly after rainfall and reducing the risk of fungal infections. Wider spacing also reduces root competition, promoting healthy growth and minimizing the spread of root rot pathogens. In nurseries, spacing practices are particularly critical due to the high density of seedlings. Overcrowded seedbeds are highly susceptible to damping-off and other diseases, necessitating regular thinning and transplanting to maintain optimal conditions. Incorporating these practices into forest management plans enhances seedling survival and long-term forest health.

4.4 Nutrient and Water Management for Healthy Tree Growth

Effective nutrient and water management is essential for promoting healthy tree growth and reducing susceptibility to diseases. Nutrient deficiencies weaken trees, impairing their ability to produce defensive compounds and repair damaged tissues. Over-fertilization, on the other hand, can create lush, nitrogen-rich foliage that is highly susceptible to foliar pathogens and insect pests. Similarly, improper water management, whether through overwatering or drought, creates stress conditions that predispose trees to infections.

Nutrient management begins with soil testing to determine nutrient availability and deficiencies. Balanced fertilization, tailored to the specific needs of tree species, ensures optimal growth without creating conditions favorable for pathogens. For example, adequate levels of potassium and calcium strengthen cell walls, making them more resistant to pathogen penetration. Slow-release fertilizers and organic amendments are preferred over synthetic fertilizers to avoid nutrient leaching and promote soil health.

Water management focuses on maintaining consistent soil moisture levels that support tree growth while minimizing the risk of pathogen proliferation. Overwatering, particularly in poorly drained soils, promotes the activity of waterborne pathogens like *Phytophthora* and *Pythium*. Drip irrigation systems, which deliver water directly to the root zone, reduce water wastage and minimize the risk of foliar diseases caused by splashing water. During drought conditions, mulching helps retain soil moisture and regulate temperature, protecting roots from stress. Incorporating moisture sensors and automated irrigation systems enhances the efficiency of water management, ensuring that trees receive adequate hydration without excess. These technologies are particularly valuable in nurseries, where precise water management is critical for seedling survival and disease prevention.

5. Chemical Control Methods

Chemical control methods play a significant role in the management of tree diseases, particularly in nurseries, urban landscapes, and commercial forestry. These methods involve the use of synthetic or natural chemicals to prevent, suppress, or eradicate pathogens, providing rapid and targeted solutions for disease outbreaks. However, the application of chemical controls requires careful consideration of their efficacy, environmental impact, and the potential for resistance development.

5.1 Fungicides

Fungicides are the most widely used chemicals in tree disease management, targeting fungal pathogens responsible for a broad spectrum of diseases, including root rots, leaf spots, cankers, and vascular wilts. These compounds are categorized based on their mode of action, chemical structure, and application methods, offering diverse options for controlling fungal diseases.

Types of Fungicides

Fungicides are broadly classified into two categories: contact and systemic.

» **Contact fungicides** remain on the surface of plant tissues, forming a protective barrier that prevents fungal spores from germinating or penetrating the plant. Common examples include copper-based fungicides, such as copper oxychloride and copper sulfate, which are effective against foliar pathogens like anthracnose and rusts. These fungicides are preventive in nature and must be applied before infection occurs.

» **Systemic fungicides** are absorbed by the plant and translocated to various tissues, providing both preventive and curative action. These fungicides, such as phosphonates and triazoles, are particularly effective against diseases like root rots caused by *Phytophthora* and vascular wilts caused by *Verticillium*. Systemic fungicides often target specific biochemical pathways in fungi, making them highly effective against targeted pathogens.

Fungicides can also be classified based on their chemical groups, such as dithiocarbamates, benzimidazoles, and strobilurins, each with distinct mechanisms of action. The selection of an appropriate fungicide depends on the pathogen, disease severity, and environmental conditions.

Application Strategies

The effectiveness of fungicides depends on proper application techniques, timing, and dosages. Fungicides can be applied as sprays, soil drenches, seed treatments, or trunk injections, depending on the target pathogen and site of infection.

- **Spray applications** are commonly used for foliar diseases, ensuring even coverage of leaves and stems. Proper calibration of equipment and adherence to recommended dosages are critical to avoid under- or over-application.
- **Soil drenches** are effective against root rot pathogens, delivering fungicides directly to the root zone. This method is often used in nurseries to protect seedlings from damping-off and root infections.
- **Trunk injections** are employed for vascular diseases, allowing fungicides to reach the xylem and phloem where pathogens like *Ophiostoma* (Dutch elm disease) and *Fusarium* reside.

Timing is another critical factor in fungicide application. Preventive applications are most effective when applied during periods of high disease risk, such as early in the growing season or before expected weather conditions favorable for pathogen development. Repeated applications may be necessary for long-lasting protection, particularly in areas with persistent pathogen pressure.

5.2 Bactericides and Their Role in Tree Disease Control

Bactericides are chemical agents designed to manage bacterial pathogens that cause diseases such as blights, cankers, and leaf spots in trees. While less commonly used than fungicides, bactericides play a crucial role in preventing and mitigating bacterial infections, particularly in nurseries and urban forestry.

Types of Bactericides

Most bactericides fall into two primary categories:

- **Copper-based bactericides**, such as copper hydroxide and copper sulfate, are widely used for managing bacterial pathogens like *Pseudomonas syringae* (bacterial canker) and *Xanthomonas* (bacterial leaf spot). Copper ions disrupt bacterial cell membranes and metabolic processes, providing broad-spectrum activity.
- **Antibiotic bactericides**, such as streptomycin and oxytetracycline, are used more selectively, often in high-value settings like fruit orchards or nurseries. These bactericides inhibit bacterial protein synthesis, effectively halting pathogen proliferation.

Application Strategies

Bactericides are typically applied as foliar sprays or trunk injections, depending on the disease and site of infection. For example, copper sprays are commonly used to manage bacterial blights in nurseries, while trunk injections are effective for systemic bacterial infections like fire blight in certain tree species. The timing of bactericide application is critical for success. Preventive applications, made before bacterial populations reach high levels, are more effective than curative treatments. Bactericides are often applied during periods of high moisture, as wet conditions promote bacterial multiplication and dispersal. While bactericides are effective in controlling bacterial diseases, their use is limited by concerns about environmental toxicity and the development of antibiotic resistance. Copper-based bactericides, in particular, pose risks to non-target organisms and soil health when used excessively. Integrated management approaches, combining bactericides with cultural and biological controls, are recommended to minimize these risks.

5.3 Challenges of Chemical Resistance and Environmental Impact

The use of chemical controls in tree disease management is not without challenges. The development of pathogen resistance to fungicides and bactericides is a significant concern, reducing the long-term efficacy of these chemicals and necessitating the use of alternative strategies. Additionally, the environmental impact of chemical applications poses risks to non-target organisms, soil and water quality, and overall ecosystem health.

Chemical Resistance

Pathogens can develop resistance to fungicides and bactericides through genetic mutations, rendering these chemicals less effective over time. Resistance is particularly common in systemic fungicides that target specific biochemical pathways, such as strobilurins and benzimidazoles. For example, resistance to benzimidazole fungicides has been documented in *Botrytis* and *Fusarium* species, limiting their use in disease management.

To mitigate resistance development, it is essential to implement resistance management strategies, including:

- **Rotating fungicides** with different modes of action to reduce selective pressure on pathogens.
- **Mixing fungicides** with complementary modes of action to enhance efficacy and delay resistance.
- **Limiting the use of chemical controls** to periods of high disease risk, reducing unnecessary exposure.

Environmental Impact

The environmental impact of chemical controls is another critical concern. Excessive or improper use of fungicides and bactericides can lead to the accumulation of toxic residues in soil and water, affecting beneficial microbes, aquatic organisms, and wildlife. Copper-based chemicals, while effective, are particularly problematic due to their persistence in the environment and toxicity to soil biota. To minimize environmental risks, forest managers should adopt integrated approaches that prioritize cultural and biological controls over chemical interventions. When chemicals are necessary, precision application techniques, such as targeted sprays or trunk injections, can reduce off-target effects. Additionally, the development and use of eco-friendly chemicals, such as biopesticides derived from natural compounds, offer promising alternatives with lower environmental impact.

6. Biological Control Approaches

Biological control approaches represent a sustainable and environmentally friendly alternative to chemical control methods for managing tree diseases. These strategies utilize living organisms, including beneficial microorganisms, mycorrhizal fungi, and natural predators, to suppress pathogens and enhance tree health. Biological controls work through mechanisms such as competition, parasitism, induced resistance, and antagonism, targeting pathogens while minimizing negative impacts on non-target organisms and ecosystems.

6.1 Use of Beneficial Microorganisms: Fungi, Bacteria, and Viruses

Beneficial microorganisms, including fungi, bacteria, and viruses, play a pivotal role in biological control by directly suppressing pathogens or promoting the health and resilience of host trees. These microorganisms are integral to the soil microbiome and can be introduced or enhanced through inoculation to mitigate tree diseases.

Fungi

Certain fungi, such as *Trichoderma* and *Gliocladium* species, are widely used in biological control due to their antagonistic properties. These fungi suppress pathogens through various mechanisms, including:

- **Competition**: Outcompeting pathogens for nutrients and space in the rhizosphere.
- **Parasitism**: Producing enzymes that degrade pathogen cell walls, effectively destroying fungal pathogens like *Fusarium* and *Phytophthora.*
- **Induced resistance**: Stimulating host plants to activate their own defense mechanisms.For example, *Trichoderma harzianum* has demonstrated efficacy against root rot and damping-off pathogens, enhancing seedling survival in nurseries and plantations.

Bacteria

Beneficial bacteria, such as *Bacillus subtilis* and *Pseudomonas fluorescens*, are also effective biological control agents. These bacteria suppress pathogens through:

- **Antibiosis**: Producing antimicrobial compounds that inhibit pathogen growth.
- **Colonization**: Forming protective biofilms on root surfaces, creating a barrier against pathogen invasion.
- **Nutrient cycling**: Enhancing nutrient availability and uptake, promoting overall tree health.
- For instance, *Pseudomonas fluorescens* has been used to control bacterial blight and fungal diseases in nurseries, while *Bacillus subtilis* is effective against leaf spots and foliar blights.

Viruses

Viruses targeting specific fungal or bacterial pathogens are emerging as a promising area of biological control. Mycoviruses, which infect and inhibit fungal pathogens, are being explored for their potential to suppress diseases like chestnut blight caused by *Cryphonectria parasitica.* While research in this area is still in its early stages, the specificity and self-replicating nature of viral agents make them a valuable tool for targeted disease control. The successful application of beneficial microorganisms depends on understanding their ecological interactions, optimal environmental conditions, and appropriate delivery methods. Inoculation techniques, such as seed treatments, soil amendments, and foliar sprays, ensure that these microorganisms establish and persist in the desired environment.

6.2 Mycorrhizal Associations in Disease Resistance

Mycorrhizal fungi form mutualistic associations with tree roots, providing a range of benefits, including enhanced nutrient uptake, improved water relations, and increased resistance to soilborne pathogens. These fungi establish a symbiotic relationship with their host, extending the root system through a network of hyphae that explore the soil for nutrients and water.

Types of Mycorrhizal Associations

» **Arbuscular mycorrhizal (AM) fungi**: Penetrate root cortical cells and form arbuscules, facilitating nutrient exchange. AM fungi are particularly effective in improving phosphorus uptake and protecting against pathogens like *Phytophthora*.

» **Ectomycorrhizal (ECM) fungi**: Form a sheath around roots and extend hyphal networks into the soil, enhancing nitrogen and water uptake. ECM fungi, such as *Pisolithus* and *Rhizopogon* species, are commonly associated with forest trees like pines and oaks.

Mechanisms of Disease Resistance

Mycorrhizal fungi enhance disease resistance through multiple pathways:

» **Physical barrier**: The fungal sheath or biofilm formed around roots acts as a physical barrier against pathogen penetration.

» **Competition**: Mycorrhizal fungi outcompete pathogens for root colonization sites and soil resources.

» **Induced systemic resistance (ISR)**: Mycorrhizal associations trigger host defense mechanisms, increasing resistance to a broad range of pathogens.

» **Antimicrobial production**: Certain mycorrhizal fungi produce bioactive compounds that inhibit pathogen growth.

For example, AM fungi have been shown to reduce root rot caused by *Fusarium* and *Pythium*, while ECM fungi protect against *Armillaria* and other soilborne pathogens. The use of mycorrhizal inoculants in nurseries and reforestation projects enhances seedling establishment and resilience, particularly in degraded or pathogen-prone soils.

6.3 Natural Predators and Pathogen Antagonists

Natural predators and antagonists, including insects, nematodes, and parasitic fungi, contribute to biological control by directly targeting pathogens or their vectors. These organisms play an important role in regulating pathogen populations, particularly in forest ecosystems where chemical controls may be impractical.

Insect Predators

Certain insects act as natural predators of fungal spores or parasitic plant seeds. For example, the dodder moth (*Cuscuta epilobii*) feeds on dodder vines, reducing their spread and impact in nurseries and forests. Similarly, beetles that prey on pathogen vectors, such as bark beetles carrying blue-stain fungi, indirectly limit the spread of diseases like Dutch elm disease.

Nematodes

Entomopathogenic nematodes, which infect and kill insect vectors, are valuable tools for managing diseases transmitted by pests. For instance, nematodes targeting weevils and other soil-dwelling pests reduce the spread of root rot pathogens. These nematodes are environmentally friendly and can be applied as soil treatments in nurseries or plantations.

Parasitic Fungi

Fungi that parasitize pathogens or their reproductive structures are highly effective biological control agents. For example:

- ***Gliocladium* species** target fungal pathogens by producing enzymes that degrade their cell walls.
- ***Ampelomyces quisqualis*** parasitizes powdery mildew fungi, reducing their spore production and viability.
- These fungi are often applied as biocontrol products, either as soil amendments or foliar sprays, to manage specific diseases.

The success of natural predators and antagonists in disease control depends on maintaining ecological balance and minimizing disruptions to their habitats. Practices such as reducing pesticide use, conserving natural vegetation, and introducing compatible species enhance the effectiveness of these biological controls.

7. Integrated Disease Management (IDM)

Integrated Disease Management (IDM) is a comprehensive approach to controlling tree diseases that combines multiple strategies, including preventive, cultural, biological, and chemical methods, to achieve sustainable disease management. IDM emphasizes the use of environmentally friendly and economically viable practices tailored to specific pathogen-host-environment interactions. This approach integrates scientific knowledge, community engagement, and adaptive management to reduce disease impacts while maintaining ecosystem health and productivity.

7.1 Combining Preventive, Cultural, and Chemical Strategies

The cornerstone of IDM lies in its multifaceted approach, which combines preventive, cultural, and chemical strategies to minimize disease risks and manage outbreaks effectively. By addressing multiple aspects of the disease triangle pathogen, host, and environment IDM ensures comprehensive coverage against diverse threats.

Preventive Strategies

Preventive measures are the first line of defense in IDM, aiming to reduce the introduction and establishment of pathogens in nurseries and forests. These strategies include:

- **Sanitation practices**: Removing infected plant material, sterilizing tools, and maintaining clean nursery environments to minimize pathogen reservoirs.
- **Quarantine measures**: Restricting the movement of potentially infected planting materials to prevent the spread of invasive pathogens.
- **Resistant varieties**: Planting disease-resistant tree species and rootstocks to limit susceptibility to common pathogens.

Cultural Practices

Cultural practices enhance tree health and create unfavorable conditions for pathogens. These include:

- **Soil management**: Improving soil health through organic amendments, proper drainage, and maintaining soil structure to reduce root rot risks.
- **Mixed-species planting**: Diversifying species in plantations to reduce the spread of host-specific pathogens and increase ecosystem resilience.
- **Timing and spacing**: Planting during favorable seasons and maintaining optimal spacing to reduce pathogen load and improve airflow.

Chemical Controls

Chemical interventions, including fungicides and bactericides, are employed judiciously within an IDM framework. These controls are typically used as a supplementary measure, targeting specific outbreaks or high-risk areas. The integration of chemical controls must consider:

- **Rotating active ingredients**: Preventing resistance development by alternating fungicides with different modes of action.
- **Precision application**: Using targeted methods, such as trunk injections or soil drenches, to minimize environmental impact.
- **Combining with other methods**: Enhancing efficacy by pairing chemical treatments with cultural and biological practices.

By integrating these strategies, IDM reduces the reliance on any single method, enhancing the overall effectiveness of disease management while minimizing risks to non-target organisms and ecosystems.

7.2 Role of Community Participation in IDM Programs

Community participation is a critical element of IDM, particularly in large-scale forest management and reforestation programs. The success of IDM relies not only on technical interventions but also on the active involvement of local communities, stakeholders, and policymakers in disease prevention and control.

Education and Awareness

Raising awareness about tree diseases and their impacts is essential for fostering community engagement. Educational programs can provide communities with knowledge about disease symptoms, pathogen spread, and management practices. For example, workshops and field demonstrations can teach nursery operators and forest workers about sanitation practices, the importance of resistant varieties, and the benefits of integrated management strategies.

Collaborative Monitoring and Reporting

Communities play a vital role in early detection and monitoring of tree diseases. Local observations and traditional ecological knowledge can complement scientific methods, providing valuable insights into disease outbreaks and environmental changes. Collaborative monitoring programs, supported by mobile apps or community-based reporting systems, enable rapid detection and response to emerging threats.

Participatory Management

Involving communities in decision-making processes ensures that IDM strategies are context-specific and culturally appropriate. Participatory approaches, such as stakeholder consultations and joint management plans, enhance the adoption and sustainability of IDM practices. For instance, community-led initiatives to remove infected trees, implement reforestation projects, or adopt sustainable silvicultural practices contribute significantly to disease management efforts.

Shared Responsibility and Incentives

Establishing a sense of shared responsibility among stakeholders fosters collective action in disease control. Providing incentives, such as subsidies for disease-resistant planting materials or training programs, encourages communities to adopt IDM practices. Additionally, recognizing and integrating traditional forest management practices can strengthen community involvement and ensure the long-term success of IDM programs.

7.3 Adaptive Management in the Context of Emerging Diseases

Emerging diseases pose significant challenges to forest health, driven by factors such as climate change, globalization, and habitat fragmentation. Adaptive management is a core principle of IDM, emphasizing flexibility and responsiveness to address the uncertainties and dynamics of disease emergence.

Monitoring and Risk Assessment

Adaptive management begins with robust monitoring systems that track disease outbreaks, pathogen distribution, and environmental changes. Tools such as remote sensing, molecular diagnostics, and predictive modeling enable forest managers to assess risks and anticipate potential threats. Regular risk assessments help prioritize interventions and allocate resources effectively.

Research and Innovation

Continuous research is essential for understanding emerging diseases and developing innovative solutions. Studies on pathogen biology, host resistance, and environmental drivers provide insights into disease dynamics and inform management strategies. For example, research on the genetic diversity of *Phytophthora* species has guided the development of resistant rootstocks and targeted fungicide treatments.

Scenario Planning and Flexibility

Adaptive management involves scenario planning to prepare for multiple disease outcomes. By considering different climatic, ecological, and socio-economic scenarios, forest managers can develop contingency plans for disease outbreaks. Flexibility in management practices ensures that strategies can be adjusted based on new information or changing conditions.

Integration of Climate Adaptation

Climate change exacerbates the risks of emerging diseases by altering pathogen lifecycles, expanding host ranges, and increasing environmental stress on trees. Adaptive management incorporates climate-resilient practices, such as selecting species suited to future climatic conditions, enhancing genetic diversity, and modifying planting schedules to mitigate climate impacts.

Stakeholder Engagement

Engaging stakeholders, including policymakers, researchers, and communities, is vital for adaptive management. Collaborative efforts facilitate knowledge exchange, resource sharing, and coordinated responses to emerging threats. Public-private partnerships and international cooperation further strengthen the capacity to address transboundary pathogens and large-scale disease challenges.

8. Role of Biotechnology in Disease Control

Biotechnology is revolutionizing the field of tree disease management by providing precise and innovative tools to enhance resistance, suppress pathogens, and improve overall forest health. The integration of biotechnological advances into forest management strategies has become increasingly important, particularly as emerging pathogens, climate change, and global trade amplify disease threats. Modern biotechnological methods such as genetic modification, CRISPR-based gene editing, and RNA interference (RNAi) have expanded the possibilities for developing disease-resistant trees and targeting pathogens with unparalleled specificity. These approaches offer sustainable alternatives to traditional chemical controls, aligning with global efforts to minimize environmental impact and preserve biodiversity.

8.1 Genetically Modified Trees with Enhanced Disease Resistance

Genetically modified (GM) trees represent a major milestone in biotechnological approaches to tree disease control, enabling the introduction of specific traits that enhance resistance to pathogens. Unlike conventional breeding, which relies on the natural genetic variability within a species, genetic modification allows for the precise insertion of genes from other organisms to confer desired traits. This technology is particularly valuable for addressing diseases that threaten forest ecosystems and commercial plantations, where natural resistance is limited or insufficient.

Mechanisms of Genetic Modification

Genetic modification involves the transfer of resistance genes, often from other plants, microbes, or even animals, into the genome of a target tree species. These genes may encode proteins with antimicrobial properties, enzymes that degrade pathogen cell walls, or signaling molecules that activate the host's defense mechanisms. For example:

- **Antimicrobial peptides (AMPs):** Genes encoding AMPs, such as defensins, have been inserted into trees to provide broad-spectrum resistance against fungal and bacterial pathogens.
- **Chitinase genes:** Derived from fungi or bacteria, these genes enable trees to produce enzymes that degrade the chitin in fungal cell walls, reducing infections caused by pathogens like *Fusarium* and *Armillaria*.

- **Pathogenesis-related (PR) proteins:** Genes encoding PR proteins, such as glucanases, enhance the tree's ability to break down pathogen-derived molecules and limit their spread.

Applications in Forest Disease Control

GM trees have been developed to combat some of the most devastating diseases in forestry. For example:

- **American chestnut restoration:** The introduction of a wheat-derived gene for oxalate oxidase into American chestnuts (*Castanea dentata*) has provided resistance to chestnut blight (*Cryphonectria parasitica*), a fungal pathogen that nearly eradicated this species in North America.
- **Poplar plantations:** GM poplars with enhanced resistance to leaf rust and canker pathogens have been created to improve the productivity and sustainability of commercial plantations.
- **Eucalyptus species:** Genes conferring resistance to *Phytophthora* and other root pathogens have been inserted into eucalyptus trees to mitigate losses in tropical and subtropical regions.

8.2 CRISPR and Gene Editing for Pathogen Resistance

CRISPR (Clustered Regularly Interspaced Short Palindromic Repeats) and associated gene-editing technologies have emerged as powerful tools for engineering disease-resistant trees with unprecedented precision. Unlike traditional genetic modification, which often involves the insertion of foreign DNA, CRISPR allows for targeted modifications to the tree's own genome, enhancing or suppressing specific genes to improve resistance.

How CRISPR Works

CRISPR technology relies on the Cas9 enzyme, guided by RNA sequences, to locate and edit specific DNA sequences within the genome. This system can:

- **Knock out susceptibility genes**: Disabling genes that make trees vulnerable to pathogens. For example, knocking out susceptibility genes involved in fungal infection pathways can significantly reduce disease incidence.
- **Enhance defense-related genes**: Increasing the expression of genes associated with immune responses, such as those encoding PR proteins or enzymes involved in secondary metabolite production.
- **Target pathogen genes**: Directly editing pathogen DNA in cases where pathogens integrate into the host genome, such as certain viruses.

Applications in Forestry

CRISPR-based gene editing is being explored for several forest species to address critical disease challenges:

- **Ash dieback:** Editing genes in ash trees (*Fraxinus excelsior*) to enhance resistance to *Hymenoscyphus fraxineus*, a fungal pathogen responsible for widespread ash mortality in Europe.
- **Pine pitch canker:** Modifying genes in pines to increase resistance to *Fusarium circinatum*, a significant threat to conifer plantations worldwide.

» **Dutch elm disease:** Targeting susceptibility pathways in elms to reduce the impact of *Ophiostoma* species.

Advantages of CRISPR

CRISPR offers several advantages over traditional genetic modification:

» **Precision**: Allows for specific edits without introducing foreign DNA, addressing public and regulatory concerns.

» **Speed**: Accelerates the breeding process by bypassing the need for multiple generations of crossing and selection.

» **Flexibility**: Can be used to edit multiple genes simultaneously, enabling complex trait improvements.

8.3 RNA Interference (RNAi) for Targeted Pathogen Suppression

RNA interference (RNAi) is a natural biological process that regulates gene expression by silencing specific RNA molecules. In the context of tree disease control, RNAi is used to suppress the expression of genes essential for pathogen survival, growth, or virulence. This technology offers a highly targeted approach to managing tree diseases without affecting non-target organisms or environmental processes.

Mechanism of RNAi

RNAi involves the introduction of double-stranded RNA (dsRNA) molecules into the host tree or pathogen. These dsRNA molecules are processed into small interfering RNAs (siRNAs) that bind to complementary RNA sequences, leading to their degradation and preventing protein synthesis. RNAi can target:

» **Pathogen genes**: Silencing genes required for infection or reproduction. For example, targeting fungal effector genes that suppress host immune responses.

» **Vector genes**: Suppressing genes in insect vectors that transmit pathogens, such as bark beetles carrying blue-stain fungi.

» **Host susceptibility genes**: Silencing host genes that facilitate pathogen entry or colonization.

Applications in Tree Disease Management

RNAi has shown promise in controlling various tree diseases and pests:

» **Bark beetle-associated pathogens:** Silencing genes in bark beetles to disrupt their ability to carry fungal pathogens responsible for Dutch elm disease and other vascular infections.

» **Fungal pathogens:** Targeting genes in *Phytophthora* species to inhibit spore production and infection.

» **Viral diseases:** Using RNAi to degrade viral RNA in trees affected by viral pathogens.

Advantages of RNAi

» **Specificity**: Targets only the intended pathogen or vector without affecting beneficial organisms.

» **Environmental safety**: Degrades naturally and does not persist in the environment, reducing ecological risks.

» **Versatility**: Can be applied as topical sprays, soil treatments, or genetic modifications to trees.

Conclusion

Tree disease control is an indispensable aspect of forest management, addressing the complex interplay of pathogens, hosts, and environmental factors that threaten forest health and productivity. Forests, as critical ecosystems, provide essential services such as carbon sequestration, biodiversity conservation, and water regulation. Tree diseases, driven by both biotic pathogens and abiotic stressors, significantly disrupt these functions, leading to ecological imbalance, economic losses, and reduced resilience to environmental changes. Effective disease control strategies are therefore vital for maintaining the health and sustainability of forests, particularly in the face of mounting challenges posed by climate change, invasive species, and anthropogenic pressures. The importance of tree disease control lies in its ability to mitigate the direct and cascading impacts of forest diseases. Pathogens such as fungi, bacteria, viruses, and nematodes, along with stressors like pollution and climate variability, weaken trees, reduce biodiversity, and disrupt forest dynamics. Preventing disease outbreaks and minimizing their spread requires a multifaceted approach that integrates preventive measures, early detection, cultural practices, chemical and biological control, and cutting-edge biotechnology. These strategies work together to protect forest ecosystems, enhance their resilience, and ensure their continued provision of ecological and economic benefits.

Preventive measures are fundamental to disease control, emphasizing the importance of sanitation, quarantine, and the use of resistant tree varieties. These practices reduce the likelihood of pathogen introduction and establishment, safeguarding both natural and managed forests. For instance, quarantine measures and biosecurity protocols prevent the spread of invasive pathogens, while selecting resistant rootstocks enhances the inherent defense mechanisms of tree populations. Additionally, climate-adapted silvicultural practices help mitigate the impacts of environmental stressors, creating conditions that reduce tree vulnerability to disease.

Early detection and monitoring are critical components of effective disease control, enabling timely interventions to contain outbreaks. Advances in molecular diagnostics, such as PCR and next-generation sequencing, have revolutionized pathogen detection, offering rapid and precise identification of even cryptic pathogens. Complementing these tools are remote sensing and GIS technologies, which provide large-scale surveillance and spatial analysis of disease dynamics. Together, these methods allow forest managers to anticipate disease trends and allocate resources effectively.

Cultural practices also play a significant role in disease management by promoting healthy tree growth and reducing pathogen pressure. Site preparation, soil health management, mixed-species planting, and optimized planting schedules are effective strategies for creating resilient forest systems. Proper nutrient and water management further enhances tree defenses, ensuring robust growth and reducing susceptibility to pathogens. These practices are particularly valuable for integrating disease control into sustainable forestry operations. Chemical and biological control methods provide additional tools for managing tree diseases. While fungicides and bactericides remain effective for targeted interventions, concerns over resistance development and environmental impacts necessitate their judicious use. Biological control approaches, such as the use of beneficial microorganisms, mycorrhizal fungi, and natural predators, offer sustainable alternatives that complement other management strategies. These methods harness ecological processes to suppress pathogens and enhance forest health. Biotechnology represents the frontier of tree disease control, offering innovative solutions to longstanding challenges. Genetically modified trees with enhanced disease resistance, CRISPR-based gene editing, and RNA interference technologies are transforming how forests can be protected against pathogens. These advancements provide opportunities to develop resilient tree populations capable of withstanding emerging threats while minimizing environmental impacts.

Chapter - 7

Protection Against Injuries from Insects and Pests

1. Insect and Pest Damage in Forests

Insect and pest damage in forests represents a significant challenge for forest management and conservation, with wide-ranging ecological and economic implications. Forests, as dynamic and interconnected ecosystems, are home to a diverse array of insect species, many of which play beneficial roles in pollination, nutrient cycling, and decomposition. However, certain insects and pests become harmful when their populations grow unchecked, leading to direct and indirect damage to trees. These infestations threaten forest health by weakening trees, reducing productivity, altering species composition, and facilitating the spread of pathogens. Effective management of insect and pest damage is therefore critical for maintaining the ecological integrity and economic value of forests, particularly in the face of climate change and human-induced environmental pressures.

The importance of managing insect and pest damage in forests stems from the need to preserve ecosystem services, sustain biodiversity, and mitigate economic losses. Pests can cause widespread defoliation, disrupt water and nutrient transport, and directly kill trees through boring, girdling, or root feeding. Additionally, insects such as bark beetles often act as vectors for diseases, compounding the effects of infestations. Forest ecosystems play an essential role in carbon sequestration, water regulation, and habitat provision, all of which are jeopardized by unchecked pest activity. For industries reliant on forest products, such as timber and paper, pest outbreaks translate into reduced yields, higher management costs, and economic instability. These considerations underscore the need for proactive and integrated pest management strategies that balance ecological preservation with economic demands. The types of injuries caused by insects and pests are varied and depend on the feeding behaviors and life cycles of the species involved. Defoliators, such as caterpillars, beetles, and leaf miners, strip foliage, reducing photosynthetic capacity and weakening trees. Borers, including bark beetles, wood borers, and ambrosia beetles, attack trunks and branches, creating galleries that disrupt water and nutrient transport. Sap-sucking insects like aphids, scale insects, and cicadas drain plant sap, often leading to leaf discoloration, stunted growth, and secondary infections. Root feeders, such as white grubs, nematodes, and termites, attack belowground structures, impairing water and nutrient uptake and destabilizing trees. Seed and cone pests, including moths, weevils, and beetles, target reproductive structures, affecting forest regeneration and genetic diversity. Understanding these mechanisms of injury is essential for devising targeted control measures.

DOI: 10.1201/9781042011063-7

Pest infestations have far-reaching ecological and economic impacts on forest ecosystems. Ecologically, severe pest outbreaks can lead to widespread tree mortality, altering species composition and disrupting food webs. The loss of dominant tree species creates opportunities for invasive species to establish themselves, potentially leading to long-term shifts in forest structure. Additionally, pests that serve as vectors for diseases introduce a new dimension of complexity, as pathogen infections further exacerbate tree decline. Economically, the impacts of pest infestations are equally significant. Reduced timber yields, increased costs for pest control and salvage operations, and disruptions to industries reliant on forest resources represent tangible losses. Furthermore, the indirect effects of pest outbreaks, such as increased vulnerability to wildfires and reduced carbon storage capacity, have global implications for climate regulation and environmental health. The mechanisms of insect damage extend beyond physical injuries to trees. Chewing, boring, and girdling directly weaken structural tissues, while sap feeding drains vital nutrients. Indirectly, pests can disrupt essential physiological processes, such as photosynthesis and water transport, by introducing pathogens or creating wounds that serve as entry points for secondary infections. These disruptions often have cumulative effects, as weakened trees become more susceptible to additional stressors, including drought, pollution, and extreme weather events. Monitoring and early detection are critical for managing pest infestations effectively. Field surveys and visual inspections remain essential tools for identifying early signs of pest activity, such as frass deposits, discoloration, and dieback. Advances in technology, including pheromone traps and light traps, have enhanced the ability to monitor pest populations and predict outbreaks. Remote sensing and GIS technologies provide large-scale surveillance capabilities, allowing forest managers to track pest spread and assess infestation severity. These tools enable timely interventions, reducing the impact of pest outbreaks and optimizing resource allocation.

Preventive measures are central to forest pest management, focusing on reducing the risk of infestations before they occur. Sanitation practices, such as removing infested material and maintaining forest hygiene, minimize breeding grounds for pests. Mixed-species planting diversifies the forest composition, reducing the likelihood of severe outbreaks by interrupting pest-host relationships. Quarantine and biosecurity measures, including the regulation of plant material movement and monitoring trade pathways, are essential for preventing the introduction of invasive pests. These strategies create a proactive framework for minimizing pest risks and preserving forest health. Biological control offers environmentally sustainable solutions for managing forest pests by leveraging natural mechanisms. Beneficial insects, such as predators, parasitoids, and entomopathogens, play a critical role in suppressing pest populations. For example, lady beetles and parasitic wasps target aphids and caterpillars, while entomopathogenic fungi and nematodes provide effective control against borers and root feeders. Conservation of natural enemies within forest ecosystems enhances the resilience of forests, reducing dependency on chemical treatments and promoting ecological balance.

Cultural and silvicultural practices complement biological control by creating unfavorable conditions for pests. Proper timing of planting and harvesting can disrupt pest life cycles, while thinning and pruning reduce pest habitats and improve airflow, limiting the spread of infestations. Soil and nutrient management strengthen tree health, enhancing their ability to resist and recover from pest attacks. These practices integrate pest control into sustainable forestry operations, ensuring long-term forest productivity and health. Chemical control methods, including insecticides and systemic treatments, remain effective tools for managing severe pest outbreaks. Insecticides are available in various formulations, targeting specific pests at different life stages. Systemic insecticides and trunk injections provide targeted protection, delivering active ingredients directly to affected tissues. However, the reliance on chemical treatments raises concerns about resistance development, non-target effects, and environmental contamination. These challenges highlight the importance of integrating chemical methods with other control strategies to achieve sustainable pest management.

Integrated Pest Management (IPM) provides a comprehensive framework for managing forest pests by combining biological, cultural, and chemical methods. IPM emphasizes the importance of monitoring and early detection, enabling informed decision-making and minimizing unnecessary interventions. Community involvement in pest management programs fosters collaboration and ensures effective implementation. Adaptive management, which incorporates emerging knowledge and responds to changing pest dynamics, is critical for addressing new threats, particularly those linked to climate change and globalization. Advances in pest management technologies are revolutionizing how forest pests are monitored and controlled. Genetically modified trees with insect resistance offer long-term solutions to pest challenges, reducing the need for chemical treatments. RNA interference (RNAi) and gene editing technologies provide targeted approaches to suppress specific pest populations, minimizing collateral damage to non-target organisms. Remote monitoring and predictive modeling enhance the ability to anticipate outbreaks, enabling proactive management and reducing the impact of infestations.

1.1 Importance of Managing Insect and Pest Damage

The importance of managing insect and pest damage in forests lies in their ability to disrupt the ecological and economic functions of these vital systems. Forests provide an array of ecosystem services, including carbon sequestration, water regulation, soil stabilization, and habitat provision. When insect populations reach outbreak levels, they can significantly impair these functions, leading to cascading effects on biodiversity, climate regulation, and human livelihoods. For example, large-scale defoliation events caused by pests like gypsy moths (*Lymantria dispar*) can reduce photosynthetic capacity, leading to decreased growth rates, increased tree mortality, and altered forest composition. Pests also represent a direct threat to the economic value of forests. Timber, pulp, and non-timber forest products are critical components of global and local economies. Insect infestations can reduce wood quality by causing structural defects, discoloration, or fungal infections, leading to financial losses for forest owners and industries. In addition to direct damage, pests increase the costs associated with forest management, as efforts to monitor, suppress, and recover from infestations require substantial resources.

The urgency of managing insect and pest damage has intensified with the advent of climate change and globalization. Warmer temperatures and altered precipitation patterns are expanding the ranges and activity periods of many insect species, enabling them to colonize new areas or produce multiple generations within a single season. Global trade has facilitated the accidental introduction of invasive species, such as the emerald ash borer (*Agrilus planipennis*), which has caused extensive damage to ash trees in North America. Effective pest management is thus essential for maintaining forest health, protecting biodiversity, and ensuring the continued provision of ecosystem services and economic benefits.

1.2 Types of Injuries Caused by Insects and Pests

Insects and pests cause various types of injuries to trees, depending on their feeding habits, life cycles, and ecological niches. These injuries can be broadly categorized as defoliation, boring, sap-sucking, and root damage, each with distinct mechanisms and consequences for tree health.

Defoliation

Defoliating insects, such as caterpillars, beetles, and leaf miners, consume or damage foliage, reducing the tree's photosynthetic capacity. Severe or repeated defoliation weakens trees, delaying growth, reducing reproductive output, and increasing susceptibility to secondary stressors like drought or pathogens. Examples of defoliators include the spruce budworm (*Choristoneura fumiferana*), which targets conifers, and the gypsy moth, which affects a wide range of hardwood species.

Boring and Girdling

Boring insects, such as bark beetles and wood borers, tunnel into the bark, cambium, or wood of trees, causing structural damage and disrupting vascular transport systems. This can lead to girdling, where the flow of water and nutrients is completely obstructed, resulting in tree death. Borers often serve as vectors for pathogenic fungi, compounding the damage. Notable examples include the mountain pine beetle (*Dendroctonus ponderosae*), responsible for widespread pine mortality in North America, and the Asian longhorned beetle (*Anoplophora glabripennis*), which infests hardwoods.

Sap-Sucking Injuries

Sap-sucking insects, such as aphids, scale insects, and cicadas, extract nutrients directly from the phloem or xylem, depleting the tree's resources and introducing pathogens through feeding wounds. These pests can cause stunted growth, leaf yellowing, and premature leaf drop, particularly in young or stressed trees. Hemlock woolly adelgid (*Adelges tsugae*) is a prominent example, decimating hemlock populations across eastern North America.

Root Damage

Root-feeding insects and nematodes, such as white grubs, termites, and root weevils, target the below-ground portions of trees, disrupting water and nutrient uptake. Root damage is often less visible than other forms of injury, but its consequences are severe, including reduced stability, diminished growth, and heightened vulnerability to drought. Subterranean termites are particularly destructive in tropical and subtropical regions, undermining both natural and plantation forests. The cumulative effects of these injuries depend on the severity, duration, and frequency of pest activity, as well as the resilience of the affected tree species and ecosystems. Effective management requires a detailed understanding of pest biology and behavior to mitigate these impacts.

1.3 Ecological and Economic Impacts of Pest Infestations

Pest infestations exert profound ecological and economic impacts, altering forest dynamics, reducing biodiversity, and imposing significant costs on forestry operations and communities. These impacts can be immediate and localized or long-term and widespread, depending on the nature of the infestation and the management responses implemented.

Ecological Impacts

Ecologically, pest infestations disrupt the balance of forest ecosystems by altering species composition, successional patterns, and nutrient cycles. Outbreaks of pests like the spruce budworm or the mountain pine beetle can lead to large-scale tree mortality, transforming dense, mature forests into open landscapes dominated by pioneer species. Such shifts affect not only vegetation dynamics but also the fauna that depend on specific tree species for food, shelter, or breeding sites. Pests also influence nutrient cycling and carbon storage in forests. The decomposition of infested and dead trees releases stored carbon, contributing to greenhouse gas emissions. The replacement of mature forests with younger stands further reduces the forest's capacity to sequester carbon, exacerbating the effects of climate change. Additionally, the accumulation of dead wood increases the risk of wildfires, compounding the ecological and economic consequences of pest outbreaks.

Economic Impacts

The economic costs of pest infestations are substantial, encompassing direct losses in timber value, increased management expenditures, and indirect costs related to ecosystem services. Infestations that compromise wood quality or cause widespread tree mortality reduce the availability of marketable timber, affecting industries reliant on forestry products. For example, the emerald ash borer has caused billions of dollars in losses by destroying ash trees used in furniture, flooring, and sports equipment manufacturing. Management costs include the expenses associated with monitoring, suppressing, and recovering from infestations. These efforts often involve labor-intensive activities such as removing and disposing of infested trees, applying chemical treatments, and replanting affected areas. Urban forestry programs face additional costs related to tree removal, replacement, and public safety measures in pest-affected areas. The loss of ecosystem services, such as water filtration, flood control, and recreational opportunities, represents another significant economic burden. Forest-dependent communities may experience reduced income and livelihood opportunities as pest outbreaks disrupt tourism, hunting, or non-timber forest product harvesting. Mitigating these impacts requires a proactive and integrated approach to pest management, combining early detection, prevention, and targeted interventions. By understanding the ecological and economic ramifications of pest infestations, forest managers and policymakers can develop strategies to minimize losses and enhance forest resilience.

2. Types of Forest Insect Pests

Forest ecosystems face a diverse range of insect pests, each with distinct feeding habits and ecological impacts. These pests exploit different parts of trees, including leaves, stems, roots, seeds, and cones, often causing significant damage that affects tree growth, reproduction, and survival. Understanding the biology, behavior, and impacts of these pests is critical for developing effective management strategies.

2.1 Defoliators: Caterpillars, Beetles, and Leaf Miners

Defoliators are among the most conspicuous and damaging forest pests, feeding on the leaves and needles of trees. These pests include caterpillars, beetles, and leaf miners, which consume or damage foliage, reducing photosynthetic capacity and weakening trees. Defoliation can lead to stunted growth, reduced reproductive output, and increased susceptibility to secondary stressors such as drought or pathogens.

Caterpillars

Lepidopteran larvae, or caterpillars, are common defoliators in forest ecosystems. Species such as the gypsy moth (*Lymantria dispar*), spruce budworm (*Choristoneura fumiferana*), and forest tent caterpillar (*Malacosoma disstria*) are notorious for their large-scale defoliation events. Gypsy moth larvae feed on over 300 tree species, with a preference for oaks, causing widespread defoliation that weakens trees and alters forest composition. Similarly, the spruce budworm targets conifers, including spruce and fir, periodically causing extensive defoliation and tree mortality in North America.

Beetles

Defoliating beetles, such as the elm leaf beetle (*Xanthogaleruca luteola*) and pine sawfly (*Neodiprion* spp.), feed on tree leaves and needles during both larval and adult stages. These pests are particularly destructive in monoculture plantations, where their populations can grow unchecked. Repeated defoliation by beetles can lead to significant growth loss and predispose trees to secondary infections.

Leaf Miners

Leaf miners, including species of moths, flies, and beetles, feed within the leaf tissue, creating distinctive tunnels or blotches. While individual infestations are often localized, severe outbreaks can reduce photosynthetic efficiency and weaken young trees. For example, the birch leaf miner (*Fenusa pusilla*) is a major pest of birch trees, creating extensive leaf damage that diminishes their aesthetic and ecological value.

2.2 Borers: Bark Beetles, Wood Borers, and Ambrosia Beetles

Borers are insects that tunnel into the bark, cambium, or wood of trees, causing structural damage and disrupting vascular transport. These pests often act as vectors for pathogenic fungi, compounding their destructive effects. Borers include bark beetles, wood borers, and ambrosia beetles, each with unique ecological roles and management challenges.

Bark Beetles

Bark beetles, such as the mountain pine beetle (*Dendroctonus ponderosae*), southern pine beetle (*Dendroctonus frontalis*), and spruce beetle (*Dendroctonus rufipennis*), attack the cambium layer beneath the bark, cutting off the flow of water and nutrients. These pests often target stressed or weakened trees but can also cause widespread mortality during outbreaks. For instance, the mountain pine beetle has killed millions of hectares of pine forests in North America, driven by climate change and forest management practices.

Wood Borers

Wood borers, including longhorned beetles and metallic wood-boring beetles, feed on the xylem tissue of trees, causing structural weakening and reducing timber quality. The Asian longhorned beetle (*Anoplophora glabripennis*), an invasive species, poses a significant threat to hardwood trees, including maples, birches, and elms, in North America and Europe.

Ambrosia Beetles

Ambrosia beetles are a group of bark beetles that cultivate symbiotic fungi within their galleries. These fungi serve as a food source for the beetles but often infect the surrounding wood, causing staining and decay. Species like the redbay ambrosia beetle (*Xyleborus glabratus*), which vectors the laurel wilt pathogen, have caused extensive damage to redbay and avocado trees in the southeastern United States.

2.3 Sap-Sucking Insects: Aphids, Scale Insects, and Cicadas

Sap-sucking insects feed on the phloem or xylem of trees, extracting nutrients and water while introducing pathogens through feeding wounds. These pests can cause wilting, stunted growth, and premature leaf drop, particularly in young or stressed trees. Common sap-sucking pests include aphids, scale insects, and cicadas.

Aphids

Aphids, such as the green spruce aphid (*Elatobium abietinum*), feed on the sap of needles and leaves, causing chlorosis, necrosis, and defoliation. Aphid infestations are often exacerbated by their ability to reproduce rapidly and form dense colonies. The honeydew excreted by aphids also promotes the growth of sooty mold, further impairing tree health.

Scale Insects

Scale insects, including armored scales (*Diaspididae*) and soft scales (*Coccidae*), attach themselves to tree stems, branches, and leaves, where they extract sap and weaken the host. The hemlock woolly adelgid (*Adelges tsugae*) is a notable scale insect that has decimated hemlock populations in eastern North America, altering entire forest ecosystems.

Cicadas

Periodic cicadas, such as *Magicicada* species, feed on tree roots during their nymphal stages and cause damage to twigs and branches when adults lay eggs. While their activity is often localized and cyclical, high populations can significantly weaken young trees. Management of sap-sucking insects involves promoting tree vigor, applying horticultural oils or insecticides, and introducing natural enemies like lady beetles and lacewings to control pest populations.

2.4 Root Feeders: White Grubs, Nematodes, and Termites

Root feeders target the below-ground portions of trees, disrupting water and nutrient uptake and compromising stability. These pests are particularly damaging in nurseries, young plantations, and degraded forests where roots are more exposed and vulnerable.

White Grubs

White grubs, the larvae of scarab beetles, feed on fine roots, reducing nutrient absorption and causing stunted growth. Severe infestations can lead to wilting and tree death, particularly in sandy soils where grubs are more prevalent.

Nematodes

Plant-parasitic nematodes, such as root-knot nematodes (*Meloidogyne* spp.), invade root tissues, causing galls, lesions, and reduced root function. Nematode infestations are often associated with secondary infections by soilborne fungi or bacteria, compounding tree stress.

Termites

Subterranean termites feed on woody roots and can cause extensive damage to trees, particularly in tropical and subtropical regions. Their activity weakens structural integrity and increases susceptibility to windthrow and other stressors. Root feeder management includes soil treatments with nematicides or biological agents, such as entomopathogenic nematodes and fungi, as well as maintaining soil health through organic amendments and proper irrigation.

2.5 Seed and Cone Pests: Moths, Weevils, and Beetles

Seed and cone pests target the reproductive structures of trees, reducing seed viability and regeneration potential. These pests are particularly problematic in seed orchards and natural forests that rely on seed dispersal for regeneration.

Moths

Moth larvae, such as those of the pine coneworm (*Dioryctria* spp.), feed on the seeds and cones of conifers, destroying reproductive capacity. These pests are often associated with fungal infections, further reducing seed quality.

Weevils

Cone and seed weevils, such as the white pine cone beetle (*Conophthorus coniperda*), bore into cones to feed on seeds. Infestations can decimate cone crops, particularly in managed seed orchards.

Beetles

Beetles, including seed chalcids and nut borers, attack seeds and nuts, causing physical damage and reducing germination rates. Their activity can disrupt natural regeneration cycles and affect wildlife that depend on seeds as a food source.

3. Mechanisms of Insect Damage

Insects and pests inflict damage on forest trees through various mechanisms that directly and indirectly impair tree health, reduce growth, and sometimes lead to mortality. Understanding these mechanisms is crucial for developing effective management strategies to mitigate their impacts. The primary ways insects damage trees include physical damage through feeding activities such as chewing, boring, and girdling; indirect effects by vectoring pathogens that cause diseases; and disrupting vital physiological processes like photosynthesis, water transport, and overall growth. Each of these mechanisms can have profound effects on individual trees and forest ecosystems as a whole, especially when pest populations reach outbreak levels.

3.1 Physical Damage: Chewing, Boring, and Girdling

Physical damage caused by insects is one of the most immediate and visible forms of injury to trees. This damage results from the feeding activities of various insect pests that chew leaves, bore into stems and branches, or girdle trunks and twigs. Such physical disruptions can severely impair a tree's structural integrity, physiological functions, and ability to defend itself against other stressors.

Chewing Damage

Chewing insects, including many species of caterpillars, beetles, and grasshoppers, consume foliage, flowers, buds, and sometimes bark. Defoliation, the removal of leaves and needles, reduces the tree's photosynthetic capacity, thereby limiting its ability to produce energy for growth and maintenance. For instance, caterpillars of the gypsy moth (*Lymantria dispar*) can defoliate vast areas of hardwood forests during outbreak periods, consuming nearly all the leaves on affected trees. While healthy trees may recover from a single defoliation event by producing a second flush of leaves, repeated defoliations can deplete stored energy reserves, leading to branch dieback, reduced growth, increased susceptibility to other pests and diseases, and ultimately, tree mortality. Chewing damage is not limited to foliage. Some insects target reproductive structures, such as flowers and seeds, which can affect a tree's reproductive success and the regeneration of forests. Additionally, certain beetles and weevils chew on bark and cambial tissues, creating wounds that can serve as entry points for pathogens and other pests.

Boring Damage

Boring insects penetrate the bark and tunnel into the cambium, sapwood, or heartwood of trees. This group includes bark beetles, wood borers, and ambrosia beetles. Their feeding and tunneling activities disrupt the flow of water and nutrients within the tree's vascular system, weaken structural support, and can introduce fungal pathogens.

Bark beetles, such as the mountain pine beetle (*Dendroctonus ponderosae*), attack conifers by boring through the bark into the phloem tissue, where they construct galleries for laying eggs. The feeding larvae further expand these galleries, severing the conductive tissues responsible for transporting nutrients. Mass attacks by bark beetles can overwhelm a tree's defenses, leading to rapid decline and death. The mountain pine beetle has caused extensive mortality of lodgepole pine forests in North America, altering ecosystem dynamics and increasing fuel loads for wildfires. Wood borers, including longhorn beetles and metallic wood-boring beetles, infest both hardwoods and softwoods. The larvae tunnel into the sapwood and heartwood, compromising the mechanical strength of the wood, making trees more susceptible to windthrow and breakage. The emerald ash borer (*Agrilus planipennis*), an invasive species from Asia, bores into ash trees, disrupting the transport of water and nutrients and causing widespread mortality of ash species in North America. Ambrosia beetles bore into wood and cultivate symbiotic fungi in their galleries. While they primarily infest stressed or dying trees, some species can attack healthy trees, and the fungal growth can degrade the wood, reducing its commercial value.

Girdling

Girdling occurs when insects chew or bore in a continuous ring around a stem, branch, or trunk, effectively severing the vascular tissues that transport water and nutrients between the roots and the crown. This action can quickly lead to wilting, dieback, and death of the affected part or the entire tree. Some weevils and beetles, such as the twig girdler (*Oncideres cingulata*), girdle small branches or twigs to create suitable sites for egg-laying. The larvae develop in the dead or dying wood, but the girdling process can cause significant damage to young trees or saplings by disrupting growth and causing deformities. Girdling can also occur as a result of the cumulative feeding damage from sap-sucking insects that weaken the vascular tissues. The loss of connectivity in the vascular system impairs the tree's ability to transport essential resources, leading to stress and increased vulnerability to other pests and environmental factors.

Implications of Physical Damage

The physical damage caused by chewing, boring, and girdling has several implications for tree health and forest ecosystems:

» **Reduced Growth and Vigor**: Physical injuries divert the tree's resources towards wound healing and defense, reducing the energy available for growth and reproduction.

» **Increased Susceptibility**: Wounds serve as entry points for pathogens such as fungi and bacteria, leading to secondary infections. Physically damaged trees are also more attractive to other pests.

» **Altered Forest Structure**: Extensive damage can lead to tree mortality, changing species composition, and successional trajectories in forests.

» **Economic Losses**: Damage to timber quality from borers and other insects reduces the commercial value of wood products.

3.2 Indirect Effects: Vectoring of Pathogens

Beyond the direct physical damage they cause, many insect pests have significant indirect effects on tree health by acting as vectors for pathogens. These insects transmit fungi, bacteria, viruses, and nematodes that cause diseases, leading to complex interactions that exacerbate tree decline and mortality. The vectoring of pathogens by insects is a critical component of many forest disease epidemics and presents substantial challenges for forest management.

Mechanisms of Pathogen Transmission

Insect vectors transmit pathogens through various mechanisms:

» **Feeding Activities**: Insects introduce pathogens into trees during feeding when their mouthparts penetrate plant tissues. For example, sap-sucking insects like aphids and leafhoppers can transmit viruses and phytoplasmas as they feed on phloem sap.

» **Oviposition**: Some insects deposit eggs within plant tissues, introducing pathogens carried on their ovipositors. For instance, certain species of beetles and wasps can inoculate trees with fungal spores during egg-laying.

» **Physical Carriage**: Insects can carry pathogens on their bodies, transferring them to trees as they move and interact with plant surfaces.

Notable Insect-Pathogen Complexes

Several significant forest diseases involve insect vectors:

Dutch Elm Disease

Dutch elm disease, caused by the fungal pathogens *Ophiostoma ulmi* and *Ophiostoma novo-ulmi*, has devastated elm populations in Europe and North America. The fungus is primarily transmitted by elm bark beetles, such as the European elm bark beetle (*Scolytus multistriatus*) and the native elm bark beetle (*Hylurgopinus rufipes*). The beetles carry fungal spores on their bodies and introduce them into healthy trees when feeding or breeding. The fungus spreads through the tree's vascular system, causing wilting, yellowing of leaves, and eventual death.

Pine Wilt Disease

Pine wilt disease, caused by the pinewood nematode (*Bursaphelenchus xylophilus*), is transmitted by pine sawyer beetles (*Monochamus* species). The nematodes reside in the beetles' tracheal system and are introduced into pine trees as the beetles feed on young shoots. The nematodes multiply within the tree's resin canals, disrupting water transport and causing rapid wilting and mortality. Pine wilt disease has caused significant losses in pine forests in Asia and Europe.

Laurel Wilt Disease

Laurel wilt disease affects members of the Lauraceae family, including redbay and avocado trees. The causal agent is the fungus *Raffaelea lauricola*, transmitted by the redbay ambrosia beetle (*Xyleborus glabratus*). As the beetle bores into trees, it introduces the fungus, which colonizes the xylem and induces a defensive response that blocks water transport, leading to wilting and death.

Sudden Oak Death

While not primarily insect-vectored, insects can play a role in the spread of *Phytophthora ramorum*, the pathogen responsible for sudden oak death. Insects may carry spores on their bodies or facilitate the entry of the pathogen through wounds caused by feeding or boring.

Implications of Vectoring Pathogens

The vectoring of pathogens by insects has several significant implications:

- **Synergistic Effects**: The combination of insect damage and pathogen infection can have synergistic effects, leading to more rapid and severe tree decline than either agent alone.
- **Epidemic Spread**: Insect vectors can rapidly disseminate pathogens across large geographic areas, facilitating epidemic outbreaks that are difficult to contain.
- **Management Challenges**: Controlling diseases that involve insect vectors requires integrated approaches that address both the insect and the pathogen. This complexity increases the difficulty and cost of management efforts.
- **Biodiversity Loss**: Diseases vectored by insects can lead to widespread mortality of specific tree species, resulting in significant losses of biodiversity and alterations of ecosystem functions.

Strategies to Mitigate Indirect Effects

Effective management of insect-vectored diseases involves:

- **Monitoring and Early Detection**: Regular surveillance to detect early signs of insect activity and disease symptoms allows for prompt action to prevent spread.
- **Control of Vector Populations**: Reducing the populations of insect vectors through biological control agents, pheromone traps, or targeted insecticide applications can limit pathogen transmission.
- **Quarantine Measures**: Restricting the movement of potentially infested wood and plant materials helps prevent the introduction of vectors and pathogens to new areas.
- **Resistant Varieties**: Breeding and planting tree varieties that are resistant to both the insect vector and the pathogen can provide long-term solutions.

3.3 Disruption of Photosynthesis, Water Transport, and Growth

Insect pests disrupt critical physiological processes in trees, including photosynthesis, water transport, and growth. These disruptions can be direct, through feeding activities that damage tissues essential for these functions, or indirect, by inducing stress responses that alter normal physiological pathways. The cumulative effects of these disruptions weaken trees, reduce their ability to cope with environmental stressors, and can lead to mortality.

Impact on Photosynthesis

Photosynthesis is the process by which trees convert light energy into chemical energy, producing carbohydrates needed for growth and maintenance. Insects affect photosynthesis in several ways:

- **Defoliation**: Chewing insects that consume leaves and needles reduce the photosynthetic surface area, directly decreasing the tree's capacity to produce energy. Severe defoliation forces trees to mobilize stored carbohydrates to produce new foliage, depleting energy reserves.
- **Leaf Mining and Skeletonization**: Leaf miners and skeletonizers damage leaf tissues, impairing photosynthetic efficiency even when leaves remain attached. The damaged areas may become necrotic, further reducing functional leaf area.
- **Chlorosis and Necrosis**: Sap-sucking insects can cause chlorosis (yellowing) and necrosis (death) of leaf tissues by extracting sap and injecting toxins or enzymes. This damage disrupts chlorophyll production and reduces photosynthetic activity.
- **Sooty Mold Development**: Honeydew excreted by sap-sucking insects promotes the growth of sooty mold fungi on leaf surfaces, blocking sunlight and interfering with gas exchange necessary for photosynthesis.

Disruption of Water Transport

Water transport from roots to leaves is essential for nutrient distribution, turgor maintenance, and transpiration cooling. Insects disrupt water transport through:

- **Vascular Tissue Damage**: Boring insects that tunnel through xylem vessels create physical blockages and introduce air embolisms that impair water flow. Bark beetles and wood borers can cause extensive damage to the vascular system.

» **Girdling**: As previously discussed, girdling interrupts the continuity of vascular tissues, severing the transport pathways for water and nutrients between roots and shoots.

» **Induced Defensive Responses**: Some insect feeding induces the tree to produce resins, gums, or tyloses (outgrowths into xylem vessels) as defensive measures. While these responses can impede insect progress, they may also obstruct water transport.

» **Transpiration Loss**: Defoliation reduces the leaf area available for transpiration, which can alter the tree's water balance and affect nutrient uptake.

Effects on Growth and Development

Insect-induced disruptions to photosynthesis and water transport have cascading effects on tree growth and development:

» **Reduced Biomass Accumulation**: Limited energy production and impaired nutrient transport slow down the formation of new tissues, leading to reduced diameter and height growth.

» **Delayed Reproduction**: Energy deficits may force trees to allocate resources away from reproduction, resulting in fewer flowers, cones, or seeds. This can impact forest regeneration and the availability of food for wildlife.

» **Altered Carbon Allocation**: Trees may shift carbon allocation towards defensive compounds (e.g., phenolics, terpenoids) at the expense of growth. While this can enhance resistance to pests, it further limits growth potential.

» **Increased Susceptibility to Stressors**: Weakened trees are less able to withstand environmental stressors such as drought, temperature extremes, and competition. This increased vulnerability can lead to a decline spiral, where multiple factors contribute to mortality.

Physiological Stress Responses

In addition to direct damage, insects can induce stress responses that alter hormonal balances and physiological processes:

» **Hormonal Disruption**: Feeding by insects can affect the production of plant hormones like auxins, gibberellins, and jasmonates, which regulate growth and defense responses. Disruptions in hormonal signaling can lead to abnormal growth patterns or reduced defenses.

» **Stress Metabolites**: Trees under insect attack may accumulate stress metabolites like reactive oxygen species (ROS), leading to oxidative stress and cellular damage if not properly managed by the tree's antioxidant systems.

Implications for Forest Ecosystems

The disruption of physiological processes by insect pests has broader implications for forest ecosystems:

» **Forest Productivity**: Reduced growth rates diminish the productivity of forests, impacting timber yields and carbon sequestration capacity.

» **Species Composition**: Differential impacts on tree species can alter species composition and competitive dynamics, potentially leading to shifts in forest structure and function.

» **Wildlife Habitat**: Changes in tree growth and reproduction affect the availability of habitat and food resources for wildlife, influencing biodiversity.

» **Ecosystem Resilience**: Weakened trees contribute to decreased resilience of forest ecosystems to disturbances such as wildfires, storms, and climate change.

4. Monitoring and Early Detection of Pest Infestations

Monitoring and early detection are critical components of forest pest management, enabling timely interventions to prevent the escalation of infestations and minimize damage to trees and ecosystems. Effective monitoring involves the systematic collection of data on pest populations, their dynamics, and the environmental factors influencing their activity. Early detection techniques allow forest managers to identify outbreaks at their nascent stages, significantly reducing the economic and ecological costs associated with large-scale pest damage.

4.1 Field Surveys and Visual Inspections

Field surveys and visual inspections are the most fundamental methods of monitoring pest infestations, relying on direct observation to detect signs of pest activity, damage, and tree health. These methods are critical for establishing baseline data, identifying pest hotspots, and guiding the deployment of more advanced monitoring techniques.

Field Surveys Field surveys involve systematic observation of forested areas to assess the presence and severity of pest activity. These surveys can be conducted using grid patterns, transects, or random sampling to ensure representative coverage of the forest. Surveyors look for visual indicators of pest presence, such as chewed leaves, boreholes, frass (insect excrement), resin flow, galls, or signs of defoliation. For instance, bark beetle infestations may be identified by pitch tubes on tree trunks, while caterpillar activity is often indicated by webbing or silk threads in the canopy.

Visual Inspections Visual inspections focus on identifying individual pest species, their life stages, and associated damage. Inspectors examine leaves, stems, branches, and roots for characteristic symptoms, such as discoloration, lesions, or deformities. For example, scale insects like the hemlock woolly adelgid produce cottony masses on twigs, while sap-sucking aphids may leave behind honeydew deposits and sooty mold. Inspections may also involve collecting samples of insects, damaged tissues, or frass for laboratory analysis to confirm pest identity.

Advantages and Limitations Field surveys and visual inspections are cost-effective and accessible, requiring minimal equipment and training. However, they are labor-intensive and time-consuming, particularly for large or remote forest areas. Additionally, these methods may miss early-stage infestations that lack obvious symptoms, underscoring the importance of integrating them with other monitoring techniques for comprehensive detection.

4.2 Pheromone Traps and Light Traps for Insect Monitoring

Pheromone and light traps are widely used tools for monitoring insect populations, offering an efficient and species-specific approach to detecting pest activity. These traps provide valuable data on pest abundance, distribution, and seasonal dynamics, supporting early warning systems and targeted management efforts.

Pheromone Traps Pheromone traps exploit the chemical communication systems of insects by mimicking the sex or aggregation pheromones produced by pests. These traps are designed to attract specific species, allowing for precise monitoring of target populations. For example:

» **Bark beetles**: Traps baited with aggregation pheromones are used to monitor species like the mountain pine beetle (*Dendroctonus ponderosae*) and southern pine beetle (*Dendroctonus frontalis*).

» **Gypsy moths**: Traps baited with synthetic sex pheromones effectively attract male gypsy moths, providing data on population density and spread.

Pheromone traps are typically deployed in a grid pattern or along forest edges to capture a representative sample of the pest population. The captured insects are counted and identified to assess infestation levels and guide management decisions.

Light Traps

Light traps attract insects that are phototactic (drawn to light), such as moths, beetles, and flies. These traps use ultraviolet or visible light to lure pests, which are then captured in a collection chamber or adhesive surface. Light traps are effective for monitoring a broad range of nocturnal pests, including defoliating caterpillars and seed-eating moths.

Applications and Benefits

Pheromone and light traps are valuable for:

» **Early Detection**: Capturing low-density populations before visible damage occurs.

» **Population Dynamics**: Tracking seasonal activity and migration patterns of pests.

» **Impact Assessment**: Evaluating the effectiveness of control measures based on changes in trap captures.

However, pheromone traps are species-specific and may not provide comprehensive data on multi-pest systems. Light traps, while versatile, may attract non-target insects, requiring careful interpretation of results.

4.3 Use of Remote Sensing and GIS in Pest Detection

Remote sensing and geographic information systems (GIS) represent cutting-edge technologies for monitoring pest infestations across large and inaccessible forested areas. These tools leverage satellite imagery, aerial photography, and spatial data analysis to provide real-time insights into pest activity and forest health.

Remote Sensing

Remote sensing involves the collection of data from sensors mounted on satellites, drones, or aircraft to detect changes in forest vegetation caused by pest infestations. Advanced imaging technologies, such as hyperspectral and multispectral cameras, can identify subtle variations in tree health by analyzing light reflectance across different wavelengths.

» **Hyperspectral Imaging**: Detects specific stress indicators, such as reduced chlorophyll content or water stress, which are early signs of pest damage. For example, foliage affected by defoliators like spruce budworm exhibits distinct spectral signatures that can be detected remotely.

» **Thermal Imaging**: Captures temperature variations in tree canopies, identifying stressed or dying trees that may host borers or bark beetles.

» **Lidar Technology**: Measures forest structure and canopy density, providing insights into defoliation severity and pest-induced changes in forest composition.

Geographic Information Systems (GIS)

GIS integrates spatial data from remote sensing, field surveys, and other sources to create detailed maps of pest distribution and forest health. These maps allow forest managers to:

» **Identify Hotspots**: Pinpoint areas of active infestations for targeted interventions.

» **Analyze Spread**: Track the movement of pest populations over time and predict future outbreaks.

» **Model Risk**: Combine environmental data, such as temperature, humidity, and host availability, to assess the risk of pest outbreaks under different scenarios.

Applications of Remote Sensing and GIS

These technologies have been successfully applied to monitor and manage forest pests:

» **Bark Beetle Infestations**: Satellite imagery has been used to map tree mortality caused by mountain pine beetle outbreaks in North America.

» **Defoliator Activity**: Hyperspectral data has detected defoliation by gypsy moths and spruce budworm, enabling early intervention.

» **Emerging Pests**: GIS models have predicted the spread of invasive pests like the emerald ash borer, informing quarantine and management efforts.

Advantages and Challenges

Remote sensing and GIS provide large-scale, non-invasive, and real-time monitoring capabilities, making them indispensable for modern pest management. However, these technologies require significant investment in equipment, expertise, and data processing infrastructure. Interpretation of remote sensing data can be complex, necessitating collaboration between ecologists, entomologists, and geospatial analysts.

5. Preventive Measures Against Insect and Pest Damage

Preventive measures are the foundation of sustainable pest management in forests, aiming to minimize the risk of infestations and their associated damage by proactively addressing conditions that favor pest proliferation. Unlike reactive strategies, which are employed after pests have already established and caused harm, preventive measures focus on creating an environment that is inherently less susceptible to pest activity. These measures encompass a range of practices, including sanitation to remove infested material, mixed-species planting to deter pests, and stringent quarantine and biosecurity protocols to prevent the introduction and spread of invasive species. Implementing these strategies in an integrated and adaptive framework enhances forest resilience, reduces economic losses, and supports long-term ecological stability.

Sanitation practices play a pivotal role in maintaining forest health by removing or treating infested material that serves as a reservoir for pests. This involves the prompt removal of dead or dying trees, logging debris, and any visibly infested wood or plant parts. Such material often harbors pest populations, including bark beetles, wood borers, and fungal pathogens that rely on weakened or dead tissues for reproduction and spread. For example, bark beetles like the mountain pine beetle (*Dendroctonus ponderosae*) thrive in recently felled or weakened pines, using these trees as breeding sites before spreading to adjacent healthy stands. By removing these reservoirs through practices such as salvage logging or controlled burning, managers can disrupt pest life cycles and limit their ability to establish and proliferate. Additionally, the proper disposal of infested material, whether through chipping, incineration, or deep burial, is critical to ensure that pests do not escape and reinfest surrounding areas. In urban forestry and nurseries, sanitation measures extend to the sterilization of tools, containers, and equipment to prevent the inadvertent spread of pests and pathogens. While effective, these practices require regular monitoring, labor-intensive efforts, and careful coordination, especially in large or remote forest areas.

Mixed-species planting is another effective preventive measure that leverages biodiversity to reduce the risk and impact of pest infestations. Monoculture plantations, characterized by the dominance of a single tree species, are particularly vulnerable to pests due to the uniform availability of host material. In such systems, pests that specialize in attacking a specific tree species can spread rapidly, leading to widespread defoliation, mortality, and economic losses. Mixed-species planting disrupts this dynamic by introducing a variety of tree species with different levels of resistance or attractiveness to pests. This diversification reduces the continuity of host material, making it more difficult for pests to locate and colonize their preferred hosts. For instance, interplanting resistant or non-preferred species among susceptible ones has been shown to slow the progression of pests like the spruce budworm (*Choristoneura fumiferana*) and gypsy moth (*Lymantria dispar*). Beyond pest deterrence, mixed-species forests enhance overall ecosystem resilience by promoting soil health, supporting beneficial organisms, and mitigating the impacts of environmental stressors such as drought or storms. However, the successful implementation of mixed-species planting requires careful planning, including the selection of compatible species that thrive under similar ecological conditions, appropriate spacing to minimize competition, and long-term monitoring to ensure that the desired effects are achieved.

Quarantine and biosecurity measures are essential for preventing the introduction and spread of invasive pests, which pose significant threats to native forests and global biodiversity. The globalization of trade and travel has facilitated the accidental transport of pests across continents, leading to the establishment of non-native species in areas where they lack natural predators or controls. Examples include the emerald ash borer (*Agrilus planipennis*), which was introduced to

North America from Asia and has since caused widespread mortality of ash trees, and the Asian longhorned beetle (*Anoplophora glabripennis*), which threatens hardwoods in both Europe and North America. Quarantine regulations aim to intercept such pests at points of entry, such as ports, airports, and border checkpoints, by inspecting and treating imported wood, plants, and soil. Measures include heat treatments, fumigation, and the use of pest-free certification systems for traded goods. Biosecurity protocols extend these efforts to domestic settings, ensuring that nurseries, plantations, and urban forests implement practices to minimize the risk of pest introduction and establishment. For example, restricting the movement of firewood, cleaning vehicles and equipment before entering pest-free zones, and monitoring high-risk areas for early signs of infestation are critical components of a robust biosecurity strategy. Public awareness campaigns also play an important role in encouraging compliance with quarantine measures and fostering a sense of shared responsibility for forest protection. Preventive measures against insect and pest damage require a combination of proactive strategies tailored to local ecological conditions and pest pressures. By integrating sanitation practices, mixed-species planting, and stringent quarantine protocols, forest managers can create resilient systems that are less prone to pest outbreaks and better equipped to recover from disturbances. These approaches not only protect forest health and productivity but also contribute to broader efforts to conserve biodiversity, support ecosystem services, and mitigate the impacts of climate change.

6. Biological Control of Insect Pests

Biological control offers a sustainable and environmentally friendly approach to managing insect pests in forests by leveraging the natural enemies of these pests to regulate their populations. Unlike chemical methods, which may have non-target effects and contribute to resistance development, biological control emphasizes ecological balance, reducing pest populations through predators, parasitoids, pathogens, entomopathogenic fungi, and nematodes. Additionally, conserving and enhancing the activity of natural enemies within forest ecosystems plays a crucial role in maintaining long-term pest suppression.

6.1 Beneficial Insects: Predators, Parasitoids, and Pathogens

Beneficial insects are among the most important agents of biological control, directly targeting pest populations through predation, parasitism, or pathogen transmission. These insects provide a natural mechanism for regulating pest outbreaks and are particularly effective in diverse forest ecosystems where ecological interactions are complex and self-sustaining.

Predators

Predatory insects actively hunt and consume pest species, often feeding on multiple individuals during their lifetimes. Lady beetles (*Coccinellidae*), for instance, are voracious predators of aphids, scale insects, and other sap-sucking pests. Similarly, lacewing larvae (*Chrysopidae*) prey on eggs, larvae, and nymphs of various pest species, including mites and caterpillars. Ground beetles (*Carabidae*) and ants also play significant roles as generalist predators, feeding on soil-dwelling larvae and pupae of defoliators and root pests. Predators provide rapid pest suppression, especially during outbreaks, and contribute to overall forest health by keeping pest populations in check.

Parasitoids

Parasitoids are insects, primarily wasps and flies, that lay their eggs inside or on pest hosts. The developing larvae feed on the host, ultimately killing it. Examples include parasitic wasps in the families *Braconidae* and *Ichneumonidae*, which target caterpillars, sawflies, and borers. For instance, the parasitoid *Cotesia congregata* attacks the tobacco hornworm (*Manduca sexta*), reducing its impact on forest vegetation. Tachinid flies (*Tachinidae*) are another important group, parasitizing a range of defoliators and sap-feeding insects. The specificity of parasitoids to their hosts makes them highly effective and environmentally safe tools for biological control, minimizing non-target effects.

Pathogens

Insect pathogens, including bacteria, viruses, and fungi, cause diseases in pest populations, leading to significant mortality during outbreaks. *Bacillus thuringiensis* (Bt), a bacterium that produces insecticidal toxins, is widely used to control caterpillars, beetles, and flies. Viral pathogens, such as nucleopolyhedroviruses (NPVs), are species-specific agents used against pests like gypsy moths (*Lymantria dispar*) and spruce budworms (*Choristoneura fumiferana*). These pathogens are environmentally safe, degrade naturally, and are compatible with integrated pest management (IPM) programs. The success of beneficial insects as biological control agents depends on their ability to establish and persist in the target environment. Factors such as habitat quality, climate, and pest-host availability influence their effectiveness. Introducing or augmenting populations of these natural enemies through releases or habitat enhancements can significantly improve pest management outcomes.

6.2 Use of Entomopathogenic Fungi and Nematodes

Entomopathogenic fungi and nematodes are specialized biological control agents that infect and kill insect pests, offering targeted and sustainable pest management solutions. These organisms exploit specific vulnerabilities in pest physiology, often delivering high mortality rates without harming non-target species or the environment.

Entomopathogenic Fungi

Fungi such as *Beauveria bassiana*, *Metarhizium anisopliae*, and *Lecanicillium lecanii* are widely used in biological control due to their ability to infect a broad range of pest species. These fungi infect insects by penetrating their cuticle, proliferating within their body, and ultimately killing them through systemic infection. *Beauveria bassiana*, for example, is effective against defoliators, aphids, and beetles, while *Metarhizium anisopliae* targets soil-dwelling pests like root grubs and weevils. Fungal biopesticides are typically applied as sprays, soil treatments, or seed coatings, depending on the target pest. Their advantages include high specificity, minimal environmental impact, and compatibility with other pest management methods. However, their efficacy can be influenced by environmental conditions such as humidity and temperature, requiring careful planning and application.

Entomopathogenic Nematodes

Entomopathogenic nematodes, such as *Steinernema* and *Heterorhabditis* species, are microscopic worms that parasitize and kill insect pests. These nematodes enter the host through natural openings, release symbiotic bacteria that cause septicemia, and reproduce within the dead insect. They are particularly effective against soil-dwelling pests, including root weevils, white grubs, and cutworms. Nematodes are applied as soil drenches or sprays, making them suitable for managing pests in nurseries, plantations, and forested areas. They are safe for non-target organisms, decompose naturally, and can persist in the soil for extended periods under favorable conditions. Their integration with other biological and cultural control measures enhances their effectiveness in forest pest management programs.

6.3 Conservation of Natural Enemies in Forest Ecosystems

Conserving natural enemies within forest ecosystems is a cornerstone of biological control, focusing on maintaining and enhancing the populations of predators, parasitoids, and pathogens that naturally regulate pest populations. This approach aligns with ecological principles, promoting biodiversity and resilience while reducing reliance on external inputs.

Habitat Management

Creating and maintaining habitats that support natural enemies is crucial for their conservation. Practices such as preserving native vegetation, maintaining forest edges, and planting cover crops provide shelter, alternative prey, and nectar resources for beneficial insects. For example, flowering plants attract pollinators and parasitoids, enhancing their activity in forest ecosystems. Maintaining a mosaic of habitats within and around forests fosters diverse predator and parasitoid populations, increasing their effectiveness in pest suppression.

Minimizing Pesticide Use

Reducing or eliminating the use of broad-spectrum pesticides is essential for conserving natural enemies, as these chemicals often harm beneficial organisms along with target pests. Integrated pest management (IPM) programs emphasize the selective use of pest-specific chemicals, minimizing non-target effects and preserving ecological balance.

Encouraging Biological Interactions

Promoting interactions between natural enemies and pests enhances the overall efficacy of biological control. For instance, introducing or augmenting specific parasitoids during pest outbreaks can complement the activity of existing predators. Similarly, using microbial agents like fungal pathogens in conjunction with insect predators creates synergistic effects, improving pest suppression outcomes.

Climate Adaptation and Resilience

Climate change poses challenges to the conservation of natural enemies by altering their habitats, life cycles, and interactions with pests. Adaptive management strategies, such as selecting climate-resilient species and enhancing habitat connectivity, help mitigate these impacts and support the long-term persistence of beneficial organisms. By integrating habitat management, pesticide reduction, and adaptive practices, forest managers can conserve and enhance natural enemies, creating self-regulating ecosystems that minimize pest outbreaks and maintain forest health. This approach not only reduces the costs and environmental impacts of pest management but also contributes to broader conservation goals, supporting biodiversity and ecosystem services.

7. Cultural and Silvicultural Practices

Cultural and silvicultural practices form the backbone of sustainable forest management, providing proactive strategies to enhance tree health and resilience against pest infestations. These practices, which focus on the manipulation of environmental and structural factors within forest ecosystems, aim to reduce the availability of suitable habitats for pests, limit the spread of infestations, and promote the overall vigor of trees. By addressing both the causes and conditions that allow pests to thrive, cultural and silvicultural interventions minimize the reliance on chemical controls and align with broader ecological and conservation goals.

7.1 Timing of Planting and Harvesting

The timing of planting and harvesting plays a critical role in mitigating pest pressures by reducing the overlap between tree vulnerability and pest activity. Careful scheduling of these operations helps forest managers minimize stress on trees, enhance establishment success, and disrupt pest life cycles, thereby lowering the risk of infestations.

Planting Timing

Planting trees during optimal environmental conditions promotes healthy establishment and reduces susceptibility to pests. Young seedlings are particularly vulnerable to pest attacks, as their limited resources make them less capable of mounting effective defenses. For example, planting during cooler, wetter periods reduces stress on seedlings and enhances root development, making them more resistant to sap-sucking pests like aphids and scale insects. Conversely, planting during peak pest activity, such as during the emergence of defoliators or root-feeding larvae, increases the likelihood of damage. In regions prone to specific pests, aligning planting schedules with pest phenology is crucial. For instance, delaying planting until after the flight period of bark beetles can reduce the risk of infestation in susceptible species. Similarly, avoiding planting in the late growing season, when drought stress and pest activity are higher, helps mitigate pest impacts.

Harvesting Timing

The timing of harvesting operations influences pest dynamics by altering habitat availability and resource distribution. Harvesting during periods of low pest activity can reduce the attraction of pests to logging residues or stumps, which often serve as breeding sites. For example, felling trees during winter months, when bark beetles and wood borers are less active, minimizes the risk of these pests colonizing fresh material. Post-harvest practices, such as prompt removal of felled trees and proper disposal of logging debris, further reduce pest habitat and breeding opportunities. Delays in these operations can exacerbate pest problems, particularly for species like ambrosia beetles and long horned beetles, which are attracted to freshly cut wood.

7.2 Thinning and Pruning to Reduce Pest Habitat

Thinning and pruning are essential silvicultural practices that enhance forest structure and reduce pest habitat by improving air circulation, light penetration, and tree spacing. These interventions create unfavorable conditions for pest proliferation, limit the spread of infestations, and promote the growth of healthy, vigorous trees.

Thinning

Thinning involves selectively removing trees within a stand to reduce competition for resources such as light, water, and nutrients. Dense forests with overcrowded canopies create humid microclimates that favor the development of pests like defoliators and fungal pathogens. By increasing spacing between trees, thinning improves airflow and sunlight penetration, which reduces humidity levels and disrupts pest life cycles. For example, thinning in coniferous forests has been shown to reduce populations of pests like spruce budworm (*Choristoneura fumiferana*) by limiting the availability of suitable host material. Thinning also enhances the resilience of remaining trees by redirecting resources toward their growth and defense mechanisms. Healthier trees are better equipped to produce defensive compounds, such as resins and tannins, that deter pests like bark beetles. However, care must be taken to avoid over-thinning, which can stress trees and make them more susceptible to pests. The timing of thinning is equally important; conducting operations during pest dormancy or low activity periods minimizes disturbance and reduces the risk of colonization.

Pruning

Pruning involves the removal of lower branches and diseased or damaged limbs to improve tree health and reduce pest habitat. Many pests, such as borers and canker-causing fungi, exploit wounds or weakened tissues for entry. By removing these vulnerable parts, pruning helps eliminate potential entry points and breeding sites. For instance, pruning infected branches can contain the spread of rusts and anthracnose diseases, while removing galled or infested limbs disrupts the life cycles of pests like gall wasps. Pruning also facilitates better access for monitoring and treatment applications, making it easier to detect and manage infestations. As with thinning, proper timing is critical. Pruning should be conducted during dormant periods to minimize stress on trees and reduce the risk of attracting pests to fresh wounds. Wound treatments, such as the application of fungicidal or insecticidal sealants, can further protect pruned areas from colonization.

7.3 Soil and Nutrient Management to Enhance Tree Health

Healthy soil is the foundation of resilient forests, as it provides the nutrients and water necessary for tree growth and defense against pests. Soil and nutrient management practices aim to optimize these resources, creating conditions that enhance tree vigor and reduce susceptibility to pest infestations.

Soil Management

Maintaining soil health involves preserving its physical structure, chemical balance, and biological activity. Compacted or degraded soils restrict root growth, reduce water infiltration, and impair nutrient uptake, making trees more vulnerable to root-feeding pests and drought stress. Practices such as minimizing heavy machinery use, avoiding soil disturbance, and incorporating organic amendments improve soil structure and promote root development. Mulching is another effective soil management practice that regulates soil temperature, retains moisture, and suppresses weeds, all of which contribute to healthier root systems. Organic mulches, such as wood chips or compost, also enhance soil microbial activity, fostering beneficial organisms that suppress soilborne pests and pathogens. In nurseries and young plantations, soil sterilization or solarization can be used to eliminate pests like root-knot nematodes and damping-off fungi, providing a clean start for seedlings.

Nutrient Management

Balanced fertilization ensures that trees receive the essential nutrients needed for growth and defense. Nutrient deficiencies, particularly of nitrogen, phosphorus, and potassium, weaken trees and increase their vulnerability to pests. For example, potassium is critical for strengthening cell walls and enhancing resistance to borers and sap-sucking pests. Conversely, excessive nitrogen fertilization can stimulate lush, tender growth that attracts defoliators and sap feeders. Soil testing is a key step in nutrient management, allowing forest managers to identify deficiencies and tailor fertilization programs to the specific needs of the site. Slow-release fertilizers and organic amendments are preferred over synthetic fertilizers to maintain soil health and prevent nutrient leaching. In degraded or nutrient-poor soils, the addition of mycorrhizal fungi or other beneficial microorganisms can improve nutrient uptake and support tree health.

Water Management

Proper water management is integral to soil and nutrient management, as water facilitates nutrient absorption and maintains physiological processes in trees. Overwatering can create waterlogged conditions that favor root rot pathogens and pests, while drought stress weakens trees and makes them more susceptible to borers and defoliators. Techniques such as drip irrigation, rainwater harvesting, and the use of moisture sensors help optimize water delivery and prevent stress. By prioritizing soil health and nutrient availability, forest managers can strengthen tree defenses, enhance growth, and reduce the risk of pest infestations. Combined with thinning, pruning, and strategic planting and harvesting, these practices form a holistic approach to cultural and silvicultural pest management.

8. Chemical Control of Insect Pests

Chemical control is one of the most commonly employed strategies in pest management, providing rapid and effective solutions for mitigating insect infestations in forests, plantations, and nurseries. The use of insecticides, including both contact and systemic formulations, targets specific pest populations and disrupts their life cycles to minimize damage. However, the deployment of chemical controls requires careful consideration of application methods, environmental impacts, and the potential for resistance development among pest populations.

8.1 Insecticides: Types and Application Strategies

Insecticides are chemical agents designed to kill or inhibit the development of insect pests, offering targeted and immediate relief from infestations. These compounds are classified based on their mode of action, chemical composition, and method of application, providing forest managers with a range of options to address specific pest threats.

Types of Insecticides

» **Contact Insecticides**: These formulations kill pests upon direct contact by disrupting their nervous systems or causing physiological damage. Examples include pyrethroids, organophosphates, and carbamates. Pyrethroids, derived from chrysanthemum flowers, act as neurotoxins that paralyze and kill pests such as aphids, caterpillars, and beetles. While highly effective, contact insecticides require precise application to reach target pests and may have non-target effects on beneficial insects.

- **Stomach Poisons**: These insecticides are ingested by pests during feeding, targeting their digestive and metabolic systems. Compounds like spinosad and *Bacillus thuringiensis* (*Bt*) toxins are particularly effective against leaf-feeding insects, such as caterpillars and beetles. *Bt* a naturally occurring bacterium produces proteins that disrupt the gut lining of specific pests, providing a highly targeted approach with minimal environmental impact.
- **Systemic Insecticides**: These compounds are absorbed by plants and translocated through their vascular systems, allowing pests to ingest the toxins when feeding on treated tissues. Systemic insecticides, such as imidacloprid and dinotefuran, are effective against sap-sucking pests like aphids, scale insects, and whiteflies. Their systemic nature ensures long-lasting protection but requires careful monitoring to avoid overuse and environmental contamination.

Application Strategies

Effective application of insecticides depends on understanding the pest's biology, behavior, and habitat, as well as the timing of interventions:

- **Foliar Sprays**: Contact and stomach insecticides are typically applied as sprays to cover leaves, stems, and branches. This method is particularly effective for defoliators and sap-sucking pests but requires precision to avoid non-target exposure and drift.
- **Soil Treatments**: Systemic insecticides can be applied to the soil, where they are absorbed by tree roots and transported to above-ground tissues. This method is commonly used for root-feeding pests and sap-sucking insects, providing prolonged protection.
- **Aerial Spraying**: Large-scale infestations, such as those caused by gypsy moths or spruce budworms, often necessitate aerial application of insecticides. While efficient for covering extensive areas, aerial spraying poses higher risks of drift and non-target effects, requiring strict adherence to regulatory guidelines.
- **Bait Stations**: For pests like termites and ants, insecticides can be delivered through bait stations that attract and kill foraging individuals. This method minimizes environmental exposure and targets pest colonies with precision.

The success of chemical applications relies on accurate pest identification, proper dosage, and adherence to safety protocols. Integrating insecticides with other pest management strategies, such as biological and cultural controls, enhances their efficacy while reducing reliance on chemical inputs.

8.2 Systemic Insecticides and Trunk Injections

Systemic insecticides and trunk injections provide targeted methods for delivering chemical treatments directly into trees, offering effective control of pests that are difficult to reach with conventional sprays. These techniques are particularly valuable for managing pests that reside within the bark, cambium, or vascular tissues, such as borers and sap-sucking insects.

Systemic Insecticides

Systemic insecticides, when applied as soil drenches, trunk injections, or foliar treatments, are absorbed by the plant and distributed through its vascular system. This ensures that all parts of the tree are protected, including those that are inaccessible to external sprays. Imidacloprid, a widely used systemic insecticide, is effective against pests like emerald ash borer (*Agrilus planipennis*), hemlock woolly adelgid (*Adelges tsugae*), and whiteflies.

- **Soil Drenches**: Soil-applied systemic insecticides are taken up by the roots and translocated to other tissues. This method is ideal for treating sap-sucking pests in large trees or areas where spray applications are impractical. However, soil applications must be managed carefully to prevent leaching into water bodies or harm to non-target organisms.
- **Trunk Injections**: Trunk injection involves injecting insecticides directly into the vascular tissues of trees, delivering concentrated doses to specific areas. This method is highly effective for pests like bark beetles, wood borers, and aphids. Trunk injections reduce the risk of off-target exposure and environmental contamination, making them suitable for urban and high-value forestry applications.

Advantages of Systemic Insecticides and Injections

- **Precision**: Systemic treatments target pests without affecting surrounding flora and fauna.
- **Longevity**: Systemic insecticides provide long-lasting protection, often persisting for several months or even years.
- **Reduced Drift**: Trunk injections and soil treatments minimize the risk of pesticide drift, protecting non-target species and nearby habitats.

Limitations and Considerations

- **Timing**: The effectiveness of systemic insecticides depends on applying treatments during the appropriate season, often before pest activity peaks.
- **Tree Health**: Trunk injections can cause minor wounds, which may lead to secondary infections if not managed properly.
- **Environmental Risks**: Overuse or misapplication of systemic insecticides can contaminate groundwater and harm pollinators, necessitating strict adherence to guidelines.

8.3 Challenges of Resistance and Environmental Impact

The widespread use of chemical controls in pest management has led to significant challenges, including the development of resistance among pest populations and the unintended environmental consequences of pesticide applications. Addressing these challenges requires a balanced approach that minimizes reliance on chemicals while prioritizing sustainability and ecological integrity.

Resistance Development

Pests can develop resistance to insecticides through genetic mutations or behavioral adaptations, rendering treatments less effective over time. Resistance is particularly problematic with systemic insecticides and broad-spectrum sprays that target multiple pest species. For instance, resistance to pyrethroids and neonicotinoids has been documented in aphids, whiteflies, and other sap-sucking pests, limiting the efficacy of these compounds.

- **Mechanisms of Resistance**: Pests may develop resistance by detoxifying insecticides through enzyme activity, altering their target sites to reduce susceptibility, or avoiding treated areas altogether.

- **Factors Driving Resistance**: Frequent and repeated use of the same insecticide class, high pest densities, and incomplete coverage during application increase the likelihood of resistance development.

Strategies to Manage Resistance

- **Insecticide Rotation**: Alternating insecticides with different modes of action reduces selective pressure on pests, slowing resistance development.
- **Combination Treatments**: Using insecticides in combination with biological controls or synergists enhances efficacy and reduces the risk of resistance.
- **Targeted Applications**: Precision application techniques minimize unnecessary exposure and focus treatments on pest hotspots.

Environmental Impacts

The environmental consequences of chemical controls include non-target effects, pollution, and disruption of ecosystem services. Broad-spectrum insecticides, for example, may harm beneficial insects like pollinators, predators, and parasitoids, reducing natural pest control and biodiversity. Runoff from soil-applied insecticides can contaminate water bodies, affecting aquatic organisms and compromising water quality.

Mitigation Measures

- **Integrated Pest Management (IPM)**: Combining chemical, biological, and cultural controls minimizes chemical inputs and their environmental impacts.
- **Eco-Friendly Alternatives**: Developing and adopting biopesticides, natural oils, and microbial agents reduces reliance on synthetic chemicals.
- **Regulation and Monitoring**: Adhering to pesticide regulations, conducting regular environmental monitoring, and promoting best practices help mitigate risks.

Balancing the benefits of chemical controls with their potential drawbacks is essential for sustainable pest management. By integrating these methods into a broader framework, forest managers can protect tree health while minimizing resistance and environmental harm. Let me know if further elaboration is needed!

9. Integrated Pest Management (IPM) in Forests

Integrated Pest Management (IPM) is a holistic and sustainable approach to controlling pest populations in forest ecosystems. Unlike conventional methods that rely heavily on chemical treatments, IPM emphasizes the integration of biological, cultural, and chemical strategies to minimize environmental impacts and ensure long-term forest health. This comprehensive approach combines preventive measures, community involvement, and adaptive strategies to address the complex challenges of managing pests in diverse forest landscapes. With emerging threats such as climate change and the increasing prevalence of invasive species, IPM has become an indispensable tool for forest management.

9.1 Combining Biological, Cultural, and Chemical Methods

The cornerstone of IPM lies in its emphasis on combining biological, cultural, and chemical methods to achieve effective and sustainable pest control. Each approach offers unique strengths and limitations, and their integration maximizes efficacy while minimizing ecological disruptions.

Biological Methods

Biological control is central to IPM, leveraging natural enemies such as predators, parasitoids, pathogens, and beneficial microbes to regulate pest populations. For example, introducing predatory insects like lady beetles (*Coccinellidae*) to manage aphids or employing parasitic wasps to target defoliating caterpillars provides targeted pest suppression. Similarly, the use of entomopathogenic fungi such as *Beauveria bassiana* and nematodes like *Steinernema carpocapsae* effectively controls soil-dwelling and leaf-feeding pests. Biological methods are environmentally friendly and compatible with other control measures, but they require careful planning to ensure the compatibility of introduced agents with the native ecosystem. Monitoring is critical to assess their impact and effectiveness, particularly in areas where introduced species may interact with local biodiversity.

Cultural Methods

Cultural practices enhance the resilience of forest ecosystems and reduce the availability of suitable habitats for pests. Techniques such as mixed-species planting, thinning, pruning, and proper nutrient management improve tree vigor and create unfavorable conditions for pest establishment. Timing planting and harvesting operations to avoid peak pest activity further reduces pest pressure. For example, planting pest-resistant tree varieties or interplanting non-host species disrupts pest life cycles and limits the spread of infestations. Cultural methods are preventive, reducing the likelihood of pest outbreaks before they occur. When combined with biological controls, these practices create a robust framework for maintaining forest health.

Chemical Methods

Chemical treatments, including insecticides and fungicides, are used selectively within IPM to manage high pest densities or contain outbreaks. These methods are deployed as a last resort, targeting specific pests while minimizing non-target effects. For example, systemic insecticides like imidacloprid are applied to manage sap-sucking pests such as the hemlock woolly adelgid, while pheromone traps are used to monitor and suppress populations of bark beetles. The judicious use of chemicals complements biological and cultural strategies, ensuring effective pest control while reducing the risks of resistance development and environmental contamination. Combining these methods requires detailed knowledge of pest biology, tree health, and environmental conditions, as well as regular monitoring to adjust interventions as needed.

9.2 Community Involvement in Pest Management Programs

Community involvement is an integral aspect of IPM, particularly in forested landscapes where stakeholders include local communities, private landowners, industries, and government agencies. The active participation of these groups enhances the effectiveness of pest management programs and fosters a sense of shared responsibility for forest health.

Education and Awareness

Raising awareness about the importance of pest management and the principles of IPM is essential for gaining community support. Educational programs, workshops, and outreach initiatives can inform stakeholders about the impacts of pests, the benefits of IPM, and practical measures they can implement. For example, teaching nursery operators and forest workers about sanitation practices, early detection methods, and the importance of biodiversity promotes proactive pest management.

Participatory Monitoring

Engaging communities in monitoring efforts improves the detection of pest outbreaks and ensures timely interventions. Local knowledge and observations complement scientific monitoring, providing valuable insights into pest activity and environmental changes. Mobile applications, citizen science programs, and community-based reporting systems facilitate data collection and dissemination, enabling collaborative decision-making.

Stakeholder Collaboration

Effective IPM programs require collaboration among diverse stakeholders, including forest managers, researchers, policymakers, and community members. Joint planning and management initiatives ensure that pest control measures are tailored to local conditions and priorities. For instance, community-led efforts to remove infested material, restore degraded areas, or adopt sustainable silvicultural practices contribute significantly to pest suppression and forest resilience.

Incentives and Support

Providing financial and technical support encourages community participation in IPM programs. Subsidies for pest-resistant planting materials, access to biological control agents, and training programs help stakeholders adopt IPM practices. Recognizing and incorporating traditional ecological knowledge further strengthens community engagement and ensures culturally appropriate interventions.

9.3 Adapting to Emerging Pest Threats and Climate Change

The dynamic nature of pest infestations, driven by factors such as climate change, globalization, and habitat fragmentation, presents significant challenges for IPM. Adapting to these emerging threats requires flexible and innovative strategies that account for changing pest dynamics, environmental conditions, and socio-economic factors.

Climate Change and Pest Dynamics

Climate change alters the distribution, behavior, and life cycles of many forest pests, increasing the frequency and severity of outbreaks. Warmer temperatures and prolonged growing seasons allow pests to expand their ranges, produce multiple generations, and colonize new areas. For example, the mountain pine beetle (*Dendroctonus ponderosae*) has extended its range into higher elevations and previously inhospitable regions, causing extensive damage to pine forests. Changing precipitation patterns and extreme weather events further exacerbate pest problems by stressing trees and creating favorable conditions for pest proliferation. Drought-stressed trees, for instance, are more susceptible to borers and bark beetles, while excessive rainfall promotes fungal diseases.

Emerging Pests and Invasive Species

Global trade and travel have facilitated the introduction of invasive pests, which often lack natural enemies in their new environments. Species like the emerald ash borer (*Agrilus planipennis*) and Asian longhorned beetle (*Anoplophora glabripennis*) have caused significant ecological and economic losses in North America and Europe, underscoring the need for proactive measures to prevent introductions and manage established populations.

Adaptive Management Strategies

Adaptive management is a key principle of IPM, emphasizing the need for flexible, evidence-based approaches to address emerging challenges. This involves:

» **Enhanced Monitoring**: Utilizing remote sensing, GIS, and predictive modeling to track pest distributions, assess risk, and identify emerging threats.

» **Research and Innovation**: Investing in research to understand pest biology, host resistance, and environmental interactions, as well as developing novel control methods such as RNAi and genetic engineering.

» **Climate-Resilient Practices**: Implementing silvicultural practices that enhance forest resilience, such as planting diverse and climate-adapted species, improving soil health, and managing water resources.

» **International Collaboration**: Strengthening global networks for pest surveillance, information sharing, and coordinated responses to transboundary threats.

By incorporating adaptive strategies, IPM programs can effectively respond to the uncertainties of climate change and emerging pest threats, ensuring the long-term health and sustainability of forest ecosystems.

10. Advances in Pest Management Technologies

The rapidly evolving field of pest management has witnessed significant advancements in technology, enabling forest managers to adopt more precise, efficient, and sustainable methods for controlling insect pests. These advances integrate cutting-edge scientific tools with practical applications to address the increasing challenges posed by climate change, invasive species, and evolving pest dynamics. Genetically modified trees, RNA interference (RNAi), gene editing technologies, remote monitoring, and predictive modeling are revolutionizing how forests are protected against pest threats. By enhancing precision and minimizing ecological impacts, these technologies are paving the way for a new era in integrated pest management (IPM).

10.1 Genetically Modified Trees with Insect Resistance

Genetically modified (GM) trees are among the most promising advancements in pest management, offering a proactive solution to reduce pest-related damage. Through genetic engineering, trees are modified to express traits that enhance their resistance to specific pests, reducing reliance on chemical controls and fostering long-term forest health.

Mechanisms of Genetic Modification

Genetically modified trees are engineered using techniques that incorporate insecticidal genes, pathogen resistance genes, or regulatory elements that enhance the tree's natural defenses. The most widely used genes are derived from *Bacillus thuringiensis* (*Bt*), a bacterium that produces proteins toxic to certain insect pests. These Bt proteins are highly specific, targeting the digestive systems of pests like caterpillars and beetles without affecting non-target organisms or humans. For example, *Bt*-modified poplar trees have shown increased resistance to leaf-feeding insects such as gypsy moths and forest tent caterpillars. Other approaches involve introducing genes that encode anti-feeding proteins, such as protease inhibitors, which disrupt the digestive processes of pests. Additionally, genes for systemic signaling molecules like jasmonic acid can be engineered to enhance the tree's ability to activate its immune responses against multiple threats.

Applications in Forests

GM trees have been developed to combat significant pest challenges, including:

- **American chestnut restoration:** Introducing oxalate oxidase genes to confer resistance to chestnut blight and associated fungal infections.
- **Eucalyptus plantations:** Engineering resistance to *Leptocybe invasa*, a gall-inducing wasp, and leaf-eating pests in tropical forests.
- **Pine species:** Developing Bt pines to resist bark beetles and other conifer-specific pests.

Challenges and Considerations

While GM trees hold immense potential, their deployment is accompanied by challenges, including regulatory hurdles, public perception, and potential ecological risks. Concerns about gene flow to wild relatives, impacts on non-target organisms, and the adaptability of pests to engineered traits must be carefully addressed through rigorous field trials and monitoring programs.

10.2 RNAi and Gene Editing Technologies for Pest Control

RNA interference (RNAi) and gene editing technologies, including CRISPR-Cas systems, represent transformative tools for targeted pest management. These technologies allow for precise manipulation of genetic pathways in pests or host plants, offering unprecedented specificity and effectiveness in controlling pest populations.

RNA Interference (RNAi)

RNAi is a natural cellular process in which double-stranded RNA (dsRNA) molecules silence specific genes by degrading their messenger RNA (mRNA). In pest management, RNAi is used to suppress genes critical for pest survival, reproduction, or feeding behavior. For example:

- Targeting genes involved in chitin synthesis to disrupt the formation of the exoskeleton in beetles.
- Silencing digestive enzymes in caterpillars to inhibit nutrient absorption. RNAi treatments can be delivered as topical sprays, root drenches, or through genetically modified trees that produce dsRNA molecules. The high specificity of RNAi minimizes non-target effects, making it an eco-friendly alternative to traditional pesticides.

Gene Editing (CRISPR-Cas Systems)

CRISPR-Cas technologies enable precise modifications to the DNA of pests or plants, allowing for enhanced resistance or suppression of pest populations. Applications include:

- **Knockout of susceptibility genes:** Disabling host plant genes that facilitate pest colonization or feeding.
- **Gene drives:** Engineering pest populations to carry traits that reduce their fertility or survival, effectively controlling their numbers over generations.
- **Pathogen-targeted edits:** Modifying plant genomes to resist pest-vectored pathogens, such as viruses transmitted by aphids or whiteflies.

Advantages and Future Potential

RNAi and gene editing technologies offer several advantages, including high precision, reduced environmental impact, and compatibility with IPM strategies. However, challenges such as delivery efficiency, regulatory approval, and public acceptance need to be addressed for widespread adoption. Continued research is critical to refine these technologies and ensure their safe implementation in forest pest management.

10.3 Remote Monitoring and Predictive Modeling for Pest Outbreaks

Remote monitoring and predictive modeling are advancing pest management by providing real-time data and forecasting tools to detect and anticipate pest outbreaks. These technologies integrate satellite imagery, drones, environmental sensors, and computational models, enabling forest managers to adopt proactive and data-driven strategies.

Remote Monitoring

Remote monitoring systems use advanced imaging and sensing technologies to detect changes in forest health that indicate pest activity. Key tools include:

- **Satellite Imagery**: High-resolution satellite data can identify defoliation, discoloration, and canopy thinning caused by pests. For example, hyperspectral imaging detects subtle changes in chlorophyll levels, allowing early identification of infestations.
- **Drone Surveillance**: Unmanned aerial vehicles (UAVs) equipped with cameras and sensors provide detailed, site-specific information on pest activity, tree health, and forest conditions. Drones are particularly useful for monitoring remote or inaccessible areas.
- **IoT Sensors**: Internet of Things (IoT) devices, such as pheromone traps with real-time data transmission, provide continuous monitoring of pest populations. These sensors can alert managers to changes in pest density or behavior, facilitating rapid responses.

Predictive Modeling

Predictive models use environmental data, pest life cycles, and historical outbreak patterns to forecast pest activity and assess risks. These models help forest managers prioritize areas for intervention and allocate resources effectively. Key features include:

- **Climate-Based Models**: Incorporating temperature, humidity, and precipitation data to predict pest emergence and activity. For example, models have been developed to forecast bark beetle outbreaks based on warming trends and drought conditions.
- **Spread Simulation**: Mapping the potential spread of invasive species, such as the emerald ash borer, based on host availability and transportation networks.
- **Decision Support Systems**: Integrating monitoring data with predictive algorithms to provide actionable recommendations for pest management.

Applications and Benefits

Remote monitoring and predictive modeling have been successfully applied to manage pests like gypsy moths, spruce budworms, and bark beetles. These technologies enable early detection, reduce response times, and improve the precision of interventions, ultimately minimizing pest impacts and management costs.

Challenges and Future Directions

Despite their potential, these technologies face challenges such as high costs, data complexity, and the need for technical expertise. Expanding access to affordable tools, improving data integration, and enhancing user-friendly interfaces will be critical for broader adoption. Collaboration between researchers, technology developers, and forest managers is essential to optimize these systems for practical use.

Conclusion

Managing insect and pest damage in forests is a critical aspect of forest conservation and management, given the significant ecological and economic impacts of pest infestations. Forest ecosystems provide vital services, including carbon sequestration, biodiversity conservation, and water regulation. However, pest outbreaks can weaken these systems, leading to tree mortality, changes in forest composition, and reduced ecosystem functionality. Effective pest management strategies are therefore indispensable for maintaining forest health, productivity, and resilience, particularly in the context of climate change and human activities that exacerbate pest pressures. The ecological and economic implications of pest damage underscore the need for comprehensive management. Ecologically, pests disrupt forest dynamics by directly damaging trees and indirectly altering food webs and species interactions. Defoliators, borers, sap-sucking insects, root feeders, and seed pests each contribute to tree decline through distinct mechanisms, such as defoliation, boring, sap extraction, and root damage. These injuries compromise photosynthesis, water transport, and nutrient uptake, reducing tree growth and increasing susceptibility to other stressors. In many cases, pests also act as vectors for pathogens, introducing secondary infections that further amplify damage. Economically, the impacts of pest infestations are seen in reduced timber yields, increased management costs, and losses to industries dependent on forest resources, such as construction, paper, and tourism.

Preventive measures form the foundation of effective pest management, focusing on reducing the likelihood of infestations. Sanitation practices, such as removing infested material and maintaining clean forest environments, are essential for minimizing pest breeding grounds. Mixed-species planting disrupts pest-host relationships and enhances biodiversity, creating a natural buffer against severe outbreaks. Quarantine and biosecurity measures, including the regulation of plant material movement and trade, are critical for preventing the introduction and spread of invasive pests. These strategies represent proactive approaches to safeguarding forests against pest threats. Early detection and monitoring are pivotal for controlling pest infestations before they escalate. Traditional methods, such as field surveys and visual inspections, remain essential for identifying early signs of damage. Advances in pheromone traps and light traps have improved the ability to monitor pest populations and predict outbreaks. Remote sensing and GIS technologies have further enhanced pest surveillance, enabling large-scale monitoring and spatial analysis of infestations. These tools ensure timely interventions, optimizing resource allocation and minimizing damage.

Biological control offers sustainable alternatives to chemical treatments, leveraging natural enemies to suppress pest populations. Beneficial insects, such as predators, parasitoids, and entomopathogens, play crucial roles in controlling pests like aphids, caterpillars, and borers. Conservation of these natural enemies within forest ecosystems enhances ecological balance and reduces the need for chemical interventions. Additionally, entomopathogenic fungi and nematodes provide effective solutions for managing soilborne and stem-boring pests, further diversifying biological control options. Cultural and silvicultural practices complement biological control by creating conditions unfavorable to pests. Thinning and pruning reduce pest habitats, while proper timing of planting and harvesting disrupts pest life cycles. Soil and nutrient management strengthen tree defenses, improving resilience against pest attacks. These practices integrate pest management into sustainable forestry operations, ensuring long-term forest health and productivity.

While chemical control methods, such as insecticides and systemic treatments, remain effective for severe infestations, their use requires careful consideration. Issues such as resistance development, non-target impacts, and environmental contamination necessitate integrating chemical treatments with other strategies. This balanced approach minimizes negative impacts while addressing acute pest challenges. Integrated Pest Management (IPM) provides a holistic framework for addressing pest damage, combining biological, cultural, chemical, and preventive measures. IPM emphasizes monitoring, community participation, and adaptive strategies to respond to emerging pest threats and changing environmental conditions. Advances in technology, including genetically modified trees, RNA interference (RNAi), and predictive modeling, are transforming pest management, offering innovative solutions for long-term control.

Chapter - 8

Methods of Pest Control in Forest Ecosystems

Introduction

Biological, cultural, chemical, and ecological control methods represent the pillars of sustainable pest management, each offering unique approaches to mitigating the damage caused by pests in forest ecosystems. The challenge of pest management is multifaceted, encompassing a wide array of pests, such as defoliators, borers, sap-suckers, and root feeders, each with distinct behaviors, life cycles, and ecological impacts. Effective pest control requires an integration of these strategies into cohesive management plans that minimize environmental damage, enhance ecosystem resilience, and safeguard the economic and ecological value of forests. Central to this effort is the goal of achieving long-term pest suppression while maintaining the balance and biodiversity of forest ecosystems.

Biological control stands as one of the most environmentally sustainable approaches to pest management, leveraging the natural relationships between predators, parasitoids, pathogens, and their prey. Predators such as lady beetles, lacewings, and spiders play a crucial role in controlling pest populations by preying on species like aphids, caterpillars, and borers. Their effectiveness lies in their ability to consume large quantities of prey, often regulating pest populations below damaging thresholds. Parasitoids, such as parasitic wasps and flies, provide target-specific pest control by laying their eggs within or on pest hosts, leading to the eventual death of the host. These agents are particularly effective in controlling pests like moth larvae and aphids while minimizing non-target effects. Pathogens, including fungi, bacteria, and viruses, offer another avenue for biological control, often leading to pest mortality through infections that spread naturally or via deliberate application. Conservation and augmentation of natural enemies through habitat enhancement, reduced pesticide use, and biological agent release are essential strategies for strengthening biological control in forest ecosystems.

Cultural and silvicultural practices complement biological control by altering environmental conditions to reduce pest habitats and limit infestations. Sanitation techniques, such as removing infested plant material and debris, reduce breeding sites for pests and interrupt their life cycles. Mixed-species planting creates diverse habitats that disrupt pest-host relationships, reducing the likelihood of pest outbreaks that often occur in monocultures. Thinning and pruning help manage pest habitats by improving airflow and sunlight penetration, thereby creating unfavorable conditions

DOI: 10.1201/9781042011063-8

for many pests. Timing of planting and harvesting allows forest managers to avoid periods of peak pest activity, further minimizing the risk of damage. These practices align pest management with sustainable forestry principles, ensuring long-term productivity and ecological stability.

Chemical control methods remain a critical component of pest management, especially during severe outbreaks where rapid intervention is necessary. Contact insecticides, applied directly to pests, provide immediate control and are widely used against defoliators and other visible pests. Systemic insecticides, absorbed and distributed throughout the plant, are particularly effective against sap-sucking insects and borers that are otherwise difficult to target. The use of pheromones for pest monitoring and control has revolutionized chemical strategies, enabling species-specific trapping and mating disruption techniques that reduce reliance on broad-spectrum insecticides. However, chemical control poses challenges, including the development of pesticide resistance, contamination of non-target species, and environmental impacts. These limitations underscore the importance of integrating chemical methods with other strategies for a balanced and sustainable approach.

Physical and mechanical control methods offer targeted, environmentally friendly solutions for pest removal. Trapping systems, such as pheromone traps and light traps, effectively capture specific pest species, aiding both monitoring and control efforts. Barriers, such as tree wrapping and netting, prevent pests from accessing vulnerable tree parts, while manual removal of pest colonies provides immediate relief in localized infestations. Soil preparation techniques, including tilling and mulching, disrupt pest life cycles by exposing larvae and pupae to predators or unsuitable conditions. These methods, though labor-intensive, are invaluable for integrated pest management programs that emphasize minimal environmental disruption.

Integrated Pest Management (IPM) serves as the cornerstone of modern pest control, combining biological, cultural, chemical, and mechanical methods into a comprehensive framework. IPM emphasizes early monitoring and risk assessment to ensure timely and targeted interventions. By relying on a combination of strategies, IPM minimizes the overuse of any single method, reducing the risk of resistance development and environmental harm. Community-based approaches in IPM foster collaboration among stakeholders, including forest managers, researchers, and local communities, ensuring broader implementation and success. The adaptability of IPM makes it particularly effective in addressing emerging pest threats and adapting to changing environmental conditions, including those driven by climate change. Advances in technology are transforming pest management, offering innovative tools for monitoring, prediction, and control. RNA interference (RNAi) and gene editing technologies provide precise methods for suppressing specific pest populations without harming non-target species. Remote monitoring and predictive modeling, powered by satellite imagery and machine learning algorithms, enable real-time tracking of pest movements and outbreak forecasting. Genetically modified trees with pest resistance, developed through biotechnological advancements, offer long-term solutions to persistent pest problems. Artificial intelligence and machine learning are increasingly being used to analyze pest behavior, optimize management strategies, and automate surveillance systems, making pest control more efficient and effective.

Quarantine and regulatory measures are critical for preventing the spread of invasive pests, which often cause severe damage to forest ecosystems. International standards, such as those set by the International Plant Protection Convention (IPPC), guide the inspection and certification of plant material to minimize pest risks in global trade. Domestic quarantine protocols for high-risk zones prevent the movement of infested material within countries, reducing the spread of established pests. Public awareness campaigns and compliance with quarantine practices play a vital role in ensuring the success of these measures, as human activities often serve as vectors for pest introduction and distribution. Ecological approaches to pest control focus on restoring and enhancing natural ecosystem functions to suppress pest populations. Biodiversity, including the presence of diverse plant and animal species, creates ecological checks and balances that limit pest outbreaks. Microbial communities in soil and on plant surfaces act as natural suppressants, outcompeting or directly antagonizing pest populations. Rewilding practices, which involve restoring natural habitats and reintroducing native species, contribute to ecosystem balance and reduce the likelihood of pest outbreaks. These approaches have long-term benefits, promoting sustainable forest management while minimizing the need for external interventions.

1. Biological Control

Biological control is a cornerstone of sustainable pest management in forest ecosystems, utilizing natural predators, parasitoids, pathogens, and other biological agents to suppress pest populations. By leveraging the ecological balance of predator-prey interactions, biological control minimizes the need for chemical pesticides, reduces environmental impact, and preserves biodiversity. This approach is rooted in the principles of integrated pest management (IPM) and aligns with the broader goals of ecological conservation and forest sustainability.

1.1 Role of Predators in Controlling Pest Populations

Predators play a critical role in regulating pest populations by directly consuming their prey, thereby preventing outbreaks and maintaining ecological balance. Predatory insects, arachnids, birds, and even mammals are integral components of forest ecosystems, acting as natural checks on herbivorous pests that threaten tree health. These predators often exhibit generalist feeding habits, allowing them to target a wide range of pests and respond dynamically to population fluctuations.

Insect Predators

Insects are among the most effective biological control agents in forests, with species such as lady beetles (*Coccinellidae*), ground beetles (*Carabidae*), lacewings (*Chrysopidae*), and predatory stink bugs (*Pentatomidae*) playing significant roles. Lady beetles are renowned for their voracious appetite for aphids, scale insects, and whiteflies. Their larvae and adults actively hunt and consume pests across a variety of forested habitats, reducing sap-sucking insect populations that weaken tree health. Similarly, ground beetles patrol the forest floor, preying on soil-dwelling larvae, pupae, and other pests, including root-feeding grubs and caterpillars. Lacewings are particularly valuable in managing pests during the larval stage, as their larvae, often referred to as "aphid lions," exhibit insatiable predatory behavior. These insects are commonly used in biocontrol programs to manage aphid outbreaks on young forest saplings. Predatory stink bugs, on the other hand, target larger pests such as caterpillars and leaf beetles, delivering effective pest suppression in both natural and managed forest ecosystems.

Arachnid Predators

Spiders are ubiquitous predators in forest ecosystems, preying on a wide range of insect pests. Orb-weaver spiders construct intricate webs that capture flying insects, while wolf spiders actively hunt ground-dwelling prey. By targeting defoliators, sap-sucking insects, and other pests, spiders contribute to the overall health and resilience of forests.

Birds and Mammals

Birds, particularly insectivorous species, are highly effective at controlling pest populations in forest canopies. Woodpeckers, for example, specialize in extracting bark beetles, borers, and other wood-dwelling insects, reducing their impact on tree health. Songbirds, such as warblers and chickadees, feed on caterpillars, leafhoppers, and other defoliators, while larger birds like crows and magpies target larger pest species, including grasshoppers. Small mammals, such as shrews and bats, also contribute to pest control by consuming ground-dwelling insects and nocturnal pests. Bats, in particular, are efficient at managing flying pests such as moths, whose larvae can cause defoliation.

Integration with Forest Management

The effectiveness of predators in biological control depends on the availability of suitable habitats, food resources, and minimal disturbance from human activities. Maintaining diverse forest ecosystems with varied vegetation structures supports predator populations and enhances their impact on pest suppression. Practices such as preserving native vegetation, reducing pesticide use, and providing nesting sites for birds and mammals foster a thriving predator community.

1.2 Use of Parasitoids

Parasitoids are a unique group of biological control agents that parasitize pest hosts during their developmental stages, ultimately killing them. Unlike predators, which consume multiple prey throughout their lifetimes, parasitoids rely on a single host to complete their life cycle. This specificity makes parasitoids highly effective at targeting specific pests without affecting non-target organisms, making them invaluable tools in integrated pest management.

Types of Parasitoids

Parasitoids are primarily insects, including wasps (families *Braconidae* and *Ichneumonidae*) and flies (family *Tachinidae*), that parasitize various life stages of their hosts.

» **Egg Parasitoids**: These parasitoids, such as *Trichogramma* species, lay their eggs inside the eggs of pests. The developing parasitoid larvae consume the host egg contents, preventing the emergence of pest larvae. *Trichogramma* wasps are widely used to control defoliators like gypsy moths and spruce budworms.

» **Larval Parasitoids**: Many braconid and ichneumonid wasps target caterpillars, sawflies, and other larval pests. For example, *Cotesia congregata* parasitizes hornworms, significantly reducing their populations in forest ecosystems.

» **Pupal Parasitoids**: These parasitoids attack pests during the pupal stage, often delivering highly effective suppression by preventing the emergence of adult pests. Tachinid flies, for instance, parasitize the pupae of moths and beetles.

- **Adult Parasitoids**: Some parasitoids, such as *Anastatus* wasps, target adult pests, particularly those responsible for egg laying, thereby reducing future pest generations.

Biological Control Programs

Parasitoids have been successfully used in numerous biological control programs worldwide. In forests, their deployment focuses on managing key pests that cause economic and ecological damage. For instance, parasitoids like *Oobius agrili* have been introduced to control the emerald ash borer (*Agrilus planipennis*), an invasive pest that has devastated ash trees in North America. Similarly, parasitoid wasps have been employed to manage bark beetles, reducing their impact on coniferous forests.

Advantages of Parasitoids

The specificity of parasitoids minimizes non-target effects and ensures that their activity is confined to pest populations. Additionally, many parasitoids can establish self-sustaining populations in forests, providing long-term pest suppression without the need for repeated releases. However, their effectiveness can be influenced by environmental factors such as temperature, humidity, and host availability.

1.3 Pathogens as Natural Pest Suppressants

Pathogens, including bacteria, fungi, viruses, and nematodes, naturally infect and kill insect pests, serving as important agents of biological control. These microorganisms often cause epizootics (disease outbreaks) in pest populations, leading to rapid and widespread declines in their numbers.

Bacterial Pathogens

Bacillus thuringiensis (*Bt*) is the most widely known bacterial pathogen used in pest management. Bt produces crystalline proteins that, when ingested by pests, disrupt their gut lining, leading to starvation and death. Different Bt strains target specific pests, such as lepidopteran caterpillars, coleopteran beetles, and dipteran flies. *Bt*-based biopesticides are extensively used in forestry to manage pests like gypsy moths and spruce budworms.

Fungal Pathogens

Entomopathogenic fungi, such as *Beauveria bassiana* and *Metarhizium anisopliae*, infect pests by penetrating their cuticle and proliferating within their bodies. These fungi are effective against a wide range of pests, including aphids, beetles, and borers. Fungal pathogens are particularly valuable in damp and humid forest environments, where their spores thrive.

Viral Pathogens

Nucleopolyhedroviruses (NPVs) and granuloviruses are species-specific pathogens that target caterpillars and other pests. These viruses are ingested during feeding and cause systemic infections, leading to high mortality rates. NPVs are widely used in forest pest management, particularly against defoliators like gypsy moths and armyworms.

Nematodes

Entomopathogenic nematodes, such as *Steinernema* and *Heterorhabditis* species, infect soil-dwelling pests by releasing symbiotic bacteria that kill the host. These nematodes are effective against root grubs, cutworms, and other subterranean pests.

Integration in IPM

Pathogens are integral to IPM programs, providing environmentally friendly alternatives to chemical pesticides. Their application as biopesticides is compatible with other control methods, enhancing the overall effectiveness of pest management strategies.

1.4 Conservation and Augmentation of Natural Enemies

Conserving and augmenting natural enemies is critical for maintaining the efficacy of biological control in forest ecosystems. This involves creating favorable conditions for predators, parasitoids, and pathogens to thrive and exert their suppressive effects on pest populations.

Conservation of Habitats

Preserving natural habitats, such as forest edges, understory vegetation, and water sources, supports diverse communities of natural enemies. Providing shelter, food resources, and breeding sites ensures their persistence and activity. For example, planting flowering plants attracts parasitoids and predatory insects by providing nectar and pollen.

Reduction of Disturbances

Minimizing human-induced disturbances, such as pesticide application and deforestation, protects natural enemy populations. Selective use of pest-specific chemicals and adoption of ecological forestry practices enhance the survival of beneficial organisms.

Augmentation Techniques

Augmentation involves the deliberate release of natural enemies to boost their populations in pest-affected areas. This can be done through:

- **Inoculative Releases**: Introducing a small number of natural enemies to establish self-sustaining populations.
- **Inundative Releases**: Releasing large numbers of natural enemies to achieve immediate pest suppression.

Monitoring and Research

Regular monitoring of natural enemy populations and their interactions with pests is essential for adaptive management. Research on the ecology, behavior, and effectiveness of natural enemies informs targeted interventions and enhances the success of biological control programs. By integrating predators, parasitoids, pathogens, and conservation efforts, biological control provides a sustainable framework for managing pest populations in forests. These strategies not only protect tree health but also contribute to the ecological integrity of forest ecosystems.

2. Cultural and Silvicultural Practices

Cultural and silvicultural practices are foundational to forest pest management, focusing on modifying environmental conditions and forest structures to minimize pest habitats, interrupt pest life cycles, and enhance the overall resilience of trees. These methods emphasize proactive measures, targeting the ecological factors that facilitate pest outbreaks rather than relying solely on reactive interventions. By integrating sanitation, mixed-species planting, thinning and pruning, and carefully timed planting and harvesting, forest managers can mitigate pest pressures and support sustainable forest health.

2.1 Sanitation Techniques to Reduce Pest Habitats

Sanitation is one of the most direct and effective strategies for reducing pest populations by removing or treating materials that serve as breeding grounds or refuges for pests. Dead or dying trees, infested wood, and logging debris are often hotspots for pest activity, especially for species like bark beetles, wood borers, and fungal pathogens. Removing these materials disrupts the life cycles of pests, preventing their spread to healthy trees. For example, in forests impacted by the mountain pine beetle (*Dendroctonus ponderosae*), sanitation involves identifying and promptly removing infested trees. These trees are often cut down, debarked, or treated with heat to kill any beetles or larvae present. Similarly, salvage logging after storm damage or disease outbreaks is critical to prevent pests from exploiting weakened or downed trees. In urban and nursery settings, sanitation extends to sterilizing tools, containers, and equipment to prevent the inadvertent spread of pests and pathogens. Proper disposal of infested material is crucial for effective sanitation. Methods such as chipping, burning, or deep burial ensure that pests do not escape and reinfest nearby areas. Additionally, leaving untreated debris in forests can create microhabitats that favor pest proliferation. Sanitation must also include regular monitoring to detect early signs of infestation and take prompt action before pests reach outbreak levels.

2.2 Mixed-Species Planting to Prevent Pest Monocultures

Mixed-species planting is a proactive strategy that leverages biodiversity to reduce the vulnerability of forests to pest outbreaks. Monoculture plantations, where a single tree species dominates, create ideal conditions for specialized pests to thrive, as they provide a continuous and abundant food source. Mixed-species forests, by contrast, introduce ecological complexity that disrupts pest behavior, reduces host availability, and enhances the presence of natural pest enemies. In mixed-species systems, pests often find it more challenging to locate their preferred hosts due to the physical and chemical diversity of the forest. For example, interplanting pest-resistant species among susceptible ones can create natural barriers that slow the spread of pests like the spruce budworm (*Choristoneura fumiferana*) or gypsy moth (*Lymantria dispar*). Mixed-species planting also promotes a balanced ecosystem by supporting beneficial insects, birds, and other organisms that act as natural pest control agents. This practice is particularly valuable in regions prone to climate change or environmental stressors, as diverse forests are more adaptable to fluctuating conditions. Selecting species with complementary growth habits and resource requirements ensures that trees thrive together without excessive competition. However, implementing mixed-species planting requires careful planning, including site analysis, species compatibility, and long-term monitoring to evaluate its effectiveness.

2.3 Thinning and Pruning for Pest Habitat Management

Thinning and pruning are essential silvicultural practices that optimize forest structure, reduce pest habitats, and improve tree health. These techniques aim to create conditions that are less favorable for pests while enhancing the resilience of individual trees and forest stands. Thinning involves selectively removing trees within a stand to reduce competition for sunlight, water, and nutrients. Dense forests with overcrowded canopies create humid microclimates that encourage the proliferation of pests like sap-sucking insects, defoliators, and fungal pathogens. By increasing spacing between trees, thinning improves air circulation and light penetration, disrupting the conditions pests need to thrive. For instance, thinning in coniferous forests has been shown to reduce bark beetle infestations by limiting the availability of stressed or weakened trees that these pests prefer. Pruning complements thinning by removing lower branches, dead wood, and other vulnerable parts of trees that pests frequently exploit. Many pests, such as wood borers and canker-causing fungi, enter through damaged or decayed branches. Pruning these parts minimizes entry points for pests and helps contain existing infestations. Additionally, pruning facilitates better access for pest monitoring and treatment applications, making it easier for forest managers to implement integrated pest management strategies. Timing is critical for both thinning and pruning operations. Conducting these activities during dormant seasons or periods of low pest activity reduces stress on trees and minimizes the risk of attracting pests to fresh wounds. Proper wound treatments, such as fungicidal or insecticidal sealants, further protect pruned areas from colonization.

2.4 Timing of Planting and Harvesting to Avoid Pest Peaks

The timing of planting and harvesting operations is a crucial factor in reducing pest pressures and ensuring successful forest regeneration. Aligning these activities with pest life cycles and environmental conditions helps minimize tree vulnerability and disrupt pest activity patterns. Planting during optimal environmental conditions, such as cool and wet seasons, promotes healthy seedling establishment and reduces susceptibility to pests. Young trees are particularly vulnerable to pests like sap-sucking insects and defoliators, as they lack the energy reserves and defenses of mature trees. Delaying planting until after the flight periods of bark beetles or the emergence of caterpillars can protect seedlings from early infestations. Conversely, planting late in the growing season, when drought stress or frost risks are higher, can exacerbate pest impacts and compromise tree health. Harvesting timing also plays a significant role in pest management. Freshly felled trees and logging residues attract pests like ambrosia beetles, wood borers, and bark beetles, providing them with breeding sites and food sources. Scheduling harvesting during low pest activity periods, such as winter months, reduces the risk of infestations. Post-harvest practices, including debris removal, chipping, or controlled burning, further limit pest habitats and prevent the spread of infestations to surrounding stands.

3. Chemical Control

Chemical control remains a pivotal component of pest management strategies in forest ecosystems, offering rapid and effective solutions for controlling pest populations. This method involves the use of chemical agents, such as contact and systemic insecticides, to directly target pests or disrupt their life cycles. Additionally, semiochemicals like pheromones are employed in innovative ways to monitor and suppress pest activity. However, chemical control is not without challenges, including the risks of resistance development and environmental contamination. To maximize effectiveness and sustainability, chemical controls are best integrated into broader pest management frameworks.

3.1 Contact Insecticides: Types and Application

Contact insecticides are a traditional and widely used method in forest pest management, offering immediate action against target pests. These compounds work by directly interacting with the insect's body, causing rapid physiological disruptions that result in death. Contact insecticides are classified based on their chemical composition and mode of action, with pyrethroids, organophosphates, and carbamates being the most common types. Pyrethroids, synthetic derivatives of natural insecticidal compounds found in chrysanthemum flowers, are highly effective against a broad spectrum of pests. These compounds act as neurotoxins, overstimulating the nervous system and causing paralysis. Pyrethroids are favored for their quick knockdown effects and relative safety for mammals and birds. They are widely used to control defoliators, aphids, and other sap-sucking insects. Organophosphates, another class of contact insecticides, inhibit acetylcholinesterase, an enzyme essential for proper nerve function, leading to similar paralytic effects. These compounds, such as malathion, are used against a range of pests, including caterpillars and beetles, but require careful handling due to their higher toxicity to non-target species. Carbamates, which also target acetylcholinesterase, are similar in action to organophosphates but generally have shorter environmental persistence, making them suitable for applications where rapid degradation is desired. The application methods for contact insecticides include foliar sprays, which are used to manage pests like caterpillars and aphids, and soil treatments to target root-dwelling insects. While effective, contact insecticides pose risks of non-target effects and environmental contamination if not applied judiciously. Precision application techniques, including localized sprays and calibrated nozzles, are critical for minimizing drift and reducing harm to beneficial organisms.

3.2 Systemic Insecticides for Sap-Sucking and Boring Pests

Systemic insecticides provide an advanced and targeted approach to pest management, particularly for pests that feed within plant tissues, such as sap-suckers and borers. These compounds are absorbed by the tree and translocated through its vascular system, effectively reaching all parts of the plant, including those inaccessible to surface sprays. Systemic insecticides offer long-lasting protection and are particularly valuable in high-value forestry, where pest damage can lead to significant economic losses. Neonicotinoids, such as imidacloprid and dinotefuran, are among the most widely used systemic insecticides. These compounds act on nicotinic acetylcholine receptors in the insect nervous system, causing paralysis and death. They are particularly effective against sap-sucking pests like aphids, scale insects, and mealybugs, as well as borers like the emerald ash borer (*Agrilus planipennis*).

Systemic insecticides can be applied using several methods. Soil drenches involve applying the compound around the base of the tree, allowing the roots to absorb it. This method is commonly used for young trees or those with extensive root systems. Trunk injections, where the insecticide is directly injected into the vascular tissue, are effective for managing pests like bark beetles and borers. This approach ensures precise delivery to infested areas, minimizing environmental exposure. While systemic insecticides offer numerous advantages, including reduced reliance on frequent applications, they must be used cautiously. Overuse can lead to contamination of groundwater and harm non-target organisms, including pollinators and soil microbes, making integration with other pest management methods essential

3.3 Use of Pheromones in Pest Monitoring and Control

Pheromones represent a sophisticated and environmentally friendly tool in pest management, leveraging the chemical communication systems of insects to monitor and control pest populations.

These semiochemicals are highly species-specific, making them ideal for targeted applications without affecting non-target organisms. Pheromones are primarily used in three contexts: monitoring, mass trapping, and mating disruption. In monitoring programs, pheromone-baited traps are deployed to assess pest density, distribution, and activity. For instance, traps baited with aggregation pheromones are used to monitor bark beetles like the mountain pine beetle (*Dendroctonus ponderosae*), while traps with sex pheromones are employed to detect gypsy moths (*Lymantria dispar*). This information is critical for early detection and timely intervention. Mass trapping involves deploying a large number of pheromone traps to capture and reduce pest populations. This method is particularly effective for localized outbreaks, such as those caused by defoliating caterpillars or fruit pests. Mating disruption, another application, involves releasing synthetic pheromones into the environment to confuse male insects, preventing them from locating females. This technique has been successfully used to manage pests like the spruce budworm (*Choristoneura fumiferana*) and codling moth (*Cydia pomonella*). Pheromone-based strategies are non-toxic, environmentally safe, and compatible with other pest management methods. However, they require species-specific formulations and careful deployment to achieve desired results.

3.4 Challenges of Resistance and Environmental Contamination

Despite their effectiveness, chemical controls face significant challenges, including the development of resistance among pest populations and the potential for environmental contamination. Resistance occurs when pest populations adapt to chemical treatments through genetic mutations or behavioral changes, reducing the efficacy of insecticides. Factors contributing to resistance include overuse, improper application, and reliance on a single class of chemicals. For example, widespread use of neonicotinoids has led to resistance in pests like whiteflies and aphids, necessitating the development of alternative control methods. Environmental contamination is another critical concern, as chemical residues can persist in soil, water, and non-target organisms. Broad-spectrum insecticides, in particular, may harm beneficial insects, birds, and aquatic life, disrupting ecological balance. For instance, runoff from soil-applied systemic insecticides can contaminate nearby water bodies, affecting fish and amphibians.

4. Physical and Mechanical Control

Physical and mechanical control methods are essential components of pest management strategies, offering direct and environmentally safe approaches to mitigate pest infestations. These methods involve the use of physical barriers, traps, manual interventions, and soil management techniques to reduce pest populations and prevent their spread. Unlike chemical controls, physical and mechanical methods are non-toxic, often targeting specific pests without harming non-target species or the environment. Their integration into broader pest management frameworks enhances the sustainability and efficacy of forest pest control efforts.

4.1 Trapping Systems for Targeted Pest Removal

Trapping systems are widely used in forest pest management to monitor, suppress, and sometimes eradicate pest populations. These systems leverage the behavior and biology of pests to capture them efficiently, often using bait, attractants, or mechanical designs tailored to specific species.

Pheromone Traps

Pheromone traps are highly effective tools for targeting specific pests, as they mimic the chemical signals used by insects for communication, such as sex or aggregation pheromones. For instance, traps baited with synthetic sex pheromones are commonly used to monitor and control gypsy moths (*Lymantria dispar*) and bark beetles like the mountain pine beetle (*Dendroctonus ponderosae*). These traps attract male insects, preventing them from mating and thereby reducing subsequent generations.

Light Traps

Light traps exploit the phototactic behavior of nocturnal pests, such as moths and beetles, by luring them to a light source and capturing them in a collection chamber. These traps are particularly effective during peak activity periods and are often deployed in nurseries or forest edges where pest activity is concentrated.

Sticky Traps

Sticky traps use adhesive-coated surfaces to capture flying or crawling insects. These traps are simple to deploy and are often used for monitoring sap-sucking pests, such as aphids and scale insects, in young plantations or nursery stock.

Pitfall and Ground Traps

For soil-dwelling pests, such as root-feeding grubs and beetle larvae, pitfall traps are used to intercept crawling insects. These traps consist of containers buried flush with the soil surface, often baited with attractants to lure pests. Trapping systems are invaluable for pest monitoring and control, providing real-time data on pest activity and population density. However, their effectiveness depends on strategic placement, appropriate attractants, and regular maintenance to prevent trap saturation and ensure consistent results.

4.2 Use of Barriers and Tree Wrapping to Prevent Pest Entry

Barriers and tree wrapping are preventive measures designed to protect trees from pests by physically blocking their access to host tissues. These methods are particularly effective for pests that crawl, climb, or bore into trees, such as cankerworms, bark beetles, and borers.

Tree Wrapping

Tree wrapping involves encircling tree trunks with materials that prevent pests from ascending or boring into the tree. For example, sticky bands or tanglefoot barriers are applied to trunks to intercept climbing insects like cankerworms and gypsy moth larvae. Burlap wraps can be used to trap pests like the emerald ash borer (*Agrilus planipennis*), which seek shelter in crevices during certain life stages. Tree wraps also protect young saplings from rodent and mammal damage.

Exclusion Fences and Nets

Exclusion fences or netting are deployed around nurseries, plantations, or individual trees to prevent access by larger pests, such as deer, rabbits, or invasive insects. Fine mesh nets can also be used to shield young trees from flying pests, such as leaf miners or moths, without impeding sunlight or airflow.

Trunk Guards and Shields

Trunk guards made of metal or plastic are installed to prevent borers and other pests from accessing vulnerable tree parts. These guards are particularly useful in high-value forestry or urban landscapes, where aesthetic and economic considerations are paramount. While barriers and tree wrapping are effective for targeted protection, they require regular inspection and maintenance to ensure functionality and avoid unintended consequences, such as moisture buildup or fungal growth beneath the wraps.

4.3 Manual Removal and Destruction of Pest Colonies

Manual removal and destruction of pest colonies are labor-intensive but highly effective methods for managing localized pest infestations. These techniques involve physically removing pests or their habitats from trees and surrounding areas, disrupting their life cycles and preventing further damage.

Handpicking Pests

Handpicking is a direct method used to remove visible pests, such as caterpillars, larvae, and egg masses, from tree foliage, bark, or soil. This approach is particularly useful for controlling small infestations in young plantations or nurseries, where early intervention can prevent outbreaks.

Pruning Infested Branches

Pruning is employed to remove branches and stems infested with pests like borers, scale insects, or canker-causing fungi. Proper disposal of pruned material, through chipping, burning, or burial, ensures that pests do not escape and reinfest nearby trees.

Destroying Pest Nests and Galleries

For pests that build nests or galleries, such as tent caterpillars or ambrosia beetles, manual destruction of these structures disrupts their reproductive cycles and reduces population density. Removing and destroying bark infested by borers or beetles is another effective measure. Manual removal is best suited for small-scale applications or as a supplementary measure in integrated pest management programs. While effective, it requires skilled labor, regular monitoring, and prompt action to address infestations before they spread.

4.4 Soil Preparation Techniques to Disrupt Pest Life Cycles

Soil preparation techniques play a vital role in managing soil-dwelling pests, such as root grubs, nematodes, and wireworms, by altering their habitats and interrupting their development. These methods improve soil health while reducing the likelihood of pest outbreaks.

Tillage and Soil Disturbance

Tillage disrupts the soil profile, exposing pests and their life stages to environmental conditions and natural enemies. For example, plowing or disking can destroy pupal chambers or burrows of pests like cutworms and white grubs, reducing their survival rates.

Soil Solarization

Soil solarization involves covering the soil with transparent plastic sheeting to trap solar heat, raising the temperature to levels lethal for pests and pathogens. This technique is particularly effective for managing nematodes and soilborne fungal spores in nurseries or reforestation sites.

Organic Amendments

Incorporating organic matter, such as compost or manure, enhances soil microbial activity, which can suppress pest populations through natural competition. Certain organic amendments, like neem cake or biochar, have pest-repellent or insecticidal properties that further reduce pest density.

Flooding and Moisture Management

In some cases, controlled flooding of soil can drown soil-dwelling pests, particularly those with low mobility, such as nematodes. Proper water management also prevents conditions conducive to pest proliferation, such as waterlogging or drought stress. Soil preparation techniques are integral to creating a healthy growing environment for trees, enhancing their resilience to pests while promoting long-term forest health.

5. Integrated Pest Management (IPM)

Integrated Pest Management (IPM) represents a holistic and sustainable approach to controlling pest populations in forest ecosystems. Unlike conventional pest control methods that may rely heavily on chemical applications, IPM integrates biological, cultural, mechanical, and chemical strategies to minimize pest impacts while safeguarding environmental health and biodiversity. This approach emphasizes prevention, early detection, and the judicious use of control measures tailored to the specific ecological and socio-economic contexts of forests. The dynamic nature of IPM allows it to adapt to emerging threats, including invasive species and the effects of climate change.

5.1 Combining Biological, Cultural, and Chemical Methods

The cornerstone of IPM is the seamless integration of biological, cultural, and chemical methods to achieve effective pest control with minimal ecological disruption. Each of these approaches offers distinct advantages and limitations, and their combination allows forest managers to address pest problems comprehensively and sustainably.

Biological Methods

Biological control is a central pillar of IPM, leveraging natural enemies such as predators, parasitoids, and pathogens to suppress pest populations. For instance, predatory insects like lady beetles and lacewings prey on aphids and scale insects, while parasitoid wasps target the larvae of defoliating caterpillars. Pathogens, including *Bacillus thuringiensis* (*Bt*) and nucleopolyhedroviruses, provide species-specific control of pests like gypsy moths and spruce budworms. These biological agents are environmentally friendly and self-regulating, often establishing long-term pest suppression in forest ecosystems. However, their efficacy depends on factors such as habitat quality, pest density, and environmental conditions.

Cultural Practices

Cultural practices aim to create conditions that reduce pest populations and enhance tree resilience. Techniques such as mixed-species planting disrupt pest-host interactions, while sanitation measures remove infested material that serves as pest reservoirs. Thinning and pruning improve air circulation and light penetration, creating an unfavorable environment for pests and fungal pathogens. Timing planting and harvesting to avoid peak pest activity further reduces the likelihood of infestations. These preventive measures complement biological control, reducing the need for chemical interventions.

Chemical Methods

Chemical control serves as a supplementary tool within IPM, deployed judiciously to manage severe pest outbreaks or protect high-value forest stands. The use of targeted insecticides, such as systemic compounds for sap-sucking pests or pheromone-based strategies for mating disruption, minimizes non-target effects and environmental risks. Integrating chemical methods with biological and cultural approaches ensures that pesticides are used sparingly and strategically, reducing the potential for resistance development and ecological harm. The integration of these methods requires careful planning, informed by a thorough understanding of pest biology, forest ecology, and management objectives. Regular monitoring and adaptive management are critical to fine-tuning interventions and ensuring the long-term success of IPM programs.

5.2 Role of Early Monitoring and Risk Assessment

Early monitoring and risk assessment are fundamental to the success of IPM, enabling forest managers to detect pest problems at their inception and implement timely interventions. These proactive measures reduce the economic and ecological costs associated with large-scale pest outbreaks, ensuring that control efforts are both efficient and effective.

Field Surveys and Visual Inspections

Field surveys provide essential baseline data on pest populations, their distribution, and associated damage. Surveyors use systematic sampling methods, such as transects or grid patterns, to assess tree health and identify early signs of pest activity, such as defoliation, discoloration, or resin flow. Visual inspections of foliage, bark, and soil help detect specific pests, such as bark beetles, caterpillars, or root grubs, and their characteristic symptoms.

Pheromone Traps and Monitoring Devices

Pheromone traps are invaluable tools for monitoring the activity of specific pest species. These traps attract male insects using synthetic pheromones, providing data on pest density, emergence patterns, and seasonal dynamics. For instance, pheromone traps are widely used to monitor gypsy moths, bark beetles, and defoliators, facilitating early detection and guiding control measures. Light traps and sticky traps complement pheromone-based monitoring by capturing a broader range of pests.

Remote Sensing and Predictive Modeling

Advances in technology have enhanced the scope and precision of pest monitoring. Remote sensing, using satellite imagery or drones, detects changes in forest health, such as canopy thinning or discoloration, that may indicate pest activity. Geographic Information Systems (GIS) integrate spatial data with environmental variables to map pest distribution and assess risk. Predictive models, incorporating pest biology, weather patterns, and host availability, forecast potential outbreaks and inform decision-making.

Risk Assessment and Decision Thresholds

Risk assessment involves evaluating the potential impact of pest populations on forest health and productivity. Decision thresholds, defined as the pest density or damage level at which intervention becomes necessary, guide the timing and intensity of control measures. These thresholds vary depending on management objectives, economic considerations, and environmental factors. Regular monitoring ensures that interventions are deployed at the optimal time, preventing unnecessary treatments and minimizing disruptions to the ecosystem.

5.3 Community-Based Approaches in Pest Control

Community involvement is a critical component of IPM, particularly in forests that intersect with human settlements, agricultural areas, or recreational sites. Engaging local communities fosters collaboration, enhances the effectiveness of pest management programs, and promotes a sense of shared responsibility for forest health.

Education and Outreach

Raising awareness about pest management and the principles of IPM is essential for gaining community support. Educational initiatives, such as workshops, field demonstrations, and informational campaigns, empower stakeholders with the knowledge and skills needed to participate in pest control efforts. For instance, training programs for nursery operators and landowners can focus on early detection, sanitation practices, and biological control techniques.

Participatory Monitoring

Involving communities in monitoring activities enhances data collection and ensures timely detection of pest outbreaks. Citizen science programs and mobile applications enable residents to report observations, such as pest sightings or tree damage, contributing to a collective understanding of pest dynamics. These efforts complement professional monitoring, creating a more comprehensive and responsive management system.

Collaborative Management Initiatives

Community-based approaches emphasize collaboration among stakeholders, including forest managers, local governments, non-governmental organizations, and private landowners. Joint planning and decision-making ensure that pest control measures are tailored to local needs and conditions. For example, community-led efforts to remove infested material, restore degraded areas, or implement mixed-species planting can significantly enhance the effectiveness of IPM programs.

Incentives and Support Mechanisms

Providing financial and technical support encourages community participation in pest management. Subsidies for pest-resistant planting materials, access to biological control agents, and assistance with monitoring equipment reduce barriers to adoption. Recognizing and incorporating traditional ecological knowledge further strengthens community engagement and ensures culturally appropriate interventions.

5.4 Adapting IPM Strategies to Emerging Threats

The dynamic nature of forest ecosystems and the increasing prevalence of challenges such as climate change, invasive species, and habitat fragmentation necessitate adaptive IPM strategies. These strategies emphasize flexibility, innovation, and collaboration to address emerging threats effectively.

Climate Change and Pest Dynamics

Climate change is altering pest behavior, distribution, and life cycles, leading to more frequent and severe outbreaks. Warmer temperatures and longer growing seasons allow pests to expand their ranges, produce multiple generations, and exploit stressed trees. For example, the mountain pine beetle has extended its range into higher altitudes and northern latitudes, causing unprecedented damage to pine forests. Adaptive IPM strategies must account for these changes by incorporating climate-resilient practices, such as selecting drought-tolerant tree species and enhancing forest diversity.

Invasive Species Management

The globalization of trade and travel has facilitated the introduction of invasive pests, which often lack natural enemies in their new environments. Species like the emerald ash borer and Asian longhorned beetle have caused significant ecological and economic losses in North America and Europe. Early detection, stringent quarantine measures, and targeted biological control are critical components of IPM strategies for managing invasive pests.

Technological Innovations

Advances in biotechnology, such as RNA interference (RNAi) and gene editing, offer new opportunities for pest control. These technologies enable precise targeting of pest genes, reducing their survival or reproductive capabilities. Remote monitoring, using drones, sensors, and AI-driven models, enhances the accuracy and efficiency of pest surveillance, supporting proactive interventions.

International Collaboration

Emerging threats often transcend regional boundaries, requiring coordinated responses at national and international levels. Collaborative research, information sharing, and harmonized regulations are essential for addressing transboundary pest challenges and ensuring the long-term success of IPM programs.

6. Advances in Technology for Pest Control

Advancements in technology have revolutionized pest control strategies in forest ecosystems, offering innovative, precise, and sustainable solutions to mitigate pest impacts. As traditional methods face limitations in the context of evolving pest dynamics, climate change, and growing environmental concerns, the integration of cutting-edge technologies has become essential for effective pest management. From molecular tools like RNA interference (RNAi) and gene editing to the application of artificial intelligence (AI), remote monitoring, and genetically modified pest-resistant trees, these technologies promise to transform pest control practices.

6.1 RNAi and Gene Editing Technologies

RNA interference (RNAi) and gene editing technologies, such as CRISPR-Cas systems, have emerged as transformative tools in pest management, offering highly specific and efficient methods for targeting pests at the genetic level. These molecular approaches allow for the precise disruption of pest biology while minimizing non-target effects and environmental risks.

RNA Interference (RNAi)

RNAi is a natural cellular process in which double-stranded RNA (dsRNA) molecules silence specific genes by degrading their corresponding messenger RNA (mRNA). In pest management, RNAi is utilized to disrupt genes critical to pest survival, reproduction, or feeding behavior. For example, RNAi-based treatments can target genes responsible for chitin synthesis, weakening the exoskeleton of insects like beetles and borers, or silence digestive enzyme genes in caterpillars to inhibit nutrient absorption. These treatments can be delivered through topical sprays, soil applications, or transgenic trees that produce dsRNA internally. RNAi is particularly effective because of its high specificity, allowing researchers to design dsRNA molecules that affect only the target pest species while leaving non-target organisms unharmed. This precision reduces the environmental impact associated with broad-spectrum chemical pesticides and aligns with the principles of integrated pest management (IPM).

Gene Editing (CRISPR-Cas Systems)

Gene editing technologies like CRISPR-Cas systems enable precise modifications to the DNA of pests or host plants, offering novel solutions for pest control. For instance, CRISPR can be used to knock out susceptibility genes in trees, enhancing their resistance to pests, or modify pest populations by reducing their fertility or viability through gene drives. Gene drives are particularly promising for controlling invasive pest species, as they spread specific genetic traits rapidly through populations, effectively reducing their numbers over generations. These technologies are also being explored for engineering resistance in trees against pest-vectored pathogens. For example, modifying tree genomes to resist viruses transmitted by aphids or whiteflies provides dual protection against direct and indirect pest damage. While RNAi and gene editing offer immense potential, challenges such as delivery mechanisms, regulatory approval, and public acceptance must be addressed for widespread adoption.

6.2 Remote Monitoring and Predictive Modeling

Remote monitoring and predictive modeling represent significant advancements in pest surveillance, enabling forest managers to detect, track, and anticipate pest activity with unprecedented accuracy and efficiency. These technologies integrate data collection, analysis, and visualization tools to inform proactive management strategies.

Remote Sensing Technologies

Remote sensing uses satellite imagery, drones, and aerial photography to monitor forest health and detect pest-induced changes, such as defoliation, discoloration, or canopy thinning. Hyperspectral and multispectral imaging identify subtle variations in light reflectance that indicate stress caused by pests, often before visible symptoms appear. For instance, thermal imaging detects temperature changes associated with pest infestations, such as the heat generated by bark beetles tunneling under tree bark.

IoT Sensors and Traps

The Internet of Things (IoT) has revolutionized pest monitoring by enabling real-time data collection and transmission. IoT-enabled pheromone traps, for example, automatically count captured pests and transmit the data to central databases for analysis. Soil sensors monitor pest activity underground, while environmental sensors track temperature, humidity, and other factors that influence pest dynamics.

Predictive Modeling and GIS Integration

Predictive models combine biological, climatic, and spatial data to forecast pest outbreaks and assess risk. These models use algorithms to simulate pest population dynamics, migration patterns, and interactions with environmental variables. Geographic Information Systems (GIS) integrate these models with spatial data to create detailed risk maps, highlighting areas vulnerable to pest infestations. For example, models predicting gypsy moth outbreaks use temperature and precipitation data to forecast larval emergence and feeding periods, guiding timely interventions. Remote monitoring and predictive modeling enhance the precision and efficiency of pest management, enabling early detection, rapid response, and informed decision-making. However, these technologies require investment in infrastructure, technical expertise, and data management systems.

6.3 Development of Genetically Modified Pest-Resistant Trees

Genetically modified (GM) trees offer a proactive and sustainable solution to pest management, enabling forests to withstand pest pressures through enhanced natural defenses. By incorporating pest-resistant traits into tree genomes, GM trees reduce the need for chemical controls and support long-term forest health.

Mechanisms of Genetic Modification

GM trees are engineered to express traits that deter pests, inhibit their feeding, or enhance the tree's immune responses. Common approaches include introducing genes for toxic proteins, such as those derived from *Bacillus thuringiensis* (Bt), which target specific pest species. For example, Bt poplar trees have been developed to resist defoliators like caterpillars, reducing the economic and ecological damage caused by these pests. Other modifications focus on strengthening structural defenses, such as lignin synthesis, or enhancing chemical defenses, such as the production of secondary metabolites that deter pests. Additionally, GM trees can be engineered to produce RNAi molecules internally, targeting pests at the genetic level.

Applications and Success Stories

GM trees have been successfully developed for various species and pest challenges. For instance, Bt cottonwood trees resist leaf-feeding pests in North American forests, while modified eucalyptus trees exhibit resistance to gall-forming insects in tropical regions. Efforts to restore American chestnut trees have involved genetic modifications to resist fungal blight and associated pests.

7. Quarantine and Regulatory Measures

Quarantine and regulatory measures are critical components of pest management, aiming to prevent the introduction, establishment, and spread of invasive and harmful pests in forest ecosystems. These measures encompass a wide range of activities, including international and domestic regulations, inspection protocols, and public compliance initiatives. They are especially vital in a globalized world where the movement of goods, plants, and materials across borders increases the risk of pest translocation. Effective quarantine systems not only protect biodiversity but also safeguard the economic and ecological value of forests.

7.1 International Standards for Preventing Pest Spread

International standards play a pivotal role in harmonizing pest management practices across borders, ensuring a coordinated response to the global challenge of pest translocation. The movement of plants, wood products, and other forest-related materials poses a significant risk of introducing invasive pests into new regions. International standards established by organizations such as the International Plant Protection Convention (IPPC) and regional bodies like the European and Mediterranean Plant Protection Organization (EPPO) provide a framework for mitigating these risks.

The Role of the IPPC

The IPPC, a treaty under the Food and Agriculture Organization (FAO), sets global guidelines for plant health, including the management of pests associated with international trade. Its primary instrument, the International Standards for Phytosanitary Measures (ISPMs), outlines protocols for the safe movement of plant materials. For example, ISPM 15 mandates the treatment and marking of wood packaging materials to eliminate pests like wood borers and bark beetles.

Harmonized Inspection Standards

International standards promote uniform inspection and treatment protocols, facilitating trade while minimizing pest risks. This harmonization reduces disputes between trading partners and ensures that exported and imported goods meet agreed-upon phytosanitary requirements.

Collaboration and Information Sharing

International organizations facilitate collaboration among countries, enabling the exchange of information on pest outbreaks, emerging threats, and best practices. Platforms like the IPPC's Phytosanitary Capacity Evaluation tool help countries assess and strengthen their plant health systems. Despite their benefits, implementing international standards requires significant resources, technical expertise, and compliance monitoring, especially in developing countries. Strengthening global partnerships and capacity-building efforts is essential for ensuring the effectiveness of these measures.

7.2 Inspection and Certification of Plant Material

Inspection and certification are integral to quarantine systems, ensuring that plant materials, seeds, and forest products are free from pests before they are moved across borders or within domestic regions. These processes are designed to detect and mitigate pest risks at critical control points, such as ports, nurseries, and storage facilities.

Pre-Export Inspection

Exporters are required to conduct thorough inspections of plant materials to certify their pest-free status. This involves visual examinations, laboratory testing, and the application of treatments such as fumigation, heat, or irradiation. For instance, plant nurseries producing seedlings for export must comply with strict phytosanitary standards to prevent the spread of soilborne pathogens like nematodes and *Phytophthora* species.

Phytosanitary Certification

Phytosanitary certificates, issued by national plant protection organizations (NPPOs), attest to the pest-free status of plant materials. These certificates are a prerequisite for international trade and serve as a guarantee that the shipment meets the importing country's quarantine requirements.

Post-Import Inspection

At points of entry, plant materials undergo rigorous inspection to detect any signs of pest contamination. Advanced tools like X-ray scanners, DNA-based diagnostics, and sensor technologies enhance the accuracy and efficiency of these inspections. Intercepted pests are identified, and appropriate measures, such as destruction or re-export, are taken to prevent their entry. Inspection and certification systems are vital for maintaining the integrity of quarantine measures. However, their effectiveness depends on the availability of skilled personnel, state-of-the-art facilities, and robust regulatory frameworks.

7.3 Domestic Quarantine Protocols for High-Risk Zones

Domestic quarantine protocols are essential for containing pest outbreaks and preventing their spread within a country. These measures target high-risk zones, such as areas adjacent to ports, nurseries, or regions with known pest infestations, and impose restrictions on the movement of plant materials, wood products, and equipment.

Establishment of Quarantine Zones

Quarantine zones are designated in response to pest detections or outbreaks, with restrictions on the movement of potentially infested materials. For example, areas affected by the emerald ash borer (*Agrilus planipennis*) may be subject to regulations prohibiting the transport of ash logs, firewood, or nursery stock.

Monitoring and Surveillance

Regular monitoring and surveillance within quarantine zones help detect new infestations and assess the effectiveness of control measures. Trapping systems, visual inspections, and remote sensing technologies are commonly employed to monitor pest activity.

Treatment and Eradication Protocols

Infested materials within quarantine zones are subjected to treatments such as fumigation, chipping, or incineration. In cases of severe outbreaks, eradication measures, including the removal of host trees, are implemented to eliminate pest populations.

Enforcement and Compliance

Enforcement of quarantine protocols requires collaboration between regulatory authorities, local governments, and stakeholders. Non-compliance, whether intentional or accidental, can undermine the effectiveness of quarantine measures, highlighting the need for stringent penalties and public awareness campaigns.

7.4 Public Awareness and Compliance in Quarantine Practices

Public awareness and compliance are vital for the success of quarantine measures, as human activities are a significant factor in pest introductions and spread. Educating the public about the importance of quarantine practices fosters a sense of responsibility and encourages adherence to regulations.

Outreach Campaigns

Public awareness campaigns use various media, including social media, posters, workshops, and educational programs, to inform people about quarantine rules and the risks of non-compliance. For example, campaigns like "Don't Move Firewood" educate the public about the dangers of transporting firewood, which can harbor pests like bark beetles and borers.

Engaging Stakeholders

Involving stakeholders, such as forest managers, nursery operators, and the transportation industry, in quarantine efforts enhances compliance. Stakeholder workshops and training sessions provide practical guidance on implementing quarantine measures and detecting pest signs.

Community Involvement

Community participation in pest monitoring, reporting, and control activities strengthens quarantine measures. Citizen science initiatives, such as apps for reporting pest sightings, enable early detection and rapid response to pest threats.

Incentives for Compliance

Offering incentives, such as subsidies for pest-free certification or financial assistance for implementing quarantine measures, encourages voluntary compliance. Recognizing and rewarding good practices also motivates stakeholders to adhere to regulations. Public awareness and compliance are the backbone of successful quarantine systems. By fostering a culture of shared responsibility, these efforts enhance the effectiveness and sustainability of pest management programs.

8. Ecological Approaches to Pest Control

Ecological approaches to pest control emphasize the restoration and maintenance of natural processes to manage pest populations sustainably. These methods align with the principles of conservation biology and ecosystem management, seeking to harness ecological interactions, enhance biodiversity, and restore balance within forest ecosystems. By reducing dependency on chemical interventions, ecological pest control minimizes environmental risks and promotes resilience against emerging threats, such as climate change and invasive species.

8.1 Enhancing Biodiversity to Suppress Pest Populations

Biodiversity plays a critical role in regulating pest populations by supporting a wide array of natural enemies and disrupting pest-host relationships. Diverse ecosystems provide multiple ecological niches, fostering a balanced community of predators, parasitoids, and competitors that collectively suppress pest species.

Role of Predators and Parasitoids

Biodiversity enhances the abundance and diversity of predators and parasitoids, which are natural enemies of pests. Predators such as birds, spiders, and predatory insects actively hunt and consume pests, while parasitoids like wasps lay their eggs inside or on pest hosts, leading to their eventual death. Mixed-species forests, which include a variety of tree and understory species, provide habitat complexity and resources that sustain these natural enemies throughout their life cycles. For instance, flowering plants in forest understories offer nectar and pollen, attracting parasitoid wasps that target caterpillar pests.

Disruption of Pest-Host Dynamics

Monoculture forests are highly susceptible to pest outbreaks because they provide an uninterrupted food source for specialized pests. In contrast, mixed-species planting disrupts pest-host dynamics by interspersing non-host species, making it more challenging for pests to locate and infest their preferred trees. This strategy has proven effective against pests like spruce budworm (*Choristoneura fumiferana*) and gypsy moth (*Lymantria dispar*).

Ecosystem Services and Resilience

Biodiverse ecosystems are inherently more resilient to pest outbreaks due to their ability to absorb and recover from disturbances. For example, diverse forests maintain critical ecosystem services such as pollination, nutrient cycling, and soil stability, which indirectly contribute to pest management by promoting tree health and vigor. Enhancing biodiversity requires deliberate management practices, including the protection of native species, habitat restoration, and the introduction of complementary plant and animal species. While these measures often take time to yield results, their long-term benefits for pest suppression and ecosystem health are substantial.

8.2 Role of Microbial Communities in Pest Suppression

Microbial communities in soil, plants, and forest environments play a pivotal role in suppressing pest populations through direct and indirect mechanisms. These communities include bacteria, fungi, viruses, and nematodes that interact with pests, their hosts, and the broader ecosystem.

Pathogens as Biocontrol Agents

Many microbial organisms act as pathogens that infect and kill pests. For example, entomopathogenic fungi such as *Beauveria bassiana* and *Metarhizium anisopliae* are effective against a wide range of pests, including aphids, beetles, and caterpillars. These fungi penetrate the pest cuticle, proliferate within the host, and ultimately cause death. Similarly, bacterial agents like *Bacillus thuringiensis* (Bt) produce insecticidal toxins that disrupt the digestive systems of pests, particularly caterpillars and beetles.

Soil Microbial Interactions

Healthy soil microbial communities contribute to pest suppression by enhancing plant vigor and directly competing with soilborne pests. Mycorrhizal fungi, for instance, form symbiotic associations with tree roots, improving nutrient uptake and increasing resistance to root-feeding pests. Beneficial bacteria in the rhizosphere produce antimicrobial compounds that inhibit the growth of nematodes and pathogenic fungi.

Virus-Based Pest Control

Viruses such as nucleopolyhedroviruses (NPVs) are highly specific to certain pest species, making them valuable tools for targeted pest control. These viruses infect the cells of caterpillars and other pests, causing systemic infections that lead to high mortality rates. NPVs are particularly effective against pests like gypsy moths and armyworms. Maintaining and enhancing microbial diversity requires sustainable soil management practices, such as minimizing chemical inputs, adding organic matter, and avoiding soil compaction. These practices support robust microbial communities that naturally suppress pests and promote forest health.

8.3 Rewilding Practices to Restore Ecosystem Balance

Rewilding involves restoring natural processes and reintroducing native species to rebuild ecological balance and resilience. In the context of pest management, rewilding aims to reestablish predator-prey dynamics, enhance ecosystem stability, and reduce the likelihood of pest outbreaks.

Reintroducing Keystone Species

Keystone species, such as apex predators and herbivores, play a crucial role in regulating ecosystem dynamics. Their reintroduction can have cascading effects that suppress pest populations indirectly. For instance, reintroducing carnivores like wolves or foxes can reduce populations of pest-carrying rodents, while herbivores like deer can shape forest structure in ways that influence pest habitats.

Restoring Natural Vegetation

Rewilding efforts often focus on restoring native vegetation to create habitats that support diverse communities of natural enemies. For example, replanting native understory plants can provide food and shelter for predatory insects and birds, enhancing their effectiveness in pest control.

Promoting Ecological Succession

Allowing forests to undergo natural succession processes can help rebuild ecosystem complexity, reducing vulnerability to pest outbreaks. Early successional forests often harbor diverse species that contribute to pest suppression, while mature forests provide stable habitats for long-term ecosystem health. Rewilding practices require careful planning and monitoring to ensure compatibility with existing land uses and ecological conditions. While they may face challenges such as public acceptance and funding, their potential to restore ecosystem function and resilience makes them a valuable component of ecological pest control.

8.4 Long-Term Effects of Ecological Pest Control

Ecological pest control methods offer significant long-term benefits for forest ecosystems by promoting sustainability, resilience, and biodiversity. Unlike chemical interventions, which often provide short-term solutions, ecological approaches address the underlying causes of pest problems, leading to enduring pest suppression and ecosystem stability.

Reduction in Pest Outbreak Frequency

By enhancing natural pest regulation through biodiversity, microbial activity, and ecosystem restoration, ecological approaches reduce the likelihood and severity of pest outbreaks. This stability benefits forest productivity, carbon sequestration, and other ecosystem services.

Minimization of Environmental Impacts

Ecological methods minimize the environmental risks associated with chemical pesticides, such as contamination of soil and water, harm to non-target species, and disruption of ecological interactions. This reduction in environmental impact aligns with global conservation goals and supports the health of adjacent ecosystems.

Adaptation to Emerging Threats

Ecological approaches enhance the adaptive capacity of forests to withstand emerging threats, including climate change and invasive pests. By fostering diverse and resilient ecosystems, these methods provide a buffer against unpredictable environmental changes and pest dynamics.

Economic and Social Benefits

Sustainable pest management practices reduce long-term management costs by decreasing the reliance on expensive chemical treatments and minimizing economic losses from pest damage. Additionally, these practices promote community engagement and support for conservation initiatives, fostering a shared sense of stewardship for forest resources. Implementing ecological pest control requires a shift in management philosophy, prioritizing long-term ecosystem health over immediate gains. However, the cumulative benefits of these methods far outweigh the initial investments, making them an essential component of sustainable forest management.

Conclusion

The management of pest damage in forest ecosystems is a critical component of sustainable forestry, requiring a combination of biological, cultural, chemical, physical, and ecological strategies. Pests such as defoliators, borers, sap-suckers, and root feeders threaten forest health by causing direct damage, weakening trees, and facilitating the spread of secondary pathogens. These impacts extend beyond individual trees, disrupting forest structure, reducing biodiversity, and compromising the ecosystem services forests provide, including carbon sequestration, water regulation, and habitat preservation. Effective pest control is essential for maintaining the resilience and productivity of forests, particularly in the context of climate change and increasing human influence on forest ecosystems. Biological control methods, which leverage natural predators, parasitoids, and pathogens, play a pivotal role in sustainable pest management. Predators such as lady beetles, lacewings, and spiders reduce pest populations through direct predation, while parasitoids like parasitic wasps offer target-specific control by exploiting pest hosts. Pathogens, including fungi, bacteria, and viruses, naturally suppress pest populations, often leading to widespread mortality. Conservation and augmentation of these natural enemies strengthen ecological resilience, reducing reliance on chemical pesticides and aligning pest management with environmental preservation goals.

Cultural and silvicultural practices complement biological control by modifying forest conditions to make them less hospitable to pests. Sanitation measures, such as the removal of infested materials, interrupt pest life cycles and reduce breeding sites. Mixed-species planting enhances biodiversity, disrupting pest-host relationships and preventing monoculture-associated outbreaks. Thinning and pruning improve airflow and sunlight penetration, creating unfavorable conditions for many pests. Timing of planting and harvesting allows managers to avoid periods of peak pest activity, further reducing the risk of infestations. These practices integrate pest control into broader forest management strategies, supporting long-term ecological and economic goals.

Chemical control methods remain vital for addressing severe infestations and protecting high-value forest areas. Contact insecticides provide immediate relief by targeting visible pests, while systemic insecticides are effective against sap-suckers and borers that are otherwise challenging to manage. Pheromones, used for monitoring and mating disruption, represent a significant advancement in chemical control, allowing for species-specific interventions with minimal non-target effects. However, the challenges of resistance development, environmental contamination, and non-target impacts necessitate careful application and integration with other pest management strategies. Physical and mechanical control methods offer targeted and environmentally friendly options for pest suppression. Trapping systems, such as pheromone traps and light traps, effectively capture specific pest species, aiding both monitoring and control efforts. Barriers, tree wrappings, and manual removal of pest colonies provide localized solutions that can significantly reduce pest pressures. Soil preparation techniques disrupt pest life cycles, particularly for root-feeding pests, further enhancing control measures. While labor-intensive, these methods contribute to integrated pest management frameworks by minimizing chemical reliance and environmental impacts.

Integrated Pest Management (IPM) combines these approaches into a holistic strategy, emphasizing monitoring, prevention, and adaptive management. By integrating biological, cultural, chemical, and mechanical methods, IPM minimizes the overuse of any single approach, reducing risks of resistance and environmental harm. Community-based programs within IPM frameworks foster collaboration and collective action, enhancing the efficacy of pest management efforts. Adaptability to emerging threats and changing environmental conditions ensures that IPM remains effective in the face of evolving challenges.

Chapter - 9

Addressing Forest Threats

1. Introduction to Forest Threats

Forests, as dynamic ecosystems, are vital for maintaining global biodiversity, regulating climate, and providing essential resources and ecosystem services. However, these ecosystems face a multitude of threats that endanger their health, functionality, and longevity. Forest threats encompass a diverse array of biotic and abiotic factors, as well as emerging challenges influenced by human activity and environmental changes. Understanding these threats and addressing their interconnected nature is critical for safeguarding forest ecosystems in a rapidly changing world. The growing complexity of forest threats, driven by climate change, globalization, and unsustainable practices, underscores the need for proactive and integrated management strategies that ensure the resilience and sustainability of forests for future generations. Defining forest threats in a global context involves recognizing their multifaceted nature and the scale at which they operate. Forests around the world are under pressure from invasive species, pathogens, pests, deforestation, climate change, pollution, and habitat fragmentation. Each of these threats poses unique challenges while often interacting with others to exacerbate their impact. For example, climate change not only stresses forest ecosystems directly through rising temperatures and altered precipitation patterns but also creates conditions conducive to the proliferation of invasive species, pests, and pathogens. Similarly, deforestation and habitat fragmentation reduce forest resilience, making them more vulnerable to abiotic stressors and biotic invasions.

The importance of proactive management strategies in addressing forest threats cannot be overstated. Reactive approaches often come too late, resulting in significant ecological and economic losses. Proactive measures, such as early detection systems, ecological restoration, and policy interventions, offer the potential to mitigate threats before they escalate into crises. These strategies not only reduce the direct impacts of forest threats but also enhance the adaptive capacity of forest ecosystems, enabling them to cope with future challenges. By adopting a forward-looking approach, forest managers and policymakers can safeguard the ecological integrity and socio-economic value of forests. One of the defining characteristics of forest threats is their interconnected nature. Biotic and abiotic threats rarely operate in isolation; rather, they interact in complex ways to create cascading effects that amplify their impact. For instance, invasive species often outcompete native flora and fauna, disrupting ecosystem balance and increasing susceptibility to pests and diseases. Climate change, by altering

DOI: 10.1201/9781042011063-9

temperature and precipitation patterns, accelerates these processes while introducing new stressors, such as droughts and wildfires. These interactions highlight the need for integrated management strategies that address the root causes of forest threats while considering their synergistic effects. Biotic threats are among the most pressing challenges facing forests, with invasive species, pathogens, and herbivorous pests causing widespread damage. Invasive species, often introduced through global trade and human movement, disrupt native ecosystems by competing for resources, altering habitat structure, and displacing native species. Pathogens and disease outbreaks, such as sudden oak death and ash dieback, further exacerbate forest decline by weakening trees and reducing biodiversity. Herbivorous pests, including defoliators and borers, inflict direct damage on trees, impairing growth and increasing mortality rates. Addressing these biotic threats requires a combination of prevention, monitoring, and control measures tailored to specific contexts.

Abiotic threats, including climate change, deforestation, and pollution, compound the challenges of forest management. Climate change alters forest dynamics by increasing the frequency and severity of extreme weather events, shifting species ranges, and disrupting phenological cycles. Deforestation and habitat fragmentation reduce forest cover, disrupt wildlife corridors, and increase edge effects, making forests more vulnerable to biotic and abiotic stressors. Pollution, particularly soil and air contamination, further degrades forest health by impairing nutrient cycling and soil fertility. These abiotic threats necessitate large-scale interventions, such as reforestation, habitat restoration, and sustainable land-use practices. Emerging threats, driven by globalization and urbanization, present new challenges for forest ecosystems. Global trade and human movement facilitate the rapid spread of novel pests and pathogens, often outpacing the capacity of existing management systems to respond. Urbanization exerts additional pressure on forests by encroaching on natural habitats, fragmenting ecosystems, and introducing anthropogenic stressors. These emerging threats require innovative solutions, including enhanced biosecurity measures, urban planning that prioritizes green spaces, and international cooperation to address transboundary challenges.

Integrated response strategies are essential for mitigating the impacts of forest threats and building ecosystem resilience. Early detection and rapid response systems enable timely interventions, reducing the spread and impact of invasive species, pests, and pathogens. Policy frameworks, both at national and international levels, provide the regulatory foundation for coordinated action and resource allocation. Community engagement and public awareness initiatives foster local stewardship, ensuring that forest conservation efforts are inclusive and sustainable. By integrating these components, forest managers can address threats more effectively and equitably. Technological innovations offer powerful tools for mitigating forest threats and enhancing management practices. Biotechnology and genetic engineering are revolutionizing forest resilience by enabling the development of pest- and disease-resistant tree species. Remote sensing and predictive modeling enhance monitoring capabilities, providing real-time data on forest health and enabling proactive decision-making. Ecological restoration practices, including reforestation with native species and habitat enhancement, contribute to long-term ecosystem recovery while addressing the root causes of forest decline. These technologies, when combined with traditional knowledge and practices, create a robust foundation for sustainable forest management.

Building resilience in forest ecosystems is a long-term goal that requires a holistic and adaptive approach. Enhancing forest diversity and connectivity strengthens ecological networks, enabling forests to withstand and recover from disturbances. Adaptive management, which involves continuously refining strategies based on monitoring and feedback, ensures that management practices remain effective in the face of changing environmental conditions. Ecosystem-based approaches, which prioritize the health and functionality of entire ecosystems, provide a comprehensive framework for achieving sustainability. By focusing on resilience, forest managers can ensure that forests continue to provide essential services and benefits for future generations.

1.1 Defining Forest Threats in a Global Context

Forest threats encompass a wide range of biotic and abiotic factors that adversely affect forest ecosystems. Biotic threats include invasive species, pests, and pathogens that disrupt ecological balance and degrade forest health. Abiotic threats, on the other hand, stem from environmental stressors such as climate change, deforestation, pollution, and extreme weather events. These threats are not only local but also global in scope, with their impacts transcending geographical boundaries and requiring coordinated international efforts to address them. The rapid globalization of trade and human movement has exacerbated the spread of invasive species and pathogens, introducing foreign pests into ecosystems where they lack natural predators or controls. For example, the emerald ash borer (*Agrilus planipennis*) and the Asian longhorned beetle (*Anoplophora glabripennis*) have caused significant damage to North American forests after being introduced through international trade routes. Similarly, climate change has emerged as a dominant abiotic threat, altering temperature and precipitation patterns, which in turn affect pest behavior, tree physiology, and overall forest dynamics. Defining forest threats also involves recognizing their economic, ecological, and social dimensions. Forest degradation impacts livelihoods, disrupts ecosystem services, and contributes to greenhouse gas emissions. Consequently, the need to classify, prioritize, and respond to these threats has become a critical component of global forest management.

1.2 Importance of Proactive Management Strategies

Proactive management strategies are essential for mitigating forest threats before they escalate into severe crises. Historically, many forest management efforts have been reactive, addressing threats only after significant damage has occurred. However, the increasing complexity and scale of modern threats necessitate a shift toward prevention, early detection, and rapid response. Proactive management begins with robust monitoring systems that enable the early detection of emerging threats. Technological advancements, such as remote sensing, Geographic Information Systems (GIS), and artificial intelligence (AI), have enhanced the ability to track changes in forest health and predict potential risks. For instance, satellite imagery can detect early signs of deforestation or pest outbreaks, while predictive models can forecast the spread of invasive species under varying climatic scenarios. In addition to technological tools, proactive management requires policy frameworks that prioritize forest health and resilience. International agreements, such as the Convention on Biological Diversity (CBD) and the United Nations Framework Convention on Climate Change (UNFCCC), emphasize the importance of sustainable forest management in achieving global conservation and climate goals. National and regional policies, including quarantine measures and reforestation programs, play a crucial role in implementing these objectives on the ground. Furthermore, proactive strategies emphasize the integration of ecological approaches, such as enhancing biodiversity and restoring degraded landscapes, to build forest resilience against threats. By addressing the root causes of vulnerability, these strategies ensure long-term sustainability while reducing the need for costly interventions.

1.3 Interconnected Nature of Forest Threats

Forest threats are often interconnected, with one type of threat exacerbating or triggering others. For example, climate change not only stresses trees but also creates favorable conditions for pests and pathogens, leading to compounded impacts on forest health. Droughts associated with climate change weaken tree defenses, making them more susceptible to bark beetle infestations. Similarly, deforestation and habitat fragmentation increase the vulnerability of forest ecosystems to invasive species and reduce their capacity to recover from disturbances. The interconnected nature of forest threats underscores the need for integrated management approaches that consider the broader ecological and socio-economic context. Addressing a single threat in isolation may yield temporary benefits but fail to achieve long-term solutions. For instance, the eradication of an invasive species without restoring the underlying ecosystem processes that support native biodiversity may lead to new invasions or other unintended consequences. Human activities further amplify the interconnections between threats. Unsustainable logging practices, agricultural expansion, and urbanization not only degrade forests but also contribute to greenhouse gas emissions, accelerating climate change. The loss of forest cover reduces the capacity of ecosystems to regulate water cycles and buffer against extreme weather events, creating a feedback loop that intensifies abiotic stressors. Recognizing these interconnections requires a systems-based perspective that integrates forest management with broader land-use planning, climate adaptation, and biodiversity conservation efforts. Collaborative approaches involving multiple stakeholders, from local communities to international organizations, are essential for addressing the complex web of interactions that shape forest threats.

2. Biotic Threats

Biotic threats to forests arise from living organisms that disrupt ecological balance, damage forest health, and diminish the resilience of ecosystems. These threats include invasive species, pathogens, and herbivorous pests, which collectively contribute to significant ecological and economic losses. Unlike abiotic stressors, biotic threats often have the capacity to proliferate and adapt, making their management particularly challenging. As globalization and climate change intensify, the frequency and severity of biotic threats are expected to rise, necessitating proactive and integrated management approaches.

2.1 Invasive Species: Impact on Native Biodiversity

Invasive species are among the most significant biotic threats to forest ecosystems. These non-native organisms are introduced, intentionally or accidentally, into regions outside their natural range, where they often establish, proliferate, and displace native species. Invasive species can include plants, animals, and microorganisms, all of which can disrupt ecological processes, alter habitats, and reduce biodiversity.

Ecological Impact

Invasive species compete with native flora and fauna for resources, often outcompeting them due to the absence of natural predators or pathogens in their introduced range. For example, invasive tree species like *Ailanthus altissima* (tree of heaven) dominate forest understories, suppressing native plants and altering nutrient cycling. Similarly, invasive vines such as kudzu (*Pueraria montana*) smother native vegetation and disrupt forest structure. Invasive animals also pose significant threats. For instance, the emerald ash borer (*Agrilus planipennis*), introduced to North America from Asia, has devastated ash tree populations, leading to cascading ecological effects on species that rely on these trees for habitat or food. In aquatic forested wetlands, invasive species like Asian carp disrupt food webs and affect water quality, further impacting surrounding ecosystems.

Economic Consequences

Invasive species impose substantial economic costs on forestry, agriculture, and conservation. The costs associated with managing invasive species, restoring affected areas, and compensating for lost ecosystem services amount to billions of dollars annually. For example, the management of invasive species like the gypsy moth (*Lymantria dispar*), which defoliates trees and weakens forest stands, incurs significant expenses for monitoring, control measures, and reforestation efforts.

Management Challenges

Controlling invasive species is challenging due to their rapid reproduction, wide dispersal, and adaptability to diverse environments. Early detection and rapid response (EDRR) systems are crucial for preventing their establishment, while long-term management often involves a combination of mechanical, chemical, and biological methods. Public awareness campaigns and international collaboration are also vital to addressing the global nature of invasive species.

2.2 Pathogens and Disease Outbreaks in Forest Ecosystems

Pathogens, including fungi, bacteria, viruses, and nematodes, are significant biotic threats that cause widespread disease outbreaks in forest ecosystems. These outbreaks can lead to extensive tree mortality, reduced productivity, and long-term ecosystem degradation.

Notable Disease Outbreaks

Pathogens have caused some of the most devastating forest disease epidemics in history. The chestnut blight (*Cryphonectria parasitica*), introduced to North America in the early 20th century, nearly eradicated American chestnut trees, altering forest composition and ecosystem dynamics. Similarly, sudden oak death (*Phytophthora ramorum*), caused by an invasive oomycete, has decimated oak and tanoak populations in the western United States, with ripple effects on associated species and habitats.

Mechanisms of Spread and Impact

Pathogens spread through various pathways, including airborne spores, water movement, and human activities such as logging and plant trade. The introduction of infected plant materials or soil can facilitate the rapid establishment of pathogens in new regions. Once established, pathogens weaken tree defenses, reduce growth rates, and increase susceptibility to secondary pests. Climate change exacerbates the impacts of pathogens by altering their distribution, virulence, and interactions with hosts. For example, warming temperatures and increased precipitation create favorable conditions for the proliferation of fungal pathogens like *Fusarium* and *Armillaria*.

Management Strategies

Addressing forest diseases requires an integrated approach that combines early detection, quarantine measures, and the deployment of resistant tree varieties. Advances in molecular diagnostics, such as

DNA-based tools, enable rapid identification of pathogens, improving the effectiveness of monitoring and control programs. Biological control agents, including antagonistic fungi and bacteria, are also being explored as eco-friendly alternatives to chemical treatments.

2.3 Herbivorous Pests

Herbivorous pests, such as insects and mammals, pose a persistent threat to forest health by feeding on leaves, bark, roots, and other tree tissues. Their feeding activity weakens trees, disrupts physiological processes, and predisposes forests to secondary threats like pathogens and drought.

Insect Pests

Defoliators, bark beetles, and sap-sucking insects are among the most damaging herbivorous pests. Defoliators, such as gypsy moths and forest tent caterpillars, strip trees of their foliage, reducing photosynthesis and growth. Repeated defoliation can lead to tree mortality and increased vulnerability to secondary pests. Bark beetles, including the mountain pine beetle (*Dendroctonus ponderosae*), tunnel into tree bark to lay eggs, disrupting vascular tissues and introducing fungal pathogens that further compromise tree health. Sap-sucking insects, such as aphids and scales, drain nutrients from trees, weakening their overall vigor.

Mammalian Herbivores

Large herbivores like deer and rodents also contribute to forest degradation. Overbrowsing by deer can prevent forest regeneration, particularly in young plantations or recovering areas. Rodents, including voles and squirrels, girdle tree stems or consume seeds, reducing the reproductive success of certain species.

Control Challenges and Strategies

Controlling herbivorous pests is complicated by their reproductive potential, mobility, and adaptability. For insect pests, chemical controls, such as insecticides, provide rapid relief but can have negative environmental consequences. Biological controls, including the release of natural predators, parasitoids, or pathogens, offer a more sustainable alternative. Cultural practices, such as mixed-species planting and maintaining tree health through proper silvicultural techniques, can reduce susceptibility to herbivorous pests.

3. Abiotic Threats

Abiotic threats are non-living factors that adversely affect forest ecosystems, leading to declines in biodiversity, productivity, and ecological stability. Unlike biotic threats, which arise from living organisms such as pests or pathogens, abiotic threats stem from environmental changes and anthropogenic activities. Climate change, deforestation, habitat fragmentation, pollution, and soil degradation represent some of the most pervasive and challenging abiotic threats. These factors not only harm forests directly but also exacerbate biotic pressures, creating compounding impacts that undermine forest resilience.

3.1 Climate Change: Shifting Dynamics in Forest Health

Climate change is one of the most significant and far-reaching abiotic threats to forests. It alters the environmental conditions that govern forest ecosystems, including temperature, precipitation, and seasonal patterns, leading to profound changes in forest health and dynamics.

Impact on Forest Physiology and Growth

Changes in temperature and precipitation patterns directly affect tree physiology, altering growth rates, reproduction, and survival. For example, prolonged droughts associated with climate change reduce water availability, weakening tree defenses against pests and diseases. Conversely, excessive rainfall and flooding can cause root asphyxiation and increase the spread of soilborne pathogens like *Phytophthora*. Warming temperatures also extend the growing season in some regions, potentially enhancing tree growth in temperate forests but stressing species adapted to cooler climates.

Pest and Pathogen Proliferation

Climate change exacerbates biotic threats by creating favorable conditions for pests and pathogens. Warmer winters reduce the mortality of overwintering pest populations, while extended growing seasons allow for multiple pest generations per year. For instance, the mountain pine beetle (*Dendroctonus ponderosae*) has expanded its range into higher altitudes and latitudes due to warming temperatures, causing widespread destruction of pine forests in North America.

Shifts in Forest Composition and Distribution

Climate change drives shifts in forest composition and geographic distribution as species migrate to adapt to changing conditions. Some tree species may move toward higher elevations or latitudes in search of suitable habitats, while others may face local extinction if they cannot adapt or migrate quickly enough. These shifts disrupt existing ecosystems, altering food webs, nutrient cycling, and ecosystem services.

Management Strategies

Addressing the impacts of climate change requires adaptive management approaches. Strategies include selecting climate-resilient tree species for reforestation, enhancing forest connectivity to facilitate species migration, and implementing silvicultural practices that improve water use efficiency and soil health. Monitoring systems, such as remote sensing and predictive modeling, are essential for tracking changes in forest health and informing proactive interventions.

3.2 Deforestation and Habitat Fragmentation

Deforestation and habitat fragmentation are critical abiotic threats driven primarily by human activities, including agriculture, urbanization, logging, and infrastructure development. These processes result in the loss and degradation of forest cover, leading to significant ecological, economic, and social consequences.

Loss of Biodiversity

Deforestation eliminates habitats for countless plant and animal species, leading to local extinctions and a decline in biodiversity. Habitat fragmentation compounds this effect by isolating populations, reducing genetic diversity, and disrupting ecological interactions. For example, forest-dependent species such as orangutans and jaguars face severe population declines due to the loss and fragmentation of their habitats.

Disruption of Ecosystem Services

Forests provide critical ecosystem services, including carbon sequestration, water regulation, and soil stabilization. Deforestation disrupts these services, contributing to climate change, altering hydrological cycles, and increasing the risk of soil erosion and landslides. Fragmented forests are also less effective at regulating local climates, exacerbating temperature extremes and altering precipitation patterns.

Edge Effects

Fragmentation creates "edge effects," where the conditions at forest edges differ significantly from interior areas. Edges are often hotter, drier, and more susceptible to invasive species, pests, and human disturbances. These changes reduce the overall quality and resilience of the forest ecosystem.

Management Strategies

Efforts to combat deforestation and fragmentation include afforestation and reforestation projects, sustainable logging practices, and the establishment of protected areas. Restoring connectivity through wildlife corridors and buffer zones helps mitigate fragmentation and supports species migration. International agreements, such as REDD+ (Reducing Emissions from Deforestation and Forest Degradation), incentivize conservation and sustainable forest management by linking these efforts to climate change mitigation goals.

3.3 Pollution and Soil Degradation

Pollution and soil degradation are pervasive abiotic threats that compromise forest health by altering soil properties, water quality, and air composition. These factors directly impact tree growth, biodiversity, and ecosystem functionality.

Air Pollution

Airborne pollutants, including sulfur dioxide, nitrogen oxides, and particulate matter, contribute to acid rain and ozone formation, which harm forests. Acid rain alters soil chemistry by leaching essential nutrients like calcium and magnesium while increasing the availability of toxic aluminum ions. This nutrient imbalance weakens trees and reduces their resistance to pests and diseases. Ground-level ozone, a byproduct of industrial emissions, damages leaf tissues, impairing photosynthesis and growth.

Soil Degradation

Soil degradation encompasses a range of processes, including erosion, compaction, nutrient depletion, and contamination. Unsustainable agricultural practices, deforestation, and industrial activities accelerate soil erosion, reducing the availability of fertile land for forest regeneration. Soil compaction, caused by heavy machinery or overgrazing, restricts root growth and water infiltration, further limiting tree health. Pollution from industrial and agricultural runoff introduces heavy metals, pesticides, and other toxic substances into forest soils and waterways. These contaminants disrupt microbial communities, reduce nutrient availability, and harm aquatic ecosystems within forested landscapes.

Water Pollution

Forests adjacent to agricultural or industrial areas often suffer from water pollution, which affects both terrestrial and aquatic ecosystems. Excessive nutrients from fertilizers cause eutrophication in water bodies, leading to oxygen depletion and the loss of aquatic biodiversity. Pollutants such as oil and heavy metals can infiltrate forest soils, reducing tree health and productivity.

Management Strategies

Addressing pollution and soil degradation requires a combination of restoration and prevention measures. Rehabilitating degraded soils through organic amendments, reforestation, and erosion control techniques helps restore ecosystem functionality. Reducing industrial emissions and adopting sustainable agricultural practices, such as agroforestry, minimizes the introduction of pollutants into forest systems. Policies regulating pollutant discharge and promoting cleaner technologies are critical for mitigating the impacts of air and water pollution.

4. Emerging Threats

Emerging threats to forest ecosystems represent a dynamic and often unpredictable set of challenges that have escalated in recent decades. These threats are closely linked to human activity, global trade, urbanization, and environmental changes, which together disrupt forest health and stability. Emerging threats often involve new or intensifying interactions between biotic and abiotic factors, such as the introduction of novel pests and pathogens, increased fragmentation of forest habitats, and the expansion of urban areas.

4.1 Impact of Global Trade and Human Movement

Global trade and human movement are among the primary drivers of emerging threats to forests, facilitating the introduction and spread of invasive species, pests, and pathogens. The interconnectedness of the global economy, coupled with advancements in transportation, has dramatically increased the volume and speed of goods and people crossing international borders. While these developments have brought numerous economic benefits, they have also created pathways for the inadvertent transfer of forest threats, often with devastating consequences.

Invasive Species and Pest Introduction

The globalization of trade has significantly increased the risk of invasive species infiltrating forest ecosystems. Wood packaging materials, shipping containers, and plant imports are common vectors for pests and pathogens. For example, the emerald ash borer (*Agrilus planipennis*), an invasive beetle native to Asia, was introduced to North America through infested wood packaging materials. This pest has caused widespread devastation of ash trees across the United States and Canada, leading to severe ecological and economic losses. Similarly, the Asian longhorned beetle (*Anoplophora glabripennis*), another invasive species introduced through trade, has targeted hardwood species such as maple, birch, and elm in North America and Europe. Its spread has required costly eradication programs, including the removal of thousands of infested trees, which has further fragmented forests and reduced urban canopy cover.

Pathogen Spread through Trade

Pathogens often hitchhike on nursery plants, seeds, and soil transported across borders. For instance, sudden oak death (*Phytophthora ramorum*), an oomycete pathogen responsible for significant oak and tanoak mortality in the western United States, is believed to have been introduced through infected nursery stock. The pathogen's spores are easily dispersed by rain, wind, and human activity, complicating containment efforts.

Human Movement and Recreational Activities

Human movement, including tourism and recreational activities, also contributes to the spread of forest threats. Hikers, campers, and vehicles can inadvertently transport invasive plant seeds, fungal spores, or pest eggs into pristine forest areas. Firewood transported for camping has been a significant vector for pests like bark beetles and borers. Public awareness campaigns, such as "Don't Move Firewood," aim to reduce these risks by educating people about the importance of sourcing firewood locally.

Management and Policy Responses

Mitigating the impacts of global trade and human movement requires robust international and national policies. The International Plant Protection Convention (IPPC) establishes phytosanitary measures to reduce the risk of pest introductions through trade. Protocols such as ISPM 15 mandate the treatment of wood packaging materials to eliminate pests. National inspection programs, coupled with early detection systems, are essential for identifying and intercepting threats before they establish. Public awareness and compliance with regulations also play a critical role in minimizing human-mediated introductions.

4.2 Novel Pests and Pathogens

The emergence of novel pests and pathogens poses unanticipated challenges to forest ecosystems, often with devastating impacts. These threats arise from the introduction of previously unknown species, the evolution of new virulent strains, or the spread of pests and pathogens into areas where they have no natural enemies or controls. The complexity of these challenges is compounded by the difficulty in predicting their occurrence and the lack of immediate management solutions.

Characteristics of Novel Threats

Novel pests and pathogens often exhibit aggressive behavior in their new environments, exploiting ecological niches and host species that are ill-equipped to resist them. For example, the fungus-like pathogen *Hymenoscyphus fraxineus*, responsible for ash dieback in Europe, was introduced from Asia, where it coexists with native ash species. In Europe, however, the pathogen has caused widespread mortality in ash populations due to the lack of co-evolved resistance mechanisms. Similarly, the arrival of the pinewood nematode (*Bursaphelenchus xylophilus*) in European pine forests has raised significant concerns. Native to North America, this nematode is spread by beetles and causes a lethal disease in pine trees, particularly under stressed conditions.

Rapid Proliferation and Impact

Novel threats often spread rapidly due to their ability to exploit vulnerable host populations and favorable environmental conditions. The proliferation of these threats leads to direct tree mortality, reduced forest productivity, and cascading ecological effects. For instance, the laurel wilt disease, caused by the fungal pathogen *Raffaelea lauricola*, has decimated populations of native laurel species in the southeastern United States, affecting both forest ecosystems and the avocado industry.

Compounding Effects of Climate Change

Climate change exacerbates the impact of novel pests and pathogens by altering their geographic ranges, life cycles, and interactions with host species. Warmer temperatures and shifting precipitation patterns create conditions conducive to the establishment and spread of new threats. For example, the southern pine beetle (*Dendroctonus frontalis*), previously restricted to the southeastern United States, has expanded its range into northeastern states due to milder winters.

Challenges in Management

Managing novel pests and pathogens is particularly challenging due to the lack of pre-existing knowledge and control measures. Early detection and rapid response are critical but require advanced diagnostic tools, robust monitoring systems, and coordinated action. Research into host resistance, biological control agents, and targeted treatments is essential for developing long-term solutions.

4.3 Urbanization and Its Pressure on Forest Ecosystems

Urbanization exerts significant pressure on forest ecosystems, contributing to habitat loss, fragmentation, and the introduction of new threats. As cities expand, they encroach on forested areas, altering the natural landscape and creating environments that facilitate pest and pathogen spread. Urban forests, while providing essential ecosystem services, are particularly vulnerable to these pressures due to their proximity to human activities and fragmented nature.

Habitat Loss and Fragmentation

Urban expansion often leads to the conversion of forested land into residential, commercial, or industrial areas. This process reduces the extent of forest cover and isolates remaining patches, making them more susceptible to environmental stressors and invasions. Fragmented forests experience edge effects, where increased sunlight, wind exposure, and human activity alter microclimates, favoring invasive species and pests.

Urban Heat Island Effect

The urban heat island effect, characterized by elevated temperatures in urban areas compared to surrounding rural regions, exacerbates stress on urban and peri-urban forests. Higher temperatures can increase the activity and reproduction rates of pests such as aphids, bark beetles, and caterpillars. For instance, warming conditions in urban areas have been linked to increased infestations of the hemlock woolly adelgid (*Adelges tsugae*), which targets hemlock trees.

Introduction of Exotic Species

Urban areas serve as hubs for the introduction of exotic species, as they concentrate human activity, trade, and transportation. Ornamental plants, landscaping materials, and urban green spaces often act as reservoirs for non-native pests and pathogens. For example, the introduction of *Phytophthora* species into urban landscapes has facilitated their spread into nearby natural forests.

Pressure on Urban Forests

Urban forests face unique challenges, including soil compaction, pollution, and vandalism, which weaken tree health and increase susceptibility to pests and diseases. The removal of trees for infrastructure development further reduces canopy cover, limiting the ability of urban forests to provide shade, regulate temperature, and improve air quality.

Management and Mitigation Strategies

Addressing the pressures of urbanization requires integrating urban forestry into broader land-use planning and conservation efforts. Strategies include preserving existing forest patches, establishing green corridors to enhance connectivity, and planting diverse, pest-resistant tree species. Community engagement in urban forestry initiatives, such as tree planting programs and citizen science monitoring, is essential for building resilience and fostering a sense of stewardship among residents.

5. Integrated Response Strategies

Integrated response strategies are essential for managing the complex and interconnected threats facing forest ecosystems today. These strategies aim to combine technological advancements, policy frameworks, and community involvement to create a comprehensive and adaptive approach to forest threat management. By integrating early detection, international cooperation, and public awareness, forest managers and policymakers can address biotic and abiotic challenges proactively and sustainably.

5.1 Early Detection and Rapid Response Systems

Early detection and rapid response (EDRR) systems are critical for preventing the establishment and spread of invasive species, pests, and pathogens in forest ecosystems. These systems focus on identifying potential threats at their inception and implementing swift measures to mitigate their impacts before they escalate into large-scale crises.

Monitoring and Surveillance

Effective EDRR systems rely on robust monitoring and surveillance networks that provide real-time data on forest health and pest activity. Technologies such as remote sensing, Geographic Information Systems (GIS), and Internet of Things (IoT)-enabled sensors are increasingly being deployed to track changes in forest ecosystems. For example, drones equipped with hyperspectral imaging can detect early signs of pest infestations or disease outbreaks, such as canopy discoloration or reduced photosynthetic activity. Ground-based monitoring systems, including pheromone traps and soil sensors, complement aerial surveillance by providing localized data on pest density and behavior.

Rapid Identification and Diagnostics

Accurate and rapid identification of pests and pathogens is essential for EDRR systems. Advances in molecular diagnostics, such as DNA barcoding and polymerase chain reaction (PCR) techniques, allow for the precise identification of invasive species and pathogens at an early stage. Portable diagnostic tools, such as handheld PCR devices, enable on-site identification, reducing the time required to initiate control measures.

Emergency Response Protocols

Rapid response involves deploying targeted interventions to contain and eradicate identified threats.

These interventions may include quarantine measures, chemical treatments, or the removal of infested trees. For example, in the case of an outbreak of the Asian long horned beetle (*Anoplophora glabripennis*), rapid response efforts typically involve identifying and removing infested trees within a designated radius, followed by intensive monitoring to prevent re-infestation.

Challenges and Limitations

While EDRR systems are highly effective, their success depends on adequate funding, technical expertise, and interagency coordination. Additionally, the sheer scale of some forest ecosystems can make comprehensive monitoring challenging, underscoring the need for strategic prioritization and the use of predictive models to identify high-risk areas.

5.2 Policy Frameworks and International Cooperation

Policy frameworks and international cooperation are critical for addressing forest threats that transcend national borders. The interconnected nature of global trade, climate change, and pest dynamics necessitates collaborative approaches to forest management, supported by robust legal and institutional structures.

National and Regional Policies

Effective national policies provide the foundation for forest threat management, outlining regulations, incentives, and enforcement mechanisms to protect forest ecosystems. For example, quarantine laws and phytosanitary regulations are essential for preventing the introduction of invasive species through trade. Additionally, forest management plans that integrate pest control, habitat restoration, and climate adaptation ensure a coordinated response to multiple threats.

At the regional level, transboundary initiatives facilitate the sharing of resources, expertise, and data among neighboring countries. For instance, the European Union's Forest Strategy emphasizes cross-border collaboration to address shared challenges such as invasive species and deforestation.

International Agreements and Standards

International agreements and standards play a pivotal role in harmonizing pest management practices and promoting global cooperation. The International Plant Protection Convention (IPPC) establishes guidelines for preventing the spread of pests through trade, including the International Standards for Phytosanitary Measures (ISPMs). ISPM 15, which mandates the treatment of wood packaging materials, is a prime example of how international standards mitigate pest risks. Multilateral agreements such as the Convention on Biological Diversity (CBD) and the United Nations Framework Convention on Climate Change (UNFCCC) also emphasize the importance of sustainable forest management in achieving global conservation and climate goals.

Data Sharing and Research Collaboration

International cooperation facilitates the sharing of data, research findings, and best practices. Platforms such as the Global Invasive Species Database (GISD) and the Global Forest Watch provide valuable resources for tracking forest threats and informing policy decisions. Collaborative research initiatives, such as those supported by the Food and Agriculture Organization (FAO) and the International Union for Conservation of Nature (IUCN), enhance the collective capacity to address emerging challenges.

5.3 Community Engagement and Public Awareness

Community engagement and public awareness are integral to the success of integrated response strategies, as local stakeholders often serve as the first line of defense against forest threats. Empowering communities to participate in monitoring, reporting, and management efforts enhances the effectiveness of pest control measures and fosters a sense of stewardship for forest resources.

Role of Local Communities

Local communities play a crucial role in detecting and responding to forest threats, particularly in remote or under-monitored areas. Community-based monitoring programs, supported by training and resources, enable residents to identify early signs of pest infestations, diseases, or habitat degradation. For example, citizen science initiatives, such as tree health monitoring apps, allow individuals to report observations and contribute to data collection efforts.

Educational Campaigns

Public awareness campaigns are essential for educating people about the importance of forest health and the actions they can take to mitigate threats. These campaigns often target specific behaviors, such as avoiding the transport of firewood, practicing sustainable land use, or complying with quarantine regulations. Programs like "Don't Move Firewood" have successfully reduced the spread of invasive pests like emerald ash borer and gypsy moth by encouraging responsible practices among campers and homeowners.

Collaborative Decision-Making

Involving communities in decision-making processes ensures that pest management strategies are locally relevant and culturally appropriate. Collaborative approaches, such as participatory planning workshops, foster trust and cooperation between forest managers, policymakers, and local stakeholders. This inclusivity enhances the likelihood of successful implementation and long-term compliance with management measures.

Incentives for Participation

Providing incentives, such as financial compensation for reporting pests or subsidies for adopting sustainable practices, encourages community involvement. Recognizing and rewarding contributions through certification programs or public acknowledgment further motivates individuals and organizations to support forest health initiatives.

6. Innovative Tools for Threat Mitigation

The evolving complexity of threats to forest ecosystems necessitates the development and deployment of innovative tools for threat mitigation. These tools combine cutting-edge technology, advanced scientific research, and sustainable ecological practices to address both immediate and long-term challenges. Biotechnology and genetic engineering are transforming the genetic landscape of forests to enhance their resilience, while technological innovations in monitoring and prediction allow for precise and proactive threat management.

6.1 Biotechnology and Genetic Engineering for Forest Resilience

Biotechnology and genetic engineering are revolutionizing forest management by enabling the development of trees with enhanced resistance to pests, diseases, and environmental stressors. These tools provide a powerful means to address biotic and abiotic threats while preserving ecosystem functionality.

Genetically Modified (GM) Trees

Genetically modified trees are engineered to express specific traits that confer resistance to pests or pathogens. For example, *Populus* species have been genetically modified to produce *Bacillus thuringiensis* (Bt) toxins, which are highly effective against leaf-feeding caterpillars and other defoliators. Similarly, transgenic American chestnut trees have been developed to resist the fungal blight caused by *Cryphonectria parasitica*. These GM trees hold significant potential for restoring species that have been decimated by invasive pathogens while reducing reliance on chemical pesticides.

Gene Editing with CRISPR-Cas Systems

CRISPR-Cas systems have revolutionized genetic engineering by allowing precise editing of tree genomes. This technology enables researchers to enhance traits such as drought tolerance, pest resistance, and rapid growth. For instance, gene editing can be used to knock out susceptibility genes in trees, making them resistant to pathogens like *Phytophthora* or *Hymenoscyphus fraxineus*. Unlike traditional genetic modification, CRISPR-based approaches are highly specific and minimize unintended changes to the genome, making them more acceptable from an environmental and regulatory perspective.

RNA Interference (RNAi)

RNA interference is another promising tool for forest resilience. RNAi-based technologies target specific genes in pests or pathogens, disrupting their ability to reproduce, feed, or infect trees. This approach has been successfully demonstrated in controlling pests like bark beetles and borers, offering a species-specific and environmentally friendly alternative to broad-spectrum insecticides.

6.2 Technological Innovations in Monitoring and Prediction

Technological advancements in monitoring and predictive modeling are transforming the way forest threats are detected, tracked, and managed. These tools provide unprecedented accuracy, speed, and scale in assessing forest health and anticipating future risks.

Remote Sensing and Aerial Surveillance

Remote sensing technologies, including satellite imagery and drone-based systems, enable large-scale monitoring of forest conditions. Hyperspectral and multispectral imaging detect subtle changes in canopy color, structure, and moisture content, which may indicate early signs of pest infestations, diseases, or nutrient deficiencies. Thermal imaging is particularly effective for identifying stressed trees, as they exhibit higher temperatures due to reduced transpiration rates.

6.3 Ecological Restoration Practices

Ecological restoration is a cornerstone of sustainable forest management, aiming to repair degraded ecosystems and enhance their resilience to future threats. Restoration practices are grounded in ecological principles and often involve the active participation of local communities and stakeholders.

Afforestation and Reforestation

Afforestation (planting trees in areas where there were no previous forests) and reforestation (restoring forests in deforested areas) are vital for reversing habitat loss and mitigating climate change. These practices not only sequester carbon but also provide habitat for biodiversity and stabilize soil. Selecting native and climate-resilient tree species is critical for ensuring the long-term success of these efforts.

Enhancing Biodiversity

Restoration projects often prioritize the reintroduction of native species to rebuild ecological complexity. Mixed-species planting creates diverse habitats that support pollinators, natural pest predators, and other beneficial organisms. Biodiversity also enhances ecosystem functions, such as nutrient cycling and water regulation, which are essential for forest health.

Soil Rehabilitation

Restoring soil health is a fundamental aspect of ecological restoration. Techniques such as adding organic amendments, implementing erosion control measures, and reintroducing mycorrhizal fungi help rebuild soil structure and fertility. Healthy soils improve water retention, support robust tree growth, and suppress soilborne pests and pathogens.

Hydrological Restoration

Restoring hydrological systems within forests is critical for mitigating the impacts of drought and ensuring water availability for trees and wildlife. This may involve re-establishing natural watercourses, constructing wetlands, or planting riparian vegetation to stabilize streambanks and improve water quality.

Community-Led Restoration

Community participation is essential for the success and sustainability of restoration projects. Engaging local stakeholders in activities such as tree planting, monitoring, and maintenance fosters a sense of ownership and ensures that restoration efforts align with local needs and knowledge. Incentive-based programs, such as payment for ecosystem services (PES), encourage long-term community involvement.

Long-Term Monitoring and Maintenance

Restoration is a long-term process that requires ongoing monitoring and adaptive management. Regular assessments of tree survival, soil health, and biodiversity help identify challenges and inform corrective actions. Maintenance activities, such as controlling invasive species and preventing fire, are critical for ensuring the success of restoration projects.

7. Building Resilience in Forest Ecosystems

Resilience in forest ecosystems refers to their ability to absorb disturbances, recover from adverse events, and maintain their essential functions over time. As forests face mounting threats from climate change, invasive species, habitat fragmentation, and human activities, building resilience has become a cornerstone of sustainable forest management. Resilient forests are better equipped to adapt to environmental changes, support biodiversity, and continue providing critical ecosystem services such as carbon sequestration, water regulation, and soil stabilization.

7.1 Enhancing Forest Diversity and Connectivity

Diversity and connectivity are fundamental attributes of resilient forest ecosystems. Diverse forests, composed of a variety of species and genetic material, are less vulnerable to pest outbreaks, diseases, and environmental stressors. Connectivity, meanwhile, ensures the movement of species, nutrients, and genetic material across landscapes, supporting ecological processes and reducing the risks associated with habitat fragmentation.

Biodiversity as a Buffer Against Threats

Species diversity enhances the ecological stability of forests by distributing risks across a wider range of organisms. When forests are dominated by a single species or a narrow range of species, they are more susceptible to threats such as pest infestations or climate-induced stress. Mixed-species forests, by contrast, create complex habitats that support a variety of organisms, including pollinators, decomposers, and natural pest predators. For example, planting a combination of broadleaf and conifer species can reduce the risk of monoculture-targeting pests like bark beetles while promoting mutual benefits among species.

Genetic Diversity for Adaptive Potential

Genetic diversity within tree populations is essential for adapting to environmental changes. Trees with diverse genetic traits are more likely to survive stressors such as drought, disease, or extreme temperatures. Conservation programs often prioritize the preservation of genetic material through seed banks, ex situ conservation, and selective breeding of climate-resilient species.

Enhancing Connectivity Through Landscape Management

Habitat connectivity is crucial for maintaining ecological functions and facilitating species movement, especially in fragmented landscapes. Wildlife corridors, riparian buffer zones, and stepping-stone habitats are effective strategies for connecting isolated forest patches. For example, corridors allow pollinators, seed dispersers, and predators to move freely, maintaining biodiversity and ecosystem health. Connectivity also supports the migration of tree species and wildlife in response to climate change, enabling forests to adapt over time.

7.2 Adaptive Management in a Changing Climate

Adaptive management is a dynamic and iterative approach to forest management that emphasizes flexibility, learning, and responsiveness to change. In the context of a changing climate, adaptive management provides a framework for addressing uncertainties, incorporating new knowledge, and adjusting practices to enhance forest resilience.

Monitoring and Feedback Loops

Adaptive management begins with robust monitoring systems that track changes in forest health, pest activity, and environmental conditions. Continuous data collection and analysis allow managers to identify emerging trends, evaluate the effectiveness of interventions, and refine management strategies. For example, monitoring soil moisture levels and tree growth rates during droughts provides insights into the effectiveness of water management practices.

Scenario Planning for Uncertainty

Given the uncertainty of climate change impacts, scenario planning is a key component of adaptive management. This approach involves developing and testing multiple management strategies under different climate scenarios to identify the most effective options. For instance, scenarios may explore the potential impacts of increased temperature variability on pest populations or the effectiveness of mixed-species planting in mitigating drought stress.

Incorporating Climate-Resilient Practices

Adaptive management emphasizes the adoption of climate-resilient practices to buffer forests against environmental stressors. Strategies include planting drought-tolerant tree species, enhancing soil health through organic amendments, and implementing water conservation measures. Thinning and pruning are also used to reduce competition among trees, improving their ability to withstand stress.

7.3 Long-Term Sustainability Through Ecosystem-Based Approaches

Ecosystem-based approaches to forest management focus on maintaining the integrity and functionality of ecosystems as a whole rather than addressing individual threats in isolation. These approaches emphasize the interdependence of species, processes, and environmental factors, fostering resilience and sustainability over the long term.

Holistic Forest Management

Ecosystem-based management integrates ecological, economic, and social objectives to achieve sustainability. Practices such as selective logging, agroforestry, and controlled burns mimic natural processes, maintaining biodiversity and ecosystem functions while supporting local livelihoods. For example, selective logging avoids clear-cutting and preserves habitat complexity, reducing the risk of erosion and biodiversity loss.

Restoring Ecosystem Services

Restoration projects are central to ecosystem-based approaches, focusing on repairing degraded ecosystems and enhancing their capacity to provide services such as carbon storage, water filtration, and soil stabilization. For instance, reforestation efforts in watershed areas improve water quality and reduce the risk of flooding, benefiting both ecosystems and human communities.

Mitigating Cumulative Impacts

Cumulative impacts, such as habitat loss, pollution, and invasive species, require integrated strategies that address multiple stressors simultaneously. For example, combining reforestation with invasive species management and soil rehabilitation creates synergies that amplify the effectiveness of each intervention.

Climate Change Mitigation and Adaptation

Ecosystem-based approaches play a dual role in mitigating and adapting to climate change. Forests sequester significant amounts of carbon, reducing greenhouse gas concentrations, while also providing natural buffers against climate impacts such as storms and heatwaves. Adaptive measures, such as maintaining diverse and connected ecosystems, further enhance forests' ability to cope with changing conditions.

Conclusion

The threats facing forest ecosystems today are diverse, interconnected, and increasingly complex, posing significant challenges to their health, resilience, and sustainability. Forests provide indispensable ecological, economic, and social benefits, including biodiversity conservation, carbon sequestration, and cultural value. However, biotic threats such as invasive species, pathogens, and herbivorous pests, alongside abiotic stressors like climate change, pollution, and deforestation, have escalated the risks to these vital ecosystems. Emerging threats, driven by globalization and urbanization, further complicate forest management by introducing novel challenges, including the spread of new pests and pathogens and increased habitat fragmentation. Addressing these threats is critical for maintaining the integrity of forests and the services they provide. Biotic threats, including invasive species, pathogens, and pests, remain among the most pressing challenges for forest ecosystems. Invasive species disrupt native biodiversity and ecosystem balance, often leading to long-term ecological shifts. Pathogens, such as fungi, bacteria, and viruses, exacerbate forest decline by causing disease outbreaks that weaken trees and reduce biodiversity. Herbivorous pests, including defoliators and borers, directly damage trees, impairing their growth and increasing mortality rates. The interaction of these biotic factors with abiotic stressors, such as drought and temperature extremes, further magnifies their impact, underscoring the need for integrated management approaches that address multiple threats simultaneously.

Abiotic threats, particularly those associated with climate change, pollution, and deforestation, compound the challenges facing forests. Climate change alters forest dynamics by shifting species ranges, increasing the frequency of extreme weather events, and disrupting natural cycles, such as flowering and migration. Pollution, including soil contamination and air pollution, degrades forest health by impairing nutrient cycles and soil fertility. Deforestation and habitat fragmentation not only reduce forest cover but also weaken ecological connectivity, making forests more vulnerable to other stressors. These abiotic threats require large-scale interventions, such as reforestation, habitat restoration, and sustainable land-use practices, to mitigate their effects and enhance forest resilience.

Emerging threats, driven by global trade, human movement, and urbanization, add new layers of complexity to forest conservation. Globalization accelerates the introduction and spread of invasive species, pests, and pathogens, outpacing traditional management systems. Urbanization encroaches on natural habitats, fragmenting ecosystems and introducing anthropogenic pressures, such as pollution and heat islands. These emerging threats demand innovative solutions, including enhanced biosecurity measures, urban planning that integrates green infrastructure, and international cooperation to address transboundary challenges. Effective responses to these threats require an integrated and proactive approach. Early detection and rapid response systems are crucial for identifying and addressing threats before they escalate. Policy frameworks, both national and international, provide the structure for coordinated action, while community engagement and public awareness initiatives foster local stewardship and long-term commitment to forest conservation. Technological innovations, such as remote sensing, predictive modeling, and biotechnology, enhance the ability to monitor, predict, and mitigate threats. The integration of these strategies ensures a comprehensive response to the multifaceted challenges facing forests.

Building resilience in forest ecosystems is essential for their long-term sustainability. Enhancing biodiversity and ecological connectivity strengthens the natural defenses of forests, enabling them to recover from disturbances. Adaptive management, which incorporates continuous learning and refinement of strategies, ensures flexibility in addressing new and evolving threats. Ecosystem-based approaches, which prioritize the health and functionality of entire ecosystems, provide a holistic framework for achieving sustainability. By focusing on resilience, forest managers can secure the ecological and economic benefits forests provide while preserving their value for future generations.

Chapter - 10

Basics of Logging and Its Scope

1. Introduction to Logging

Logging, the process of harvesting trees for timber and other forest products, has been a cornerstone of human development and forest management for centuries. From its humble beginnings as a manual labor-intensive practice to its evolution into a mechanized, globally integrated industry, logging has played a pivotal role in shaping economies, cultures, and landscapes. Today, logging continues to serve as a critical component of forest resource management, providing essential raw materials for construction, paper production, furniture, and countless other industries. However, its impacts on forest ecosystems, biodiversity, and climate have raised significant concerns, necessitating the adoption of more sustainable and regulated practices. Understanding the complexities of logging, including its methods, economic contributions, environmental implications, and future directions, is crucial for balancing the demands of forest resource utilization with the need for conservation and ecological stability.

The definition of logging extends beyond the simple act of tree felling to encompass the entire process of extracting timber and non-timber products from forests. Historically, logging began as a rudimentary activity using basic tools such as axes and saws, primarily for local needs. Over time, technological advancements transformed logging into a mechanized and industrialized operation, enabling large-scale extraction to meet growing global demand. The historical evolution of logging reflects its dual role as both an economic driver and an environmental stressor, illustrating the need for responsible management to mitigate its adverse effects. The importance of logging in forest resource management lies in its ability to provide renewable resources while supporting economic development. Timber, as a renewable resource, can be harvested and replenished through sustainable practices, contributing to long-term forest productivity. Logging also plays a role in forest health by removing dead or diseased trees, reducing fuel loads that exacerbate wildfires, and facilitating regeneration. However, unsustainable logging practices can lead to deforestation, habitat destruction, and biodiversity loss, emphasizing the need for carefully planned and regulated operations that align with conservation goals.

Global trends in logging practices reveal a shift toward more sustainable and technologically advanced methods. The growing demand for timber, driven by urbanization and industrial development,

DOI: 10.1201/9781042011063-10

has prompted the exploration of efficient and environmentally friendly techniques. Reduced Impact Logging (RIL) and selective logging have gained prominence as strategies to minimize ecological damage, while certification schemes such as the Forest Stewardship Council (FSC) have emerged to promote responsible logging practices. These trends highlight the increasing recognition of the need to balance economic benefits with environmental preservation. Logging can be categorized into several types, each with distinct applications, benefits, and challenges. Selective logging involves the removal of specific trees, preserving the overall structure and function of the forest. While beneficial for reducing environmental impact, selective logging requires skilled planning to avoid collateral damage to the remaining vegetation. Clear-cutting, a more intensive approach, involves the removal of all trees in a designated area and is often used in plantation forestry. Despite its efficiency, clear-cutting has significant environmental consequences, including soil erosion, habitat loss, and reduced carbon sequestration. Shelterwood and seed tree systems offer a middle ground, allowing for natural regeneration while maintaining forest cover. Reduced Impact Logging emphasizes techniques that minimize soil disturbance, canopy damage, and waste, representing a shift toward ecological sensitivity in logging operations.

Methods and techniques in logging have evolved alongside technological advancements, ranging from traditional manual logging to modern mechanical and aerial systems. Manual logging, though labor-intensive, remains relevant in small-scale and community-based operations. Mechanical logging, utilizing advanced machinery such as harvesters and skidders, has revolutionized efficiency and safety in large-scale operations. Cable logging and aerial systems, including helicopters and drones, enable access to remote and rugged terrains, reducing environmental impact. Innovations in robotics and automation are further enhancing precision and reducing human labor requirements, paving the way for a more efficient and sustainable future in logging.

Economically, logging is a major industry with significant contributions to global and local markets. The timber trade drives substantial revenue, supporting industries such as construction, paper production, and furniture manufacturing. Value addition through downstream processing, including sawmilling and veneer production, enhances economic output and creates employment opportunities. In addition to timber, logging also provides access to non-timber forest products such as resins, nuts, and medicinal plants, diversifying the economic potential of forests and supporting local livelihoods. However, the environmental and ecological impacts of logging present substantial challenges. Unsustainable logging practices contribute to deforestation, leading to habitat loss, soil erosion, and reduced water quality. The removal of trees impacts the carbon cycle, increasing greenhouse gas emissions and exacerbating climate change. Biodiversity loss, driven by habitat destruction and fragmentation, disrupts ecological balance and threatens the survival of countless species. Mitigation strategies, including afforestation, reforestation, and the adoption of sustainable logging practices, are essential for addressing these impacts and preserving forest ecosystems.

Legal and regulatory frameworks play a critical role in ensuring sustainable logging practices. International standards, such as those established by the FSC and the Programme for the Endorsement of Forest Certification (PEFC), promote responsible forest management and provide market incentives for certified products. National policies and forest certification schemes regulate logging operations, ensuring compliance with environmental and social standards. Combatting illegal logging, a pervasive issue in many regions, requires robust enforcement mechanisms, community-based monitoring, and international cooperation to address the underlying drivers of deforestation and forest degradation. The scope of sustainable logging extends beyond resource extraction to integrate conservation goals, agroforestry, and technological advancements. By aligning logging with ecosystem-based management, it is possible to balance economic needs with ecological preservation. Agroforestry systems, which combine agriculture with sustainable forestry, provide alternative livelihoods while maintaining forest cover. Advances in logging technology, including precision machinery and data-driven planning, enhance sustainability by reducing waste and environmental impact. These approaches highlight the potential for logging to coexist with conservation efforts, contributing to the broader goals of sustainable development.

The social and cultural dimensions of logging underscore its importance to communities and economies worldwide. Logging provides employment and livelihood opportunities, particularly in rural and forest-dependent regions. However, it also impacts indigenous communities, whose cultural and spiritual ties to forests may be disrupted by logging activities. Public perception and ethical considerations play a growing role in shaping logging practices, with increased demand for transparency, sustainability, and equitable benefit-sharing. Education and training in the logging industry are essential for equipping workers with the skills and knowledge needed to implement sustainable practices and adapt to evolving technologies. Looking ahead, the future of logging lies in balancing emerging opportunities with pressing challenges. Global markets are increasingly demanding sustainably sourced timber, creating incentives for adopting environmentally responsible practices. Innovations in machinery, robotics, and gene editing offer new possibilities for efficiency and sustainability in logging operations. Policy shifts toward sustainable forest management, driven by international agreements and public awareness, provide a framework for addressing deforestation and promoting conservation. However, challenges such as climate change, illegal logging, and the need for equitable benefit-sharing require ongoing efforts and adaptive strategies to ensure the long-term viability of logging as a sustainable industry.

1.1 Definition and Historical Evolution of Logging

Logging refers to the process of cutting down trees, extracting logs, and transporting them for commercial use. Traditionally, logging was a labor-intensive activity that relied on manual tools such as axes and saws. Over time, technological advancements revolutionized the practice, introducing mechanized equipment like chainsaws, skidders, and logging trucks, which significantly increased efficiency and scale. Logging encompasses various methods, including selective logging, clear-cutting, and reduced-impact logging, each tailored to specific ecological and economic contexts. The origins of logging can be traced back to ancient civilizations that used wood for fuel, construction, and tools. For instance, in ancient Egypt, timber was harvested for shipbuilding, while in Mesopotamia, wood was essential for constructing irrigation systems. During the Roman Empire, logging expanded significantly, as wood became a critical resource for building infrastructure, heating homes, and fueling industries like metallurgy. However, this increased demand also led to deforestation and soil degradation in some regions, highlighting the environmental consequences of unchecked logging.

The Industrial Revolution marked a turning point in the history of logging. With the rise of machinery, railroads, and sawmills, logging operations scaled up to meet the growing demand for timber in construction and manufacturing. The development of mechanized tools like steam-powered saws and logging trains enabled the extraction of timber from previously inaccessible forests. By the

20th century, logging had become a global industry, driving economic growth in many countries but also raising concerns about deforestation, habitat loss, and climate change. Modern logging practices continue to evolve, integrating advanced technologies such as drones, GIS mapping, and automated machinery to enhance efficiency and sustainability. Despite these advancements, the historical legacy of logging underscores the need for responsible management to mitigate its ecological impact while supporting economic and social development.

1.2 Importance of Logging in Forest Resource Management

Logging plays a pivotal role in forest resource management, serving as a primary means of utilizing forest resources to meet human needs. It provides essential materials for industries such as construction, furniture making, and paper production, contributing significantly to local and global economies. Beyond its economic value, logging also has implications for forest health, land-use planning, and sustainable development.

Forests are a vital resource, and the judicious use of timber can support sustainable economic growth. Logging provides livelihoods for millions of people worldwide, from small-scale operators in rural communities to large-scale industrial firms. For example, the forestry sector is a major employer in countries like Canada, Brazil, and Indonesia, generating income and supporting ancillary industries such as transportation, manufacturing, and trade. Additionally, timber is a renewable resource when harvested sustainably, offering an alternative to non-renewable materials like steel and concrete. In the context of forest resource management, logging is not solely about timber extraction. It also involves maintaining a balance between utilization and conservation. For instance, selective logging practices focus on removing specific trees while preserving the overall structure and biodiversity of the forest. This approach minimizes environmental damage and supports natural regeneration, making it an integral part of sustainable forest management plans. Logging can also serve as a tool for controlling forest density, reducing competition among trees, and mitigating the risk of wildfires by clearing excessive undergrowth.

However, the importance of logging extends beyond economic and ecological considerations. It plays a role in land-use dynamics, often paving the way for agricultural development, urban expansion, and infrastructure projects. While these changes can drive economic development, they also raise concerns about deforestation and habitat fragmentation. Effective forest resource management must therefore reconcile the competing demands of logging with the need to conserve forests for their ecological, cultural, and recreational values. In recent years, the focus has shifted toward integrating logging practices into broader frameworks of sustainable development. Initiatives such as forest certification programs, including those by the Forest Stewardship Council (FSC) and the Programme for the Endorsement of Forest Certification (PEFC), promote responsible logging by setting standards for environmental, social, and economic sustainability. These programs aim to ensure that timber harvesting aligns with principles of forest conservation, climate mitigation, and community well-being.

1.3 Global Trends in Logging Practices

Logging practices vary widely across regions, reflecting differences in forest types, management objectives, and socio-economic contexts. Understanding global trends in logging is essential for addressing the challenges and opportunities associated with forest utilization.

Expansion of Industrial Logging in Tropical Forests

Tropical forests in regions such as the Amazon, Congo Basin, and Southeast Asia have become focal points for industrial logging due to their rich timber resources. However, the expansion of logging in these areas has sparked concerns about deforestation, biodiversity loss, and carbon emissions. Illegal logging remains a significant challenge in many tropical countries, undermining conservation efforts and contributing to forest degradation. To address these issues, international initiatives like the European Union's Forest Law Enforcement, Governance and Trade (FLEGT) program and the United Nations REDD+ framework aim to promote legal and sustainable logging practices.

Shift Toward Sustainable Logging Practices

Sustainability has become a central theme in modern logging practices, driven by growing awareness of environmental issues and consumer demand for eco-friendly products. Reduced-impact logging (RIL) techniques, which minimize damage to surrounding vegetation and soil, are gaining traction as a viable alternative to traditional methods. For example, RIL practices include directional felling to avoid damage to nearby trees, the use of pre-determined skid trails, and the implementation of low-impact harvesting equipment.

Technological Innovations in Logging

The integration of advanced technologies is transforming the logging industry, enhancing efficiency and reducing environmental impact. Drones and remote sensing tools are increasingly used to map forests, assess tree health, and plan logging operations. Automated machinery, such as harvesters and forwarders, streamlines the extraction process while reducing labor intensity. Geographic Information Systems (GIS) enable precise mapping and monitoring of logging activities, ensuring compliance with sustainability standards.

Growth of Certification Programs and Ethical Consumerism

Forest certification programs have emerged as key drivers of responsible logging. Consumers in developed markets increasingly demand certified wood products, prompting companies to adopt sustainable practices to maintain market access. The FSC and PEFC are leading certification schemes that verify the sustainability of logging operations, promoting transparency and accountability across supply chains.

2. Types of Logging

Logging, a central activity in forest utilization, entails the harvesting of trees for various industrial, commercial, and local purposes. This activity plays a pivotal role in satisfying the world's timber demands and supporting livelihoods across the globe. Logging practices, however, have far-reaching implications on ecological balance, biodiversity conservation, and the global climate. Consequently, understanding the types of logging and their implications has become an essential focus for forestry scholars and policymakers alike. Different methods of logging have been developed, each tailored to suit specific ecological contexts, economic objectives, and conservation priorities.

2.1 Selective Logging

Selective logging, often regarded as a compromise between economic gain and ecological conservation, involves the targeted removal of specific trees based on criteria such as species, size, or quality. Unlike more intensive methods, selective logging aims to minimize ecological disturbances while meeting market demands. This practice is widely adopted in tropical forests where biodiversity conservation is critical, yet the need for valuable hardwood species like mahogany, teak, and rosewood persists. One of the primary benefits of selective logging lies in its potential to preserve forest structure and biodiversity. By selectively harvesting mature or economically valuable trees, the practice allows younger trees to continue growing, maintaining the forest canopy, and preventing extensive soil erosion. This method often provides a habitat for numerous species, as the overall forest cover remains intact. Additionally, selective logging can support long-term timber production by fostering natural regeneration and ensuring the availability of commercially desirable species for future harvests. However, selective logging is not without its challenges. Poorly planned or executed selective logging can result in significant ecological disruptions, including soil compaction, reduced water infiltration, and microhabitat destruction. The selective removal of high-value species often leads to "high-grading," where the genetic diversity of these species declines, compromising their future adaptability to environmental changes. Moreover, selective logging operations frequently necessitate the construction of logging roads, which fragment habitats and open previously inaccessible forest areas to illegal logging and hunting. From an economic perspective, selective logging can be resource-intensive, requiring skilled labor and careful planning to identify and harvest suitable trees without damaging the surrounding vegetation. While selective logging is often promoted as a sustainable practice, its implementation demands rigorous monitoring and enforcement to prevent overharvesting and ensure compliance with ecological guidelines. These complexities underscore the need for integrating selective logging into broader forest management frameworks that prioritize conservation and community engagement.

2.2 Clear-Cutting

Clear-cutting, one of the most controversial logging practices, involves the complete removal of all trees within a designated area. This method is primarily employed in regions where timber production is the primary objective, as it facilitates large-scale harvesting and simplifies forest management operations. Clear-cutting is particularly prevalent in boreal and temperate forests, where it supports industries such as paper production, construction, and biomass energy generation. The primary advantage of clear-cutting lies in its economic efficiency. By harvesting trees uniformly, clear-cutting minimizes the costs associated with selective tree removal and allows for mechanized operations that maximize productivity. Additionally, this method often accelerates forest regeneration in species that rely on full sunlight for growth, such as pines and birches. In managed forests, clear-cutting can be integrated with replanting programs to ensure the continuous supply of timber resources while maintaining economic viability. Despite its economic benefits, clear-cutting poses significant environmental risks. The complete removal of tree cover disrupts the forest ecosystem, leading to habitat loss for countless species. Soil erosion, nutrient depletion, and water cycle alterations are common consequences of clear-cutting, particularly in sloped or sensitive landscapes. The method also contributes to carbon emissions, as the removal and decomposition of trees release stored carbon into the atmosphere, exacerbating climate change. Clear-cutting's visual impact on the landscape often evokes public concern, as it transforms verdant forests into barren expanses. These aesthetic and ecological concerns have led to growing advocacy for alternative logging practices and the adoption of stricter regulations to mitigate clear-cutting's environmental footprint. Sustainable clear-cutting models, which involve smaller clearings and replanting efforts, have been proposed to balance economic objectives with ecological preservation. However, the long-term viability of these models remains subject to ongoing scientific scrutiny and policy development.

2.3 Shelterwood and Seed Tree Systems: Sustainable Approaches

Shelterwood and seed tree systems represent innovative approaches to logging that emphasize sustainability and ecological resilience. Both methods aim to mimic natural forest regeneration processes while ensuring the continued availability of timber resources. These systems have gained prominence in forestry management as viable alternatives to clear-cutting, particularly in regions where conservation priorities are high. The shelterwood system involves the gradual removal of mature trees in multiple stages, allowing younger trees to establish under the protective canopy of the remaining mature trees. This method fosters a more natural forest structure, providing habitat continuity and reducing the risk of soil erosion. By maintaining partial canopy cover, shelterwood systems also mitigate microclimatic changes, ensuring favorable conditions for seed germination and seedling growth. Over time, as the younger trees mature, the remaining canopy trees are harvested, completing the regeneration cycle. The seed tree system, on the other hand, focuses on retaining a select number of mature, seed-producing trees during the initial harvest. These "seed trees" facilitate natural regeneration by dispersing seeds across the harvested area. Once the new generation of trees has established, the seed trees may be harvested or left standing to enhance biodiversity and aesthetic value. The seed tree system is particularly effective for species that exhibit prolific seed production and thrive in open conditions. Both shelterwood and seed tree systems offer significant ecological benefits, including the preservation of genetic diversity, improved soil stability, and enhanced habitat connectivity. However, their implementation requires careful planning and monitoring to ensure successful regeneration. Factors such as seed viability, site conditions, and competition from invasive species must be considered to optimize outcomes. Additionally, these systems may be less economically efficient than clear-cutting or selective logging in the short term, as they often necessitate longer rotation periods and more intensive management efforts.

2.4 Reduced Impact Logging (RIL)

Reduced Impact Logging (RIL) represents a paradigm shift in forestry, prioritizing ecological sustainability and minimizing environmental damage. This method involves a suite of techniques designed to reduce the negative impacts of logging activities, including soil disturbance, habitat destruction, and carbon emissions. RIL has emerged as a cornerstone of sustainable forest management, particularly in tropical regions where logging often threatens biodiversity hotspots.

The principles of RIL include pre-harvest planning, directional felling, and the use of specialized equipment to minimize ground damage. Pre-harvest planning involves the careful mapping of trees to be harvested, identifying skid trails, and avoiding ecologically sensitive areas. Directional felling ensures that trees are felled in a manner that minimizes damage to surrounding vegetation and facilitates efficient extraction. By employing low-impact machinery and techniques, RIL reduces soil compaction and prevents excessive disruption to forest ecosystems. One of the key ecological benefits of RIL is its ability to maintain forest carbon stocks. By limiting collateral damage and preserving non-target trees, RIL significantly reduces the carbon emissions associated with traditional logging methods. This makes RIL an essential strategy for mitigating climate change and promoting carbon sequestration in forest landscapes. Additionally, RIL supports biodiversity conservation by preserving critical habitats and fostering the recovery of logged forests.

Despite its advantages, RIL faces several challenges in implementation. The upfront costs associated with training personnel, acquiring specialized equipment, and conducting pre-harvest assessments can deter adoption, particularly in low-income regions. Moreover, enforcing RIL practices requires robust regulatory frameworks and monitoring systems to ensure compliance. Addressing these challenges necessitates collaborative efforts among governments, industry stakeholders, and local communities to promote the widespread adoption of RIL as a standard logging practice. The growing recognition of RIL's ecological and economic potential underscores its importance in shaping the future of forestry. As global efforts to combat deforestation and climate change intensify, RIL offers a viable pathway for balancing timber production with environmental stewardship. Through continued research, capacity-building initiatives, and policy support, RIL can play a pivotal role in advancing sustainable forest management and fostering resilience in forest ecosystems worldwide.

3. Methods and Techniques in Logging

The methods and techniques used in logging have undergone a remarkable transformation over centuries, reflecting advances in technology, evolving economic needs, and growing environmental concerns. From traditional manual methods to cutting-edge robotic systems, each technique offers unique advantages and challenges depending on the ecological and industrial contexts in which it is applied. Logging methods can broadly be categorized based on their level of mechanization and the technological sophistication involved.

3.1 Manual Logging: Traditional Tools and Labor

Manual logging represents the oldest form of timber harvesting, relying on human labor and simple tools like axes, hand saws, machetes, and wedges. This method has been practiced for centuries and continues to be prevalent in certain parts of the world, especially in rural or remote areas where access to advanced machinery is limited. Manual logging is often associated with small-scale operations, subsistence forestry, or traditional practices passed down through generations in forest-dependent communities.

One of the defining features of manual logging is its reliance on precision and skill. Unlike mechanical methods, manual logging allows loggers to carefully select and fell trees with minimal collateral damage to the surrounding vegetation. This makes it particularly suitable for selective logging operations in biodiversity-rich environments, where preserving ecological integrity is paramount. Moreover, manual logging is often seen as a culturally significant activity, providing employment opportunities and sustaining traditional knowledge systems in rural communities.

The benefits of manual logging extend to its relatively low cost of operation, particularly in regions where labor is inexpensive. It also minimizes the infrastructure requirements typically associated with mechanized logging, such as roads and heavy equipment. Additionally, manual methods are often less intrusive to the environment, as they avoid soil compaction and habitat destruction caused by heavy machinery.

However, manual logging is not without its limitations. It is labor-intensive and time-consuming, making it unsuitable for large-scale commercial operations. The physical demands of the work also pose significant risks to loggers, including musculoskeletal injuries, fatigue, and accidents involving falling trees. Furthermore, the productivity of manual logging is significantly lower compared to mechanized methods, which can impact the economic viability of the practice in competitive markets. In recent years, efforts have been made to modernize manual logging through the introduction of ergonomic tools and training programs designed to improve efficiency and worker safety. By combining traditional methods with contemporary innovations, manual logging remains a vital component of forest management in contexts where environmental sensitivity and cultural heritage take precedence over industrial efficiency.

3.2 Mechanical Logging: Modern Machinery and Efficiency

The advent of mechanical logging marked a transformative shift in the timber industry, driven by the need to meet growing global demands for wood products. This method employs advanced machinery such as chainsaws, skidders, harvesters, forwarders, and feller-bunchers, each designed to perform specific tasks with high precision and speed. Mechanical logging is the backbone of industrial-scale forestry operations, enabling the efficient harvesting of large volumes of timber in relatively short periods.

Chainsaws, introduced in the mid-20th century, revolutionized logging by providing a powerful yet portable tool for felling and bucking trees. Skidders are used to drag felled logs to centralized collection points, facilitating transport and processing. Feller-bunchers combine multiple functions by cutting trees, gathering them into bundles, and depositing them in predetermined locations. Harvesters and forwarders represent the pinnacle of mechanization, capable of performing tasks such as cutting, delimbing, and loading logs with minimal human intervention. The primary advantage of mechanical logging lies in its efficiency and scalability. By automating key aspects of the logging process, mechanical methods significantly reduce the time and labor required for timber harvesting. This not only lowers operational costs but also enhances worker safety by minimizing direct exposure to hazardous tasks. Additionally, mechanical equipment can be adapted to diverse terrains and forest types, from plantations to natural forests, ensuring versatility and adaptability in various logging scenarios.

Despite its numerous benefits, mechanical logging is often criticized for its environmental impact. The use of heavy machinery can result in soil compaction, erosion, and damage to non-target vegetation, particularly in sensitive ecosystems. Logging roads and skid trails, essential for machine access, can fragment habitats and disrupt wildlife movement. Furthermore, the high initial costs associated with acquiring and maintaining logging equipment can limit the accessibility of mechanical methods to small-scale operators. To mitigate these challenges, the logging industry has embraced innovations such as low-impact machinery, hybrid and electric-powered equipment, and sustainable operational practices. These advancements aim to reduce the environmental footprint of mechanical logging while maintaining its economic viability. As forestry practices evolve, mechanical logging continues to play a critical role in balancing the demands of industrial production with the principles of sustainability.

3.3 Cable Logging and Aerial Systems for Remote Areas

Cable logging and aerial systems are specialized techniques designed to address the challenges of harvesting timber in remote or rugged terrains. These methods employ advanced engineering and technology to transport logs from felling sites to collection points without the need for extensive ground infrastructure, making them particularly suitable for mountainous regions, steep slopes, and inaccessible forests. Cable logging systems utilize a network of cables and winches to move logs across challenging landscapes. Skyline systems, for example, involve suspending a stationary cable between two anchor points, enabling a carriage to transport logs along the cable line. High-lead systems use a single cable for hauling logs uphill or downhill, while yarders combine multiple cables to enhance operational flexibility and efficiency. These systems minimize soil disturbance and reduce the need for logging roads, preserving the ecological integrity of the forest. Aerial logging, commonly known as heli-logging, employs helicopters or drones to lift and transport logs from the felling site to designated drop zones. This method is particularly effective in extracting high-value timber from remote or environmentally sensitive areas where ground-based equipment cannot operate. Helicopters can maneuver through dense forests and steep terrains with minimal ecological impact, avoiding soil compaction and habitat destruction.

The environmental advantages of cable and aerial logging make them attractive options for sustainable forestry. By reducing ground disturbance, these methods help maintain soil stability, protect water resources, and preserve biodiversity. However, the high costs associated with equipment, infrastructure, and skilled operators pose significant challenges to their widespread adoption. Aerial logging, in particular, is one of the most expensive logging techniques, often limited to operations involving premium timber species or areas with strict conservation priorities. Technological advancements in cable and aerial systems, including automated controls and precision mapping, have improved their efficiency and cost-effectiveness. These innovations, coupled with integrated forest management strategies, offer promising opportunities for the responsible utilization of remote forest resources while minimizing environmental harm.

3.4 Innovations in Logging

The integration of robotics and automation into logging represents a revolutionary development in the forestry sector, promising to enhance productivity, safety, and environmental sustainability. These technologies encompass a wide range of applications, from autonomous harvesting machines and robotic tree climbers to drone-based monitoring and mapping systems. Autonomous harvesters equipped with advanced sensors, machine learning algorithms, and GPS technology can identify, fell, and process trees with minimal human intervention. These machines are capable of operating in complex forest environments, optimizing timber yield while reducing waste. By analyzing data in real-time, autonomous systems can make informed decisions about tree selection and processing, ensuring sustainable resource utilization.

Robotic tree climbers are designed for specialized tasks such as pruning, thinning, and selective harvesting in challenging forest conditions. These machines can navigate tree trunks and branches with precision, performing operations that are labor-intensive or hazardous for human workers. Their use enhances worker safety, reduces operational costs, and minimizes damage to surrounding vegetation. Drones have emerged as versatile tools in modern forestry, revolutionizing planning and monitoring processes. Equipped with high-resolution cameras, LiDAR sensors, and thermal imaging, drones can map forest areas, assess tree health, and identify logging sites with unprecedented accuracy. These capabilities enable forest managers to optimize resource allocation, monitor compliance with environmental regulations, and detect illegal logging activities.

While robotics and automation hold immense potential, their adoption faces challenges such as high initial costs, technical complexity, and the need for specialized training. Ethical considerations, including job displacement and the potential for overexploitation of resources, also warrant careful deliberation. Nonetheless, continued research and development in this field are likely to address these issues, paving the way for a more efficient and sustainable logging industry. The fusion of traditional practices with modern innovations underscores the dynamic nature of forestry, reflecting a balance between economic imperatives and ecological stewardship. By embracing cutting-edge technologies alongside proven methods, the logging sector can navigate the complexities of contemporary forest management and contribute to the global pursuit of sustainability.

4. Economic Scope of Logging

The economic scope of logging encompasses its role as a significant industrial activity that drives global timber markets, supports national economies, and fosters regional development. Logging forms the foundation of a vast economic network, contributing not only through direct timber production but also through downstream industries and the utilization of non-timber forest products (NTFPs). The multifaceted economic implications of logging require examining its contributions as an industry, its role in the global timber trade, the value-added processes involved in downstream production, and the diverse opportunities presented by non-timber resources.

4.1 Logging as an Industry

Logging is a cornerstone of the forestry industry, playing a critical role in supporting national and regional economies worldwide. As a labor-intensive sector, logging provides direct employment opportunities for millions of individuals, including loggers, machine operators, transporters, and supervisors. The industry also generates indirect employment in ancillary sectors such as equipment manufacturing, logistics, and forest management services, making it an integral part of rural economies where alternative job opportunities are often limited. In many countries, logging contributes significantly to gross domestic product (GDP) through the production of raw timber, which is subsequently utilized in construction, furniture manufacturing, paper production, and bioenergy generation. Countries with extensive forest cover, such as Canada, Russia, Brazil, and the United States, derive substantial revenue from logging and related industries, which support infrastructure development, education, and public health systems. Additionally, logging often serves as a major source of foreign exchange earnings, particularly for developing nations that export timber and wood products to global markets. The economic impact of logging extends to government revenues through taxation and royalties. Logging companies are typically required to pay fees for forest concessions, timber harvest permits, and export duties, which contribute to national and regional budgets. These revenues can be reinvested in sustainable forest management, community development, and conservation programs, creating a cyclical benefit for both the economy and the environment. However, the logging industry's economic contributions are not without challenges. Unsustainable practices, illegal logging, and market volatility can undermine the long-term viability of the sector. Overexploitation of forest resources may lead to declining timber yields, increased costs, and social conflicts over land use. To address these issues, governments and industry stakeholders are increasingly adopting sustainable logging practices, certification systems, and community engagement models to balance economic objectives with environmental and social responsibilities.

4.2 Global Timber Trade and Market Dynamics

The global timber trade is a complex and dynamic system that connects producers and consumers across continents, shaping the economic scope of logging on an international scale. Timber and wood products, including logs, sawn timber, plywood, and paper, are among the most traded commodities worldwide, driven by demand from the construction, furniture, and packaging industries. The timber trade is influenced by various factors, including forest resource availability, market demand, international trade policies, and technological advancements. Major timber-exporting countries, such as Canada, Russia, Brazil, and Indonesia, dominate global supply chains, leveraging their extensive forest resources and established infrastructure. Importing nations, including China, the United States, and the European Union, account for significant timber consumption, driven by robust industrial sectors and growing populations. The interplay between exporting and importing regions creates a dynamic market environment characterized by competition, price fluctuations, and regulatory challenges.

Market dynamics in the timber trade are shaped by several key trends, including the increasing demand for certified wood products. Certification schemes, such as the Forest Stewardship Council (FSC) and the Programme for the Endorsement of Forest Certification (PEFC), have gained prominence as mechanisms to ensure sustainable logging practices and ethical trade. Consumers and businesses are increasingly prioritizing sustainably sourced timber, prompting producers to adopt certification to access lucrative markets. Global timber trade is also influenced by geopolitical factors, including trade agreements, tariffs, and sanctions. For instance, trade disputes between major economies can disrupt supply chains and impact timber prices, creating uncertainty for producers and consumers alike. Additionally, climate change policies and carbon markets are emerging as influential factors in the timber trade, as forest products are increasingly recognized for their role in carbon sequestration and renewable material production.

Despite its economic significance, the global timber trade faces challenges such as illegal logging, which undermines legitimate markets and contributes to deforestation and biodiversity loss. Efforts to combat illegal timber trade include stricter enforcement of international regulations, such as the EU Timber Regulation (EUTR) and the US Lacey Act, as well as increased transparency and traceability in supply chains. By addressing these challenges, the global timber trade can continue to drive economic growth while supporting sustainable development and forest conservation.

4.3 Value Addition through Downstream Processing

Downstream processing in the logging industry involves transforming raw timber into value-added products, significantly enhancing the economic scope of logging. This segment of the industry encompasses a wide range of activities, including sawmilling, kiln drying, veneer production, plywood manufacturing, and furniture making. Value addition not only increases the economic returns from forest resources but also diversifies employment opportunities and stimulates industrial growth. The sawmilling sector represents the initial stage of downstream processing, where logs are converted into sawn timber for use in construction, furniture, and other applications. Advanced technologies, such as computer-controlled saws and laser-guided cutting systems, have improved the efficiency and precision of sawmilling operations, reducing waste and maximizing output. Kiln drying, another critical process, enhances the durability and stability of timber by reducing its moisture content, making it suitable for high-quality applications.

The production of engineered wood products, such as plywood, particleboard, and laminated veneer lumber (LVL), exemplifies the potential for value addition in the timber industry. These products combine the strength and versatility of natural wood with the durability and uniformity required for modern construction and manufacturing. Engineered wood products are increasingly favored for their environmental benefits, as they utilize wood residues and lower-grade materials that might otherwise go to waste. Furniture-making and artisanal woodcraft represent high-value segments of downstream processing, catering to domestic and international markets. These industries rely on skilled craftsmanship and innovative design to create products that command premium prices, contributing to local economic development and cultural heritage preservation. In addition to traditional furniture, the demand for modular and custom-designed wood products has created new opportunities for small and medium-sized enterprises in the forestry sector. Challenges in downstream processing include the need for capital investment in technology and infrastructure, skilled labor shortages, and competition from alternative materials such as metal and plastics. Additionally, the environmental impact of processing operations, including energy consumption and waste generation, must be carefully managed to ensure sustainability. By investing in advanced technologies, workforce development, and environmental management systems, the timber industry can enhance its value-added potential while addressing these challenges.

4.4 Opportunities in Non-Timber Forest Products

Non-timber forest products (NTFPs) offer a diverse range of economic opportunities that extend beyond traditional timber harvesting. NTFPs include a wide array of resources such as medicinal plants, edible fruits and nuts, resins, latex, essential oils, and fibers, which contribute to local livelihoods and global markets. These products are particularly important in rural and indigenous communities, where they provide supplementary income and support traditional knowledge systems. Medicinal plants, for instance, have long been valued for their therapeutic properties and play a crucial role in traditional medicine systems such as Ayurveda, traditional Chinese medicine, and indigenous healing practices. The global demand for herbal remedies and natural health products has created significant economic opportunities for forest-dependent communities and small-scale enterprises. Similarly, edible NTFPs, including wild fruits, nuts, mushrooms, and honey, are increasingly sought after for their nutritional and culinary value, driving the growth of niche markets such as organic and specialty foods. Resins, latex, and essential oils derived from forest species are widely used in industries such as cosmetics, pharmaceuticals, and adhesives. For example, natural rubber obtained from Hevea brasiliensis is a critical raw material for tire manufacturing and other industrial applications. Essential oils from species like sandalwood, cedar, and eucalyptus are highly valued for their aromatic and therapeutic properties, contributing to the global fragrance and wellness industries. In addition to direct economic benefits, NTFPs play a vital role in ecosystem services and biodiversity conservation. Sustainable harvesting of

NTFPs can promote forest preservation by providing alternative income streams that reduce pressure on timber resources. However, the commercialization of NTFPs also poses challenges, including overharvesting, habitat degradation, and inequitable benefit-sharing. To realize the full potential of NTFPs, it is essential to develop robust value chains, market access mechanisms, and regulatory frameworks that support sustainable production and equitable trade. Investments in research, capacity building, and community-based management can enhance the contribution of NTFPs to local and global economies, creating a resilient and diversified forestry sector that extends beyond timber alone.

5. Environmental and Ecological Impacts

Logging, while economically significant, poses substantial environmental and ecological challenges. The extraction of timber from forests affects the delicate balance of ecosystems, leading to far-reaching consequences for biodiversity, soil health, water systems, and global climate dynamics. The extent and severity of these impacts depend on factors such as the intensity of logging, the methods used, and the effectiveness of forest management strategies. Understanding these effects is critical for developing sustainable practices that balance the economic benefits of logging with the imperative to conserve ecosystems and mitigate environmental degradation.

5.1 Deforestation and Habitat Loss

Deforestation, often driven by logging activities, represents one of the most pressing environmental challenges of the modern era. It entails the large-scale removal of forest cover, disrupting ecosystems and leading to the permanent loss of critical habitats for countless plant and animal species. While logging is not the sole driver of deforestation, agriculture, mining, and urbanization also contribute significantly, it remains a key factor, particularly in regions with high-value timber species.

The scale of deforestation caused by logging varies depending on the method employed. Clear-cutting, for example, involves the complete removal of forest cover within a designated area, leading to immediate and dramatic habitat destruction. This method often leaves landscapes barren, making it difficult for many species to survive or recolonize. Selective logging, while less visually impactful, can also contribute to habitat degradation by altering forest structure, fragmenting ecosystems, and reducing the availability of resources for forest-dwelling species. The consequences of habitat loss are profound, particularly in biodiversity-rich regions such as tropical rainforests. Species that rely on specific microhabitats or have limited ranges are particularly vulnerable to extinction. The loss of keystone species can have cascading effects on ecosystems, disrupting food webs and ecological processes such as pollination, seed dispersal, and nutrient cycling. Furthermore, habitat loss often forces wildlife into closer contact with human settlements, increasing the risk of human-wildlife conflicts and the spread of zoonotic diseases. Efforts to address deforestation and habitat loss include the establishment of protected areas, reforestation and afforestation initiatives, and the promotion of sustainable logging practices. Certification schemes such as the Forest Stewardship Council (FSC) and the Programme for the Endorsement of Forest Certification (PEFC) aim to ensure that timber is harvested responsibly, minimizing environmental impacts. Additionally, policies such as zero-deforestation commitments by corporations and international agreements like the United Nations' REDD+ program (Reducing Emissions from Deforestation and Forest Degradation) play a crucial role in mitigating deforestation and promoting forest conservation.

5.2 Soil Erosion and Water Quality Degradation

Logging activities can have severe consequences for soil health and water quality, particularly when conducted without adequate safeguards. The removal of trees disrupts the protective canopy and root systems that stabilize soil, increasing the risk of erosion, nutrient depletion, and sedimentation in nearby water bodies. These impacts are especially pronounced in sloped terrains and regions with heavy rainfall, where exposed soils are more susceptible to being washed away. Soil erosion caused by logging can lead to the loss of fertile topsoil, reducing the productivity of forestlands and hindering natural regeneration. Over time, this degradation can transform once-productive ecosystems into barren landscapes, with far-reaching implications for local agriculture, forestry, and biodiversity. The loss of soil also contributes to increased sedimentation in rivers, streams, and reservoirs, disrupting aquatic ecosystems and reducing water storage capacity.

Water quality degradation is another significant impact of logging, driven by increased sediment loads, chemical runoff, and changes in hydrological cycles. Sedimentation can smother aquatic habitats, reducing oxygen levels and disrupting the life cycles of fish and other aquatic organisms. Nutrient runoff from disturbed soils can lead to algal blooms and eutrophication, further compromising water quality. Additionally, the construction of logging roads and skid trails often alters natural drainage patterns, exacerbating the risk of flooding and waterlogging in downstream areas. Mitigating soil erosion and water quality degradation requires the adoption of best management practices, such as reduced-impact logging (RIL) techniques, buffer zone maintenance, and reforestation. Contour logging, which involves felling trees along the natural contours of the land, can minimize soil disturbance and reduce erosion risks. The use of sediment control measures, such as silt fences and retention basins, can also help prevent sedimentation in water bodies. By integrating these practices into forest management plans, the environmental impacts of logging on soil and water systems can be significantly reduced.

5.3 Carbon Emissions and Climate Change

The relationship between logging and climate change is multifaceted, encompassing both the release of stored carbon and the potential for carbon sequestration through sustainable forestry practices. Forests act as carbon sinks, absorbing carbon dioxide (CO2) from the atmosphere and storing it in biomass and soils. Logging disrupts this process, releasing stored carbon into the atmosphere and reducing the forest's capacity to sequester CO2. Carbon emissions from logging occur through multiple pathways, including the decomposition of harvested trees, soil disturbance, and the burning of forest residues. Clear-cutting and slash-and-burn practices are particularly detrimental, as they release large quantities of carbon in a short period. Additionally, logging operations often involve the use of fossil-fuel-powered machinery and transportation, further contributing to greenhouse gas emissions. The cumulative impact of logging on carbon dynamics is significant, particularly in tropical regions where forests contain high levels of biomass. Deforestation and forest degradation account for approximately 10-15% of global greenhouse gas emissions, making them a major driver of climate change. The loss of forest cover also reduces the planet's capacity to absorb CO2, exacerbating the effects of global warming and increasing the frequency and severity of climate-related events such as droughts, storms, and wildfires. Addressing the carbon impacts of logging requires a shift toward sustainable forestry practices and the integration of carbon considerations into forest management. Reduced-impact logging techniques, reforestation efforts, and the promotion of longer rotation periods can help mitigate carbon emissions and enhance carbon sequestration. Additionally, carbon offset programs, such as those under the REDD+ framework, provide financial incentives for countries and companies to reduce deforestation and invest in forest conservation. By recognizing the critical role of forests in climate regulation, policymakers and industry stakeholders can work together to align logging activities with global climate goals.

5.4 Biodiversity Loss and Mitigation Strategies

Biodiversity loss is one of the most profound ecological impacts of logging, particularly in regions with high species richness and endemism. Forests provide habitat, food, and shelter for a vast array of organisms, ranging from microorganisms and insects to large mammals and birds. Logging disrupts these ecosystems, leading to population declines, species extinctions, and changes in community dynamics. The impacts of logging on biodiversity vary depending on the intensity and method of extraction. Clear-cutting has the most severe effects, as it eliminates entire habitats and creates fragmented landscapes that are inhospitable to many species. Selective logging, while less destructive, can still reduce habitat quality by altering forest structure and removing key resources such as nesting sites and food sources. Habitat fragmentation caused by logging roads and trails further exacerbates biodiversity loss, as it isolates populations and impedes genetic exchange.

Mitigation strategies for biodiversity loss focus on minimizing the ecological footprint of logging and promoting habitat restoration. Sustainable forest management practices, such as the retention of buffer zones, the preservation of keystone species, and the maintenance of habitat corridors, are essential for supporting biodiversity in logged areas. Reforestation and afforestation initiatives can also help restore degraded habitats and provide new opportunities for wildlife to thrive. Additionally, biodiversity offsets and conservation programs can complement sustainable logging practices by protecting high-value habitats and investing in ecological research. Community-based forest management approaches, which involve local stakeholders in decision-making and benefit-sharing, can further enhance conservation outcomes by aligning economic incentives with biodiversity goals. By prioritizing the integration of ecological considerations into logging practices, the forestry sector can contribute to the preservation of global biodiversity while meeting the demands of timber production.

6. Legal and Regulatory Frameworks

The logging industry operates within a complex web of legal and regulatory frameworks designed to balance economic development, environmental conservation, and social equity. These frameworks encompass international agreements, national policies, and local governance systems that aim to promote sustainable forest management, combat illegal logging, and ensure compliance with environmental standards. The success of these frameworks depends on their design, implementation, and enforcement, as well as the participation of diverse stakeholders, including governments, private companies, and local communities.

6.1 International Standards for Sustainable Logging

International standards for sustainable logging have emerged as critical tools for guiding forestry practices and fostering global collaboration on forest conservation. These standards aim to ensure that logging activities are conducted in a manner that minimizes environmental damage, protects biodiversity, and respects the rights of local communities. Prominent international frameworks include the Forest Stewardship Council (FSC) and the Programme for the Endorsement of Forest Certification (PEFC), both of which provide certification schemes for sustainable forest management.

The FSC, established in 1993, sets rigorous environmental, social, and economic criteria for responsible forest management. Its certification process involves independent audits to verify compliance with principles such as the maintenance of ecological integrity, respect for indigenous rights, and sustainable yield management. The FSC label is widely recognized by consumers and markets as a symbol of sustainability, incentivizing companies to adopt responsible practices. The PEFC, founded in 1999, operates as an umbrella organization that endorses national certification systems based on globally accepted sustainability benchmarks. By adapting international standards to local contexts, the PEFC ensures that diverse forest management practices are aligned with global conservation goals. Both the FSC and PEFC contribute to the traceability of wood products through chain-of-custody certification, enabling consumers and businesses to make informed choices about sourcing.

International agreements such as the United Nations Forum on Forests (UNFF) and the Convention on Biological Diversity (CBD) also play a pivotal role in setting the global agenda for sustainable forestry. These agreements encourage member states to implement sustainable logging practices, develop national forest programs, and report progress toward conservation targets. Additionally, initiatives like the REDD+ program (Reducing Emissions from Deforestation and Forest Degradation) provide financial incentives for countries to reduce deforestation and invest in sustainable forest management. Despite these efforts, challenges persist in ensuring universal adherence to international standards. Limited enforcement capacity, variations in national implementation, and the prevalence of illegal logging undermine the effectiveness of global frameworks. Strengthening international cooperation, enhancing transparency, and fostering partnerships between governments, industry, and civil society are essential for achieving the objectives of sustainable logging.

6.2 National Policies and Forest Certification Schemes

National policies and forest certification schemes serve as foundational components of the legal framework governing logging activities within individual countries. These policies are shaped by the unique ecological, economic, and social contexts of each nation, reflecting priorities such as resource conservation, economic development, and community welfare. Forest certification schemes complement national policies by providing voluntary mechanisms for verifying compliance with sustainability standards. In many countries, national forest policies establish the legal basis for regulating logging activities, including the issuance of logging permits, the designation of protected areas, and the enforcement of environmental laws. These policies often include provisions for sustainable forest management, reforestation, and monitoring of logging operations. For example, Brazil's Forest Code requires landowners to maintain a certain percentage of native vegetation on their properties, while Canada's Forest Act mandates sustainable harvesting practices and reforestation in managed forests.

Forest certification schemes, such as the FSC and PEFC, operate alongside national policies to promote sustainable logging practices. These schemes provide market-based incentives for companies to comply with environmental and social standards, enabling certified products to access premium markets and enhance their reputation. Certification also facilitates compliance with international trade regulations, such as the EU Timber Regulation (EUTR) and the US Lacey Act, which prohibit the import of illegally harvested timber. The integration of certification schemes into national policies can enhance their effectiveness by aligning voluntary standards with legal requirements. For instance, countries such as Indonesia and Malaysia have developed national certification systems, such as the Indonesian Timber Legality Assurance System (SVLK) and the Malaysian Timber Certification Scheme (MTCS), that incorporate elements of international frameworks while addressing local challenges. These systems help bridge the gap between global sustainability goals and national implementation. However, the effectiveness of national policies and certification schemes is often hindered by factors such as inadequate enforcement, corruption, and lack of resources. Ensuring the success of these frameworks requires strengthening institutional capacity, enhancing stakeholder engagement, and addressing systemic barriers to compliance. By fostering collaboration between governments, industry, and civil society, national policies and certification schemes can play a vital role in promoting sustainable logging and forest conservation.

6.3 Combatting Illegal Logging: Challenges and Solutions

Illegal logging, defined as the harvesting, transportation, and sale of timber in violation of national laws, poses a significant threat to forests, economies, and communities worldwide. It undermines sustainable forestry practices, contributes to deforestation and biodiversity loss, and deprives governments of revenue from legal timber operations. Combatting illegal logging is a complex challenge that requires a multifaceted approach involving legal, technological, and community-based strategies. One of the primary challenges in addressing illegal logging is the lack of effective enforcement mechanisms. Weak governance, corruption, and inadequate monitoring systems allow illegal activities to persist, particularly in regions with vast forest cover and limited regulatory capacity. Illegal logging is often linked to organized crime networks that exploit legal loopholes and evade detection, making enforcement efforts more challenging.

Technological innovations offer promising solutions for combating illegal logging. Tools such as satellite imagery, drones, and remote sensing enable real-time monitoring of forest cover and logging activities, allowing authorities to detect and respond to illegal operations more effectively. Blockchain technology is also being explored as a means of ensuring transparency and traceability in timber supply chains, reducing the risk of illegal timber entering the market. International cooperation plays a critical role in addressing illegal logging. Agreements such as the EUTR and the Lacey Act require importing countries to verify the legality of timber products, creating disincentives for illegal operations. Bilateral and multilateral initiatives, such as the Forest Law Enforcement, Governance, and Trade (FLEGT) program, support capacity-building efforts in timber-producing countries and promote sustainable trade practices. Community involvement is another essential component of combating illegal logging. Empowering local communities to monitor forests, report illegal activities, and participate in decision-making processes can enhance the effectiveness of enforcement efforts while fostering a sense of ownership and stewardship. Financial incentives, such as payments for ecosystem services (PES), can further motivate communities to engage in conservation activities and resist illegal logging pressures.

6.4 Community-Based Monitoring and Compliance

Community-based monitoring and compliance systems represent a grassroots approach to forest governance, leveraging the knowledge, skills, and commitment of local stakeholders to ensure sustainable logging practices. These systems involve the active participation of communities in monitoring forest resources, enforcing regulations, and implementing conservation initiatives, often in collaboration with government agencies and non-governmental organizations (NGOs). The role of communities in monitoring forests is particularly important in regions where government enforcement capacity is limited. Residents are often the first to detect illegal logging activities, making them valuable partners in conservation efforts. Community-based monitoring systems can be supported by tools such as mobile applications, GPS devices, and data-sharing platforms, enabling residents to document and report violations efficiently.

Compliance with logging regulations can be enhanced through community engagement in forest management planning and decision-making. Participatory approaches, such as joint forest management (JFM) and community forestry, empower local stakeholders to manage forest resources sustainably while deriving economic benefits. These models often include benefit-sharing mechanisms that allocate a portion of logging revenues to community development projects, incentivizing compliance and fostering long-term sustainability. Challenges in implementing community-based monitoring and compliance systems include capacity-building, resource constraints, and potential conflicts over land use. Addressing these challenges requires investments in education and training, the provision of technical and financial support, and the establishment of clear and equitable governance structures. By integrating community-based approaches into broader legal and regulatory frameworks, the logging industry can enhance compliance, promote sustainable practices, and strengthen the resilience of forest ecosystems.

7. Scope of Sustainable Logging

Sustainable logging encompasses the integration of timber extraction practices with principles of environmental stewardship, social responsibility, and economic viability. It recognizes the dual necessity of meeting human demands for forest products while maintaining the integrity of ecosystems for future generations. The scope of sustainable logging extends beyond conventional forest management, incorporating innovative technologies, land-use systems, and holistic approaches that prioritize conservation and community well-being.

7.1 Integrating Logging with Conservation Goals

Integrating logging activities with conservation goals represents a fundamental aspect of sustainable forestry. The primary challenge lies in balancing timber production with the preservation of biodiversity, ecosystem services, and carbon sequestration capabilities. To achieve this balance, sustainable logging practices focus on minimizing environmental impacts, protecting critical habitats, and ensuring the regeneration of harvested areas. Selective logging is a key approach that aligns with conservation objectives. By carefully choosing which trees to harvest, selective logging minimizes disruption to forest structure, preserves canopy cover, and reduces the risk of soil erosion. This method allows forests to recover naturally, maintaining ecological functions and supporting diverse species. In tropical rainforests, where biodiversity is exceptionally high, selective logging has been shown to mitigate habitat loss while providing economic benefits to local communities. Protected area networks, buffer zones, and habitat corridors are essential tools for integrating logging with conservation. Logging operations can be strategically planned to avoid ecologically sensitive areas, such as wildlife migration routes, nesting sites, and watersheds. Buffer zones, established around protected areas, serve as transitional landscapes where controlled logging can coexist with conservation efforts. Habitat corridors connect fragmented forests, enabling species to move freely and maintain genetic diversity.

International conservation initiatives, such as REDD+ (Reducing Emissions from Deforestation and Forest Degradation), provide financial incentives for countries to adopt sustainable logging practices that contribute to climate mitigation. These programs encourage the preservation of carbon-rich forests while promoting responsible timber harvesting. Additionally, certification schemes like the Forest Stewardship Council (FSC) ensure that logging operations meet rigorous environmental and social standards, fostering consumer confidence in sustainable products. Despite these advances, challenges remain in fully integrating logging with conservation goals. Economic pressures, illegal activities, and weak enforcement mechanisms can undermine sustainability efforts. Addressing these issues requires collaboration among governments, industry stakeholders, and conservation organizations, as well as investments in research and monitoring to assess the long-term impacts of logging on forest ecosystems.

7.2 Agroforestry and Sustainable Land-Use Systems

Agroforestry and sustainable land-use systems represent innovative approaches to integrating logging with broader goals of landscape management and rural development. Agroforestry combines agricultural and forestry practices on the same land, creating multifunctional landscapes that provide economic, ecological, and social benefits. By diversifying land use, agroforestry enhances the sustainability of logging operations and supports the livelihoods of local communities. In agroforestry systems, trees are cultivated alongside crops or livestock, offering multiple advantages. Trees contribute to soil fertility, water retention, and microclimatic regulation, improving agricultural productivity. They also provide timber, fuelwood, and non-timber forest products (NTFPs) such as fruits, nuts, and resins, generating additional income for farmers. Species selection is a critical component of agroforestry, as different trees serve distinct purposes, from erosion control to shade provision and carbon sequestration.

Sustainable land-use systems prioritize the efficient use of forest resources while maintaining ecological balance. These systems often involve rotational logging, where forests are divided into management units and harvested on a cycle that allows for natural regeneration. Mixed-species plantations, as opposed to monocultures, enhance biodiversity and resilience to pests and diseases. Integrating sustainable logging with reforestation and afforestation initiatives further contributes to landscape restoration and carbon capture. Agroforestry and sustainable land-use systems are particularly valuable in addressing the dual challenges of deforestation and rural poverty. By creating economic opportunities within managed landscapes, these approaches reduce the pressure on natural forests for subsistence and commercial needs. They also empower local communities to take an active role in forest management, fostering a sense of ownership and stewardship. The success of these systems depends on supportive policies, capacity-building, and access to markets. Governments and development organizations can facilitate the adoption of agroforestry through technical assistance, financial incentives, and land tenure reforms. By integrating agroforestry into national forest management plans, countries can enhance the sustainability of their logging sectors while addressing broader goals of food security, climate resilience, and biodiversity conservation.

7.3 Advances in Logging Technology for Sustainability

Technological innovation plays a pivotal role in advancing the sustainability of logging operations. From precision harvesting equipment to digital monitoring tools, new technologies enable forest managers to optimize resource use, reduce environmental impacts, and enhance operational efficiency.

The adoption of these technologies reflects a broader shift toward data-driven decision-making and adaptive management in the forestry sector. Precision forestry techniques, such as geospatial mapping and LiDAR (Light Detection and Ranging), allow forest managers to assess the composition, health, and value of forest stands with high accuracy. These tools enable the identification of optimal harvest sites, reducing the risk of overexploitation and minimizing damage to non-target trees. Geographic Information Systems (GIS) and remote sensing technologies are also used to monitor logging activities in real-time, ensuring compliance with regulations and certification standards.

Mechanized logging equipment, such as harvesters and forwarders, has been adapted to minimize soil compaction and damage to forest ecosystems. Reduced-impact logging (RIL) practices emphasize directional felling and low-impact skidding techniques, which limit collateral damage and preserve forest structure. Hybrid and electric-powered machinery further contribute to sustainability by reducing greenhouse gas emissions and fuel consumption. Digital technologies, including blockchain and Internet of Things (IoT) devices, enhance transparency and traceability in timber supply chains. Blockchain platforms enable the recording of timber provenance, ensuring that wood products are sourced from legal and sustainable operations. IoT devices, such as sensors and drones, provide real-time data on forest conditions, enabling proactive management and rapid response to illegal activities.

7.4 Balancing Economic Needs and Ecological Preservation

The balance between economic needs and ecological preservation is at the heart of sustainable logging. Forests are vital for human well-being, providing timber, fuel, and other resources that support economic development. At the same time, they play a critical role in maintaining biodiversity, regulating climate, and delivering ecosystem services. Achieving sustainability requires reconciling these competing demands through integrated management approaches. Economic incentives for sustainable logging include market access, certification premiums, and payments for ecosystem services (PES). Certified wood products often command higher prices in global markets, encouraging companies to adopt responsible practices. PES schemes, such as carbon credits and biodiversity offsets, provide financial rewards for forest conservation, creating a win-win scenario for economic and ecological objectives. Ecological preservation is facilitated by setting aside protected areas, implementing buffer zones, and adopting biodiversity-friendly logging practices. Strategies such as low-intensity harvesting, habitat restoration, and the retention of old-growth trees help maintain ecological integrity while allowing for sustainable resource use. Multi-stakeholder collaboration, involving governments, industry, and local communities, is essential for aligning economic activities with conservation priorities. The challenges of balancing economic and ecological needs are compounded by factors such as population growth, market pressures, and climate change. Overcoming these challenges requires adaptive management, policy innovation, and a commitment to long-term sustainability. By fostering a holistic understanding of forest ecosystems and their value to society, the logging industry can contribute to a sustainable future that benefits both people and the planet.

Conclusion

Logging remains a critical component of forest resource management, balancing the dual imperatives of economic development and environmental sustainability. Forests provide essential resources, including timber and non-timber products, supporting industries that contribute significantly to global and local economies. At the same time, forests are vital ecosystems, hosting rich biodiversity, regulating climate, and delivering crucial ecological services. The practice of logging, however, presents a complex challenge, as unsustainable methods can lead to severe environmental degradation, including deforestation, habitat loss, and biodiversity decline. To address these challenges, modern logging must integrate sustainable practices, innovative technologies, and regulatory frameworks that align resource extraction with conservation goals. The economic importance of logging is undeniable, as it underpins industries such as construction, paper production, and furniture manufacturing, generating significant revenue and employment. In addition to timber, logging facilitates access to non-timber forest products, enhancing economic diversity in forest-dependent regions. Downstream processing, such as sawmilling and veneer production, adds value to raw materials and strengthens local economies. However, the pursuit of economic gains through unsustainable logging practices often comes at the cost of environmental integrity. This underscores the need for sustainable logging approaches that balance short-term economic benefits with long-term ecological preservation. Environmental impacts remain one of the most pressing concerns associated with logging. Unsustainable practices contribute to deforestation, soil erosion, water quality degradation, and increased greenhouse gas emissions, exacerbating climate change and threatening ecosystem stability. Biodiversity loss is a particularly significant consequence, as habitat destruction disrupts ecological balance and imperils countless species. Mitigation strategies, including afforestation, reforestation, and reduced-impact logging, are critical for addressing these issues. These measures not only restore forest cover but also enhance carbon sequestration, protect watersheds, and support wildlife habitats, ensuring that forests continue to provide essential ecosystem services.

Regulatory frameworks play a pivotal role in promoting sustainable logging practices. International certification schemes, such as the Forest Stewardship Council (FSC) and the Programme for the Endorsement of Forest Certification (PEFC), incentivize responsible forest management by ensuring that products meet rigorous environmental and social standards. National policies and regulations provide the foundation for sustainable operations, while efforts to combat illegal logging address a significant driver of deforestation and forest degradation. Community-based monitoring and public awareness initiatives further strengthen these efforts, fostering collective responsibility for forest conservation. Advancements in technology and sustainable practices are transforming the logging industry, offering new opportunities to mitigate environmental impacts. Precision machinery and reduced impact logging techniques minimize waste and ecological damage, while robotics and automation improve efficiency and safety. Biotechnology, including the development of pest-resistant tree species, enhances forest resilience, reducing losses from pests and diseases. Innovations in remote sensing and data analytics enable more effective monitoring and planning, ensuring that logging operations are conducted in an environmentally and economically sustainable manner.

Social and cultural dimensions add complexity to logging practices, as they impact forest-dependent communities and indigenous populations. Logging provides livelihoods and economic opportunities, particularly in rural areas, but it also raises ethical questions about equity and the preservation of cultural heritage. Public perception increasingly demands transparency and sustainability in logging operations, emphasizing the need for inclusive practices that respect the rights and traditions of local communities. Education and training in sustainable logging methods are essential for equipping workers with the skills to adapt to evolving technologies and practices.

Chapter - 11

Felling Techniques and Conversion

1. Felling Techniques and Conversion

1.1 Importance of Felling in Forest Operations

Felling, the act of cutting down trees, represents the foundational step in the sequence of forest operations, encompassing the processes that transition standing timber into utilizable wood products. It is integral to forest management and industrial forestry, bridging the gap between sustainable resource extraction and economic utilization. The significance of felling lies not only in its economic implications but also in its ecological, social, and technical dimensions, making it a cornerstone of modern forestry practices. The primary importance of felling is its role in timber production. Forests are a critical resource, supplying raw materials for industries ranging from construction and furniture-making to paper production and bioenergy. Effective felling techniques ensure the extraction of high-quality timber with minimal wastage, supporting the economic viability of forestry operations. Proper execution also ensures that the felled trees meet market specifications in terms of size, shape, and quality, thereby maximizing their value in downstream processes.

Beyond its economic function, felling plays a pivotal role in forest management, particularly in maintaining ecosystem health and sustainability. Selective felling, for example, is employed to remove mature, overmature, or diseased trees, promoting the growth of younger and healthier stands. This practice not only optimizes the productivity of the forest but also enhances its biodiversity, stability, and resilience. Similarly, strategic felling is used in silvicultural systems to create openings in the canopy, allowing light to reach the forest floor and enabling the regeneration of shade-intolerant species. This fosters a balanced forest structure and supports long-term ecological goals. Felling also has significant implications for safety and operational efficiency in forestry. Proper planning and execution of felling operations minimize risks to workers, machinery, and the surrounding environment. The direction and precision of tree felling are critical to preventing accidents, ensuring the safety of forestry personnel, and reducing damage to adjacent trees or infrastructure. Furthermore, advancements in mechanized felling techniques have enhanced productivity and safety by minimizing the physical demands on workers and allowing operations in challenging terrains.

The importance of felling extends to its environmental and social dimensions, emphasizing the need for sustainable practices. Unsustainable felling, such as clear-cutting without regeneration

DOI: 10.1201/9781042011063-11

planning, can lead to deforestation, habitat loss, soil erosion, and climate change impacts. Conversely, well-regulated felling operations contribute to sustainable forest management by balancing resource extraction with conservation objectives. Community involvement and adherence to certification standards, such as those set by the Forest Stewardship Council (FSC), further underscore the ethical and ecological responsibilities associated with felling.

1.2 Conversion Processes in Logging

Conversion, the process of transforming felled trees into usable wood products, is a vital step in the forestry value chain. It encompasses a range of activities, from initial on-site processing to transportation and further refinement in sawmills or other processing facilities. The efficiency and effectiveness of conversion processes directly influence the economic value of timber, the sustainability of forestry operations, and the quality of the final products. The first stage of conversion typically involves cross-cutting or bucking, where felled trees are cut into smaller sections, known as logs, based on market requirements and processing capabilities. The dimensions and quality of these logs depend on factors such as tree species, diameter, and the intended end use. For instance, longer logs are preferred for structural timber, while shorter sections are suitable for pulpwood or fuelwood. Proper cross-cutting minimizes waste and maximizes the value of each tree, ensuring that as much of the biomass as possible is utilized. Once cross-cutting is complete, logs are transported from the felling site to processing facilities, a process known as extraction. Extraction methods vary depending on terrain, forest type, and the scale of operations. Skidding, forwarding, and cable logging are commonly used techniques, each with its advantages and limitations. Skidders, for instance, drag logs to collection points, while forwarders carry logs in a suspended manner, reducing soil disturbance. In steep or remote areas, cable systems or aerial logging using helicopters may be employed to minimize environmental impacts. The conversion process also includes the utilization of by-products, such as sawdust, bark, and wood chips, which can be repurposed for energy production, particleboard manufacturing, or mulch. This holistic approach aligns with the principles of a circular economy, minimizing waste and maximizing the value extracted from each tree. Technological innovations, such as bio-refineries, are further expanding the scope of conversion by enabling the production of biofuels, bioplastics, and other high-value products from forest biomass.

1.3 Balancing Efficiency and Sustainability in Felling Practices

Balancing efficiency and sustainability in felling practices is a central challenge in modern forestry, as it requires reconciling the economic objectives of timber production with the ecological and social imperatives of forest conservation. This balance is critical for ensuring that forestry operations meet the needs of current generations without compromising the ability of future generations to benefit from forest resources. Efficiency in felling practices is achieved through the adoption of advanced technologies, skilled labor, and strategic planning. Mechanized felling systems, such as harvesters and processors, have revolutionized forestry by increasing the speed and precision of operations. These machines are equipped with hydraulic arms, cutting heads, and computerized controls that allow for precise cutting, delimbing, and bucking of trees in a single operation. By optimizing resource use and reducing labor intensity, mechanization enhances productivity and lowers operational costs.

However, the pursuit of efficiency must be tempered by considerations of sustainability, which require minimizing environmental impacts and preserving the long-term health of forest ecosystems. Unsustainable felling practices, such as large-scale clear-cutting, can lead to soil erosion, loss of biodiversity, and disruptions in water cycles. To mitigate these impacts, forestry professionals employ sustainable felling techniques, such as selective logging, shelterwood cutting, and reduced-impact logging (RIL). These methods prioritize the removal of specific trees while preserving the overall structure and function of the forest.

The integration of technology further supports the balance between efficiency and sustainability. Geographic information systems (GIS) and remote sensing technologies enable the mapping of forest resources, identification of sensitive areas, and planning of felling operations to minimize ecological disruption. Similarly, data-driven decision-making tools allow managers to optimize harvesting schedules, ensuring that trees are felled at the appropriate stage of maturity to maximize yield and regeneration potential. Certification schemes, such as those offered by the Forest Stewardship Council (FSC) and the Programme for the Endorsement of Forest Certification (PEFC), play a crucial role in promoting sustainable felling practices. These schemes set rigorous standards for environmental, social, and economic sustainability, incentivizing forestry operations to adopt responsible practices. Consumers, in turn, support these efforts by choosing certified wood products, creating market-driven pressure for sustainability.

2. Traditional Felling Techniques

2.1 Manual Felling

Manual felling, involving the use of tools such as axes and hand saws, represents one of the earliest and most fundamental approaches to tree harvesting. This technique has been practiced for centuries, serving as the primary method of timber extraction before the advent of mechanized equipment. The simplicity and universality of these tools have made them indispensable in forestry, particularly in contexts where mechanization is impractical due to terrain, scale, or economic constraints. The axe, one of humanity's oldest tools, has been central to manual felling practices. Designed for chopping, splitting, and cutting, axes were traditionally crafted with a wooden handle and a steel or iron blade. The felling process using an axe involves successive chopping at the base of the tree to create a notch, directing the tree's fall in a controlled manner. The precision and skill of the operator are critical, as improper technique can result in dangerous and uncontrolled falls. Axes were often accompanied by wedges, which were driven into the notch to enhance control and guide the tree's fall direction.

Hand saws, introduced later in the history of felling, brought a significant improvement in efficiency and precision. Early hand saws, such as the crosscut saw, featured long blades with teeth designed for cutting across the grain of the wood. These saws required two operators, one on either end, who alternated pulling the blade to slice through the tree trunk. The crosscut saw was particularly advantageous for felling large trees, as it produced cleaner cuts and required less physical effort compared to axes. Its portability and effectiveness made it a popular choice in forestry operations worldwide. Manual felling techniques, despite their simplicity, demand a high degree of skill and physical endurance. Operators must have a deep understanding of tree anatomy, including the direction of grain, the structural integrity of the trunk, and the balance of the canopy. Additionally, they must consider environmental factors, such as wind direction and the surrounding terrain, to ensure safe and efficient operations. The craft of manual felling has historically been passed down through generations, with experienced loggers mentoring apprentices in the techniques and nuances of the trade. While manual felling has largely been supplanted by mechanized methods in industrial forestry, it remains relevant in certain contexts. In remote or steep terrains where heavy machinery cannot be deployed, manual felling offers a practical solution. It is also favored in selective logging operations, where precision and minimal environmental impact are priorities. Furthermore, the enduring use of manual tools in traditional and small-scale forestry reflects their adaptability and cultural significance.

2.2 Pit Sawing and Early Wood Processing Methods

Pit sawing, an early wood processing technique, played a pivotal role in the transition from basic felling to the production of usable timber. This method involves sawing logs into planks or beams using a long saw operated by two workers, one positioned above the log and the other in a pit below it. The pit provided the necessary elevation to manoeuvre the saw, while gravity assisted the cutting process. Pit sawing was widely practiced in pre-industrial societies, enabling the production of timber for construction, shipbuilding, and other applications.

The process of pit sawing begins with the preparation of the work site, where a log is securely positioned over a pit using supports or trestles. The sawyers, often referred to as the "top man" and the "pit man," work in tandem to guide the saw along marked lines on the log. The top man pulls the saw upward, while the pit man applies downward pressure, ensuring a continuous and even cut. This labor-intensive process requires coordination, strength, and skill to produce consistent and straight timber. Pit sawing was instrumental in the early development of wood-based industries, as it enabled the conversion of felled trees into standardized timber products. The flexibility of this method allowed sawyers to tailor the dimensions and shapes of the timber to specific needs, supporting diverse applications in construction, furniture-making, and transportation. For example, pit-sawn planks were used to construct ships during the Age of Sail, highlighting the critical role of this technique in global trade and exploration.

Despite its advantages, pit sawing had several limitations. The process was slow and physically demanding, limiting the volume of timber that could be produced. Additionally, the quality of the sawn timber depended heavily on the skill of the sawyers, as inconsistencies in cutting could result in uneven or warped planks. The advent of mechanized sawmills in the 19th century gradually rendered pit sawing obsolete in industrial contexts, although it continues to be practiced in traditional woodworking communities and regions with limited access to modern technology. Pit sawing and other early wood processing methods represent a foundational stage in the evolution of forestry practices. They reflect humanity's ingenuity in overcoming the challenges of resource extraction and utilization, laying the groundwork for the sophisticated technologies and systems that define contemporary forestry.

2.3 Advantages and Limitations of Traditional Approaches

Traditional felling techniques, including manual felling and pit sawing, offer unique advantages that have sustained their use across centuries and regions. However, they also come with inherent limitations that have driven the adoption of mechanized alternatives in modern forestry. Understanding these advantages and limitations provides valuable insights into the evolution of forestry practices and the contexts in which traditional methods remain relevant. One of the primary advantages of traditional felling techniques is their simplicity and accessibility. Tools such as axes and hand saws are relatively inexpensive, portable, and easy to maintain, making them ideal for small-scale and resource-constrained operations. These tools do not require external power sources, enabling their use in remote areas where access to machinery or fuel is limited. Traditional techniques also excel in precision and control, particularly in selective logging operations. Skilled operators can carefully choose and fell individual trees without causing significant damage to the surrounding vegetation or soil. This precision is critical in ecologically sensitive areas, where preserving the integrity of the forest ecosystem is a priority. Furthermore, the cultural and historical significance of traditional felling practices fosters a deep connection to the craft of forestry, preserving knowledge and traditions that have shaped human relationships with forests over millennia. However, the limitations of traditional approaches cannot be overlooked. The labor-intensive nature of manual felling and pit sawing makes them time-consuming and physically demanding, limiting their efficiency and scalability. The reliance on human strength and skill also introduces variability in the quality and consistency of the work, posing challenges for large-scale or standardized operations. In industrial forestry, where productivity and uniformity are paramount, traditional techniques cannot compete with the speed and precision of mechanized systems.

The safety risks associated with traditional felling methods further highlight their limitations. Manual felling requires operators to work near falling trees and sharp tools, increasing the potential for accidents and injuries. The lack of protective equipment and ergonomic designs in traditional tools exacerbates these risks, particularly in challenging terrains or adverse weather conditions. Despite these limitations, traditional felling techniques continue to hold relevance in specific contexts. Their adaptability, low environmental impact, and cultural value make them a viable choice for small-scale forestry, conservation projects, and regions with limited access to mechanization. The enduring practice of these methods reflects their resilience and significance in the broader landscape of forestry.

3. Modern Felling Techniques

3.1 Chainsaws: Revolutionizing Felling Operations

The introduction of chainsaws in the early 20th century marked a transformative moment in the history of forestry, revolutionizing felling operations by dramatically increasing efficiency, precision, and accessibility. Chainsaws replaced traditional manual tools like axes and hand saws, offering a mechanized alternative that significantly reduced the physical effort and time required to fell trees. Today, chainsaws remain a cornerstone of modern forestry, serving as versatile tools for felling, delimbing, bucking, and even pruning. Chainsaws are powered by internal combustion engines or, more recently, electric motors, which drive a cutting chain around a guide bar. The chain, equipped with sharp, rotating teeth, slices through wood with remarkable speed and accuracy. This design allows operators to cut through even the largest trees quickly, making chainsaws indispensable for large-scale forestry operations as well as small-scale and urban tree management.

The portability and adaptability of chainsaws are among their most significant advantages. Unlike heavy machinery, chainsaws can be easily transported to remote or uneven terrains, enabling their use in diverse forestry contexts. They are also capable of performing a wide range of cutting tasks, from precise directional felling to rapid bucking of logs into transportable sections. Chainsaws have become essential tools for selective logging, where careful removal of individual trees is required to minimize environmental impact. Advancements in chainsaw technology have further enhanced their efficiency and safety. Modern chainsaws feature ergonomic designs, vibration-damping systems, and safety mechanisms such as chain brakes and anti-kickback devices, reducing operator fatigue and the risk of accidents. Additionally, the development of lightweight electric chainsaws has expanded their accessibility, particularly for use in urban and residential settings where noise and emissions are a concern. However, the widespread use of chainsaws also presents challenges, particularly in terms of safety and environmental impact. Chainsaw operation requires significant skill and training to ensure safe and effective use. Accidents, often caused by improper handling or inadequate protective equipment, remain a major concern in forestry. Furthermore, the emissions from gasoline-powered chainsaws contribute to air pollution and greenhouse gas accumulation, prompting efforts to transition toward more sustainable alternatives, such as battery-powered models. Despite these challenges, chainsaws continue to play a pivotal role in modern forestry. Their ability to combine efficiency, precision, and versatility has solidified their place as indispensable tools in felling operations, bridging the gap between traditional methods and fully mechanized systems.

3.2 Mechanized Felling

The advent of mechanized felling systems, such as harvesters and feller-bunchers, represents a significant advancement in forestry, enabling large-scale operations to achieve unprecedented levels of productivity and precision. These machines, designed to handle the complexities of modern logging, have revolutionized the industry by automating key processes and reducing the reliance on manual labor. Harvesters are multifunctional machines equipped with a mechanical arm and a cutting head that can fell, delimb, and cut trees into logs in a single operation. This level of automation not only accelerates the felling process but also ensures consistent quality and dimensions of the logs, meeting the specifications of downstream processing facilities. Harvesters are particularly effective in plantation forestry and other large-scale operations, where uniformity and efficiency are paramount. Feller-bunchers, on the other hand, are specialized machines designed to fell multiple trees at once and arrange them in bunches for easier extraction. These machines use a hydraulic arm equipped with a cutting attachment, such as a disc saw or shear blade, to sever trees at their base. The ability to bunch trees as they are felled significantly reduces the time and effort required for subsequent extraction and transport. Feller-bunchers are commonly used in clear-cutting operations, where rapid removal of large volumes of timber is required.

Mechanized felling systems offer numerous advantages, including enhanced safety, reduced physical strain on workers, and the ability to operate in challenging environments. These machines are equipped with advanced features such as climate-controlled cabins, GPS navigation, and real-time monitoring systems, which improve operator comfort and efficiency. Additionally, their precision reduces collateral damage to surrounding trees and minimizes soil disturbance, supporting sustainable forestry practices. However, mechanized felling is not without its limitations. The high cost of machinery and maintenance can be prohibitive for small-scale operations, and the environmental impact of heavy equipment, such as soil compaction and fuel consumption, must be carefully managed. Furthermore, the use of mechanized systems requires specialized training for operators and support staff, highlighting the need for investment in workforce development. Despite these challenges, the adoption of mechanized felling systems continues to grow, driven by their ability to meet the demands of modern forestry. The integration of technology, such as LiDAR mapping and telematics, is further enhancing their capabilities, paving the way for more efficient and sustainable felling practices.

3.3 Precision Felling

Precision felling, which encompasses directional and controlled felling techniques, is an essential aspect of modern forestry, prioritizing safety, efficiency, and environmental stewardship. These techniques involve carefully planned and executed cuts that control the direction and manner in which a tree falls, minimizing risks to workers, equipment, and surrounding vegetation. Directional felling is widely practiced in forestry to ensure that trees fall in a predetermined direction, facilitating easier extraction and reducing damage to adjacent trees and infrastructure. This technique involves creating a notch or undercut on the side of the tree facing the desired fall direction, followed by a back cut on the opposite side. The depth, angle, and placement of these cuts are critical to achieving precise control over the tree's fall. Directional felling is particularly valuable in selective logging, urban forestry, and situations where obstacles, such as power lines or buildings, are present.

Controlled felling builds on the principles of directional felling by incorporating additional measures to enhance safety and accuracy. Tools such as wedges, felling jacks, and ropes are used to stabilize the tree and guide its fall, reducing the risk of unpredictable movements. Controlled felling is often employed in hazardous situations, such as felling diseased or structurally compromised trees, where traditional methods may be insufficient to ensure safety. Technological advancements have further refined precision felling techniques. Laser rangefinders and inclinometers are used to assess tree height, lean, and surrounding obstacles, aiding in the planning and execution of felling operations. In mechanized systems, computer-controlled cutting heads enable precise placement of cuts, ensuring consistent and predictable results. These innovations have significantly improved the efficiency and safety of precision felling, particularly in complex and high-risk environments.

The benefits of precision felling extend beyond operational efficiency to include environmental and economic advantages. By minimizing damage to non-target trees and soil, these techniques support sustainable forest management and promote ecosystem health. Additionally, the ability to direct trees toward optimal extraction paths reduces the time and cost associated with transportation and processing, enhancing the overall profitability of forestry operations. While precision felling requires a high level of skill and planning, its advantages make it an indispensable component of modern forestry. The emphasis on safety, sustainability, and efficiency aligns with the goals of contemporary forest management, ensuring that timber extraction is conducted responsibly and effectively. As technology continues to advance, precision felling is poised to become even more integral to the future of forestry, blending traditional expertise with cutting-edge innovation.

4. Environmental Considerations in Felling

4.1 Minimizing Damage to Surrounding Vegetation

Felling operations inevitably impact the surrounding vegetation, but minimizing this damage is a cornerstone of sustainable forestry. Careful planning, skilled execution, and the adoption of environmentally sensitive techniques are essential to preserving forest ecosystems during timber extraction. By reducing collateral damage to non-target trees, shrubs, and understory plants, forestry operations can maintain the ecological integrity and regenerative capacity of the forest. Selective logging is one of the most effective approaches for minimizing damage to surrounding vegetation. This technique involves carefully choosing individual trees for felling based on their maturity, health, and ecological role while leaving other trees intact to maintain canopy cover and ecosystem stability. The selective removal of trees reduces the risk of widespread disturbance, supporting biodiversity and enabling the forest to continue providing habitat, carbon storage, and other ecosystem services. The use of directional felling techniques is another critical strategy. By controlling the direction in which a tree falls, operators can avoid damage to nearby trees and minimize the disruption of the forest floor. This technique requires precise planning and execution, often involving tools such as notches, backcuts, and wedges to guide the tree's fall. In dense forests, directional felling is particularly valuable, as it prevents the domino effect of falling trees, causing unintended damage to adjacent vegetation.

Mechanized felling systems, such as harvesters, also contribute to reducing collateral damage when properly managed. These machines are designed to cut and handle trees with precision, limiting the physical impact on surrounding vegetation. Advanced models equipped with GPS and mapping technologies enable operators to navigate efficiently and avoid sensitive areas, such as patches of endangered plant species or regenerating saplings. However, careful planning is necessary to balance the benefits of mechanization with its potential drawbacks, such as soil compaction and disturbances caused by machinery. In addition to technological solutions, forestry professionals employ spatial planning and zoning techniques to protect vegetation. Buffer zones, for example, are established around ecologically sensitive areas, such as wetlands, riparian zones, and wildlife habitats, to prevent damage during felling operations. These zones are often delineated using geographic information systems (GIS) and remote sensing data, allowing managers to identify and prioritize areas for protection.

Community involvement and education play a vital role in minimizing damage to surrounding vegetation. Local stakeholders, including indigenous communities and small-scale forest owners, often possess valuable knowledge about the ecological and cultural significance of specific areas. Incorporating this knowledge into felling plans ensures that operations are conducted with respect for local biodiversity and traditions. Training programs for operators also emphasize the importance of environmental stewardship, equipping them with the skills and awareness needed to minimize impact. Minimizing damage to surrounding vegetation is not only an ecological imperative but also an economic consideration. Healthy forests support long-term productivity, ensuring that timber resources remain available for future harvests. By adopting practices that preserve vegetation and promote regeneration, forestry operations can achieve a balance between resource extraction and conservation.

4.2 Soil Compaction and Erosion Control During Felling

The physical impact of felling operations on forest soils is a critical environmental consideration, as soil compaction and erosion can have long-lasting consequences for ecosystem health and productivity. Soil compaction occurs when the weight of machinery or fallen trees compresses the soil, reducing its porosity and disrupting its structure. This can impair root growth, water infiltration, and microbial activity, ultimately affecting the regeneration of the forest. Erosion, on the other hand, involves the displacement of soil particles by wind or water, leading to the loss of topsoil, nutrient depletion, and sedimentation in nearby water bodies. To mitigate soil compaction, forestry operations prioritize the use of low-impact techniques and equipment. Mechanized systems, such as forwarders and harvesters, are often equipped with wide tires or tracks that distribute weight more evenly, reducing pressure on the soil. Additionally, operators use designated skid trails and extraction paths to confine machinery movement to specific areas, leaving the majority of the forest floor undisturbed. In sensitive environments, such as wetlands or steep slopes, cable logging systems are employed to minimize ground contact altogether. Timing is another important factor in controlling soil compaction. Felling operations are scheduled during dry seasons or periods when the soil is less saturated, as wet soils are more susceptible to compaction. By avoiding operations during adverse weather conditions, forestry managers can reduce the risk of long-term soil degradation.

Erosion control measures focus on maintaining vegetation cover and stabilizing disturbed areas. Residual vegetation, such as shrubs and ground cover, plays a crucial role in anchoring soil and preventing erosion. Forestry operations are planned to retain as much vegetation as possible, particularly in riparian zones and steep terrain. When vegetation removal is unavoidable, replanting and mulching are used to restore ground cover and stabilize the soil. Techniques such as terracing and the installation of sediment traps further reduce the risk of erosion in vulnerable areas. Buffer zones along water bodies are essential for preventing sedimentation and maintaining water quality. These zones act as natural filters, trapping sediment and nutrients before they reach streams and rivers. In addition to protecting aquatic ecosystems, buffer zones help maintain soil integrity by reducing the speed and volume of surface runoff during rainfall events.

Soil health monitoring is an integral part of sustainable forestry practices. Periodic assessments of soil compaction, erosion rates, and nutrient levels provide valuable data for evaluating the effectiveness of mitigation measures and guiding future operations. Advances in soil sensors and remote sensing technologies have improved the accuracy and efficiency of soil health monitoring, enabling real-time decision-making in forestry management. The implementation of soil-friendly felling practices aligns with broader goals of ecosystem resilience and sustainability. Healthy soils support robust tree regeneration, enhance carbon sequestration, and provide habitat for a diverse array of organisms. By prioritizing soil conservation, forestry operations contribute to the long-term productivity and ecological stability of forests.

4.3 Role of Reduced Impact Logging (RIL) in Sustainable Felling

Reduced Impact Logging (RIL) represents a paradigm shift in forestry, emphasizing practices that minimize environmental damage while maintaining economic viability. This approach integrates planning, technology, and operational techniques to reduce the ecological footprint of felling operations, particularly in sensitive and biodiverse forests. By focusing on sustainability, RIL seeks to balance the demands of timber production with the preservation of forest ecosystems.

A cornerstone of RIL is meticulous planning and pre-harvest assessment. Before operations begin, forestry professionals conduct detailed surveys to map forest resources, identify sensitive areas, and evaluate the potential impacts of logging. Geographic information systems (GIS) and remote sensing tools are often used to create detailed maps that guide the layout of skid trails, extraction routes, and landing sites. This planning reduces unnecessary disturbance to the forest and optimizes the efficiency of logging operations. Directional felling is a key technique in RIL, ensuring that trees fall in predetermined directions to minimize damage to surrounding vegetation and soil. By controlling the fall direction, operators can align trees with skid trails and reduce the need for additional ground disturbance during extraction. Advanced tools, such as wedges and felling jacks, enhance the precision of directional felling, contributing to the overall effectiveness of RIL practices.

RIL also incorporates measures to protect biodiversity and maintain ecosystem functions. For example, only a limited number of trees are harvested per hectare, leaving a sufficient number of seed trees and habitat structures for wildlife. Logging intensity is carefully calibrated to ensure that the forest retains its regenerative capacity and ecological integrity. In tropical forests, RIL has been shown to significantly reduce canopy loss and carbon emissions compared to conventional logging methods. Training and capacity building are critical components of RIL implementation. Forestry workers receive specialized training in low-impact techniques, safety protocols, and the use of advanced equipment. This not only enhances the quality of logging operations but also empowers workers with skills that contribute to their long-term professional development. The benefits of RIL extend beyond environmental considerations to include social and economic advantages. By reducing damage to forests, RIL ensures that timber resources remain available for future harvests, supporting the sustainability of forestry-dependent economies. Certification schemes, such as those offered by the Forest Stewardship Council (FSC), recognize and promote RIL practices, providing market incentives for sustainable timber products. While RIL offers a promising pathway toward sustainable forestry, its widespread adoption faces challenges, including higher upfront costs, technical complexity, and the need for strong governance and enforcement. Addressing these barriers requires collaboration among governments, industry stakeholders, and conservation organizations to build the infrastructure, knowledge, and incentives necessary for effective implementation.

5. Conversion Processes

5.1 Primary Conversion

Primary conversion refers to the initial stage of transforming felled trees into usable wood products, specifically saw timber. This process involves the cutting, debarking, and shaping of logs into standardized forms such as planks, beams, or boards. Primary conversion is a critical link in the forestry value chain, bridging the gap between raw timber extraction and its preparation for industrial and commercial use. The efficiency and precision of this stage significantly impact the quality of the final products, the sustainability of operations, and the economic returns of forestry ventures. The process begins with log preparation, where felled trees are stripped of branches and bark. Debarking serves multiple purposes: it removes the outer protective layer of the log, which often harbors pests and fungi, and prepares the wood for further cutting. In modern sawmills, debarking is typically carried out by mechanized systems that use rotating knives, drums, or high-pressure water jets to ensure consistency and efficiency. After debarking, logs are cut into smaller sections or billets based on their size and intended use. This step, known as bucking, requires precision to optimize the yield and quality of timber. Automated bucking systems use sensors and computer algorithms to determine the optimal cutting points, maximizing the value of each log while minimizing waste. The billets are then transported to sawing stations for the primary cutting process.

Sawing is the core activity in primary conversion, where logs are transformed into planks or beams. Different sawing methods are employed based on the desired product specifications, wood species, and equipment capabilities. The most common techniques include plain sawing, quarter sawing, and rift sawing. Plain sawing, the fastest and most economical method, involves cutting the log longitudinally to produce wide planks. Quarter sawing and rift sawing, while slower and more resource-intensive, yield timber with superior stability and aesthetic appeal, making them suitable for high-end applications. Advancements in sawmilling technology have significantly enhanced the efficiency and precision of primary conversion. Modern sawmills are equipped with computer numerical control (CNC) systems, which automate the cutting process and ensure uniformity in product dimensions. Laser-guided saws and 3D scanners further improve accuracy by analyzing the log's internal structure and optimizing the sawing pattern to maximize usable wood. These technologies reduce material waste and increase the profitability of forestry operations.

The by-products of primary conversion, such as sawdust, bark, and wood chips, are also valuable resources. Sawdust can be used to manufacture particleboard or pellets for bioenergy, while bark and wood chips are often repurposed as mulch or raw materials for pulp production. The utilization of these by-products aligns with the principles of a circular economy, minimizing waste and maximizing resource efficiency. Primary conversion plays a crucial role in the forestry industry, setting the foundation for downstream processes and value creation. By adopting advanced technologies and optimizing operations, primary conversion not only enhances the economic viability of forestry but also supports sustainable resource management.

5.2 Secondary Conversion

Secondary conversion involves the further processing of primary timber products into value-added goods such as furniture, flooring, doors, and decorative panels. This stage significantly enhances the economic value of timber, diversifying its applications and supporting a range of industries from construction to interior design. Secondary conversion is characterized by precision machining, surface treatments, and the integration of aesthetic and functional considerations.

The transformation of sawn timber into value-added products begins with precise cutting and shaping. Advanced woodworking machines, including CNC routers and panel saws, are used to cut timber into specific dimensions and shapes based on the desired product. These machines enable high levels of accuracy and repeatability, ensuring consistency across large production volumes. For intricate designs, such as carved furniture components or decorative moldings, CNC machines use computer-generated patterns to guide the cutting process.

Surface treatment is a critical aspect of secondary conversion, enhancing the appearance, durability, and functionality of wood products. Sanding, planing, and polishing are commonly employed to achieve a smooth finish, while staining and painting add color and protection against environmental factors. Coating technologies, such as UV-cured lacquers and water-based finishes, offer durable and environmentally friendly alternatives to traditional varnishes. Wood modification techniques, such as thermal treatment and lamination, are increasingly used to enhance the properties of value-added products. Thermal treatment involves heating timber in a controlled environment to improve its dimensional stability, resistance to decay, and color uniformity. Laminated products, such as cross-laminated timber (CLT) and laminated veneer lumber (LVL), combine multiple layers of wood bonded with adhesives to create strong and versatile materials suitable for structural applications.

Secondary conversion also incorporates assembly and integration processes for complex products. Furniture manufacturing, for example, involves joining individual components using techniques such as doweling, mortise and tenon joints, or metal fasteners. In flooring production, individual planks are milled with interlocking profiles, such as tongue-and-groove, to facilitate installation. These processes emphasize both functionality and aesthetics, meeting the diverse needs of consumers and industries. Sustainability is a key consideration in secondary conversion, with increasing emphasis on eco-friendly practices and materials. The use of certified wood from responsibly managed forests ensures that products meet environmental and ethical standards. Additionally, manufacturers are adopting energy-efficient technologies and minimizing waste through improved production planning and the repurposing of offcuts and sawdust.

5.3 Role of Mobile and On-Site Sawmills in Conversion Efficiency

Mobile and on-site sawmills have emerged as valuable tools for enhancing the efficiency and sustainability of timber conversion processes. These portable systems allow logs to be processed directly at the felling site or in remote locations, reducing transportation costs, minimizing waste, and supporting localized forestry operations. The flexibility and accessibility of mobile sawmills make them an attractive option for small-scale producers, community forestry projects, and operations in challenging terrains. One of the primary advantages of mobile sawmills is their ability to process timber close to its source. By eliminating the need to transport entire logs to centralized sawmills, mobile systems significantly reduce fuel consumption, greenhouse gas emissions, and transportation-related expenses. This is particularly beneficial in remote or mountainous regions, where infrastructure limitations and high transport costs can hinder traditional sawmilling operations. Mobile sawmills are also well-suited for urban forestry, enabling the efficient processing of trees removed from city landscapes due to maintenance or development projects.

Mobile sawmills are designed for versatility and ease of use. They can be mounted on trailers for transport and quickly set up at the felling site. Equipped with band saws, circular saws, or chainsaw attachments, these systems can handle a wide range of log sizes and species. Operators can customize the cutting pattern to produce specific dimensions, optimizing yield and meeting local market demands. Some models feature automated functions, such as hydraulic log handling and computerized controls, to enhance efficiency and precision. On-site sawmilling also contributes to resource efficiency by enabling the utilization of timber that might otherwise go to waste. Logs with irregular shapes, smaller diameters, or defects that make them unsuitable for large-scale sawmills can be processed on-site to produce beams, planks, or specialty products. This approach aligns with the principles of sustainable forestry by maximizing the value derived from each harvested tree. In addition to economic and environmental benefits, mobile and on-site sawmills offer social advantages by supporting local economies and community forestry initiatives. Small-scale producers and local communities can use these systems to generate income, create employment opportunities, and produce timber for local construction and craft industries. Mobile sawmills also empower landowners to manage their forests sustainably, converting felled trees into usable products without relying on external processing facilities.

Despite their advantages, mobile and on-site sawmills face challenges such as lower production volumes compared to industrial sawmills and the need for skilled operators. Addressing these challenges requires training programs, technical support, and the development of cost-effective models that cater to diverse user needs. The role of mobile and on-site sawmills in the conversion process highlights their potential to enhance efficiency, reduce environmental impact, and support decentralized forestry operations. As the demand for sustainable and localized solutions continues to grow, these systems are poised to play an increasingly important role in modern timber processing.

6. Technological Innovations in Felling and Conversion

6.1 Use of Robotics and Automation in Felling Operations

The integration of robotics and automation into felling operations represents a revolutionary shift in forestry practices, offering solutions that enhance precision, efficiency, and safety. By reducing the dependence on manual labor and incorporating advanced technologies, these systems address the challenges posed by traditional methods, such as human fatigue, variability, and high-risk working conditions. Robotics and automation are particularly valuable in industrial-scale logging operations, where high productivity and consistency are essential. Robotic systems in felling operations are often embedded within advanced machinery, such as harvesters and feller-bunchers. These machines are equipped with robotic arms that perform cutting, delimbing, and bucking with minimal human intervention. Automated control systems ensure that these operations are carried out with high precision, optimizing the yield and quality of timber. For instance, programmable robotic arms can adapt to variations in tree diameter, height, and species, tailoring the cutting process to maximize efficiency and reduce waste.

Automation has also enhanced the safety of felling operations by removing human operators from hazardous environments. Remote-controlled and semi-autonomous machines allow operators to manage felling activities from a safe distance, minimizing exposure to falling trees, uneven terrain, and adverse weather conditions. Fully autonomous systems, though still in the developmental stage, promise to further eliminate the need for on-site human intervention, leveraging machine learning and AI to execute complex tasks independently. One notable application of robotics in forestry is the use of drones for preliminary assessment and monitoring. Drones equipped with cameras and sensors provide real-time data on tree health, density, and terrain, informing robotic felling systems about the optimal approach for each site. This integration of aerial and ground-based robotics creates a synergistic workflow that enhances the efficiency and precision of operations.

The environmental benefits of robotics and automation in felling are significant. These systems are designed to minimize collateral damage to surrounding vegetation, soil disturbance, and carbon emissions. Precision cutting and optimized tree handling reduce waste, while the ability to operate in environmentally sensitive areas supports sustainable logging practices. Despite their advantages, the adoption of robotics and automation faces challenges, including high initial costs, technical complexity, and the need for skilled operators and maintenance personnel. Investments in research, development, and training are essential to address these barriers and unlock the full potential of robotic technologies in forestry. As advancements continue, robotics and automation are poised to redefine the future of felling operations, blending technological innovation with sustainable forestry principles.

6.2 Laser and Sensor-Based Systems for Precision Cutting

Laser and sensor-based technologies have emerged as transformative tools in forestry, enabling unprecedented precision and control in cutting operations. These systems address critical challenges in felling and conversion, such as material waste, inconsistent quality, and environmental impact, by providing real-time data and feedback to guide decision-making. Their integration into both manual and mechanized systems has enhanced the efficiency and sustainability of forestry practices.

Laser systems are commonly used to assess tree dimensions, identify cutting points, and ensure the accuracy of felling and bucking operations. Mounted on harvesters, feller-bunchers, or handheld devices, laser rangefinders and scanners measure tree height, diameter, and lean angle with millimeter-level precision. This data allows operators to plan cuts that optimize the volume and quality of timber, reducing waste and improving the economic returns of forestry operations. For example, laser-guided harvesters can adjust their cutting patterns dynamically, ensuring that logs meet specific dimensional requirements for downstream processing. In addition to guiding felling operations, laser technologies are used in sawmills to enhance the precision of primary and secondary conversion processes. Automated saws equipped with laser alignment systems produce consistent and uniform cuts, maximizing yield and minimizing material loss. Lasers are also employed in defect detection, identifying knots, cracks, or other imperfections in logs and boards. By mapping these defects, sawmills can adjust cutting patterns to exclude damaged sections, improving the quality and value of the final products.

Sensor-based systems complement laser technologies by providing additional data on wood properties, environmental conditions, and machinery performance. Infrared sensors, for instance, detect variations in wood moisture content, density, and temperature, enabling more informed decisions during cutting and processing. Proximity sensors and gyroscopes enhance the stability and accuracy of felling machines, particularly in challenging terrains or adverse weather. Advanced sensor networks also play a critical role in safety and environmental monitoring. Proximity sensors and collision avoidance systems prevent machinery from damaging non-target trees or entering restricted zones, supporting sustainable logging practices. Environmental sensors measure soil compaction, canopy cover, and biodiversity impacts, providing data that inform adaptive management strategies.

The integration of AI and machine learning has further enhanced the capabilities of laser and sensor-based systems. AI algorithms analyze data from these technologies to predict outcomes, optimize workflows, and identify opportunities for improvement. For instance, predictive models based on sensor data can forecast equipment wear and tear, enabling preventive maintenance and reducing downtime. While laser and sensor-based systems offer numerous benefits, their implementation requires significant investment in technology, training, and infrastructure. Ensuring accessibility and affordability for small-scale and community forestry operations remains a key challenge. However, as these technologies become more widespread, their potential to enhance precision, efficiency, and sustainability in forestry will continue to grow.

6.3 Advancements in Sawmill Technology and Waste Reduction

Modern sawmill technology has undergone significant advancements, driven by the dual goals of improving efficiency and reducing waste. Innovations in machinery, process automation, and material recovery have transformed the way timber is processed, aligning with the principles of sustainable forestry and the circular economy. These advancements address key challenges in sawmilling, such as material loss, energy consumption, and environmental impact, while enhancing the quality and value of wood products.

High-speed sawing systems equipped with advanced blade technologies have significantly increased the throughput of modern sawmills. Bandsaws and circular saws, designed with precision-engineered teeth and coatings, achieve faster cutting speeds with minimal material loss. Thin-kerf saws, which produce narrower cuts, further reduce waste by maximizing the usable volume of each log. These innovations not only improve efficiency but also reduce the cost of raw materials, enhancing the profitability of sawmill operations. Automation has become a cornerstone of modern sawmilling, enabling precise control over every stage of the conversion process. Computer numerical control (CNC) systems, integrated with sensors and actuators, automate tasks such as log positioning, cutting, and sorting. Real-time data from these systems ensures consistent quality and dimensional accuracy in timber products. Automated grading systems, which use cameras and machine learning algorithms, evaluate wood properties and classify products based on their intended use, reducing the reliance on manual inspection.

Waste reduction is a central focus of advancements in sawmill technology. By-products such as sawdust, bark, and wood chips, once considered waste, are now valuable resources for a range of applications. Biomass energy facilities convert sawdust and wood chips into renewable energy, reducing reliance on fossil fuels. Bark is repurposed as mulch or raw material for bio-based products, while offcuts and trimmings are used in the production of engineered wood products, such as particleboard and medium-density fiberboard (MDF). The integration of AI and IoT (Internet of Things) technologies has further enhanced waste reduction and operational efficiency in sawmills. AI-driven optimization algorithms analyze data from sensors and machinery to identify inefficiencies, suggest process improvements, and minimize material loss. IoT-enabled devices provide real-time monitoring of equipment performance, energy usage, and material flow, supporting proactive maintenance and resource management. Environmental sustainability is a key consideration in modern sawmill technology. Energy-efficient machinery, renewable energy integration, and water recycling systems are increasingly adopted to minimize the ecological footprint of sawmills. Certification schemes, such as those offered by the Forest Stewardship Council (FSC), incentivize sustainable practices by linking them to market advantages for eco-friendly products. Advancements in sawmill technology and waste reduction have redefined the conversion process, transforming timber processing into a more efficient, sustainable, and economically viable industry. By embracing these innovations, sawmills contribute to the broader goals of sustainable forestry, resource conservation, and environmental stewardship. As technology continues to evolve, the potential for further improvements in sawmilling efficiency and sustainability remains vast.

7. Safety and Ergonomics in Felling Operations

7.1 Personal Protective Equipment (PPE) and Best Practices

Safety in felling operations begins with the use of appropriate Personal Protective Equipment (PPE) and adherence to industry-recognized best practices. The inherently dangerous nature of forestry work, particularly tree felling, necessitates rigorous safety measures to protect workers from injuries caused by falling trees, flying debris, and contact with cutting equipment. PPE serves as the first line of defense against these hazards, mitigating risks and ensuring that workers are adequately prepared for the physical demands of their tasks.

The essential components of PPE for felling operations include helmets, eye and ear protection, gloves, chainsaw-resistant clothing, and steel-toed boots. Helmets protect workers from head injuries caused by falling branches or tools, while integrated visors or safety glasses shield their eyes from wood chips and dust. Ear protection, such as earplugs or earmuffs, is critical for preventing hearing loss due to prolonged exposure to the high decibel levels generated by chainsaws and machinery. Gloves provide a firm grip on tools while protecting hands from abrasions and cuts, and chainsaw-resistant clothing, made with specialized fibers, guards against lacerations in the event of accidental contact with the saw chain. Steel-toed boots with anti-slip soles further enhance safety by protecting the feet from crushing injuries and providing stability on uneven or slippery terrain.

The effectiveness of PPE depends not only on its quality but also on its proper use and maintenance. Workers must be trained to select the appropriate gear for their specific tasks, inspect it regularly for signs of wear or damage, and replace it as needed to ensure continued protection. For example, chainsaw operators should check their protective chaps or trousers for cuts or frayed fibers, as compromised gear may fail to provide adequate defense against accidents. Employers have a responsibility to supply high-quality PPE, enforce its use, and provide ongoing education on safety standards.

In addition to PPE, best practices in felling operations play a crucial role in minimizing risks. These practices include pre-operational planning, such as assessing the site for hazards, identifying the fall direction of trees, and establishing escape routes. Clear communication among team members is essential, particularly in mechanized operations where operators and ground workers must coordinate their activities to avoid accidents. Workers are advised to maintain a safe distance from falling trees and machinery, stay alert to environmental conditions such as wind or rain, and adhere to the guidelines provided by regulatory agencies and industry standards. The combination of comprehensive PPE and adherence to best practices fosters a culture of safety in felling operations. By prioritizing worker protection and risk mitigation, forestry professionals can reduce the incidence of injuries and fatalities, ensuring that operations are conducted in a manner that respects the well-being of all involved.

7.2 Training and Certification for Safe Felling Practices

Effective training and certification programs are essential components of safe felling operations, providing workers with the knowledge, skills, and confidence to perform their tasks safely and efficiently. Tree felling involves complex techniques that require a deep understanding of tree anatomy, environmental factors, and equipment handling. Without proper training, workers are at a heightened risk of accidents and injuries, which can have severe consequences for both individuals and organizations. Training programs for safe felling practices typically combine theoretical instruction with hands-on experience, covering a wide range of topics related to forestry safety. Participants learn about the proper use and maintenance of tools, such as chainsaws, wedges, and winches, as well as techniques for directional felling, cutting notches, and creating back cuts. Emphasis is placed on understanding tree behavior, including how factors like lean, weight distribution, and wind conditions influence the direction and stability of a falling tree. Additionally, training includes situational awareness, such as recognizing potential hazards in the surrounding environment and planning escape routes.

Certification programs, such as those offered by organizations like the Occupational Safety and Health Administration (OSHA), the International Society of Arboriculture (ISA), and regional forestry associations, set standardized benchmarks for competency and safety in felling operations. Certification ensures that workers meet industry-recognized standards and are equipped to handle the challenges of their roles. For example, certified chainsaw operators are trained to assess and mitigate risks, perform emergency procedures, and maintain their equipment to minimize downtime and hazards. Advanced training is particularly important for workers involved in mechanized felling operations, where the use of harvesters, feller-bunchers, and other machinery adds layers of complexity. Operators must learn to navigate challenging terrains, manage the mechanical components of their equipment, and work in coordination with ground crews. Simulated training environments, which use virtual reality or other technologies, provide valuable opportunities for workers to practice their skills in controlled settings before entering the field. Regular refresher courses and continuous education are critical for maintaining safety standards in felling operations. As technologies and industry practices evolve, workers must stay up to date on the latest tools, techniques, and regulatory requirements. Ongoing training also reinforces safety protocols and helps workers develop a proactive mindset toward risk management.

Training and certification programs not only enhance individual and organizational safety but also contribute to broader industry goals, such as reducing accident rates and promoting sustainability. Skilled and certified workers are better equipped to implement environmentally responsible practices, such as directional felling and reduced-impact logging, which minimize damage to surrounding vegetation and soil. Investing in comprehensive training and certification fosters a professional workforce that prioritizes safety, efficiency, and environmental stewardship. By empowering workers with the skills and knowledge to perform their tasks safely, the forestry industry can create a safer and more sustainable future for felling operations.

7.3 Addressing Hazards in Mechanized and Manual Felling

Felling operations, whether mechanized or manual, are fraught with hazards that require careful management to protect workers and minimize environmental impact. Addressing these hazards involves identifying potential risks, implementing preventive measures, and equipping workers with the tools and knowledge to respond effectively to emergencies. The nature and severity of hazards vary depending on the methods and equipment used, making it essential to tailor safety protocols to the specific context of each operation. In manual felling, the primary hazards stem from the physical demands of the work, the unpredictability of tree behavior, and the proximity of workers to dangerous tools and falling trees. Chainsaw operators, for example, face risks of lacerations, kickback injuries, and fatigue-related accidents. The unpredictable trajectory of falling trees adds another layer of danger, particularly in dense forests or windy conditions where trees may snag on surrounding vegetation or change direction unexpectedly. To mitigate these risks, manual felling relies heavily on the skill and vigilance of operators, as well as the use of PPE and directional felling techniques.

Mechanized felling introduces a different set of hazards, primarily related to the operation of heavy machinery and the interaction between ground crews and equipment. Harvesters and feller-bunchers, while enhancing efficiency and safety, require operators to manage complex mechanical systems and navigate challenging terrains. Equipment failure, poor visibility, and operator fatigue can lead to accidents, such as collisions with obstacles or machinery rollovers. Ground workers, who assist with extraction and cleanup, face additional risks from moving equipment, falling debris, and uneven terrain. Addressing these hazards begins with thorough site assessments and planning. Before operations commence, forestry professionals evaluate the terrain, tree characteristics, and environmental conditions to identify potential risks. This information informs the layout of work zones, escape routes, and skid trails, ensuring that operations are conducted with minimal disruption to the environment and maximum safety for workers. In mechanized felling, geospatial technologies such as GPS and LiDAR assist in mapping terrain and guiding machinery, reducing the likelihood of accidents.

Effective communication is critical in managing hazards, particularly in large-scale operations involving multiple teams and equipment. Two-way radios, hand signals, and standardized protocols ensure that workers remain aware of each other's activities and can respond quickly to changing conditions. Mechanized systems often include safety features, such as proximity sensors and collision avoidance technologies, which further enhance communication and coordination.

Emergency preparedness is another key aspect of hazard management. Workers must be trained in first aid, evacuation procedures, and the use of emergency equipment, such as fire extinguishers and harnesses. Regular safety drills reinforce these skills and build confidence in the ability to handle unforeseen events. In both manual and mechanized felling, fostering a culture of safety is paramount. Employers and supervisors play a crucial role in promoting awareness, accountability, and adherence to safety standards. By prioritizing hazard management and investing in training, equipment, and technology, the forestry industry can significantly reduce risks and ensure that felling operations are conducted safely and responsibly.

8. Economic and Market Implications

8.1 Cost Efficiency in Felling and Conversion Operations

The economic viability of forestry operations hinges on cost efficiency in both felling and conversion processes. Cost efficiency is the ability to minimize expenses while maximizing output, quality, and sustainability. Achieving this balance requires an intricate understanding of operational dynamics, resource management, technological integration, and market demands. As global forestry faces challenges such as fluctuating timber prices, environmental regulations, and the growing demand for sustainable practices, enhancing cost efficiency has become a top priority.

In felling operations, cost efficiency begins with planning and the selection of appropriate methods and equipment. Mechanized systems such as harvesters and feller-bunchers, though capital-intensive, have revolutionized cost structures in large-scale forestry. These machines can fell, delimb, and cut logs into transportable sizes in a single operation, significantly reducing labor costs and time. Their ability to operate efficiently in various terrains further contributes to cost reductions by minimizing delays and damages associated with manual felling. However, the initial investment and maintenance costs of mechanized systems necessitate careful cost-benefit analysis, particularly for smaller operations. Manual felling, while less expensive in terms of equipment, incurs higher labor costs and longer operational timelines. For small-scale and selective logging operations, however, manual felling remains a cost-effective choice, especially when combined with portable technologies such as chainsaws and mobile sawmills. The adaptability of manual techniques to diverse forest conditions and their minimal environmental impact offer additional cost advantages in certain contexts, such as community forestry or conservation projects.

Conversion processes also play a critical role in determining overall cost efficiency. Sawmills equipped with advanced technology, such as automated log scanners, thin-kerf saws, and optimized cutting systems, reduce material waste and enhance productivity. These technologies maximize the yield of high-value timber from each log, ensuring that the raw material is utilized to its fullest potential. Modern sawmills also incorporate energy-efficient systems and waste recovery mechanisms, such as biomass boilers that convert sawdust and offcuts into energy, further lowering operational costs.

Transportation logistics represent another significant cost factor in felling and conversion operations. Transporting raw logs to centralized sawmills is expensive and labor-intensive, particularly in remote areas. Mobile and on site sawmills provide a cost efficient alternative by enabling logs to be processed closer to their source, reducing fuel consumption and transportation expenses. Additionally, the reduction in bulk during primary conversion facilitates easier and more economical transport of finished products to markets. Economies of scale play a decisive role in cost efficiency. Larger forestry operations benefit from reduced per-unit costs due to high-volume production, advanced machinery, and streamlined workflows. However, smaller operations can achieve cost efficiency by focusing on niche markets, value-added products, and sustainable practices that command premium prices. The integration of technological innovations, workforce training, and effective resource management strategies further enhances cost efficiency across operations of all sizes.

8.2 Market Trends for Sawn Timber and Processed Wood Products

The market for sawn timber and processed wood products is influenced by a complex interplay of global demand, technological advancements, environmental policies, and consumer preferences. Understanding these trends is essential for forestry stakeholders seeking to align their operations with market dynamics and maximize economic returns. The demand for wood products continues to grow, driven by expanding construction, furniture, and packaging industries, as well as the increasing use of wood in bioenergy and bio-based materials.

One of the most prominent trends in the sawn timber market is the shift toward sustainable and certified wood products. Consumers and industries are increasingly prioritizing wood sourced from responsibly managed forests, as evidenced by the rising demand for certifications such as the Forest Stewardship Council (FSC) and the Programme for the Endorsement of Forest Certification (PEFC). These certifications not only assure environmental and social responsibility but also serve as market differentiators, enabling producers to command higher prices and access premium markets. The construction industry remains the largest consumer of sawn timber, with demand driven by housing developments, infrastructure projects, and the adoption of wood in sustainable building practices. Engineered wood products, such as cross-laminated timber (CLT) and glulam, are gaining popularity due to their strength, durability, and versatility. These products are used in everything from residential buildings to large-scale commercial and industrial structures, positioning wood as a renewable alternative to steel and concrete. In addition to structural applications, the furniture and interior design industries are significant drivers of demand for processed wood products. The preference for natural and aesthetically appealing materials has fueled the market for high-quality, value-added wood products, including hardwood flooring, cabinetry, and decorative panels. Customization and innovation, such as the use of thermally treated wood and digitally printed veneers, are further expanding the appeal of wood in these sectors.

The rise of bioeconomy and circular economy principles is shaping new markets for wood products. Wood-based bioenergy, bio-plastics, and bio-chemicals are emerging as sustainable alternatives to fossil fuel-derived materials. The ability to transform forestry by-products, such as sawdust and wood chips, into value-added bio-based products is creating additional revenue streams for forestry operations. This diversification is particularly relevant in regions with established forestry infrastructure and supportive policies for renewable energy. Global trade in sawn timber and processed wood products is heavily influenced by regional supply and demand dynamics, currency fluctuations, and trade regulations. Major exporters, such as Canada, Russia, and Sweden, continue to dominate the market, while emerging markets in Asia, Africa, and South America are driving growth in imports. Trade policies, including tariffs and non-tariff barriers, significantly impact market access and competitiveness, highlighting the need for strategic planning and adaptability among forestry businesses.

8.3 Value Chains and the Role of Conversion in Forest Economies

The value chain in forestry encompasses all activities involved in transforming raw timber into finished products and delivering them to end-users. Conversion processes are at the heart of this value chain, creating economic opportunities, supporting livelihoods, and driving industrial development in forest economies. By adding value to raw materials, conversion operations enhance the profitability and sustainability of forestry, making them a cornerstone of forest-based industries. Primary conversion, which transforms logs into sawn timber, sets the foundation for the value chain by producing standardized products that meet market specifications. The efficiency and quality of this stage determine the potential for downstream processing and the overall value generated from each log. Sawmills equipped with advanced technologies, such as laser-guided cutting and automated grading, maximize the recovery of high-value timber, ensuring that raw materials are utilized to their fullest extent.

Secondary conversion processes, which include the production of furniture, flooring, and engineered wood products, further enhance the value of timber by meeting diverse consumer and industrial needs. These activities support a wide range of industries, from construction and interior design to packaging and bioenergy. The versatility of wood as a material enables its use in traditional applications as well as innovative products, such as wood-based composites and bio-plastics.

The economic impact of conversion extends beyond the production of goods. Conversion operations generate employment across the value chain, from skilled workers in sawmills to craftsmen in furniture manufacturing. These jobs support local economies, particularly in rural and forest-dependent regions, where forestry is often a primary source of income. By fostering industrial development, conversion processes contribute to broader economic growth and regional development. Sustainability is an increasingly important consideration in the value chain, with a growing emphasis on circular economy principles. By utilizing by-products such as sawdust, bark, and offcuts, conversion processes minimize waste and create additional revenue streams. For example, sawdust can be transformed into pellets for bioenergy, while bark is used for mulch or bio-chemical extraction. This holistic approach enhances resource efficiency and aligns with global efforts to reduce environmental impacts.

The role of conversion in forest economies is also shaped by policy and market dynamics. Supportive policies, such as tax incentives for sustainable practices and investment in infrastructure, enable the growth of conversion industries. Access to global markets, facilitated by trade agreements and certifications, further enhances the competitiveness of forest-based economies. By integrating technological advancements, sustainability principles, and market responsiveness, the value chain in forestry maximizes the economic, social, and environmental benefits of timber resources. Conversion processes are central to this value creation, driving the development of resilient and prosperous forest economies.

Conclusion

The practices of felling and conversion in forestry form the backbone of forest resource utilization, bridging the gap between raw timber harvesting and the production of processed wood products. These operations, integral to forestry and logging industries, have evolved over centuries, moving from labor-intensive traditional methods to highly mechanized and precision-driven processes. While these advancements have improved efficiency and output, they have also raised significant environmental and ethical considerations, underscoring the need for sustainable practices that balance resource extraction with ecological preservation. Felling techniques and conversion processes hold profound implications not only for forest ecosystems but also for the economic, technological, and social dimensions of forest management.

Felling, as the initial step in forest operations, plays a pivotal role in determining the efficiency and sustainability of subsequent processes. The transition from manual felling techniques, such as the use of axes and hand saws, to chainsaws and mechanized equipment like harvesters and feller-bunchers has revolutionized logging operations. Modern precision felling methods, such as directional and controlled techniques, enhance operational safety, minimize waste, and reduce collateral damage to surrounding vegetation. However, these advancements come with challenges, including the risk of overharvesting and increased soil compaction in mechanized operations, which necessitate the integration of Reduced Impact Logging (RIL) principles to mitigate environmental harm.

Conversion processes, which transform felled logs into usable timber and value-added products, are equally critical. Primary conversion methods focus on processing logs into saw timber, while secondary conversion involves producing finished goods like furniture and flooring. The introduction of mobile and on-site sawmills has significantly enhanced efficiency, reducing transportation costs and minimizing waste. Technological innovations in sawmill operations, including laser-based systems and automated waste recovery, have further streamlined conversion processes, contributing to a more sustainable utilization of forest resources. Efficient conversion practices are vital for maximizing the economic value of harvested timber while ensuring minimal environmental impact.

Environmental considerations in felling and conversion have become central to modern forestry practices. Unsustainable felling methods can lead to deforestation, habitat destruction, soil erosion, and reduced biodiversity. Mechanized felling, if improperly managed, exacerbates soil compaction and disrupts water cycles. Adopting RIL practices minimizes these impacts by emphasizing careful planning, minimal soil disturbance, and selective harvesting. Waste management during conversion processes, including the utilization of by-products for energy and secondary products, further enhances the sustainability of forestry operations.

Safety and ergonomics are paramount in felling and conversion operations, given the inherent risks involved. Personal Protective Equipment (PPE), comprehensive training, and adherence to best practices are essential for ensuring worker safety in both manual and mechanized operations.

Certification programs and ongoing education contribute to a culture of safety, equipping workers with the skills to navigate hazards effectively. Mechanized felling introduces additional safety considerations, necessitating the development of ergonomic equipment and automated systems to minimize human risk. The economic and market implications of felling and conversion extend beyond immediate operations to influence global timber trade and local economies. Efficient felling and conversion practices reduce costs, enhance profitability, and strengthen value chains, creating opportunities for employment and economic growth. Market trends for sawn timber and processed wood products are increasingly influenced by consumer demand for sustainably sourced materials, driving the adoption of forest certification standards and environmentally responsible practices.

Integrating sustainability into felling and conversion practices is essential for the long-term health of forests and the viability of the forestry industry. Certification standards, such as those established by the Forest Stewardship Council (FSC), promote compliance with sustainable logging practices, while advances in conversion technology minimize waste and enhance resource utilization. By prioritizing waste recovery, reducing environmental footprints, and aligning with conservation goals, forestry operations can contribute to sustainable development while meeting global demand for timber products.

Chapter - 12

Transportation and Storage of Timber

1. Timber Transportation and Storage

1.1 Importance of Transportation and Storage in the Timber Supply Chain

Timber transportation and storage are critical components of the forestry supply chain, bridging the gap between timber extraction in forests and its delivery to processing facilities, markets, and end users. These processes directly influence the economic viability, environmental sustainability, and quality of timber products. Effective transportation and storage practices ensure that timber maintains its structural integrity, reduces waste, and aligns with the operational and market demands of the forestry industry. Transportation represents the largest logistical cost in the timber supply chain, accounting for a significant portion of overall operational expenses. The movement of raw logs from felling sites to sawmills, and subsequently to distribution centers or manufacturing plants, requires strategic planning to optimize efficiency and minimize costs. This process often involves multimodal logistics, incorporating road, rail, water, and even aerial transport to overcome geographical and infrastructural challenges. For instance, logging trucks are commonly used in accessible areas, while barges or railways are preferred for transporting timber over long distances or through remote regions.

Storage plays an equally important role in preserving the quality and usability of timber during various stages of the supply chain. Proper storage techniques protect timber from environmental factors such as moisture, temperature fluctuations, pests, and fungal attacks, which can compromise its structural and aesthetic properties. For example, freshly felled logs with high moisture content are prone to decay if not stored under controlled conditions, reducing their value and utility. Storing timber strategically also enables the forestry industry to balance supply and demand fluctuations, ensuring a consistent flow of raw materials to markets and processing facilities. The integration of transportation and storage into the timber supply chain enhances overall sustainability. Efficient logistics reduce fuel consumption and emissions, aligning with global efforts to combat climate change. Sustainable storage practices, such as the use of eco-friendly preservatives and renewable energy for temperature control, further minimize the environmental impact. Additionally, advanced tracking and inventory management systems improve transparency and traceability, ensuring compliance with certification standards such as the Forest Stewardship Council (FSC) and supporting ethical and sustainable forestry. Beyond operational benefits, transportation and storage contribute to the

DOI: 10.1201/9781042011063-12

economic and social dimensions of the forestry industry. They create employment opportunities in logistics, warehousing, and transportation sectors, particularly in rural and forest-dependent regions. Moreover, well-planned logistics systems support local economies by facilitating the trade of timber and wood products, both domestically and internationally.

1.2 Historical Evolution of Timber Transport and Storage Practices

The history of timber transportation and storage reflects humanity's evolving relationship with forests and wood as a resource. From the earliest methods of manual hauling to the sophisticated logistics systems of today, these practices have undergone significant transformations, driven by technological advancements, economic demands, and environmental considerations. In ancient times, timber transportation was limited by the absence of advanced tools and infrastructure. Early methods relied on human and animal labor to drag or carry logs over short distances. Sledges and simple carts were employed to ease the burden, especially in regions with dense forests and rugged terrain. Rivers and waterways served as natural transportation routes, enabling the movement of logs to downstream settlements and markets. Floating logs in rivers was a common practice, particularly in Europe, Asia, and North America, where timber-rich regions were often located near major water systems. This method, while efficient for the time, posed challenges such as log jams, loss of material, and environmental impact on aquatic ecosystems.

The Industrial Revolution marked a turning point in timber transportation. Railroads and steam-powered vehicles revolutionized the movement of logs, enabling large-scale forestry operations and expanding access to distant markets. Railways, in particular, allowed timber to be transported efficiently over long distances, connecting remote forests to industrial hubs. The invention of logging trucks in the early 20th century further enhanced mobility, making road transport the dominant mode of timber transportation in many regions. Trucks provided flexibility and adaptability, capable of navigating varied terrains and reaching areas inaccessible by rail. Storage practices also evolved during this period, with increased awareness of the need to protect timber from environmental degradation. Traditional methods involved stacking logs in open yards, often exposing them to weather conditions that accelerated decay and pest infestations. Over time, the development of covered storage facilities, seasoning techniques, and chemical treatments improved the longevity and quality of stored timber. Kiln drying, introduced in the 19th century, became a standard practice for reducing moisture content and preventing fungal growth, enhancing the durability and usability of timber products.

The latter half of the 20th century saw the integration of mechanization and automation in both transportation and storage. Forklifts, cranes, and conveyors streamlined the handling of logs, reducing labor intensity and improving efficiency. Computerized inventory systems and tracking technologies, such as GPS and RFID, provided real-time visibility into the movement and condition of timber, enabling better coordination and planning. In the 21st century, sustainability and innovation have become central to the evolution of timber transport and storage. Electric and hybrid vehicles are being adopted to reduce emissions, while drones and autonomous systems are exploring new possibilities for monitoring and transporting timber. Storage practices now emphasize eco-friendly solutions, such as renewable energy-powered facilities and biodegradable preservatives, reflecting the industry's commitment to environmental stewardship. The historical evolution of timber transportation and storage demonstrates the interplay between technological innovation, economic development, and environmental awareness. By learning from past practices and embracing future advancements, the forestry industry can continue to optimize these processes for a sustainable and prosperous future.

1.3 Challenges in Ensuring Quality and Efficiency

Despite advancements in timber transportation and storage, the forestry industry continues to face significant challenges in maintaining quality and efficiency. These challenges are shaped by factors such as infrastructure limitations, environmental conditions, regulatory requirements, and market dynamics, necessitating adaptive strategies and innovative solutions. One of the primary challenges in timber transportation is the high cost and complexity of logistics. The movement of bulky and heavy materials, often from remote forest locations to processing facilities or markets, requires substantial investments in infrastructure, fuel, and labor. Poorly maintained roads, limited access to railways, and seasonal variations in weather conditions exacerbate these difficulties, particularly in developing regions. For example, logging operations in tropical forests often contend with muddy or flooded roads during the rainy season, delaying transport and increasing costs. Environmental considerations add another layer of complexity to transportation. The carbon footprint of timber logistics is a growing concern, as the use of diesel-powered trucks and machinery contributes to greenhouse gas emissions. Addressing this issue requires the adoption of cleaner technologies, such as electric vehicles, and optimizing transport routes to reduce fuel consumption. Balancing economic efficiency with environmental responsibility remains a critical challenge for the industry.

Storage practices face similar obstacles, particularly in maintaining timber quality during prolonged periods of storage. Freshly felled logs, with their high moisture content, are susceptible to fungal attacks, decay, and pest infestations if not stored under controlled conditions. Inadequate ventilation, exposure to sunlight, and fluctuations in temperature and humidity further compromise timber quality, leading to financial losses and reduced market value. Implementing climate-controlled storage facilities, though effective, involves significant costs that may be prohibitive for smaller operations. Compliance with regulatory requirements presents additional challenges. Transporting and storing timber often involves navigating complex regulations related to certification, permits, and sustainability standards. For example, ensuring that timber meets FSC or PEFC criteria requires meticulous documentation and traceability throughout the supply chain. Non-compliance can result in financial penalties, reputational damage, and restricted market access.

Global market dynamics also influence the efficiency and cost-effectiveness of timber transportation and storage. Fluctuations in demand and supply, driven by factors such as economic conditions, trade policies, and consumer preferences, create uncertainties that complicate logistics planning. For instance, sudden surges in demand for construction-grade timber may strain transportation networks and storage capacities, leading to delays and increased costs. To address these challenges, the forestry industry is increasingly turning to technological innovations and collaborative approaches. Advanced analytics and AI-driven tools optimize logistics planning, identify bottlenecks, and reduce inefficiencies. Blockchain technology enhances traceability and compliance, providing transparent records of timber movements and storage conditions. Partnerships with governments, communities, and private stakeholders support infrastructure development and policy alignment, enabling a more resilient and sustainable supply chain.

2. Methods of Timber Transportation

2.1 Ground Transportation: Trucks, Skidders, and Forwarders

Ground transportation is the most commonly used method for moving timber, offering flexibility and adaptability across various terrains and scales of operation. This category encompasses a range of vehicles and equipment, including logging trucks, skidders, and forwarders, each tailored to specific conditions and operational requirements. Ground transportation's efficiency and accessibility make it indispensable for connecting forest sites to processing facilities or transit hubs, forming the backbone of most timber supply chains.

Logging trucks are the cornerstone of ground transportation in forestry, designed to carry substantial loads of logs over long distances. These vehicles range from conventional highway trucks fitted with trailers to specialized off-road models equipped with robust suspensions and traction systems for navigating rugged terrains. Modern logging trucks are often equipped with GPS tracking, automated loading systems, and advanced safety features, such as stability control and rollover prevention, enhancing both efficiency and operator safety. The flexibility of trucks allows them to operate across diverse landscapes, from mountainous regions to urban peripheries, ensuring timber reaches its destination regardless of geographic constraints.

Skidders play a vital role in short-distance timber transportation, particularly in the initial extraction stage within forests. These powerful machines are designed to drag logs from felling sites to collection points, known as landings, where they can be loaded onto trucks or further processed. Cable skidders, which use winches and cables to haul logs, are ideal for steep or inaccessible terrain, while grapple skidders, equipped with hydraulic claws, excel in flat or gently sloping areas. The compact design and maneuverability of skidders make them highly effective in dense forests, minimizing damage to surrounding vegetation during operations. Forwarders complement skidders by transporting logs in a more controlled and environmentally friendly manner. Unlike skidders, which drag logs across the ground, forwarders lift and carry logs in a bunk, reducing soil disturbance and preventing damage to the timber. This method is particularly advantageous in sustainable forestry practices, where minimizing environmental impact is a priority. Forwarders are often equipped with advanced technology, such as load sensors and terrain mapping systems, to optimize routes and enhance productivity.

Despite their advantages, ground transportation methods face challenges such as fuel costs, environmental impact, and infrastructure limitations. The heavy reliance on diesel-powered vehicles contributes to greenhouse gas emissions, prompting a growing interest in electric and hybrid alternatives. Additionally, poor road conditions in remote areas can hinder transportation efficiency, necessitating investments in infrastructure and maintenance. Innovations in vehicle design and logistics planning are addressing these challenges, ensuring that ground transportation remains a reliable and sustainable option for timber movement.

2.2 Aerial Transportation

Aerial transportation methods, including the use of helicopters and cable systems, are employed in forestry operations where ground-based approaches are impractical due to steep terrain, dense forests, or environmental considerations. While more specialized and cost-intensive than ground transportation, these methods offer unique advantages in terms of efficiency, precision, and reduced environmental impact, particularly in regions with challenging landscapes or strict ecological constraints. Helicopter logging, also known as heli-logging, involves the use of helicopters to lift and transport logs from felling sites to collection points or processing facilities. This method is particularly effective in mountainous regions, where steep slopes and inaccessible terrain make ground transportation difficult or impossible. Helicopters are equipped with cables and grapples to lift individual logs, which are then flown to designated drop zones. The precision of heli-logging minimizes damage to surrounding vegetation and soil, supporting sustainable forestry practices. Additionally, the speed of helicopters allows for rapid extraction, reducing the time logs spend exposed to environmental conditions that could degrade their quality. Despite its advantages, heli-logging is associated with high operational costs, driven by fuel consumption, equipment maintenance, and the need for specialized personnel. These factors limit its use to high-value timber, such as old-growth hardwoods, where the economic returns justify the investment. Advances in helicopter technology, including fuel-efficient engines and autonomous flight systems, are helping to reduce costs and expand the applicability of this method.

Cable systems, also known as skyline or cable yarding systems, are another aerial transportation method used in forestry. These systems consist of a network of cables suspended between towers or anchor points, allowing logs to be transported along a fixed path. A carriage, which holds the logs, moves along the cables, powered by gravity, winches, or engines. Cable systems are particularly suited for steep and uneven terrain, where traditional vehicles cannot operate effectively. Their ability to transport logs without dragging them across the ground reduces soil disturbance and damage to the timber. Cable yarding systems are available in various configurations, including high-lead, skyline, and grapple yarders, each tailored to specific terrain and operational needs. Modern cable systems incorporate remote controls, automated tensioning, and advanced braking mechanisms, enhancing safety and efficiency. While less expensive than heli-logging, cable systems still require significant initial investment in infrastructure and equipment, making them more suitable for large-scale or long-term forestry operations. The adoption of aerial transportation methods reflects the forestry industry's commitment to overcoming logistical challenges while minimizing environmental impact. By combining precision, efficiency, and sustainability, helicopters and cable systems play a vital role in modern timber transportation.

2.3 Water Transportation: Rafting, Barges, and Ships

Water transportation has been a fundamental method for moving timber throughout history, leveraging rivers, lakes, and seas to transport large volumes of logs over long distances. While its significance has declined in some regions due to advancements in road and rail infrastructure, water transportation remains a cost-effective and environmentally friendly option for forestry operations located near navigable waterways.

Rafting, one of the oldest forms of water transportation, involves bundling logs together into rafts and floating them downstream to processing facilities or storage areas. This method is particularly common in regions with abundant river networks, such as Canada, Scandinavia, and parts of Asia. Rafting is a low-cost and energy-efficient method, as it relies on the natural flow of water to move timber. However, it poses challenges such as log losses, environmental impact on aquatic ecosystems, and limited control over transportation speed and direction. Barges offer a more modern and controlled approach to water transportation, capable of carrying large quantities of logs or processed timber. These flat-bottomed vessels are towed or pushed by tugboats, making them ideal for navigating rivers, canals, and coastal waters. Barges provide greater stability and protection for timber compared to rafting, reducing the risk of damage during transit. They are also compatible with multimodal logistics systems, facilitating seamless transitions to road or rail transport.

Ships are used for the international transportation of timber, particularly in the form of sawn wood, wood chips, or pulp. Bulk carriers and container ships transport timber products across oceans, connecting exporters in regions such as North America, Scandinavia, and Southeast Asia to global markets. The use of containerization has enhanced the efficiency and security of timber shipments, enabling precise tracking and reducing handling times. Water transportation's environmental benefits include lower fuel consumption and emissions compared to road and air transport. However, it is limited by the availability of navigable waterways and the need for infrastructure such as ports, docks, and loading equipment. Innovations such as eco-friendly vessels and digital tracking systems are addressing these limitations, ensuring that water transportation remains a viable and sustainable option for timber logistics.

2.4 Innovations in Timber Transport Logistics

The integration of technology and innovative practices in timber transport logistics is transforming the forestry industry, enhancing efficiency, sustainability, and transparency. These advancements address key challenges such as cost reduction, environmental impact, and compliance with regulatory standards, positioning the industry to meet the demands of a dynamic global market. One of the most significant innovations is the use of GPS and telematics systems for real-time tracking and route optimization. These technologies provide detailed information on vehicle locations, fuel consumption, and delivery times, enabling managers to monitor operations and make data-driven decisions. Route optimization algorithms minimize travel distances and fuel usage, reducing operational costs and carbon emissions. For example, logging trucks equipped with GPS systems can dynamically adjust their routes to avoid traffic congestion or adverse weather conditions, ensuring timely and efficient deliveries.

Autonomous vehicles and drones are emerging as promising tools for timber transportation. Autonomous logging trucks and forwarders are being developed to operate in challenging terrains, reducing the need for human intervention and enhancing safety. Drones, on the other hand, are used for site assessments, inventory management, and small-scale timber transport, offering a cost-effective and versatile solution for remote or inaccessible areas. Blockchain technology is revolutionizing timber logistics by enhancing transparency and traceability. Blockchain systems create immutable records of timber's journey from forest to market, ensuring compliance with certification standards and providing assurance to consumers. These systems also streamline documentation processes, reducing administrative costs and errors. Sustainability-focused innovations include the development of electric and hybrid vehicles for timber transport, which reduce greenhouse gas emissions and reliance on fossil fuels. Renewable energy-powered storage facilities and eco-friendly materials for vehicle construction further align the industry with environmental goals. The adoption of these innovations reflects the forestry industry's commitment to leveraging technology to overcome logistical challenges and enhance sustainability. By integrating advanced tools and practices, timber transport logistics is evolving to meet the needs of a rapidly changing world.

3. Challenges in Timber Transportation

3.1 Accessibility in Remote and Rugged Terrains

One of the most persistent challenges in timber transportation is accessibility in remote and rugged terrains, where forests often exist far from developed infrastructure. These conditions complicate the movement of logs from felling sites to processing facilities or markets, significantly increasing operational costs, logistical complexity, and environmental risks. Overcoming these challenges requires innovative approaches, infrastructure investments, and adaptive technologies.

Remote forests, particularly in mountainous or densely forested regions, often lack roads or other infrastructure suitable for heavy vehicles. Logging operations in these areas must rely on temporary access roads, which are expensive to construct and maintain. These roads are frequently affected by seasonal changes, such as heavy rainfall or snow, leading to mudslides, erosion, and road closures that delay transportation. In tropical regions, the challenge is compounded by monsoon seasons, which make unpaved roads impassable and limit the operational windows for timber extraction. The rugged nature of these terrains poses additional difficulties. Steep slopes, rocky outcrops, and unstable ground create hazardous conditions for vehicles and machinery, increasing the risk of accidents and equipment damage. Specialized vehicles, such as off-road trucks with advanced traction systems, are required to navigate these environments, but their high cost and fuel consumption present financial burdens for forestry operators. Technological innovations, such as cable yarding systems and aerial transportation methods, offer potential solutions for these challenges. Cable yarding systems, which use suspended cables to transport logs, are particularly effective in steep and inaccessible areas. Helicopter logging, though costly, provides a viable option for high-value timber in remote locations, reducing the need for road construction and minimizing environmental impact. However, the economic feasibility of these methods remains a limiting factor, especially for small-scale operations.

Infrastructure development, supported by public and private investments, is critical to improving accessibility in remote regions. Building durable roads, bridges, and loading facilities can significantly enhance the efficiency and safety of timber transportation. However, such developments must be carefully planned to minimize environmental impact, considering the risks of deforestation, habitat disruption, and soil erosion. The challenge of accessibility underscores the need for comprehensive planning and adaptive strategies in timber transportation. By leveraging advanced technologies, sustainable infrastructure, and collaborative efforts, the forestry industry can address these obstacles while balancing economic and environmental priorities.

3.2 Environmental Impacts of Timber Transport Operations

Timber transportation operations have significant environmental impacts, including greenhouse gas emissions, habitat disruption, and soil degradation. These impacts arise from the use of heavy machinery and vehicles, the construction of access roads, and the movement of timber through ecologically sensitive areas. Addressing these issues is critical to aligning forestry practices with sustainability goals and mitigating climate change. The carbon footprint of timber transportation is a major concern, as diesel-powered trucks, skidders, and other machinery emit substantial amounts of carbon dioxide and other greenhouse gases. Long-haul transport, which often relies on road or water networks, exacerbates this issue, particularly in regions where forests are located far from processing facilities. Efforts to reduce emissions include transitioning to electric or hybrid vehicles, optimizing transport routes to minimize fuel consumption, and investing in alternative transportation methods such as rail or water, which are more energy-efficient. Habitat disruption is another pressing issue. The construction of logging roads and transportation infrastructure often requires clearing vegetation and altering landscapes, which can fragment habitats and threaten biodiversity. Wildlife corridors are particularly vulnerable, as road networks can disrupt migratory patterns and increase the risk of vehicle-wildlife collisions. Sustainable road planning, which prioritizes minimal impact on ecosystems, is essential to mitigate these effects. For example, designing roads to follow natural contours and avoid critical habitats can reduce the ecological footprint of transportation projects. Soil degradation, caused by the movement of heavy vehicles and machinery, further amplifies the environmental impact. Soil compaction reduces water infiltration, affecting local hydrology and plant regeneration. Additionally, erosion caused by poorly designed roads or steep slopes can lead to sedimentation in nearby water bodies, harming aquatic ecosystems. Implementing best practices in road construction, such as proper drainage systems and erosion control measures, is vital to addressing these challenges.

Environmental regulations and certification standards, such as those established by the Forest Stewardship Council (FSC), play a critical role in promoting sustainable timber transport practices. These standards encourage the adoption of low-impact methods, transparent reporting, and the use of renewable energy sources. Compliance with such standards not only reduces environmental impact but also enhances market access and consumer trust. The integration of technology offers additional opportunities for mitigating environmental impacts. GPS tracking and route optimization software reduce unnecessary travel distances and fuel consumption, while drones and sensors monitor environmental conditions and identify areas requiring protection. By adopting these innovations, the forestry industry can balance transportation efficiency with environmental stewardship.

3.3 Cost Management and Fuel Efficiency

Cost management and fuel efficiency are critical challenges in timber transportation, as they directly impact the profitability and sustainability of forestry operations. The high costs associated with vehicle operation, fuel consumption, and maintenance, combined with fluctuating fuel prices and market demands, create significant financial pressures for the industry. Addressing these challenges requires a strategic focus on efficiency, technology adoption, and process optimization. Fuel costs represent one of the largest expenses in timber transportation, particularly for long-haul operations. Logging trucks and machinery consume large quantities of diesel, and fuel price volatility adds an element of uncertainty to budgeting and financial planning. To mitigate these costs, the forestry industry is increasingly exploring alternatives to traditional fuel sources. Electric and hybrid vehicles, though still in the early stages of adoption, offer promising solutions for reducing fuel dependency and emissions. Advances in battery technology and the development of charging infrastructure are making these options more viable for forestry operations.

Route optimization is another key strategy for improving fuel efficiency and reducing costs. By analyzing terrain, traffic patterns, and delivery schedules, forestry managers can identify the most efficient routes for transporting timber. GPS and telematics systems play a critical role in this process, providing real-time data that allows for dynamic adjustments and better planning. For example, avoiding congested routes or steep inclines can significantly reduce fuel consumption and wear on vehicles.

Maintenance and equipment management also contribute to cost efficiency. Regular maintenance of logging trucks, skidders, and forwarders ensures optimal performance, reducing the risk of breakdowns and extending the lifespan of machinery. Predictive maintenance, enabled by IoT sensors and AI analytics, is an emerging approach that identifies potential issues before they result in costly failures. By monitoring engine performance, fuel usage, and other parameters, predictive systems help forestry operations reduce downtime and maintenance expenses. Collaboration and economies of scale offer additional avenues for cost management. Cooperative transport arrangements, where multiple forestry companies share resources such as trucks and storage facilities, reduce individual costs and improve overall efficiency. Large-scale operations benefit from economies of scale, as they can distribute fixed costs across higher volumes of timber, achieving lower per-unit transportation expenses.

Government policies and incentives also play a role in addressing cost challenges. Subsidies for fuel-efficient vehicles, tax breaks for sustainable practices, and investments in infrastructure development create a supportive environment for reducing transportation costs. However, the accessibility of such incentives varies by region, requiring forestry stakeholders to navigate local regulations and opportunities. By focusing on fuel efficiency, maintenance, and collaborative strategies, the forestry industry can address the financial challenges of timber transportation while advancing its sustainability goals. These efforts not only enhance profitability but also position the industry to thrive in an increasingly competitive and environmentally conscious market.

4. Storage of Timber in Harvesting Sites

4.1 Temporary Storage Practices: Log Decks and Staging Areas

Temporary storage of timber at harvesting sites is a critical aspect of forestry operations, serving as an intermediary step between tree felling and transportation. These storage practices, which include the establishment of log decks and staging areas, ensure that harvested timber is efficiently organized, protected, and prepared for transport to processing facilities. Effective temporary storage practices optimize logistics, preserve the quality of the timber, and support the smooth functioning of the overall supply chain. Log decks are the most commonly used form of temporary storage in harvesting operations. These are designated areas near felling sites where logs are systematically stacked after being cut and processed. The arrangement of logs in decks is crucial for both operational efficiency and timber preservation. Logs are typically organized based on their size, species, and intended use, allowing for easy identification and retrieval during loading and transportation. This sorting process minimizes delays and errors, ensuring that the right timber reaches the appropriate destinations. Additionally, the strategic placement of log decks within the harvesting site reduces the distance logs need to be transported initially, saving time and fuel costs. The design and layout of log decks are influenced by several factors, including the terrain, volume of harvested timber, and environmental conditions. In flat or gently sloping areas, log decks can be larger and more uniform, accommodating higher volumes of timber. In contrast, in steep or rugged terrains, smaller, dispersed log decks may be more practical, reducing the risk of instability or accidents. Proper site selection is essential to ensure that log decks do not interfere with natural watercourses or sensitive habitats, aligning with environmental sustainability goals.

Staging areas are another important element of temporary storage practices. These are centralized locations within the harvesting site where timber from multiple log decks is consolidated before transportation. Staging areas often serve as logistical hubs, equipped with loading equipment such as cranes, skidders, or forwarders to facilitate the efficient transfer of logs onto transport vehicles. The use of staging areas streamlines the movement of timber, reducing bottlenecks and enhancing overall operational efficiency. In large-scale operations, staging areas may also include temporary shelters or storage structures to provide additional protection for high-value timber. Temporary storage practices must also address safety considerations, as improperly stored logs can pose significant risks to workers and equipment. Logs must be stacked securely to prevent shifting or collapsing, particularly in areas prone to high winds or heavy rainfall. The use of barriers, chocks, or retaining structures can enhance stability and safety. Furthermore, clear signage and designated pathways around log decks and staging areas ensure that workers can navigate the site safely and efficiently. While temporary storage practices are primarily operational in nature, they also play a vital role in maintaining the quality of harvested timber. By minimizing the time logs spend exposed to adverse environmental conditions, such as direct sunlight or heavy rain, temporary storage helps preserve their structural integrity and market value. As the forestry industry continues to adopt advanced technologies and sustainable practices, the optimization of temporary storage processes remains a key focus area, enhancing both productivity and environmental stewardship.

4.2 Managing Moisture and Decay During On-Site Storage

The management of moisture and decay is a critical challenge in the on-site storage of timber, as exposure to environmental conditions can significantly compromise the quality, durability, and marketability of harvested logs. Timber, being an organic material, is highly susceptible to fungal growth, insect infestations, and other forms of degradation when stored under unfavorable conditions. Implementing effective moisture and decay management strategies is essential to preserve the value of the timber and ensure its suitability for downstream processing.

Moisture content is one of the most important factors influencing the preservation of timber during storage. Freshly felled logs contain high levels of moisture, making them prone to fungal attacks and decay if not properly managed. Fungi thrive in warm, humid conditions, rapidly breaking down the cellulose and lignin in wood and causing discoloration, loss of structural integrity, and reduced commercial value. To mitigate this risk, forestry operations employ various techniques to control moisture levels and promote adequate airflow around stored logs. One common method for managing moisture is the use of elevated log decks or stacking techniques that allow for natural air circulation. By raising logs off the ground using sleepers or skids, the risk of moisture accumulation from soil contact is minimized. Proper spacing between logs within the stacks further enhances ventilation, reducing the likelihood of fungal growth. In regions with high humidity or frequent rainfall, tarps, plastic sheeting, or other weather-resistant covers are often used to protect logs from direct exposure to water while still allowing for some degree of ventilation.

Chemical treatments are another effective approach to preventing decay and insect infestations during on-site storage. Logs can be treated with fungicides, insecticides, or preservatives that inhibit the growth of harmful organisms. While chemical treatments are widely used, their application must be carefully managed to avoid environmental contamination and comply with regulatory standards.

In recent years, the development of eco-friendly and biodegradable preservatives has provided more sustainable options for timber preservation, aligning with the forestry industry's commitment to environmental responsibility. In addition to preventive measures, monitoring and inspection are essential for managing moisture and decay during storage. Regular assessments of stored timber allow forestry professionals to identify signs of fungal growth, insect activity, or other issues early, enabling timely intervention. Advanced technologies, such as moisture meters and thermal imaging, provide precise and non-invasive methods for assessing the condition of logs, enhancing the accuracy and efficiency of monitoring efforts.

Seasonal and climatic variations also play a significant role in moisture and decay management. In colder climates, the natural freezing of logs during winter can slow or halt the growth of decay-causing organisms, extending the safe storage period. Conversely, in tropical or subtropical regions, where high temperatures and humidity prevail year-round, timber must be transported or processed more quickly to prevent degradation. Adapting storage practices to local environmental conditions is essential for maintaining timber quality and minimizing losses. The integration of technology and sustainable practices into moisture and decay management continues to evolve, offering new solutions to longstanding challenges. For example, smart sensors and IoT devices are being used to monitor environmental conditions in real-time, providing actionable data to optimize storage practices. By leveraging these innovations, the forestry industry can enhance the efficiency and sustainability of on-site storage, ensuring that harvested timber reaches its intended market or processing facility in optimal condition.

4.3 Role of Logging Camps in Storage Operations

Logging camps play a multifaceted role in timber storage operations, serving as hubs for organizing, preserving, and managing harvested timber. These camps, typically established near harvesting sites, provide the infrastructure, equipment, and workforce necessary to support on-site storage activities. In addition to their operational functions, logging camps contribute to the overall efficiency, safety, and sustainability of forestry operations. One of the primary functions of logging camps is to facilitate the organization and staging of harvested timber. Logs are transported from felling sites to the camp, where they are sorted, measured, and prepared for storage or transportation. The availability of centralized facilities within the camp allows for the efficient handling and processing of large volumes of timber, reducing delays and streamlining logistics. Depending on the scale of operations, logging camps may also include dedicated areas for storing different types of wood, such as hardwoods and softwoods, or logs destined for specific markets or processing facilities.

The infrastructure of logging camps is designed to support effective storage practices. Camps are equipped with tools and machinery, such as loaders, cranes, and forklifts, that enable the safe and efficient movement of logs within the storage area. Sheltered storage structures, such as sheds or warehouses, provide additional protection for high-value timber, shielding it from rain, sunlight, and other environmental factors. Camps also serve as sites for implementing moisture and decay management strategies, including the application of preservatives, installation of drainage systems, and monitoring of storage conditions.

The workforce at logging camps plays a crucial role in the success of storage operations. Skilled workers are responsible for overseeing the sorting, stacking, and monitoring of timber, ensuring that storage practices align with industry standards and quality requirements. Training programs and safety protocols are integral to logging camps, equipping workers with the knowledge and skills to handle timber efficiently and safely. The presence of trained personnel also allows for immediate responses to challenges such as pest infestations or adverse weather conditions, minimizing the risk of timber degradation. In addition to their logistical and operational functions, logging camps contribute to the sustainability of forestry practices. By centralizing storage activities, camps reduce the environmental footprint associated with transporting timber over long distances. Camps located near waterways or rail lines can facilitate the use of low-impact transportation methods, such as barges or trains, further enhancing sustainability. Additionally, logging camps often include facilities for recycling and repurposing by-products, such as wood chips or sawdust, supporting a circular economy approach.

The role of logging camps extends beyond their operational significance to include social and economic dimensions. Camps provide employment opportunities for local communities, supporting livelihoods and fostering economic development in forestry-dependent regions. They also serve as hubs for collaboration between forestry companies, regulatory agencies, and conservation organizations, promoting the exchange of knowledge and best practices. As the forestry industry continues to evolve, logging camps remain integral to the efficient and sustainable storage of timber at harvesting sites. Advances in technology, infrastructure, and workforce training are enhancing the capabilities of these camps, ensuring that they meet the demands of modern forestry while maintaining their environmental and social responsibilities. By optimizing the role of logging camps, the industry can strengthen its supply chain and contribute to the long-term sustainability of forest resources.

5. Long-Term Timber Storage

5.1 Dry Storage Techniques: Preventing Warping and Cracking

Dry storage techniques are essential for long-term timber storage, especially when the wood is intended for high-value applications such as construction, furniture-making, and flooring. These techniques focus on reducing and stabilizing the moisture content of the timber, thereby preventing warping, cracking, and other forms of deformation that can compromise its structural integrity and aesthetic value. The effectiveness of dry storage depends on the careful management of environmental conditions and the application of proven preservation methods.

Warping and cracking occur when timber loses moisture unevenly, causing stress within the wood fibers. To mitigate this, dry storage typically begins with proper stacking techniques. Logs or sawn timber are stacked with uniform spacing between each piece to promote air circulation. This spacing is achieved using wooden spacers or stickers, which maintain consistent gaps and prevent direct contact between the timber. The alignment and stability of these stacks are critical, as uneven or unstable arrangements can lead to mechanical damage and inconsistent drying. Open-air drying is one of the oldest and most cost-effective dry storage methods. In this approach, timber is stored in well-ventilated areas, protected from direct sunlight and precipitation. Elevated platforms or pallets are used to keep the wood off the ground, reducing the risk of moisture absorption from the soil. Shade structures, such as roofs or tarps, provide additional protection while allowing airflow. The drying rate in open-air storage varies depending on climate conditions, with warmer and drier environments accelerating the process. However, excessive heat or rapid drying can cause surface checking and internal stresses, requiring careful monitoring.

Kiln drying is a more controlled and efficient dry storage technique, widely used in industrial forestry operations. Kilns are specialized chambers where temperature, humidity, and airflow are carefully regulated to achieve uniform drying. This method significantly reduces drying time compared to open-air drying and allows for precise control over the final moisture content of the timber, typically between 6% and 12%, depending on the intended application. Kiln drying also sterilizes the wood, eliminating pests and fungi, which is particularly important for export markets with strict phytosanitary regulations. To further prevent warping and cracking, certain wood species are treated with stabilizing agents or coatings during dry storage. Sealants, such as wax or oil-based products, are applied to the ends of logs or planks to reduce moisture loss and prevent end-checking, a common issue where cracks develop at the ends of the wood due to rapid drying. Advanced stabilizing solutions, including chemical treatments and resin impregnation, enhance dimensional stability and protect against environmental fluctuations.

Monitoring and maintenance are integral to the success of dry storage techniques. Regular inspections of stored timber allow for the early detection of warping, cracking, or other defects, enabling timely interventions. Modern storage facilities often employ digital moisture meters, thermal imaging cameras, and automated monitoring systems to provide real-time data on wood conditions and environmental parameters. These technologies ensure that dry storage practices are optimized for both efficiency and quality preservation. Dry storage techniques, when implemented effectively, ensure that timber retains its value and usability over extended periods. By preventing warping and cracking, these methods support the production of high-quality wood products and contribute to the sustainability and profitability of the forestry industry.

5.2 Wet Storage Systems: Immersion and Sprinkler Methods

Wet storage systems are employed to maintain the quality of timber over long periods by preserving its high moisture content and preventing biological degradation. These systems, which include immersion and sprinkler methods, are particularly effective for logs that are susceptible to fungal decay, insect infestations, or checking during dry storage. Wet storage is widely used in regions with abundant water resources or where extended storage is necessary due to market conditions or logistical constraints. The immersion method involves submerging logs in water bodies, such as ponds, rivers, or artificial reservoirs, to protect them from decay and insect activity. Water immersion creates an anaerobic environment that inhibits the growth of fungi and bacteria, preserving the structural integrity and appearance of the wood. This technique is especially effective for high-value timber species, such as tropical hardwoods and large-diameter softwoods, which are prone to rapid deterioration when exposed to air.

Immersion also reduces the risk of checking, as the uniform moisture content prevents internal stresses from developing. However, the quality of the water used is a critical factor in the success of this method. Polluted or stagnant water can introduce contaminants or promote undesirable microbial growth, compromising the timber. Regular water quality assessments and proper site management are essential to mitigate these risks.

Sprinkler systems offer an alternative wet storage approach, where logs are stacked in open areas and continuously sprayed with water to maintain surface moisture. This method is particularly suitable for regions where full immersion is impractical due to water scarcity or site limitations. Sprinklers provide a controlled and uniform application of water, reducing the risk of fungal growth and insect infestations. Advanced sprinkler systems use sensors and automated controls to optimize water usage and ensure consistent coverage, enhancing the efficiency and sustainability of the method. Wet storage systems also play a role in stabilizing timber for transportation and processing. By maintaining high moisture content, these systems reduce the likelihood of splitting and checking during handling and transit. Additionally, wet-stored logs are easier to process in sawmills, as their pliability reduces wear on cutting equipment and facilitates smoother operations.

Despite their advantages, wet storage systems have environmental and operational challenges. Water use and management must be carefully monitored to prevent wastage and contamination. In immersion systems, leaching of tannins or other compounds from logs can affect water quality, necessitating measures such as filtration or periodic water replacement. Sprinkler systems, while less resource-intensive, require energy for pumping and distribution, contributing to operational costs. The adoption of wet storage systems reflects the forestry industry's commitment to preserving timber quality while addressing market and environmental considerations. Innovations in water management, energy efficiency, and monitoring technologies continue to enhance the effectiveness and sustainability of these methods.

5.3 Innovations in Climate-Controlled Warehousing

Climate-controlled warehousing represents the forefront of long-term timber storage, offering precise environmental management to preserve the quality and value of timber. These facilities, equipped with advanced technologies, provide optimal conditions for controlling temperature, humidity, and airflow, ensuring that timber remains free from decay, warping, and other forms of degradation. Climate-controlled warehousing is particularly valuable for high-value timber products and export markets with stringent quality standards.

The primary advantage of climate-controlled warehousing is its ability to maintain consistent environmental conditions, regardless of external weather variations. Temperature regulation prevents the wood from expanding or contracting, while humidity control reduces the risk of fungal growth and moisture-related defects. Air circulation systems, often integrated with filtration and dehumidification units, ensure uniform drying and protection against airborne contaminants. These features make climate-controlled warehousing ideal for storing sawn timber, veneer sheets, and engineered wood products that require precise moisture levels and dimensional stability. Innovations in climate-controlled warehousing have significantly enhanced the efficiency and sustainability of these facilities. Smart sensors and IoT-enabled systems provide real-time monitoring and automated adjustments of environmental parameters, optimizing energy use and reducing operational costs. Data collected from these sensors also enables predictive analytics, allowing operators to identify potential issues, such as equipment malfunctions or environmental fluctuations, before they affect timber quality.

Energy efficiency is a key focus in the design of modern climate-controlled warehouses. Facilities often incorporate renewable energy sources, such as solar panels or biomass boilers, to power their operations. Insulated building materials and advanced thermal management systems further reduce energy consumption, aligning with the forestry industry's commitment to sustainability.

The use of modular and scalable designs is another innovation in climate-controlled warehousing.

These designs allow facilities to be tailored to specific storage requirements, accommodating different volumes, wood types, or market demands. For instance, modular units can be configured for segregated storage of hardwoods and softwoods, each with its optimal environmental settings. This flexibility enhances the cost-effectiveness and adaptability of climate-controlled storage. Integration with logistics systems further expands the role of climate-controlled warehousing in the timber supply chain. Facilities are often located near transportation hubs or processing centers, facilitating seamless transitions between storage, transport, and manufacturing. Automated inventory management systems, linked to digital tracking platforms, provide transparency and traceability, ensuring that stored timber meets regulatory and certification standards.

While climate-controlled warehousing offers unparalleled benefits, its adoption is limited by high initial costs and operational complexity. Small-scale forestry operations may find it challenging to invest in such facilities without external support or collaboration. However, government incentives, industry partnerships, and advances in affordable technologies are gradually expanding access to these solutions. Climate-controlled warehousing represents the pinnacle of long-term timber storage, combining precision, innovation, and sustainability to address the challenges of modern forestry. As the industry continues to evolve, these facilities are set to play an increasingly vital role in preserving timber quality and meeting the demands of global markets.

6. Preservation and Quality Maintenance

6.1 Chemical Treatments and Fumigation Techniques

Chemical treatments and fumigation techniques are fundamental to the preservation of timber during storage, offering robust protection against biological and environmental threats. These methods extend the lifespan of stored timber by preventing decay, insect infestations, and fungal growth, while also enhancing its durability and suitability for downstream applications. The choice and application of treatments depend on the type of timber, intended use, and storage conditions, as well as environmental and regulatory considerations.

Chemical treatments involve the application of preservatives that protect timber from biological degradation caused by fungi, bacteria, and insects. These preservatives typically include water-soluble, oil-borne, or organic compounds, each suited for specific requirements. Water-soluble treatments, such as copper-based solutions, penetrate deeply into the wood, providing long-lasting protection against decay and termite attacks. Creosote, an oil-borne preservative, is widely used for outdoor timber applications such as railway sleepers and utility poles, offering excellent resistance to moisture and pests. Organic compounds, including borates and azoles, are increasingly favored for their lower environmental impact and compatibility with sustainable practices. The application of chemical preservatives can be carried out through various methods, including brushing, spraying, dipping, and pressure treatment. Brushing and spraying are cost-effective techniques for surface protection, commonly used in temporary storage or for timber intended for short-term use. Dipping involves immersing timber in preservative solutions, ensuring more uniform coverage, while pressure treatment uses vacuum and pressure cycles to force preservatives deep into the wood's cellular structure. Pressure treatment is particularly effective for timber stored in high-humidity environments or areas prone to pest infestations. Fumigation is another critical technique in timber preservation, targeting insects and pests that can infest stored wood. This method involves exposing timber to gaseous pesticides in enclosed spaces, ensuring that the gas penetrates all parts of the wood and eliminates hidden pests. Common fumigants include methyl bromide, sulfuryl fluoride, and phosphine, each chosen based on the pest species and storage conditions. Fumigation is widely used for timber destined for export, as many countries require pest-free certification to prevent the introduction of invasive species.

While chemical treatments and fumigation offer significant benefits, their use must be carefully managed to minimize environmental and health risks. Many traditional preservatives contain toxic compounds that can leach into soil or water, posing hazards to ecosystems and human health. To address these concerns, the forestry industry is transitioning toward eco-friendly alternatives, such as plant-based biocides and waterborne formulations, which provide effective protection with reduced toxicity. Compliance with regulatory standards, such as those set by the Environmental Protection Agency (EPA) or the European Union's REACH framework, ensures that chemical treatments and fumigation practices are safe and sustainable. The integration of chemical treatments and fumigation techniques into timber storage operations reflects the industry's commitment to preserving quality while adapting to evolving environmental and market demands. Advances in treatment formulations and application methods continue to enhance the effectiveness and sustainability of these preservation strategies.

6.2 Protection Against Insects, Fungi, and Weathering

Timber storage presents significant challenges related to biological and environmental threats, including insect infestations, fungal growth, and weathering. Addressing these threats is essential to maintaining the structural integrity, aesthetic appeal, and market value of stored timber. Effective protection strategies combine preventive measures, active treatments, and monitoring to ensure long-term quality preservation.

Insects, particularly wood-boring beetles, termites, and carpenter ants, are among the most destructive pests affecting stored timber. These insects feed on wood fibers, creating tunnels and cavities that compromise the timber's strength and appearance. Preventing infestations begins with proper site management and hygiene. Storing timber away from vegetation, soil, or standing water reduces the likelihood of attracting pests. Barriers or physical enclosures, such as netting or insect-proof containers, provide additional protection, especially for high-value timber. Chemical treatments, such as insecticides or repellents, are commonly applied to protect timber from insect damage. Surface treatments, including sprays or coatings, deter insects from settling on the wood, while deeper treatments, such as pressure-injected preservatives, ensure long-lasting protection against internal infestations. Biological control methods, which introduce natural predators or pathogens to target pest populations, offer eco-friendly alternatives to chemical treatments, aligning with sustainable forestry practices.

Fungal growth, including molds, stains, and decay fungi, poses another significant threat to stored timber. Fungi thrive in warm, humid conditions, feeding on the wood's nutrients and causing discoloration, softening, and structural breakdown. Preventing fungal growth requires controlling moisture levels and ensuring adequate ventilation. Elevating timber stacks off the ground and using spacers to promote airflow are effective practices. In high-humidity regions, dehumidifiers or climate-controlled storage facilities provide additional protection. Weathering, caused by prolonged exposure to sunlight, rain, and wind, can degrade timber during outdoor storage. UV radiation breaks down the lignin in wood, leading to surface discoloration and brittleness, while repeated wetting and drying cycles cause cracks and warping. Protective measures include the use of covers, such as tarps or sheeting, and the application of water-repellent finishes or UV-resistant coatings. These treatments shield the timber from environmental stressors while preserving its appearance and functionality.

Advances in material science and technology are enhancing the industry's ability to protect timber against biological and environmental threats. Smart coatings, for example, incorporate antimicrobial agents or moisture-regulating properties, providing active protection against fungi and insects. Integrated monitoring systems, equipped with sensors for temperature, humidity, and pest activity, enable real-time assessment and intervention, reducing the risk of damage during storage. The implementation of comprehensive protection strategies ensures that timber remains in optimal condition throughout its storage period. By addressing the challenges posed by insects, fungi, and weathering, the forestry industry can maintain the quality and value of its products, meeting the expectations of consumers and stakeholders.

6.3 Monitoring and Quality Assurance During Storage

Monitoring and quality assurance are critical components of timber storage, ensuring that preservation measures are effective and that stored timber maintains its desired characteristics. These processes involve regular inspections, data collection, and analysis to identify potential issues early and implement corrective actions. Advances in monitoring technology and quality assurance protocols are enhancing the forestry industry's ability to manage storage operations effectively and sustainably. Timber monitoring begins with regular visual inspections, during which trained personnel assess the condition of stored wood for signs of damage, decay, or infestation. Inspections focus on key indicators such as discoloration, surface irregularities, and the presence of pests or fungal growth. Any anomalies are documented and addressed promptly to prevent further deterioration. Visual inspections are complemented by tactile assessments, such as tapping or probing the wood to detect internal voids or softness.

Technological innovations are revolutionizing timber monitoring, providing precise and non-invasive methods for assessing wood quality. Digital moisture meters, for example, measure the moisture content of timber with high accuracy, enabling operators to maintain optimal drying conditions and prevent decay. Thermal imaging cameras detect temperature variations that may indicate hidden fungal activity or moisture accumulation. Advanced imaging technologies, such as X-ray or ultrasonic scanning, allow for detailed analysis of timber's internal structure, identifying defects or damage that may not be visible on the surface. Quality assurance extends beyond monitoring to include the verification of compliance with industry standards and customer specifications. Stored timber is often graded based on criteria such as strength, appearance, and moisture content, ensuring that it meets the requirements for its intended application. Certification schemes, such as those offered by the Forest Stewardship Council (FSC) or Programme for the Endorsement of Forest Certification (PEFC), provide additional assurance of the timber's sustainability and ethical sourcing. Data management and analytics play a growing role in quality assurance during storage. Integrated systems collect and analyze data from sensors, inspections, and environmental monitoring devices, providing actionable insights into storage conditions and trends. For example, predictive analytics can identify patterns that indicate an increased risk of decay or infestation, allowing operators to take preventive measures before significant damage occurs. Cloud-based platforms enable remote monitoring and centralized management of storage operations, enhancing efficiency and transparency.

Training and capacity building are essential for successful monitoring and quality assurance. Workers involved in storage operations must be equipped with the knowledge and skills to perform inspections, operate monitoring devices, and interpret data effectively. Ongoing education ensures that personnel remain up to date with the latest techniques and technologies, contributing to the overall success of preservation efforts. The integration of monitoring and quality assurance into timber storage operations ensures that preservation measures are both effective and adaptable to changing conditions. By leveraging technology, expertise, and data-driven insights, the forestry industry can maintain the quality and value of stored timber while upholding its commitments to sustainability and customer satisfaction.

7. Environmental Considerations in Transportation and Storage

7.1 Reducing Carbon Footprint in Transport Operations

Timber transport contributes significantly to the carbon footprint of forestry operations due to the extensive use of diesel-powered vehicles, heavy machinery, and energy-intensive logistics systems. Reducing this footprint is critical for aligning the industry with global sustainability goals and addressing climate change. Several strategies and innovations have been adopted to minimize emissions while maintaining efficiency in timber transport. One of the most effective approaches to reducing carbon emissions is the adoption of alternative energy sources. Electric and hybrid vehicles are gaining traction in the forestry sector, offering a cleaner alternative to traditional diesel-powered trucks and machinery. Advances in battery technology and charging infrastructure are making these vehicles more viable for both short-haul and long-haul timber transport. Renewable energy integration, such as solar-powered charging stations, further enhances the environmental benefits of these solutions.

Optimizing transport routes through GPS and advanced logistics software also plays a vital role in reducing emissions. By identifying the most efficient paths, transport systems can minimize fuel consumption and time spent on the road. These tools also help avoid congestion and delays, improving overall operational efficiency. Additionally, load optimization ensures that transport vehicles carry maximum allowable volumes without exceeding weight limits, reducing the number of trips required and, consequently, the total emissions. The use of multimodal transport systems combining road, rail, and waterways further decreases reliance on fossil fuels. Rail and water transport are significantly more fuel-efficient than road transport for large volumes of timber over long distances. Investments in infrastructure, such as intermodal terminals and port facilities, are essential to facilitate the seamless transition between these modes, maximizing their environmental benefits. Carbon offsetting and carbon-neutral initiatives are also being implemented by forward-thinking forestry companies. These programs involve planting trees, investing in renewable energy projects, or purchasing carbon credits to offset the emissions generated during transport. Transparent reporting and adherence to international standards ensure that these efforts contribute meaningfully to reducing the industry's environmental impact.

7.2 Minimizing Waste and Spillage During Handling

Waste and spillage during the transportation and storage of timber not only result in economic losses but also contribute to environmental degradation. Spilled timber fragments and by-products can pollute ecosystems, while wasted resources represent an inefficiency in the supply chain. Minimizing these losses requires careful handling practices, advanced equipment, and process optimization. In transport, proper loading and securing of timber are critical to reducing spillage. Poorly secured loads can shift during transit, leading to losses and safety hazards. Strapping systems, load sensors, and automated tie-down mechanisms ensure that timber remains stable during transport. For bulk materials, such as wood chips or sawdust, sealed containers or tarps prevent material loss due to wind or vibration.

Handling at storage and transfer points is another area where waste can occur. The use of automated loaders and precision handling equipment minimizes the risk of timber damage or spillage during loading and unloading. Technologies such as robotic arms and conveyors enhance precision, reducing manual handling errors that often result in material loss. Recycling and repurposing by-products are integral to minimizing waste. Sawdust, bark, and offcuts generated during handling can be processed into bioenergy, mulch, or engineered wood products. Establishing dedicated facilities for collecting and processing these by-products at storage sites reduces waste and creates additional revenue streams. Compliance with environmental regulations and certifications further ensures that handling practices align with sustainability goals. Organizations adhering to standards such as FSC or PEFC demonstrate a commitment to reducing waste and environmental impact across all aspects of their operations.

7.3 Sustainable Practices in Long-Term Timber Storage

Sustainable long-term storage practices are essential to maintaining timber quality while minimizing environmental impact. These practices focus on reducing resource consumption, enhancing energy efficiency, and adopting eco-friendly preservation methods.

One of the most important aspects of sustainable storage is energy-efficient climate control. Long-term storage facilities often require temperature and humidity regulation to prevent decay, warping, and fungal growth. Incorporating renewable energy sources, such as solar panels or biomass boilers, significantly reduces the carbon footprint of these facilities. Advanced insulation and thermal management systems further enhance energy efficiency, ensuring that climate control operates effectively with minimal energy input. Sustainable storage also emphasizes the use of non-toxic and biodegradable preservatives. Traditional chemical treatments can leach into soil and water, posing risks to ecosystems and human health. Eco-friendly alternatives, such as borate-based solutions or plant-derived biocides, provide effective protection against pests and fungi without harmful side effects. These treatments align with certification requirements and consumer demand for environmentally responsible products.

Innovations in water management contribute to sustainable storage practices, particularly in wet storage systems. Recycled water systems and advanced filtration technologies reduce water consumption and prevent contamination of nearby ecosystems. In immersion methods, closed-loop systems ensure that water quality remains stable while minimizing environmental impact. By integrating these sustainable practices, the forestry industry can enhance the environmental performance of its storage operations, ensuring that timber preservation aligns with broader ecological and social goals.

8. Technological Innovations in Timber Logistics

8.1 GPS and IoT Integration in Timber Tracking and Transport

The integration of GPS and IoT technologies into timber logistics has transformed the way timber is tracked, transported, and managed. These innovations provide real-time visibility, improve operational efficiency, and enhance compliance with sustainability and legal standards.

GPS tracking enables forestry companies to monitor the location and movement of timber throughout the supply chain. From harvesting sites to processing facilities, GPS devices installed on vehicles provide real-time data on transport routes, delays, and delivery times. This information allows managers to optimize routes, reduce fuel consumption, and ensure timely deliveries. For example, avoiding congested areas or identifying the shortest path to a destination minimizes emissions and operational costs.

IoT devices expand these capabilities by providing detailed data on environmental conditions and timber quality during transport and storage. Sensors embedded in trucks, containers, or storage facilities measure parameters such as temperature, humidity, and shock. Alerts are generated if conditions deviate from acceptable ranges, enabling immediate corrective action to prevent damage or degradation. Blockchain technology complements GPS and IoT integration by enhancing transparency and traceability. Blockchain systems create immutable records of timber's journey, from its source in certified forests to its final destination. This traceability ensures compliance with legal and certification requirements, builds consumer trust, and mitigates the risk of illegal logging entering supply chains.

8.2 Automation and Robotics in Storage and Handling

Automation and robotics are revolutionizing the storage and handling of timber, improving efficiency, safety, and precision. These technologies address key challenges such as labor shortages, material waste, and safety risks associated with manual handling. Robotic systems are increasingly used for loading, unloading, and stacking timber in storage facilities. Automated cranes, conveyors, and robotic arms handle timber with precision, reducing the risk of damage and spillage. These systems operate continuously, increasing throughput and reducing reliance on human labor. For instance, automated stacking systems arrange timber in optimized configurations, maximizing storage space while ensuring stability and accessibility. Advanced robotics are also used in sorting and grading timber. AI-powered vision systems analyze timber for defects, dimensions, and moisture content, ensuring consistent quality standards. This level of precision is particularly valuable for high-value timber products, where minor imperfections can significantly affect market value. Automation enhances worker safety by reducing direct contact with heavy loads and machinery. Remote-controlled or autonomous vehicles, such as forklifts and forwarders, minimize the risk of workplace accidents in storage and handling operations.

8.3 Advanced Inventory Management Systems

Advanced inventory management systems play a critical role in streamlining timber logistics, ensuring accurate tracking, and optimizing storage space. These systems integrate software, hardware, and analytics to provide real-time visibility into inventory levels, locations, and conditions. Radio-frequency identification (RFID) and barcode scanning technologies are widely used for tagging and tracking timber. Each log or batch is assigned a unique identifier, enabling precise monitoring throughout the supply chain. RFID tags, in particular, allow for remote scanning and bulk updates, saving time and labor. Cloud-based inventory management platforms enable centralized data access and collaboration among stakeholders. Forestry companies, transport providers, and processing facilities can share information seamlessly, improving coordination and reducing delays. Predictive analytics, powered by AI, further enhance these systems by forecasting demand, identifying inventory shortages, and suggesting optimal storage configurations.

Conclusion

Timber transportation and storage are critical components of the forestry supply chain, ensuring that harvested logs reach processing facilities and markets while maintaining their quality and value. These operations bridge the gap between harvesting and utilization, playing a pivotal role in the economic, logistical, and environmental dynamics of the timber industry. Effective transportation and storage practices are essential not only for maintaining the structural integrity and quality of timber but also for optimizing costs, reducing environmental impacts, and enhancing supply chain efficiency. The evolution of these practices reflects advancements in technology, logistics, and sustainability, all of which are integral to meeting the growing global demand for timber products while preserving ecological balance.

Transportation methods have diversified significantly, ranging from traditional ground-based approaches using trucks, skidders, and forwarders to aerial and waterborne systems that navigate remote and challenging terrains. Ground transportation remains the most commonly employed method due to its versatility and accessibility, but it is often constrained by terrain and infrastructure limitations. Aerial systems, including helicopters and cable systems, are invaluable in rugged landscapes, reducing environmental impact by minimizing soil disturbance. Water transportation, through rafting, barges, and ships, remains a cost-effective option for bulk timber movement, particularly over long distances. Innovations in timber transport logistics, such as GPS tracking and route optimization, have further enhanced efficiency, reducing costs and environmental footprints.

The challenges in timber transportation are multifaceted, encompassing accessibility, environmental impacts, and cost management. Remote and rugged terrains often require specialized equipment and infrastructure, increasing operational complexity. Environmental concerns, such as soil erosion, habitat disruption, and greenhouse gas emissions, necessitate the adoption of sustainable practices and low-impact technologies. Managing costs, particularly fuel efficiency and maintenance, is critical for ensuring competitiveness in the global timber market. Addressing these challenges requires a combination of technological advancements, strategic planning, and adherence to sustainable transport practices.

Storage practices, both temporary and long-term, are equally vital in maintaining timber quality and preventing degradation. On-site storage, through log decks and staging areas, facilitates efficient handling and transportation but poses risks of moisture retention and decay if not properly managed. Logging camps play a crucial role in coordinating storage operations and ensuring logistical efficiency. Long-term storage techniques, including dry storage methods that prevent warping and cracking and wet storage systems that maintain moisture levels, are essential for preserving timber integrity over extended periods. Innovations in climate-controlled warehousing and advanced inventory management systems further enhance the effectiveness of timber storage operations.

Preservation and quality maintenance are central to timber storage, as exposure to insects, fungi, and weathering can significantly reduce timber value. Chemical treatments, fumigation techniques, and protective measures are widely employed to safeguard timber against biological and environmental threats. Monitoring and quality assurance systems ensure that stored timber meets industry standards, maintaining its marketability and usability. Environmental considerations in transportation and storage have gained prominence, driven by the need to minimize the carbon footprint and ecological impact of timber operations. Strategies such as using fuel-efficient vehicles, reducing spillage and waste during handling, and adopting sustainable storage practices contribute to environmentally responsible timber logistics. Circular economy approaches, which emphasize resource efficiency and waste minimization, further align transportation and storage operations with sustainability goals.

Technological innovations have transformed timber logistics, enabling greater efficiency, accuracy, and sustainability. GPS and IoT technologies facilitate real-time tracking and monitoring, improving supply chain transparency and coordination. Automation and robotics enhance storage and handling operations, reducing labor costs, and improving safety. Advanced inventory management systems optimize stock control, ensuring timely delivery and reducing losses. Economic and market implications underscore the importance of efficient transportation and storage in the timber industry. Cost-effective logistics enhance competitiveness in global timber markets, while streamlined operations reduce waste and maximize profitability. Linking transportation and storage systems with broader supply chain management strategies ensures that timber products reach markets efficiently and sustainably, supporting the growth of the industry.

Sustainability remains a cornerstone of modern timber transportation and storage, with policies and certification standards driving the adoption of environmentally responsible practices. Circular economy approaches, which promote recycling, reuse, and resource optimization, offer a framework for reducing the environmental impact of logistics operations. Future trends, including advancements in green transport technologies and sustainable storage solutions, promise to further align timber logistics with global sustainability goals.

Chapter - 13

Forest Produce and Its Uses

1. Introduction to Forest Produce

Forests have been fundamental to human survival and development throughout history, providing a plethora of resources vital for sustenance, economic activities, and ecological balance. The term "forest produce" encompasses a wide range of goods and services derived from forests, ranging from tangible materials like timber and non-timber products to intangible ecological benefits. Forest produce is not merely an economic asset; it represents a lifeline for countless communities, a cornerstone for industrial development, and a vital component of Earth's ecological integrity.

Forests are among the most vital ecosystems on Earth, serving as the lungs of our planet and providing a myriad of resources critical to human survival and societal advancement. The utility of forests extends far beyond the provision of tangible goods, encompassing a complex interplay of ecological, economic, and social functions. Forest produce, a term that broadly encompasses all materials obtained from forests, is central to this multifaceted role. It includes timber-based products, non-timber forest products (NTFPs), minor byproducts, and even wildlife-derived resources. Each of these categories makes a unique contribution to the enrichment of human life and the preservation of global biodiversity. Furthermore, the sustainable management and utilization of forest products are imperative for addressing pressing global challenges, including climate change, biodiversity loss, and economic inequalities. The classification of forest products provides a framework for understanding the diverse resources that forests offer. Broadly, forest products can be categorized into timber and non-timber products. Timber, which is derived from trees, is a cornerstone of the construction and furniture industries, valued for its structural strength and aesthetic appeal. Non-timber forest products, on the other hand, include an array of items such as fruits, nuts, medicinal plants, resins, and essential oils, all of which have significant economic, cultural, and ecological importance. This classification not only aids in the effective management of forest resources but also highlights the vast potential for diversified forest-based industries.

The economic and ecological significance of forest products cannot be overstated. Forests act as reservoirs of carbon, mitigating climate change by sequestering carbon dioxide and releasing oxygen. Simultaneously, they serve as habitats for a wide range of species, ensuring the preservation of biodiversity. From an economic perspective, forest products underpin various industries, from paper and packaging to pharmaceuticals and cosmetics. For many rural and indigenous communities,

DOI: 10.1201/9781042011063-13

forest resources represent a primary source of income, supporting livelihoods and fostering regional development. The dual role of forests in sustaining ecosystems and economies underscores the need for balanced and sustainable utilization practices. In the context of sustainable development, forest products plays a pivotal role. The sustainable management of forests ensures that the extraction of resources does not compromise the ecological integrity or the ability of forests to regenerate. This approach aligns with global sustainability goals, including those outlined in the United Nations' Sustainable Development Goals (SDGs). By promoting sustainable forest produce utilization, societies can achieve a harmonious balance between economic growth and environmental conservation. Moreover, forests serve as natural buffers against climate extremes, protecting communities from disasters and maintaining essential ecosystem services such as water filtration and soil stabilization. Timber-based forest products represent one of the most visible and economically significant categories of forest resources. Timber, classified into hardwood and softwood varieties, is derived from a diverse range of tree species. Hardwoods, sourced from deciduous trees such as oak, teak, and mahogany, are prized for their density and durability, making them ideal for furniture and flooring. Softwoods, obtained from coniferous trees like pine, cedar, and spruce, are commonly used in construction and paper production due to their lighter weight and ease of processing. The applications of timber extend far beyond basic construction, encompassing industrial uses in pulp and paper production, packaging, and the manufacture of engineered wood products.

Non-timber forest products (NTFPs) constitute a vast and varied category of forest produce that offers immense potential for sustainable development. Edible forest products, including fruits, nuts, and mushrooms, are not only sources of nutrition but also form the basis of thriving local and international markets. Medicinal plants, long revered in traditional medicine systems, are now gaining recognition in modern pharmaceuticals for their bioactive compounds and therapeutic properties. Resins, gums, and latex extracted from forest trees have wide-ranging applications in industries such as adhesives, cosmetics, and rubber manufacturing. The versatility and renewable nature of NTFPs make them essential for promoting eco-friendly and sustainable industrial practices. Minor forest produce and byproducts, often overlooked in traditional classifications, are equally significant in their applications. Essential oils and aromatics, derived from various plant species, are cornerstones of the fragrance and wellness industries. Bamboo and rattan, known for their rapid growth and renewability, are versatile materials used in construction, handicrafts, and even bioenergy production. Traditional forest products such as natural dyes, tannins, and fibers continue to find relevance in both artisanal crafts and modern industrial processes. The utilization of these resources not only supports traditional knowledge systems but also contributes to the development of innovative green technologies.

Wildlife-derived forest produce encompass resources obtained from forest-dwelling species, underscoring the intricate link between biodiversity and forest ecosystems. Honey and other bee products, such as beeswax and royal jelly, are not only economically valuable but also vital for ecosystem health, as bees play a crucial role in pollination. Silk and lac, derived from forest species, are integral to the textile and decorative industries, supporting both local artisans and global markets. Furthermore, the role of wildlife in sustainable forest-based livelihoods highlights the importance of conserving fauna as part of forest management strategies.

Forests also hold immense cultural and recreational value, contributing to human well-being in intangible yet profound ways. They serve as sources of aesthetic inspiration and spiritual solace, forming the backdrop for artistic and cultural expressions across the world. Eco-tourism, which leverages the beauty and biodiversity of forests, has emerged as a significant driver of conservation efforts and local economic development. Additionally, forest products are deeply embedded in art, handicrafts, and cultural practices, preserving traditions and fostering community identity. The economic contributions of forest produce are vast, encompassing the generation of employment, industrial growth, and revenue streams for national and global economies. Forest-based industries, including furniture manufacturing, paper production, and pharmaceuticals, provide livelihoods for millions of people, particularly in rural and forest-adjacent communities. The export of forest products contributes significantly to the GDP of many countries, highlighting the global interdependence on these resources. Moreover, the integration of forest produce into rural development initiatives can uplift marginalized communities, reducing poverty and promoting equitable growth.

Sustainability in the utilization of forest produce is paramount to ensuring that forests continue to meet the needs of present and future generations. Certification standards, such as those established by the Forest Stewardship Council (FSC) and the Programme for the Endorsement of Forest Certification (PEFC), promote responsible forest management practices. Sustainable harvesting techniques for timber and NTFPs aim to minimize environmental impact while maximizing economic benefits. Innovations in recycling and waste utilization, such as the development of bio-based materials from forest residues, are paving the way for a circular economy in forest product industries. However, the management of forest produce is fraught with challenges. Over-exploitation and resource depletion pose significant threats to forest ecosystems, driven by unsustainable logging, illegal trade, and overharvesting of NTFPs. Market access and fair pricing remain elusive for many local communities, which often lack the resources or knowledge to compete in global markets. Furthermore, climate change is exacerbating the vulnerability of forests, altering growth patterns, and reducing productivity. Addressing these challenges requires a concerted effort involving policy interventions, community engagement, and scientific innovation.

1.1 Classification of Forest Produce: Timber and Non-Timber Products

Forest produce can be classified into two major categories: timber and non-timber forest products (NTFPs). Each category holds its unique place in both economic and ecological contexts, providing essential resources that support diverse industries, local livelihoods, and environmental functions.

Timber Products

Timber, derived from trees, has long been one of the most valuable forest products. It serves as a raw material for construction, paper production, furniture manufacturing, and countless other applications. The types of timber can be broadly classified into hardwoods and softwoods based on the tree species from which they are obtained. Hardwoods, such as teak, mahogany, sal, and oak, are known for their strength and durability. These dense woods are highly sought after for heavy construction, premium furniture, and flooring. Softwoods, including pine, cedar, fir, and spruce, are lightweight and versatile, making them suitable for pulpwood, paper, and lightweight furniture. The utility of timber extends beyond industrial applications. From ancient civilizations to modern societies, timber has played a critical role in infrastructure development, tool-making, and energy generation. Even today, timber accounts for a significant proportion of global forest trade. However, its extraction must be balanced with sustainable practices to prevent deforestation, soil erosion, and biodiversity loss.

Non-Timber Forest Products (NTFPs)

Non-timber forest products encompass all biological materials other than timber harvested from forests. These include a vast array of resources such as fruits, nuts, resins, fibers, medicinal plants, and wildlife products. Unlike timber, the extraction of NTFPs typically involves minimal ecological disturbance, allowing forests to retain their structural and functional integrity. NTFPs play a crucial role in sustaining rural livelihoods, particularly in regions where subsistence agriculture is complemented by forest-based income.

» **Edible Products:** Forests are rich sources of fruits, nuts, seeds, honey, and mushrooms, which serve as food for both humans and wildlife. For example, wild berries and chestnuts are seasonal sources of nutrition for local communities and contribute to biodiversity by supporting animal species.

» **Medicinal Plants:** Forests harbor a treasure trove of medicinal plants used in traditional and modern medicine. Herbs like neem, tulsi, and ashwagandha are widely known for their therapeutic properties. Many life-saving drugs, including quinine and taxol, are derived from forest flora.

» **Fibers and Resins:** Bamboo, rattan, gum arabic, and latex are examples of forest-derived fibers and resins that hold economic and cultural significance. Bamboo, for instance, is not only a construction material but also a source of income for artisans and small-scale industries.

» **Wildlife Products:** Products like feathers, hides, and excreta are utilized in traditional crafts, ornamental items, and fertilizers. However, the extraction of such products necessitates stringent regulation to prevent illegal poaching and biodiversity degradation.

Both timber and NTFPs highlight the diverse utility of forest produce. While timber often dominates the global trade narrative, NTFPs play a silent yet equally important role in supporting local economies, preserving cultural traditions, and promoting environmental conservation. Together, these resources reflect the immense value of forests, showcasing their ability to meet human needs sustainably.

1.2 Economic and Ecological Importance of Forest Produce

Forest produce hold unparalleled significance in both economic and ecological spheres. The intricate relationship between these dimensions underscores the vital role forests play in supporting human well-being and maintaining environmental stability.

Economic Importance

Forests contribute significantly to the global economy, providing raw materials, employment, and trade opportunities. The timber industry is a cornerstone of economic development, supplying wood for construction, furniture, and paper industries. According to global statistics, the timber trade generates billions of dollars annually, creating jobs in logging, milling, and export activities. In addition to its industrial applications, timber is a primary source of energy in many developing countries, where wood fuel is extensively used for cook ing and heating. Non-timber forest products are equally crucial in sustaining rural economies. For forest-dependent communities, NTFPs often serve as a primary source of income. The collection and sale of wild fruits, medicinal herbs, and other forest goods not only supplement agricultural livelihoods but also empower marginalized groups, including women and indigenous people. On a larger scale, the global market for NTFPs, including pharmaceuticals, cosmetics, and organic foods, has witnessed exponential growth, reflecting increased consumer demand for sustainable and ethically sourced products. Forest produce also stimulates the tourism industry, particularly in ecotourism and wildlife conservation. National parks, wildlife sanctuaries, and biosphere reserves attract millions of visitors annually, generating revenue for local communities and governments. The sustainable management of forest resources thus holds immense potential for fostering economic growth while ensuring environmental protection.

Ecological Importance

The ecological value of forest produce extends beyond its tangible utility. Forests are pivotal in maintaining the Earth's life-support systems, providing ecosystem services that include carbon sequestration, water regulation, soil conservation, and biodiversity preservation. Forests act as carbon sinks, absorbing and storing large quantities of carbon dioxide, thereby mitigating climate change. Timber products contribute to this process by storing carbon even after harvesting, especially when used in long-lasting applications like construction. The sustainable extraction of timber can promote forest regeneration and enhance carbon storage potential. The role of NTFPs in ecological conservation is equally significant. For instance, the sustainable harvesting of honey, bamboo, or medicinal plants does not compromise the forest's structural integrity, allowing ecosystems to thrive. Additionally, forests regulate the water cycle by enhancing groundwater recharge, reducing surface runoff, and maintaining river flows. These functions are particularly critical in regions prone to droughts or floods, where forest cover serves as a natural buffer. Biodiversity conservation is another vital ecological service provided by forest produce. Forests are home to more than 80% of terrestrial species, offering habitat, food, and breeding grounds. The sustainable management of forest resources ensures that these ecosystems remain vibrant and resilient, supporting both wildlife and human communities.

1.3 Role of Forest Produce in Sustainable Development

The concept of sustainable development revolves around meeting present needs without compromising the ability of future generations to meet their own. Forest produce, with its diverse applications and inherent connection to natural ecosystems, embodies this principle. Its sustainable utilization is key to addressing global challenges such as poverty, climate change, and biodiversity loss.

Promoting Livelihoods and Social Equity

Forest produce play a critical role in poverty alleviation and social equity, particularly in developing countries. For millions of rural and indigenous communities, forests are not merely a resource but a way of life. The collection and sale of NTFPs like wild fruits, honey, and medicinal plants provide essential income, helping to bridge economic disparities. Additionally, forest-based enterprises, supported by fair trade initiatives, empower marginalized groups, including women, who are often primary harvesters of NTFPs.

Sustainable forest management practices also contribute to social well-being by preserving cultural heritage. Many indigenous communities view forests as sacred spaces, with traditional knowledge systems deeply rooted in forest ecology. By integrating these traditions with modern conservation practices, sustainable development initiatives can ensure the preservation of both cultural and ecological diversity.

Mitigating Climate Change and Biodiversity Loss

The role of forest produce in combating climate change is profound. Forests absorb approximately one-third of global carbon emissions annually, making them indispensable in climate mitigation strategies. Timber products, when harvested sustainably, continue to sequester carbon, providing a renewable alternative to fossil-fuel-based materials. Similarly, NTFPs like biofuels and natural fibers offer eco-friendly solutions that reduce reliance on non-renewable resources. Biodiversity conservation is another critical aspect of sustainable development supported by forest produce. Forests harbor a vast array of species, many of which are yet to be scientifically documented. Sustainable harvesting practices ensure that forest ecosystems remain intact, supporting wildlife and maintaining ecological balance. Certification schemes like the Forest Stewardship Council (FSC) promote the ethical sourcing of forest products, encouraging consumers to make environmentally responsible choices.

Fostering Economic Growth and Environmental Resilience

The sustainable use of forest produce has immense potential to drive economic growth while enhancing environmental resilience. Agroforestry, for example, integrates tree cultivation with agriculture, providing farmers with diversified income streams and improving soil health. Similarly, community-based forest management empowers local populations to take ownership of forest resources, promoting conservation and reducing conflicts over resource use.

2. Timber-Based Forest Produce

Timber-based forest produce are one of the most significant resources derived from forests, deeply integrated into human civilization and modern industries. The term "timber" encompasses the wood obtained from trees, utilized in an array of applications ranging from construction and furniture making to industrial manufacturing. The classification of timber into hardwood and softwood varieties, its myriad uses, and its industrial applications underscore its economic and ecological importance.

2.1 Types of Timber

Timber is broadly classified into two categories: hardwood and softwood, based on the botanical characteristics of the trees from which it is sourced. While these terms do not necessarily indicate the hardness or softness of the wood, they highlight the structural and functional differences that influence their usage.

Hardwood Varieties

Hardwoods are derived from deciduous trees, which shed their leaves annually. These trees typically grow slower than conifers, resulting in denser and more durable wood. Hardwoods such as teak, oak, mahogany, sal, and maple are renowned for their strength, resistance to wear, and aesthetic appeal. Teak, for instance, is highly valued for its resistance to termites and water, making it ideal for outdoor furniture and marine applications. Mahogany is prized for its rich reddish-brown color and fine grain, commonly used in high-quality furniture and musical instruments. Hardwoods also exhibit excellent longevity, which makes them suitable for heavy-duty applications such as flooring, bridges, and railway sleepers. Due to their high density, hardwoods often require more effort to work with but provide exceptional durability and finish, making them a preferred choice for premium constructions and artistic endeavors.

Softwood Varieties

Softwoods are sourced from coniferous trees, which are typically evergreen and grow faster than deciduous trees. Examples of softwoods include pine, fir, cedar, spruce, and redwood. Unlike hardwoods, softwoods are generally lighter in weight and less dense, making them easier to process and work with. Pine, for example, is a versatile softwood widely used in furniture making, paneling, and flooring. Cedar, known for its natural resistance to decay and insects, is commonly employed in outdoor applications such as fencing and decking. Fir and spruce are extensively used in construction, particularly in framing and structural components, due to their strength-to-weight ratio. Additionally, softwoods are the primary raw material for paper production, making them indispensable in the pulp and paper industry. Both hardwoods and softwoods are vital to the timber industry, catering to diverse needs ranging from aesthetics to structural integrity. The choice between hardwood and softwood depends on the specific requirements of the application, the environmental conditions, and the desired longevity of the final product. Sustainable management of these resources is crucial to ensure their availability for future generations.

2.2 Uses of Timber in Construction and Furniture Making

Timber has been a cornerstone of construction and furniture making for centuries, offering a combination of strength, flexibility, and aesthetic appeal that few other materials can match. Its versatility makes it suitable for a wide range of applications, from traditional building techniques to modern architectural designs.

Timber in Construction

In the construction industry, timber is used extensively for structural and non-structural purposes. Its natural strength and durability make it an ideal material for framing, flooring, roofing, and cladding. In traditional architecture, timber was the primary building material for homes, bridges, and temples, with examples of timber structures enduring for centuries due to their robustness and craftsmanship.

Modern construction has embraced engineered wood products such as plywood, laminated veneer lumber (LVL), and cross-laminated timber (CLT). These innovations enhance the mechanical properties of timber, allowing it to be used in high-rise buildings and complex architectural designs. CLT, for instance, is a prefabricated material that offers exceptional strength, thermal insulation, and environmental benefits, making it a popular choice for sustainable construction. Timber is also a key material in green building practices, as it is renewable, recyclable, and has a lower carbon footprint compared to steel or concrete. When sourced from sustainably managed forests, timber can significantly reduce the environmental impact of construction projects, contributing to global efforts in combating climate change.

Timber in Furniture Making

The furniture industry relies heavily on timber for its aesthetic and functional properties. Hardwoods such as teak, mahogany, and oak are preferred for high-end furniture due to their durability, grain patterns, and resistance to wear. Softwoods like pine and cedar are used in crafting affordable yet sturdy furniture that appeals to a broad consumer base. Timber furniture is celebrated for its timeless appeal and versatility, ranging from rustic designs to modern minimalism. The adaptability of wood allows it to be shaped, carved, and finished in countless ways, enabling artisans and manufacturers to create unique and functional pieces. Moreover, timber furniture contributes to sustainable living, as it can be repaired, repurposed, and recycled, reducing waste and promoting circular economy practices. The use of timber in construction and furniture making not only highlights its functional benefits but also emphasizes its cultural and economic significance. From traditional craftsmanship to contemporary design, timber continues to play a pivotal role in shaping human habitats and lifestyles.

2.3 Industrial Applications: Pulp, Paper, and Packaging

Beyond its use in construction and furniture making, timber serves as a critical raw material in various industrial applications, particularly in the production of pulp, paper, and packaging. These industries rely on the unique properties of wood fibers to create products that are essential to modern life.

Pulp and Paper Industry

The pulp and paper industry is one of the largest consumers of softwood timber, utilizing wood fibers to produce paper, cardboard, and related products. The process begins with the mechanical or chemical pulping of wood, where fibers are separated and processed to form a slurry. This slurry is then dried and pressed into sheets, resulting in paper of various grades and qualities. Softwoods like pine and spruce are particularly suited for paper production due to their long fibers, which provide strength and durability to the final product. Hardwood fibers, on the other hand, are shorter and contribute to the smoothness and printability of paper. By combining these fibers, manufacturers can create specialized papers for printing, packaging, and hygiene products. The demand for sustainable and recyclable materials has driven innovation in the pulp and paper industry, with a focus on reducing water and energy consumption during production. Advances in technology have also enabled the use of timber by-products, such as sawdust and wood chips, in pulp production, minimizing waste and promoting resource efficiency.

Timber in Packaging

The packaging industry increasingly relies on timber-derived materials, such as paperboard, corrugated cardboard, and wooden crates, to meet global demand for eco-friendly packaging solutions. Paperboard is commonly used in food packaging, beverage cartons, and consumer goods, offering lightweight and recyclable alternatives to plastic. Corrugated cardboard, made from layers of paper pulp, is widely used for shipping and storage due to its strength and cushioning properties. Wooden packaging, including pallets and crates, remains essential in the transportation and logistics sectors, particularly for heavy or fragile goods. Timber's durability and ability to withstand environmental conditions make it a reliable choice for international trade and supply chain management.

Environmental Considerations in Industrial Applications

The industrial use of timber raises important questions about sustainability and environmental impact. Deforestation and overexploitation of forest resources pose significant challenges, emphasizing the need for responsible sourcing and management. Certification programs such as the Forest Stewardship Council (FSC) and the Programme for the Endorsement of Forest Certification (PEFC) promote sustainable forestry practices, ensuring that timber used in industrial applications is harvested in an environmentally and socially responsible manner.

3. Non-Timber Forest Products (NTFPs)

Non-timber forest products (NTFPs) represent an invaluable aspect of forest resources, encompassing all biological materials other than timber harvested from forests. These products serve a multitude of purposes, ranging from food and medicine to industrial raw materials and cultural artifacts. Unlike timber, the extraction of NTFPs typically causes minimal disruption to the forest ecosystem, making them a cornerstone of sustainable forestry and biodiversity conservation.

3.1 Edible Forest Products: Fruits, Nuts, and Mushrooms

Forests are natural repositories of a wide variety of edible products, including fruits, nuts, seeds, and mushrooms. These resources have been utilized by humans for millennia, offering nutrition, livelihood opportunities, and economic benefits.

Fruits and Berries

Forests are abundant in wild fruits and berries, which are rich sources of vitamins, minerals, and antioxidants. Species such as wild mangoes, berries, and baobab fruits are not only staples for local communities but also hold commercial potential in the global market. Wild fruits are often more nutrient-dense than their cultivated counterparts and are integral to traditional diets in many forest-dependent societies. In recent years, the demand for wild-harvested superfoods like acai berries and goji berries has surged due to their perceived health benefits, creating opportunities for sustainable harvesting and trade.

Nuts and Seeds

Forest nuts such as chestnuts, walnuts, and Brazil nuts are valuable for their nutritional content, including healthy fats, proteins, and micronutrients. These nuts serve as food for both humans and wildlife, playing a critical role in the forest ecosystem. Additionally, seeds such as those from the sal tree and neem have diverse applications in food, oil extraction, and even cosmetics. The commercial harvesting of nuts and seeds has provided significant income for forest-dwelling communities, while also promoting forest conservation through sustainable management practices.

Mushrooms and Fungi

Mushrooms are another vital edible forest product, with varieties such as morels, shiitake, and truffles being highly prized for their culinary and medicinal properties. Wild mushrooms are not only a delicacy in many cuisines but also contribute to forest ecosystems by aiding in nutrient cycling and organic matter decomposition. The collection and trade of mushrooms, particularly rare and high-value species like truffles, have become lucrative markets in certain regions, highlighting their economic potential. The sustainable harvesting of edible forest products ensures that these resources remain available for future generations while supporting local economies and biodiversity. Promoting agroforestry and community-based harvesting practices can enhance the availability and accessibility of these nutritious and economically valuable products.

3.2 Medicinal Plants and Their Pharmaceutical Applications

Medicinal plants derived from forests form the backbone of traditional medicine systems and have played a pivotal role in the development of modern pharmaceuticals. Forests harbor an immense diversity of flora with therapeutic properties, many of which remain untapped or under-researched. This section explores the significance of medicinal plants in healthcare, their pharmaceutical applications, and the challenges associated with their conservation and sustainable use.

Traditional and Modern Medicine

For centuries, medicinal plants have been integral to traditional healing systems such as Ayurveda, Traditional Chinese Medicine (TCM), and Indigenous practices. Plants like neem, tulsi, and turmeric are widely known for their antimicrobial, anti-inflammatory, and antioxidant properties. Similarly, herbs like ginseng and ashwagandha are renowned for their adaptogenic benefits, enhancing resilience to stress and promoting overall well-being. In modern medicine, forests have provided critical compounds that form the basis of numerous life-saving drugs. For instance, the bark of the cinchona tree yields quinine, a vital antimalarial drug, while the yew tree produces taxol, an essential compound in cancer treatment. Forest-derived alkaloids, flavonoids, and terpenoids have also contributed to advancements in pharmacology, demonstrating the immense potential of medicinal plants in drug discovery.

Pharmaceutical Applications

The pharmaceutical industry relies heavily on bioactive compounds sourced from medicinal plants for the development of drugs and nutraceuticals. Examples include:

» **Antibiotics:** Many antibiotics, such as penicillin, have origins in natural forest ecosystems.

» **Pain Relievers:** Plants like opium poppy and willow bark provide the basis for analgesics like morphine and aspirin.

» **Anticancer Agents:** Taxanes from yew trees and vincristine from the Madagascar periwinkle are critical in chemotherapy.

» **Cardiovascular Drugs:** Compounds such as reserpine, derived from the Indian snakeroot, are used in hypertension management.

The exploration of medicinal plants has expanded to nutraceuticals and dietary supplements, catering to the growing demand for natural and plant-based healthcare solutions. Products like herbal teas, essential oils, and supplements are increasingly popular among health-conscious consumers.

3.3 Resin, Gum, and Latex: Industrial and Commercial Uses

Resins, gums, and latex are among the most versatile and economically valuable NTFPs, with applications spanning industries such as pharmaceuticals, cosmetics, food, and manufacturing. These substances are secreted by trees and plants as part of their natural defense mechanisms, offering a renewable source of raw materials for various commercial purposes.

Resins

Resins are sticky, viscous substances secreted by certain trees, primarily conifers like pine. They are composed of complex hydrocarbons and terpenoids and are valued for their adhesive, waterproofing, and preservative properties. Resins such as rosin and turpentine are widely used in industries including:

» **Paints and Varnishes:** Rosin serves as a key component in the production of paints, varnishes, and adhesives, enhancing their durability and finish.

- » **Pharmaceuticals:** Some resins, like frankincense and myrrh, have antimicrobial and anti-inflammatory properties, making them integral to traditional and modern medicine.
- » **Perfumes and Incense:** The aromatic properties of resins are harnessed in the production of perfumes and incense, contributing to cultural and religious practices.

Gums

Natural gums, such as gum arabic, guar gum, and gum tragacanth, are water-soluble polysaccharides derived from trees and shrubs. These substances are widely used as stabilizers, thickeners, and emulsifiers in food products such as ice creams, sauces, and beverages. Additionally, gums find applications in pharmaceuticals as binders in tablet formulation and in cosmetics as natural thickeners for lotions and creams. Their non-toxic and biodegradable nature makes them ideal for eco-friendly products, aligning with global sustainability goals.

Latex

Latex is a milky fluid secreted by certain plants, notably the rubber tree (Hevea brasiliensis). It is the primary raw material for natural rubber, which has a myriad of applications in industries such as automotive, healthcare, and manufacturing. Key uses of latex include:

- » **Tires and Tubes:** Natural rubber is essential in the production of automobile tires, offering elasticity, durability, and resistance to wear.
- » **Medical Supplies:** Latex is used in the manufacture of gloves, catheters, and other medical equipment, ensuring sterility and flexibility.
- » **Adhesives and Coatings:** Latex-based adhesives and paints are valued for their strong bonding properties and environmental safety.

The extraction and processing of resins, gums, and latex must be conducted sustainably to avoid overexploitation and ensure the health of forest ecosystems. Certification schemes and fair-trade practices can promote responsible sourcing while benefiting local communities involved in their collection and trade.

4. Minor Forest Produce and Byproducts

Minor forest produces and byproducts represent an often overlooked but highly valuable component of forest resources. These products include materials that are not as prominent as timber or large-scale non-timber forest products (NTFPs) but are nonetheless critical for local economies, traditional practices, and industrial applications. Among these, essential oils and aromatics, bamboo and rattan, and dye, tannin, and fiber hold special importance due to their wide-ranging uses and sustainable potential.

4.1 Essential Oils and Aromatics

Essential oils and aromatic compounds derived from forest resources are among the most sought-after minor forest products, valued for their applications in medicine, cosmetics, and perfumery. These oils are volatile, concentrated liquids extracted from various parts of plants, including leaves, flowers, bark, and roots. The process of extraction often employs techniques such as steam distillation, cold pressing, or solvent extraction, depending on the nature of the plant material and the desired quality of the oil.

Essential oils such as eucalyptus, sandalwood, and cedarwood have long been used in traditional medicine for their therapeutic properties. Eucalyptus oil, for example, is widely recognized for its antiseptic and anti-inflammatory effects, making it a staple in remedies for respiratory conditions and muscle pain. Similarly, sandalwood oil is prized for its calming properties and has applications in aromatherapy and skincare, where it acts as an emollient and anti-aging agent. The extraction of aromatic compounds from flowers like jasmine and rose feeds into the global perfumery industry, creating luxury fragrances that are deeply embedded in cultural traditions and modern lifestyles. The commercial importance of essential oils extends to the food and beverage industry, where they are used as flavoring agents and preservatives. Lemon, peppermint, and cinnamon oils are common examples, offering both functional and sensory benefits. Furthermore, essential oils are integral to the production of natural cleaning agents and insect repellents, aligning with the growing demand for eco-friendly products. The extraction of essential oils often relies on sustainable practices to ensure the longevity of the source plants. Overharvesting of species such as sandalwood and rosewood has led to the implementation of regulations and conservation efforts, emphasizing the need for balanced utilization. Cultivation initiatives and technological advancements in extraction methods also contribute to the efficient use of forest resources, reducing waste and enhancing the overall value chain.

4.2 Bamboo and Rattan: Versatile Renewable Resources

Bamboo and rattan are among the most versatile and renewable resources obtained from forests, offering a wide range of applications in construction, furniture making, crafts, and industrial products. These fast-growing plants are not only vital for local economies but also play a significant role in environmental conservation due to their ability to sequester carbon, prevent soil erosion, and support biodiversity.

Bamboo is often referred to as "green steel" due to its exceptional strength-to-weight ratio, making it an eco-friendly alternative to traditional construction materials. Its applications in the construction industry are vast, ranging from scaffolding and flooring to prefabricated housing units and bridges. Bamboo's flexibility and durability also make it an ideal material for crafting furniture, musical instruments, and decorative items, supporting small-scale industries and artisan communities. In recent years, the use of bamboo in modern technology has expanded, with innovations such as bamboo textiles, composites, and biofuels gaining traction in sustainable markets. Rattan, another forest resource closely related to bamboo, is widely used in the production of lightweight and durable furniture, baskets, and mats. Its pliability and ease of weaving make it a preferred material for traditional and contemporary crafts, catering to both domestic and international markets. The economic significance of rattan extends to rural livelihoods, particularly in Southeast Asia, where its harvesting and processing employ thousands of people. The cultivation and management of bamboo and rattan are critical for ensuring their sustainable use. Agroforestry practices, combined with community-based management systems, have proven effective in maintaining the balance between utilization and conservation. Furthermore, certification programs like the Forest Stewardship Council (FSC) promote responsible sourcing, enabling consumers to support sustainable products while preserving forest ecosystems.

4.3 Dye, Tannin, and Fiber

Dye, tannin, and fiber derived from forest plants have been integral to human culture and industry for centuries, serving as raw materials for textiles, leather processing, and various other applications. These byproducts not only reflect the ingenuity of traditional practices but also hold significant potential in modern, sustainable industries. Natural dyes extracted from plants such as indigo, madder, and turmeric have been used for coloring textiles and crafts long before the advent of synthetic dyes. These plant-based dyes are biodegradable and non-toxic, making them a sustainable alternative to chemical dyes that often pollute water bodies. Indigo, for example, is widely used in dyeing denim, while turmeric imparts a vibrant yellow hue to fabrics and food products. The revival of natural dyeing techniques has gained momentum in recent years, driven by consumer demand for eco-friendly and artisanal goods.

Tannin, a group of water-soluble polyphenols found in tree bark, leaves, and fruits, plays a critical role in leather tanning and wine production. Oak, chestnut, and quebracho trees are primary sources of tannins, which are used to convert raw animal hides into durable leather. This process has been practiced for centuries, with traditional methods emphasizing the use of natural tannins. In the wine industry, tannins contribute to the taste, color, and aging properties of the beverage, reflecting their multifaceted applications. Fibers derived from forest plants, including jute, hemp, and coir, are widely used in textiles, ropes, mats, and brushes. These fibers are valued for their strength, biodegradability, and resistance to wear, making them essential in both traditional crafts and modern industrial applications. Jute, for example, is a key material in making sacks and bags, while coir (derived from coconut husks) is used in mattresses and upholstery. The growing interest in sustainable materials has further highlighted the importance of plant-based fibers, with innovations in biocomposites and eco-friendly packaging driving their demand. The sustainable extraction and utilization of dyes, tannins, and fibers require careful management to prevent overexploitation and habitat degradation. Supporting local artisans and implementing fair-trade practices can enhance the economic benefits of these products while ensuring their long-term availability. By integrating traditional knowledge with modern technologies, the potential of these minor forest-produce byproducts can be fully realized, contributing to both economic development and environmental sustainability.

5. Wildlife-Derived Forest Produce

Wildlife-derived forest produce refer to the range of products obtained from forest-dwelling animals and their natural processes. These products, such as honey, silk, lac, and other wildlife-dependent resources, not only contribute to local and global economies but also play a crucial role in promoting conservation and sustainable livelihoods. Wildlife and their associated produce are deeply intertwined with forest ecosystems, underscoring the need to balance their utilization with ecological preservation.

5.1 Honey and Bee Products

Honey and other bee-derived products, such as beeswax, royal jelly, and propolis, are among the most valuable resources derived from forest wildlife. These products are harvested through beekeeping and traditional honey hunting, offering substantial economic benefits while supporting forest ecosystems.

Economic Value of Honey and Bee Products

Honey, a natural sweetener with medicinal and nutritional properties, is a high-demand product both locally and internationally. Forest honey, often harvested from wild bees, is particularly prized for its unique flavors and high nutritional content, influenced by the biodiversity of the region. In addition to honey, beeswax is widely used in industries such as cosmetics, pharmaceuticals, and candle-making due to its non-toxic and versatile properties. Royal jelly, a secretion from worker bees, is highly valued in the health and wellness industry for its purported health benefits, including boosting immunity and improving skin health. Bee products also support local economies by providing a sustainable source of income for forest-dependent communities. Beekeeping, or apiculture, has been increasingly promoted as a livelihood activity, enabling communities to generate income without causing deforestation or habitat destruction. Moreover, the export potential of honey and bee products contributes significantly to national economies, especially in regions with rich forest resources.

Conservation Implications of Bee Products

Bees are vital pollinators, playing a critical role in maintaining forest biodiversity and ecosystem services. By facilitating the reproduction of flowering plants, bees contribute to the regeneration of forests and the sustenance of wildlife. The promotion of sustainable beekeeping practices not only supports conservation efforts but also mitigates the impact of declining wild bee populations caused by habitat loss, pesticide use, and climate change. Traditional honey hunting practices, while culturally significant, need to be regulated to ensure they do not harm bee populations or disrupt forest ecosystems. Community-based initiatives that integrate modern beekeeping techniques with traditional knowledge can enhance the sustainability of honey production, benefiting both people and the environment.

5.2 Silk and Lac Production from Forest Species

The production of silk and lac is deeply rooted in the ecological and economic fabric of forest-dependent communities. These products, derived from specific insects and their life processes, hold immense cultural and industrial significance.

Silk Production

Forest species such as the Antheraea assamensis (muga silkworm), Antheraea mylitta (tasar silkworm), and Philosamia ricini (eri silkworm) are integral to the production of non-mulberry silk varieties. These silks, often referred to as "wild silk," are distinct from mulberry silk and are prized for their texture, durability, and natural colors. For instance, muga silk, native to Assam in India, is renowned for its golden sheen and is considered a cultural heritage product. The process of sericulture in forests involves rearing silkworms on host plants such as oak, castor, and arjun. The cultivation of these plants not only supports silk production but also contributes to forest regeneration and biodiversity conservation. Wild silk production is labor-intensive, providing employment opportunities for forest-dependent communities, particularly women, who are often involved in rearing, spinning, and weaving.

Lac Production

Lac, a resin secreted by the insect Kerria lacca, is another valuable wildlife-derived forest product. Lac insects feed on specific host trees, such as palas, ber, and kusum, depositing resin on branches that are later harvested. This resin is processed into shellac, which is used in various industries including furniture polishing, adhesives, and cosmetics. Additionally, lac dye, derived from the resin, is a natural coloring agent used in textiles and food products. The economic importance of lac production lies in its potential to provide sustainable livelihoods for rural and tribal communities. Lac cultivation, often practiced in traditional agroforestry systems, supports environmental conservation by promoting the growth of host trees and preventing deforestation. The demand for natural and biodegradable materials has further enhanced the significance of lac in modern industries, aligning with global sustainability goals.

5.3 Role of Wildlife in Sustainable Forest-Based Livelihoods

Wildlife plays a multifaceted role in sustaining forest-based livelihoods, contributing to food security, income generation, and ecological balance. Forest-dependent communities have historically relied on wildlife and their produce for their survival, developing intricate knowledge systems and practices that harmonize resource use with conservation.

Food and Nutritional Security

Wildlife-derived produce, including bushmeat, fish, and eggs, provides essential protein and nutrition for many forest-dwelling communities. While sustainable hunting and fishing practices are critical for meeting dietary needs, they must be regulated to prevent overexploitation and ensure the conservation of wildlife populations. Initiatives such as community-managed hunting reserves and fish farming within forest landscapes can enhance food security while promoting biodiversity conservation.

Cultural and Traditional Practices

For many indigenous and tribal communities, wildlife is deeply embedded in cultural and spiritual traditions. The use of feathers, hides, and bones in rituals and crafts reflects the cultural significance of wildlife-derived products. These practices, when conducted sustainably, contribute to the preservation of cultural heritage and provide opportunities for income generation through the sale of traditional crafts and artifacts.

Eco-Tourism and Conservation-Based Livelihoods

Wildlife is a major attraction for eco-tourism, which generates significant revenue for forest-dependent communities. Activities such as birdwatching, wildlife safaris, and trekking not only create employment opportunities but also foster awareness about conservation. Community-based eco-tourism initiatives, where locals act as guides and hosts, empower communities while ensuring that tourism activities do not harm the environment.

Sustainable Harvesting and Conservation

The sustainable utilization of wildlife-derived products requires a delicate balance between meeting human needs and conserving biodiversity. Policies and practices such as wildlife conservation laws, habitat restoration, and participatory forest management play a crucial role in achieving this balance. Engaging local communities in conservation efforts, such as anti-poaching patrols and habitat monitoring, enhances the effectiveness of these initiatives while providing alternative livelihoods.

6. Cultural and Recreational Uses of Forests

Forests have long been revered for their cultural, aesthetic, and recreational significance, serving as a source of inspiration, spiritual solace, and creative expression for human societies. Beyond their ecological and economic roles, forests are deeply intertwined with cultural identity, artistic endeavors, and recreational pursuits. These uses highlight the multifaceted value of forests, showcasing their ability to nurture the human spirit while fostering environmental awareness and conservation.

6.1 Forests as Sources of Aesthetic and Spiritual Value

Forests possess an inherent aesthetic appeal that transcends cultures and time, offering a profound sense of beauty and tranquility. The lush greenery, diverse flora and fauna, and the symphony of natural sounds evoke awe and reverence, making forests a source of inspiration for art, literature, and music. Writers like Henry David Thoreau and Rabindranath Tagore have drawn upon the serene beauty of forests to craft literary masterpieces, while artists have depicted forests in paintings, sculptures, and other creative mediums to celebrate their majesty and mystery. Beyond their aesthetic value, forests hold deep spiritual significance for many cultures and religions. For indigenous communities, forests are often regarded as sacred spaces, embodying the presence of deities, ancestors, or spirits. Rituals, ceremonies, and festivals are frequently conducted in forested landscapes, underscoring their spiritual importance. Sacred groves, for instance, are patches of forests preserved for religious purposes, reflecting the intertwining of ecological and cultural values. In modern times, the spiritual connection to forests has evolved into practices like forest bathing (Shinrin-yoku), a Japanese concept that emphasizes mindful immersion in forest environments to promote mental and physical well-being. This practice has gained global recognition for its therapeutic benefits, demonstrating how forests continue to serve as sanctuaries for healing and reflection.

6.2 Eco-Tourism and Its Role in Forest Conservation

Eco-tourism, a form of responsible travel that emphasizes environmental sustainability and cultural respect, has emerged as a significant contributor to forest conservation. Forest-based eco-tourism offers visitors an opportunity to experience the natural beauty and biodiversity of forests while fostering a deeper understanding of their ecological importance.

Economic Benefits of Eco-Tourism

Eco-tourism generates revenue for forest-dependent communities and governments, creating economic incentives for conservation. Activities such as wildlife safaris, birdwatching, trekking, and camping attract millions of tourists annually, providing employment opportunities for local guides, hospitality providers, and artisans. In countries like Costa Rica and Bhutan, eco-tourism has become a cornerstone of national economic strategies, demonstrating its potential to drive sustainable development.

Environmental Education and Advocacy

Eco-tourism also plays a crucial role in raising awareness about forest conservation. By exposing visitors to the beauty and fragility of forest ecosystems, eco-tourism fosters a sense of responsibility toward environmental stewardship. Interpretive programs, guided tours, and interactive exhibits help tourists understand the interconnectedness of forests with global issues like climate change, biodiversity loss, and water security.

6.3 Forest Products in Art, Handicrafts, and Cultural Practices

Forests have served as a repository of materials and motifs for artistic and cultural expressions throughout human history. The use of forest products in art, handicrafts, and cultural practices reflects the intimate relationship between human creativity and natural resources, highlighting the cultural significance of forests.

Art and Handicrafts

Forest products such as wood, bark, leaves, and fibers are fundamental to traditional and contemporary art forms. Wood carving, for example, is a time-honored craft practiced across the globe, with artisans creating intricate sculptures, masks, and furniture from timber. The use of bamboo and rattan in weaving baskets, mats, and decorative items showcases the versatility of forest materials in handicrafts. In addition, dyes and pigments extracted from forest plants like indigo, turmeric, and madder are used in textile production and painting, preserving traditional art forms while appealing to eco-conscious markets. The global demand for eco-friendly and handmade products has revitalized interest in forest-based handicrafts, providing new opportunities for artisans and communities. Organizations and cooperatives supporting forest artisans have helped bridge the gap between traditional crafts and modern markets, ensuring the economic sustainability of these cultural practices.

Cultural Practices and Rituals

Forests are central to numerous cultural and ritualistic practices, particularly in indigenous and tribal societies. Items such as sacred wood, ceremonial plants, and wildlife products are used in rituals, symbolizing fertility, prosperity, or protection. For example, certain forest herbs are burned as incense in religious ceremonies, while specific tree species like sandalwood are associated with purity and spirituality. In addition to their ritualistic uses, forest products often feature in traditional costumes, jewelry, and musical instruments. Tribal communities, for instance, craft ornaments from seeds, shells, and feathers, reflecting their connection to the forest. Musical instruments like drums, flutes, and stringed instruments are frequently made from forest materials, contributing to the preservation of intangible cultural heritage.

Sustainable Use and Preservation of Traditions

The sustainable use of forest products in art and cultural practices is essential for maintaining the ecological balance and safeguarding traditional knowledge. Overharvesting of resources such as sandalwood, ivory, or certain medicinal plants has led to the depletion of forest biodiversity, emphasizing the need for regulation and conservation. Initiatives that combine traditional craftsmanship with sustainable harvesting practices can ensure the longevity of both cultural traditions and forest ecosystems. The incorporation of forest products into art, handicrafts, and cultural practices underscores their enduring value beyond material utility. By fostering appreciation for the cultural and recreational uses of forests, societies can cultivate a deeper respect for these vital ecosystems, inspiring collective efforts to protect and preserve them for future generations.

7. Economic Contributions of Forest Produce

Forests and their produce are indispensable to the economic framework of societies worldwide, providing raw materials, employment opportunities, and financial sustenance across local, national, and global levels. The economic contributions of forest produce extend far beyond tangible goods such as timber and non-timber products; they also include ecosystem services that underpin industries and communities. Forests contribute to economic resilience, rural development, and industrial growth while fostering sustainable livelihoods and reducing poverty in forest-dependent areas.

7.1 Forest-Based Industries and Employment Opportunities

The forest sector forms the backbone of numerous industries, ranging from construction and furniture making to paper production and pharmaceuticals. These industries rely on a diverse array of forest products, both timber and non-timber, to manufacture goods essential for modern life. Timber serves as a raw material for construction, carpentry, and the production of furniture, while non-timber forest products (NTFPs) such as resins, gums, bamboo, and medicinal plants support industries like cosmetics, food processing, and healthcare. The interconnectedness of these industries ensures a steady demand for forest produce, making it a vital economic driver.

The employment opportunities generated by forest-based industries are vast and multifaceted, encompassing a range of skill levels and professions. Logging, sawmilling, and wood processing employ millions globally, while the harvesting of NTFPs provides seasonal and supplementary income for rural and tribal populations. Moreover, downstream industries such as manufacturing, transportation, and retail create additional jobs, extending the economic impact of forests far beyond their geographic boundaries. For instance, the bamboo industry alone supports artisans, builders, and entrepreneurs in countries like India, China, and Indonesia, where bamboo is a key resource for construction, crafts, and textiles. Emerging green industries, such as bioenergy and sustainable packaging, have further expanded the scope of forest-based employment. The global shift towards eco-friendly materials has increased demand for biodegradable alternatives derived from forest resources, such as paper-based packaging and plant-based polymers. These innovations not only create new job opportunities but also contribute to environmental sustainability, reflecting the dynamic and evolving nature of forest-based industries. However, to maximize the economic potential of these industries, there must be an emphasis on capacity building, skill development, and equitable access to resources for marginalized groups, ensuring that the benefits of forest produce are shared widely.

7.2 Contribution to National and Global Economies

The economic significance of forest produce is not confined to local or regional markets; it plays a critical role in shaping national and global economies. The forest sector contributes substantially to GDP through the trade of timber, NTFPs, and forest-derived products. Countries with rich forest resources, such as Brazil, Canada, and Russia, derive significant revenue from the export of timber and forest products, establishing themselves as key players in the global market. Timber trade alone accounts for billions of dollars annually, with sawlogs, veneer sheets, and plywood being some of the most traded commodities.

Non-timber forest products also hold substantial economic value, particularly in global markets for herbal medicines, essential oils, and organic foods. The increasing demand for sustainable and natural products has expanded the international trade of forest-derived goods such as shea butter, frankincense, and forest honey. These products cater to diverse industries, including pharmaceuticals, cosmetics, and food processing, generating significant foreign exchange earnings for producing countries. In regions like Sub-Saharan Africa and Southeast Asia, the export of NTFPs contributes not only to national economies but also to rural development, as it integrates local producers into global value chains. Ecosystem services provided by forests, such as carbon sequestration, watershed protection, and biodiversity conservation, also hold immense economic value. Mechanisms like carbon credits and payments for ecosystem services (PES) have monetized these intangible benefits, enabling countries to generate income from conserving their forest resources. For example, the REDD+ (Reducing Emissions from Deforestation and Forest Degradation) initiative has facilitated financial flows to developing countries in exchange for their efforts to reduce forest loss and enhance carbon storage. These programs not only reinforce the economic importance of forests but also align with global sustainability goals, demonstrating the multifaceted contributions of forest produce to the world economy.

7.3 Role of Forest Produce in Rural Development and Livelihoods

The role of forest produce in rural development and livelihoods is profound, particularly in regions where forests serve as the primary source of sustenance and income for local communities. Forests provide food, fuel, and materials for daily living, creating a safety net for rural populations during times of economic or environmental stress. Non-timber forest products, in particular, are a lifeline for millions of forest-dependent people, offering accessible and renewable resources that support subsistence and commercial activities. Edible forest products such as fruits, nuts, mushrooms, and honey are vital for food security and nutrition, especially in remote areas with limited access to agricultural markets. These resources also generate income through local and regional trade, enabling rural households to diversify their livelihoods and reduce economic vulnerability. Medicinal plants harvested from forests contribute to healthcare and income generation, with traditional healers and herbalists playing a crucial role in rural healthcare systems. Additionally, forest fibers like bamboo, rattan, and jute support small-scale industries and crafts, empowering artisans and entrepreneurs in rural areas. The integration of forest produce into rural development initiatives has the potential to transform local economies, fostering resilience and sustainability. Community-based forest management programs, where local populations are actively involved in conserving and utilizing forest resources, have demonstrated success in improving livelihoods while protecting ecosystems. These programs often include capacity-building activities, such as training in sustainable harvesting techniques, value addition, and market access, enabling communities to maximize the benefits of forest produce. Forest-based livelihoods also contribute to social equity by empowering marginalized groups, including women and indigenous communities. Women, in particular, play a significant role in the collection and processing of NTFPs, such as wild fruits, medicinal plants, and natural dyes. By supporting women-led enterprises and cooperatives, rural development programs can enhance gender equality and ensure inclusive economic growth.

8. Sustainability in Forest Produce Utilization

The sustainable utilization of forest produce is critical for maintaining the delicate balance between human needs and the preservation of forest ecosystems. Forests provide essential resources, including timber and non-timber products (NTFPs), but their unsustainable exploitation has led to widespread deforestation, habitat loss, and climate change. Addressing these challenges requires a holistic approach that integrates economic development with environmental stewardship.

8.1 Certification Standards for Forest Products (FSC, PEFC)

Certification standards like the Forest Stewardship Council (FSC) and the Programme for the Endorsement of Forest Certification (PEFC) are cornerstone initiatives in promoting sustainable forest management and responsible sourcing of forest products. These standards provide a framework for verifying that timber, NTFPs, and other forest-derived products are harvested in ways that preserve forest ecosystems, respect local communities, and comply with legal requirements.

The Forest Stewardship Council (FSC)

The FSC, established in 1993, is one of the most widely recognized certification systems for forest products. Its principles and criteria emphasize environmental conservation, social equity, and economic viability. FSC-certified products ensure that forests are managed responsibly, protecting biodiversity, water resources, and soil quality while promoting fair labor practices and respecting indigenous rights. The FSC certification process involves rigorous assessments by accredited organizations, ensuring that certified products meet global sustainability standards. FSC labels on products such as timber, paper, and furniture provide consumers with the assurance that their purchases do not contribute to deforestation or illegal logging. This market-driven approach has incentivized companies to adopt sustainable practices, fostering a culture of accountability in the forest sector. For example, FSC-certified tropical timber from regions like the Amazon and Southeast Asia has gained premium value in global markets, encouraging sustainable forest management in these biodiversity hotspots.

The Programme for the Endorsement of Forest Certification (PEFC)

The PEFC, established in 1999, focuses on endorsing national forest certification systems that align with international sustainability benchmarks. It provides a flexible framework that accommodates regional variations in forest management practices, making it particularly effective in promoting sustainable forestry in diverse ecological and socio-economic contexts. PEFC certification covers a wide range of forest products, including timber, paper, and packaging materials, enhancing the traceability and credibility of these products in global supply chains. Both FSC and PEFC certifications play a critical role in combating illegal logging and deforestation, improving market access for sustainably managed forest products, and fostering collaboration among governments, industries, and civil society. Their widespread adoption reflects a growing recognition of the need for responsible consumption and production to safeguard forest ecosystems.

8.2 Sustainable Harvesting Practices for Timber and NTFPs

Sustainable harvesting practices are fundamental to the long-term availability of forest resources, ensuring that timber and NTFPs can be extracted without compromising the health and regenerative capacity of forest ecosystems. These practices are designed to balance resource utilization with conservation, minimizing ecological disruption and promoting biodiversity.

Timber Harvesting

Sustainable timber harvesting involves methods that maintain the structural integrity and ecological functions of forests. Selective logging, for instance, targets mature or overmature trees while preserving younger and non-commercial species, allowing forests to regenerate naturally. Reduced-impact logging

(RIL) techniques, which include careful planning of logging roads and felling directions, minimize soil erosion, habitat destruction, and carbon emissions. Reforestation and afforestation programs complement sustainable harvesting by replenishing forest cover and enhancing carbon sequestration. In addition to these practices, community forestry initiatives have emerged as effective models for sustainable timber management. By involving local communities in decision-making and benefit-sharing, these initiatives foster stewardship and reduce conflicts over resource use. For example, Nepal's community forestry program has demonstrated significant success in improving forest health and livelihoods through participatory management.

Sustainable Harvesting of NTFPs

The extraction of NTFPs such as fruits, nuts, medicinal plants, and fibers requires careful management to prevent overharvesting and ensure the sustainability of these resources. Practices such as rotational harvesting, which allows resources to regenerate between collection cycles, and quota systems, which limit the quantity harvested, are essential for maintaining NTFP availability. Additionally, cultivating high-demand NTFPs in agroforestry systems reduces pressure on natural forests while providing economic benefits to local communities.Certification schemes like FairWild focus specifically on the sustainable harvesting of wild-collected NTFPs, ensuring that these products are sourced ethically and sustainably. These certifications not only enhance market access for NTFPs but also promote biodiversity conservation and equitable benefit-sharing.

8.3 Innovations in Recycling and Waste Utilization of Forest Products

The efficient use of forest resources extends beyond sustainable harvesting to include innovations in recycling and waste utilization. These approaches not only maximize the economic value of forest products but also reduce environmental impacts by minimizing waste and promoting circular economies.

Recycling of Forest Products

Recycling is a key strategy for reducing the demand for virgin forest resources and mitigating deforestation. Paper recycling, for instance, significantly lowers the need for fresh wood pulp, conserving trees and reducing energy and water consumption in the manufacturing process. Advances in recycling technologies have improved the quality of recycled paper, enabling its use in diverse applications, from packaging to printing. Recycled wood products, such as reclaimed timber and wood composites, are increasingly being used in construction and furniture making. These materials provide a sustainable alternative to newly harvested wood, extending the lifecycle of forest products and reducing waste. For example, old wooden pallets can be transformed into flooring, furniture, or bioenergy, demonstrating the potential of recycling to create value from discarded materials.

Utilization of Forest Product Waste

The byproducts and residues generated during the processing of forest products, such as sawdust, bark, and wood chips, offer significant opportunities for waste utilization. These materials can be repurposed into biofuels, particleboard, mulch, or animal bedding, reducing landfill waste and supporting sustainable industries. Biomass energy, derived from forest residues, is a renewable source of power that contributes to energy security while reducing greenhouse gas emissions. Innovative uses of NTFP waste have also emerged, such as the conversion of fruit peels into natural dyes or the extraction of essential oils from discarded plant materials. These practices align with the principles of circular economies, where waste is minimized, and every part of the resource is utilized efficiently.

Conclusion

Forest produce embodies the profound interdependence between humanity and nature, serving as a critical resource for sustaining life, fostering economic growth, and preserving ecological balance. The wide spectrum of forest-derived products, ranging from timber and non-timber resources to wildlife-related outputs, highlights the vast utility of forests and their indispensable role in societal and environmental well-being. This intricate relationship underscores the importance of forests as multifunctional ecosystems that support biodiversity, provide livelihoods, and mitigate climate change.

Timber-based products, with their significance in construction, furniture making, and industrial applications, demonstrate the versatility and economic value of forests. The distinction between hardwood and softwood varieties further illustrates the specialized roles of different tree species in meeting diverse human needs. Similarly, non-timber forest products (NTFPs) such as fruits, medicinal plants, resins, and essential oils showcase the ecological and commercial importance of forest resources beyond timber. These products not only sustain traditional knowledge and practices but also contribute to modern industries, offering opportunities for sustainable economic development. Minor forest produces and byproducts, including bamboo, rattan, essential oils, and natural dyes, provide further evidence of forests' adaptability to meet human demands while maintaining ecological balance. These resources underscore the potential of forests to drive innovation, from traditional crafts to modern bio-based industries, thereby bridging cultural heritage with contemporary economic imperatives. Wildlife-derived Forest products such as honey, silk, and lac reinforce the interconnectedness of biodiversity and resource utilization, emphasizing the need for holistic conservation approaches that integrate flora and fauna.

Forests also transcend material contributions, offering cultural, spiritual, and recreational value. They inspire art, nurture traditions, and promote eco-tourism, fostering a deeper human connection with nature. This cultural significance amplifies the role of forests as not merely sources of economic goods but as foundational elements of human identity and well-being. Economically, forest produce supports industries, creates employment, and generates significant national and global revenue. Forest-based livelihoods are particularly crucial for rural and indigenous communities, empowering them economically while preserving their cultural and environmental knowledge. Sustainable forest management practices, supported by certification standards and innovative harvesting techniques, ensure that forest resources continue to provide economic benefits without compromising ecological health. These measures align with broader global goals, including the United Nations Sustainable Development Goals (SDGs), underscoring the importance of integrating forest produce utilization into sustainability frameworks.

However, challenges such as over-exploitation, market inequities, and climate change pose significant threats to the sustainable management of forest resources. Unsustainable practices risk depleting these resources, while climate change alters forest ecosystems, threatening their productivity and biodiversity. Addressing these challenges requires a multifaceted approach involving policy interventions, community engagement, and technological advancements. Innovations such as biotechnology and circular economy models offer promising pathways for enhancing the efficiency and sustainability of forest produce utilization. As the world moves towards a future defined by sustainability and resilience, the role of forests and their produce becomes increasingly central. Expanding markets for sustainable products, fostering green technologies, and integrating forest resources into circular economies will be critical for ensuring that forests continue to meet the needs of present and future generations. The balance between economic development and ecological preservation, achieved through sustainable forest management, will determine the trajectory of human progress and environmental health.

Chapter - 14

Minor Forest Products

1. Introduction to Minor Forest Products

Minor forest products (MFPs), often referred to as non-timber forest products (NTFPs), represent a diverse and invaluable category of resources derived from forests. Unlike major forest products such as timber, MFPs include many materials obtained from plants, animals, and other natural sources found in forest ecosystems. These resources play a significant role in the livelihoods of rural and indigenous communities, serving as a critical means of subsistence and income generation. Beyond their economic importance, MFPs also hold ecological, cultural, and medicinal value, forming an essential link between forest conservation and sustainable development.

The definition and scope of minor forest products encompass a vast array of resources that are distinct from timber in terms of their usage, extraction methods, and economic contributions. These products include edible items such as fruits, nuts, seeds, and leaves; medicinal and aromatic plants; fibers and dyes; and animal-derived resources like honey, silk, and beeswax. Resins, gums, and latex, often extracted through specialized techniques, further exemplify the utility of MFPs in industries ranging from pharmaceuticals to adhesives. Bamboo and rattan, though classified as minor products, are among the most economically significant resources due to their applications in construction, furniture, and handicrafts. Essential oils and aromatic compounds derived from forest plants contribute to wellness, cosmetics, and global trade, highlighting the international relevance of MFPs. One of the most notable aspects of MFPs is their importance in rural and indigenous economies. For millions of people living in forest-adjacent areas, these products are a primary source of food, medicine, and income. The collection and processing of MFPs often provide employment opportunities, particularly for women, thereby contributing to gender empowerment and community well-being. Indigenous communities, with their rich knowledge of forest ecosystems, have historically relied on MFPs not only for sustenance but also for maintaining their cultural heritage. The integration of traditional practices with modern market systems offers immense potential for sustainable development and poverty alleviation.

The distinction between major and minor forest products lies in their scale of utilization, economic impact, and the nature of their extraction. While major products like timber are harvested primarily for large-scale industrial and infrastructural purposes, minor products are often collected manually,

DOI: 10.1201/9781042011063-14

requiring intricate knowledge of the local ecosystem. Despite their categorization as "minor," these products contribute significantly to the global economy and biodiversity conservation. They represent the interconnectedness of forests with human livelihoods and underscore the need for holistic forest management approaches. Plant-based MFPs form a substantial segment of this category, with edible products such as fruits, nuts, seeds, and edible leaves being crucial for nutrition and food security. Medicinal and aromatic plants hold significant economic and therapeutic value, serving as raw materials for traditional medicine systems and modern pharmaceuticals. Fibers and natural dyes, derived from forest plants, find applications in handicrafts and industries, bridging the gap between traditional knowledge and contemporary markets. These plant-based resources underscore the multifaceted role of forests in supporting human life and fostering cultural identity. Animal-derived MFPs add another dimension to the utility of forests. Honey, beeswax, and other apiculture products exemplify the economic and ecological significance of animal-based resources, with bees playing a critical role in pollination and ecosystem health. Silk and lac, derived from forest ecosystems, form the basis of traditional industries that sustain local artisans and contribute to global trade. Wildlife-based products, though more contentious due to conservation concerns, highlight the interdependence of biodiversity and human livelihoods, emphasizing the importance of sustainable practices.

Resins, gums, and latex represent a specialized category of MFPs with extensive industrial applications. Tapping techniques, developed over centuries, enable the extraction of these resources without harming the parent plants. Their uses in adhesives, varnishes, and pharmaceuticals demonstrate the versatility of forest resources in meeting diverse human needs. However, challenges such as market development and resource management often limit the full potential of these products, requiring targeted interventions to enhance their utilization and sustainability. Bamboo and rattan, often referred to as the "green gold" of forests, are among the most versatile renewable resources. Their applications in construction, furniture, and handicrafts make them economically significant, while their rapid growth and ecological benefits contribute to forest regeneration and carbon sequestration. Sustainable management and cultivation of bamboo and rattan are critical for maximizing their benefits while ensuring their long-term availability. Strategies such as community-based management and value addition through processing and branding have proven effective in promoting these resources.

Essential oils and aromatics derived from forest plants further highlight the economic potential of MFPs. Extraction techniques such as steam distillation and cold pressing enable the production of high-value oils used in cosmetics, perfumery, and wellness industries. The growing global demand for natural and organic products presents significant opportunities for the expansion of this sector, particularly for forest-dependent communities. By integrating MFPs into global value chains, local producers can gain access to international markets, fostering economic growth and sustainable practices. The role of non-timber products in livelihoods and local economies is profound. They support subsistence needs while providing supplementary income, particularly for marginalized groups. The contribution of MFPs to gender empowerment is notable, as women often play a central role in their collection, processing, and trade. Community-based initiatives integrating MFPs into local and global value chains have demonstrated their potential to drive socio-economic development while preserving forest ecosystems. Sustainability and conservation are central to the long-term viability of minor forest products. Sustainable harvesting practices, informed by indigenous knowledge, are essential for maintaining resource availability and ecological balance. Addressing over-exploitation and habitat loss requires a combination of policy frameworks, community engagement, and scientific research. Integrating traditional ecological knowledge with modern conservation strategies can enhance the effectiveness of resource management. Despite their importance, MFPs face numerous challenges in management and utilization. Limited market access, unfair pricing for local collectors, and the impact of climate change on resource availability are significant barriers. Policy gaps and the lack of regulation in MFP trade further exacerbate these issues, necessitating a concerted effort to address them.

1.1 Definition and Scope of Minor Forest Products (MFPs)

Minor Forest Products, commonly referred to as MFPs, are a broad category of goods obtained from forests, excluding timber and fuelwood. These include plant-derived resources such as wild fruits, seeds, medicinal herbs, aromatic oils, resins, and fibers, as well as animal-derived products like honey, lac, and silk. Unlike timber, MFPs are typically harvested without destroying the tree or ecosystem, making them a cornerstone of sustainable forest resource management. Their utility spans subsistence, economic, cultural, and industrial applications. The scope of MFPs is vast and encompasses resources used for food, medicine, handicrafts, cosmetics, and industrial purposes. For example, wild fruits like tamarind and berries provide essential nutrition and are also processed into jams and beverages for commercial purposes. Similarly, medicinal plants such as neem, turmeric, and ashwagandha are vital components of traditional healthcare systems and modern pharmaceuticals. The aromatic properties of essential oils derived from forest plants like sandalwood and patchouli contribute to the global fragrance and wellness industries. These examples highlight the diversity of MFPs and their ability to meet both local and international demands. Ecologically, MFPs are integral to forest health and biodiversity. Products like seeds, fruits, and nuts support forest regeneration by serving as food for wildlife and aiding in seed dispersal. Pollinators such as bees, which contribute to honey production, are also crucial for the reproduction of forest plants. The sustainable management of MFPs ensures that forests continue to provide these ecosystem services while simultaneously supporting economic activities. In the global context, the demand for natural, organic, and ethically sourced products has expanded the scope of MFPs, positioning them as essential components of sustainable supply chains and green economies.

1.2 Importance of MFPs in Rural and Indigenous Economies

Minor Forest Products are a lifeline for millions of rural and indigenous communities worldwide, providing essential resources for sustenance and income generation. In many developing regions, forests are a primary source of livelihood, and MFPs play a pivotal role in bridging the economic gap between subsistence farming and market-based income. Economically, MFPs offer a critical safety net for forest-dependent populations, particularly during periods of agricultural downturn or food scarcity. Products such as wild mushrooms, berries, nuts, and honey are harvested and sold in local and regional markets, providing families with much-needed cash income. For example, the collection and trade of wild honey in Southeast Asia and Africa support thousands of households, while in India, the tendu leaf, used in bidi (traditional hand-rolled cigarettes), is a significant source of seasonal employment for tribal communities. The economic significance of MFPs is further amplified by small-scale industries that rely on these resources. Handicrafts made from bamboo, rattan, and natural fibers not only preserve traditional skills but also create employment opportunities in rural areas. Similarly, the extraction and processing of resins, gums, and essential oils form the backbone of many cottage industries. Women, in particular, play a central role in the collection, processing, and marketing of MFPs, making these activities crucial for promoting gender equity and empowering marginalized groups.

Culturally, MFPs are deeply embedded in the social and spiritual lives of indigenous communities. Products such as sacred herbs, resins like frankincense, and natural dyes are integral to rituals, festivals, and traditional medicine. The sustainable harvesting and use of these products often adhere to indigenous knowledge systems, which emphasize ecological harmony and resource conservation. This cultural dimension underscores the importance of MFPs not just as economic assets but as vital elements of heritage and identity. Despite their significance, MFPs face numerous challenges, including market access, fair pricing, and resource depletion. Addressing these issues requires policies that recognize the value of MFPs and support the livelihoods of those who depend on them. Strengthening market linkages, providing training in value addition, and ensuring equitable benefit-sharing are essential steps in maximizing the potential of MFPs for rural and indigenous economies.

1.3 Distinction Between Major and Minor Forest Products

The distinction between major and minor forest products lies primarily in their economic scale, extraction processes, and impact on forest ecosystems. Major forest products, such as timber and large-scale fuelwood, are extracted in significant quantities and are primarily used for construction, energy, and industrial purposes. These products often require intensive logging and clear-cutting, which can lead to deforestation, habitat destruction, and soil degradation if not managed sustainably. In contrast, MFPs are typically harvested in smaller quantities and involve less invasive extraction methods. For instance, collecting fruits, nuts, or medicinal plants does not require tree felling and often leaves the forest canopy intact, preserving the ecosystem's structural integrity. This distinction makes MFPs inherently more sustainable than their major counterparts, aligning them with conservation and biodiversity goals.

Economically, major forest products dominate global trade and industrial markets, with timber being one of the most valuable commodities worldwide. MFPs, on the other hand, are often localized and cater to niche markets, although their global demand is rising. For example, the international trade in essential oils, wild honey, and shea butter has gained traction in recent years, highlighting the untapped potential of MFPs. Socially and culturally, MFPs are more closely tied to the livelihoods and traditions of local communities. While timber is primarily managed by large corporations and state agencies, MFPs are often collected and processed by smallholders and indigenous groups. This distinction underscores the role of MFPs in promoting inclusive economic growth and preserving cultural heritage.

2. Plant-Based Minor Forest Products

Plant-based minor forest products (MFPs) are a critical subset of forest resources, offering a vast array of goods that contribute significantly to local economies, traditional practices, and global industries. These products, derived from the leaves, fruits, seeds, roots, and other non-timber components of forest vegetation, are central to the livelihoods of rural and forest-dependent communities. Beyond their direct uses, plant-based MFPs play an integral role in biodiversity conservation and sustainable forest management.

2.1 Edible Products: Fruits, Nuts, Seeds, and Edible Leaves

Edible products derived from forests are among the most accessible and widely utilized plant-based MFPs, encompassing a diverse range of fruits, nuts, seeds, and edible leaves that provide sustenance and nutrition to millions of people. Wild fruits such as mangoes, tamarinds, berries, and baobab are not only vital for the dietary needs of local communities but also form the basis for numerous commercial products. For instance, wild tamarind is processed into pulp for use in sauces and beverages, while baobab fruit is turned into powders and supplements rich in vitamins and antioxidants. These products are often harvested during lean agricultural seasons, acting as a crucial safety net for food security in rural areas. Nuts and seeds, such as chestnuts and acorns, serve dual roles as sources of nutrition and economic commodities. These forest products are rich in fats, proteins, and micronutrients, making them highly valued both locally and in global markets. Brazil nuts, for instance, are a significant export product from South America, harvested sustainably from the Amazon rainforest, thereby supporting the livelihoods of indigenous communities while promoting forest conservation. Seeds like sal and neem are utilized not only as food sources but also for oil extraction, with neem oil being particularly important for its medicinal and pesticidal properties.

Edible leaves, such as those from moringa and amaranth plants, are another vital category of forest-derived food resources. These leaves are nutrient-dense, providing essential vitamins and minerals that combat malnutrition, especially in developing regions. In addition to their dietary value, they are often integrated into traditional medicine and culinary practices, reflecting their multifaceted utility. Forest greens are collected sustainably, requiring minimal infrastructure and serving as a readily available resource for subsistence and trade. The collection and sale of edible forest products also contribute significantly to household incomes, particularly for women who play a central role in these activities.

2.2 Medicinal and Aromatic Plants

Medicinal and aromatic plants (MAPs) derived from forests are integral to healthcare systems, traditional medicine, and modern industries such as pharmaceuticals, cosmetics, and aromatherapy. These plants contain bioactive compounds that are used to treat a wide range of ailments, from common colds to chronic diseases. Neem, turmeric, ashwagandha, and ginseng are just a few examples of medicinal plants with well-documented therapeutic properties. Forests serve as reservoirs of biodiversity, hosting countless plant species that are yet to be fully explored for their medicinal potential.

The economic value of MAPs is immense, as these products form the raw materials for a variety of high-demand goods. For instance, the global trade in medicinal plants is worth billions of dollars annually, driven by the increasing popularity of herbal supplements, natural remedies, and plant-based pharmaceuticals. Aromatic plants like sandalwood, lavender, and patchouli are extensively cultivated and harvested for their essential oils, which are used in perfumery, skincare, and wellness industries. Sandalwood oil, in particular, is one of the most expensive essential oils globally, with its demand far exceeding supply, prompting significant investment in sustainable cultivation practices. The traditional use of MAPs is deeply rooted in indigenous knowledge systems, where these plants are used for treating ailments, conducting rituals, and maintaining well-being. For example, the bark of the cinchona tree, which yields quinine, has been used for centuries to treat malaria, while the resin of frankincense is valued for its anti-inflammatory and antimicrobial properties. These practices are now being integrated with modern pharmacological research, leading to the development of new drugs and therapies. The sustainable harvesting of medicinal plants is critical to preserving both their availability and the ecosystems they inhabit. Overharvesting, habitat loss, and climate change pose significant threats to MAPs, underscoring the need for conservation programs and cultivation initiatives.

2.3 Fibers and Natural Dyes

Fibers and natural dyes obtained from forests represent a unique category of plant-based MFPs, serving as raw materials for handicrafts, textiles, and various industrial applications. Bamboo, rattan, jute, and coir are among the most widely used forest fibers, valued for their strength, flexibility, and renewability. These fibers are the backbone of traditional crafts in many cultures, used to make baskets, mats, ropes, and furniture. For instance, bamboo, often referred to as "green steel," is not only a versatile craft material but also a sustainable alternative to conventional construction materials in many parts of the world. The applications of natural fibers extend beyond traditional uses, with advancements in technology enabling their integration into modern industries. Bamboo and hemp fibers are now being processed into textiles, offering eco-friendly alternatives to synthetic fabrics. These fibers are biodegradable, durable, and require fewer resources for cultivation, making them increasingly popular in sustainable fashion and home decor markets. Similarly, jute and coir are used in producing eco-friendly packaging materials, carpets, and geotextiles, contributing to the shift towards greener industrial practices. Natural dyes extracted from forest plants such as indigo, madder, and turmeric have been used for centuries in textile production, art, and cosmetics. These dyes are valued for their vibrant colors, biodegradability, and non-toxic nature. Indigo, for instance, is a primary dye in the production of denim, while turmeric is used to create bright yellow and golden hues in fabrics and food products. The revival of traditional dyeing techniques has gained momentum in recent years, driven by consumer demand for sustainable and artisanal goods. Natural dyes also play a crucial role in cultural preservation, as they are often integral to the traditional crafts and attire of indigenous communities.

The economic potential of fibers and natural dyes is significant, as these products cater to both local markets and international demand for sustainable and ethical goods. The collection, processing, and trade of these resources provide employment opportunities for rural and tribal populations, particularly women who are heavily involved in crafting and dyeing activities. However, the sustainability of these practices depends on regulated harvesting, equitable benefit-sharing, and market support to prevent resource depletion and ensure fair compensation for producers. By fostering innovations in fiber and dye production and promoting their integration into global markets, plant-based MFPs can continue to contribute to sustainable development while preserving the rich heritage of forest-based craftsmanship.

3. Animal-Derived Minor Forest Products

Animal-derived minor forest products (MFPs) are an essential component of forest ecosystems, encompassing resources harvested from or produced by forest-dwelling animals. These products, such as honey, beeswax, silk, lac, and wildlife-derived materials, hold immense cultural, economic, and ecological significance. They not only provide livelihoods for millions of forest-dependent communities but also play a critical role in biodiversity conservation and sustainable resource management.

3.1 Honey, Beeswax, and Other Apiculture Products

Honey, beeswax, and other apiculture products are among the most valuable animal-derived forest resources, offering a combination of nutritional, medicinal, and economic benefits. Forest honey, often harvested from wild bees, is particularly prized for its distinct flavors and health-enhancing properties, which are influenced by the biodiversity and flora of the region. Honey has been used for centuries as a natural sweetener and a remedy for various ailments, including coughs, wounds, and digestive issues. In addition to honey, products like beeswax, royal jelly, and propolis have wide-ranging applications in industries such as cosmetics, pharmaceuticals, and food processing.

The economic importance of apiculture products is profound, especially for rural and forest-dependent communities. Honey collection and beekeeping provide a sustainable source of income, with minimal environmental impact. Wild honey hunting is a traditional practice in many forested regions, supporting livelihoods and preserving cultural heritage. However, modern apiculture practices have increasingly been adopted to enhance productivity and ensure year-round harvesting. Beekeeping, or apiculture, also contributes to forest conservation, as bees are vital pollinators that support plant reproduction and biodiversity. Their activities enhance the productivity of both natural ecosystems and agricultural landscapes, creating a symbiotic relationship between forest health and human livelihoods. Sustainable apiculture practices are critical for maintaining bee populations, which face threats from habitat loss, pesticide use, and climate change. Promoting community-based beekeeping programs and integrating traditional knowledge with modern techniques can enhance the sustainability of apiculture while providing economic benefits to local populations. Certification schemes for organic and fair-trade honey further add value to these products, enabling access to premium markets and fostering ethical trade practices.

3.2 Silk and Lac Production from Forest Ecosystems

Silk and lac are two of the most significant animal-derived forest products, deeply embedded in the socio-economic and cultural fabric of many forest-dependent regions. These products are obtained from specific insect species that thrive in forested environments, and their production reflects a unique interplay between natural ecosystems and human ingenuity.

Silk Production

Silk, a luxurious natural fiber, is produced by silkworms reared on specific host plants found in forests. Non-mulberry silk varieties, such as tasar, muga, and eri, are particularly associated with forest ecosystems. Tasar silk is produced from the larvae of Antheraea mylitta, which feed on leaves of trees like Terminalia and Shorea species. Muga silk, native to Assam in India, is renowned for its golden sheen and is derived from the Antheraea assamensis silkworm, which feeds on som and sualu plants. Eri silk, also known as "peace silk," is produced from Philosamia ricini, whose larvae feed on castor plants. The economic and cultural importance of wild silk production cannot be overstated. It employs thousands of people, particularly women, in forest-dependent communities, who are involved in sericulture, spinning, and weaving. Wild silk products are highly valued in both domestic and international markets, with growing demand for eco-friendly and sustainable textiles. The cultivation of host plants for silkworms also contributes to forest conservation, as it integrates tree planting with income-generating activities.

Lac Production

Lac, a resin secreted by the lac insect (Kerria lacca), is another prominent animal-derived forest product with significant economic and industrial value. These insects are cultivated on host trees such as palas (Butea monosperma), kusum (Schleichera oleosa), and ber (Ziziphus mauritiana), which are commonly found in forested landscapes. The resin is processed into shellac, a versatile material used in industries such as furniture polishing, adhesives, and electronics. Lac dye, derived from the resin, is a natural coloring agent used in textiles, food, and cosmetics. Lac production provides livelihoods to many tribal and rural communities, particularly in India and Southeast Asia, where traditional lac cultivation practices are deeply rooted. It is an environmentally sustainable activity, as it promotes the planting and protection of host trees, thereby contributing to forest conservation. The global demand for natural and biodegradable products has further increased the economic potential of lac, with industries seeking sustainable alternatives to synthetic materials.

3.3 Wildlife-Based Products and Conservation Implications

Wildlife-based forest products encompass a diverse range of materials derived from forest-dwelling animals, including hides, feathers, bones, and other byproducts. These products have been used for centuries in traditional crafts, medicines, and rituals, reflecting their cultural and economic significance. However, the extraction and trade of wildlife-based products raise complex challenges, particularly in balancing resource utilization with biodiversity conservation. In traditional societies, wildlife-based products often serve as symbols of cultural identity and spiritual beliefs. For example, feathers from certain bird species are used in ceremonial attire, while bones and tusks are carved into ornaments and tools. These practices are often guided by indigenous knowledge systems that emphasize sustainable use and respect for nature. However, the commercialization of wildlife products has led to overexploitation and illegal trade, threatening the survival of many species.

The conservation implications of wildlife-based products are significant, as unsustainable harvesting can disrupt ecological balance and contribute to biodiversity loss. For instance, the illegal poaching of elephants for ivory or pangolins for their scales has had devastating impacts on these species' populations. Addressing these challenges requires a combination of strict enforcement of wildlife protection laws, community-based conservation initiatives, and the promotion of sustainable alternatives. Community-managed wildlife programs, where local populations are actively involved in monitoring and managing wildlife resources, have proven effective in reducing poaching and habitat destruction. Ecotourism, which generates income through wildlife observation and photography, provides an alternative to the exploitation of wildlife products while fostering awareness about conservation. By integrating economic incentives with ecological goals, wildlife-based products can be managed sustainably, ensuring that forests continue to support both biodiversity and human livelihoods.

4. Resin, Gum, and Latex

Resins, gums, and latex are among the most versatile minor forest products, offering a wide range of applications in industries such as adhesives, pharmaceuticals, varnishes, and cosmetics. These forest-derived substances are secreted by specific plants and trees, often as a defense mechanism against injury or pests. Their chemical properties make them invaluable raw materials in traditional and industrial settings. However, their sustainable extraction, market development, and resource management are critical to ensure long-term availability and ecological balance.

4.1 Tapping Techniques and Sustainable Harvesting

Tapping is the primary method used to extract resins, gums, and latex from trees. This process involves creating incisions or wounds in the bark of trees to stimulate the secretion of these substances. Proper tapping techniques are crucial for ensuring the health and longevity of the trees, as well as for maximizing yield without causing irreversible damage.

Tapping for Resins

Resins are viscous substances secreted by trees such as pine, spruce, and fir. The tapping process typically involves making V-shaped cuts on the tree trunk and collecting the exuded resin in containers. Traditional methods of resin tapping often result in excessive wounding of trees, leading to reduced productivity and increased vulnerability to pests and diseases. Sustainable techniques, such as controlled tapping intervals and minimal incision sizes, have been developed to address these issues. Additionally, resin tapping is often complemented by forest management practices, such as selective thinning and pest control, to enhance tree health and resin yields.

Tapping for Gums

Gums, such as gum arabic and gum karaya, are water-soluble polysaccharides secreted by trees like Acacia and Sterculia species. The tapping process for gums involves making shallow incisions in the bark to stimulate exudation. Over-tapping can deplete tree vitality and reduce gum quality, necessitating the adoption of sustainable practices. Rotational tapping systems, where different sections of a tree are tapped over successive seasons, allow the trees to recover and maintain their productivity. These practices are often guided by traditional knowledge, especially in regions where gum tapping has been practiced for generations.

Tapping for Latex

Latex, a milky fluid secreted by rubber trees (Hevea brasiliensis) and other plants, is the raw material for natural rubber and various other products. The tapping of rubber trees involves making spiral or herringbone-shaped cuts on the trunk and collecting the latex in small cups. Latex tapping requires precision to avoid damaging the cambium layer of the tree, which can lead to reduced yields or tree mortality. Sustainable latex tapping practices include limiting the frequency of tapping and ensuring adequate rest periods for trees, allowing them to regenerate and maintain long-term productivity. The adoption of sustainable tapping techniques is essential for preserving forest resources and ensuring the continued availability of resins, gums, and latex. Training programs for tappers, the use of improved tools, and community-based forest management initiatives are critical for promoting sustainable practices. These approaches not only enhance the economic value of these products but also contribute to forest conservation by minimizing environmental degradation.

4.2 Uses in Industries: Adhesives, Varnishes, and Pharmaceuticals

Resins, gums, and latex are integral to a wide range of industries, owing to their unique chemical and physical properties. These substances are used as raw materials in the production of adhesives, varnishes, pharmaceuticals, and numerous other products, highlighting their versatility and economic importance.

Adhesives and Sealants

Resins and gums are key components in the formulation of adhesives and sealants. Pine resin, for example, is processed into rosin, which is widely used in the production of paper adhesives, rubber-based glues, and sealants. Gum arabic, known for its excellent binding properties, is used in adhesives for stamps, envelopes, and industrial packaging. The natural tackiness and durability of these substances make them indispensable in applications requiring strong and long-lasting bonds.

Varnishes and Coatings

Resins, particularly those derived from coniferous trees, are primary ingredients in varnishes and coatings. Rosin and shellac are used to create protective finishes for wood, metal, and other surfaces, enhancing their durability and aesthetic appeal. These substances are also used in the production of paints, where they improve gloss, adhesion, and resistance to wear. Latex-based coatings, derived from rubber trees, are widely used in the construction industry for waterproofing and weatherproofing applications.

Pharmaceuticals and Personal Care Products

Gums and latex have significant applications in the pharmaceutical and personal care industries. Gum Arabic and gum karaya are used as stabilizers, emulsifiers, and binders in pharmaceutical formulations, including tablets, syrups, and ointments. Latex, particularly from plants like Papaver somniferum (opium poppy), is the source of medically important compounds such as morphine and codeine, which are used for pain management. Resins like frankincense and myrrh have antimicrobial and anti-inflammatory properties, making them valuable ingredients in traditional medicine and modern skincare products. The industrial uses of resins, gums, and latex underscore their economic value and highlight the need for sustainable sourcing practices. As demand for these products continues to grow, ensuring the ecological sustainability of their production is critical for meeting industrial needs without depleting forest resources.

4.3 Challenges in Market Development and Resource Management

Despite their economic and industrial importance, the market development and resource management of resins, gums, and latex face several challenges. These challenges include inconsistent quality, limited access to markets, resource overexploitation, and the impact of climate change on forest ecosystems.

Market Development

The market for resins, gums, and latex is often fragmented, with a significant portion of production handled by small-scale collectors and traders. The lack of organized supply chains and quality control mechanisms hampers the competitiveness of these products in both domestic and international markets. For example, variability in the quality of gum arabic and pine resin affects their usability in industrial applications, limiting their market potential. Additionally, the absence of certification schemes for sustainably sourced resins and gums restricts access to premium markets that value ethical and eco-friendly products.

Resource Overexploitation

The increasing demand for resins, gums, and latex has led to overexploitation of forest resources, particularly in regions where these products are a primary source of income for local communities. Over-tapping and unsustainable harvesting practices not only reduce tree productivity but also threaten the long-term viability of these resources. In some cases, excessive extraction has resulted in the decline of key species, disrupting forest ecosystems and biodiversity.

Impact of Climate Change

Climate change poses a significant threat to the availability and quality of resins, gums, and latex. Changes in temperature and precipitation patterns can affect tree growth, resin secretion rates, and latex yields, reducing the productivity of these forest products. Additionally, the increased frequency of extreme weather events, such as droughts and storms, can damage forest ecosystems, further exacerbating resource scarcity. Addressing these challenges requires a multi-pronged approach that includes improving market access, promoting sustainable harvesting practices, and investing in research and development. Establishing cooperatives and producer organizations can help small-scale collectors gain better access to markets and negotiate fair prices. Certification schemes and traceability systems can enhance consumer confidence and create incentives for sustainable production. Furthermore, integrating climate resilience strategies into forest management plans is essential for mitigating the impacts of climate change on resins, gums, and latex.

5. Bamboo and Rattan

Bamboo and rattan are among the most versatile and renewable forest resources, playing an integral role in construction, furniture, handicrafts, and various industrial applications. Their rapid growth, adaptability, and sustainability make them essential for rural livelihoods, ecological balance, and global markets. These resources have been utilized for centuries, providing materials for traditional practices and modern innovations alike.

5.1 Uses in Construction, Furniture, and Handicrafts

Bamboo and rattan are widely used in construction, furniture making, and handicrafts due to their strength, flexibility, and aesthetic appeal. These materials have been central to traditional practices across Asia, Africa, and Latin America and are increasingly being adopted in modern sustainable designs.

Construction Applications

Bamboo is often referred to as "green steel" for its remarkable strength-to-weight ratio, making it an eco-friendly alternative to conventional building materials like timber, steel, and concrete. It is used for structural components, scaffolding, roofing, and flooring in both traditional and contemporary architecture. Bamboo's tensile strength and resilience to earthquakes have made it a popular choice in disaster-prone regions. Modern innovations have further expanded its use, with engineered bamboo products like laminated bamboo and bamboo composites being utilized in bridges, high-rise buildings, and prefabricated housing. Rattan, though less commonly used in large-scale construction, is often employed for decorative elements and structural reinforcements in traditional architecture. Its pliability and durability make it suitable for creating intricate designs in ceilings, partitions, and screens.

Furniture and Handicrafts

Bamboo and rattan are indispensable materials in furniture making, valued for their lightweight, durability, and versatility. They are used to craft chairs, tables, beds, and cabinets that combine functionality with aesthetic appeal. Rattan, in particular, is highly sought after for its smooth texture and ability to be bent into intricate shapes, making it ideal for artisanal furniture and decor. In the realm of handicrafts, bamboo and rattan are used to create baskets, mats, utensils, and decorative items. These products are often handcrafted by skilled artisans, reflecting the cultural heritage and traditional craftsmanship of forest-dependent communities. The global demand for eco-friendly and handmade goods has further boosted the market for bamboo and rattan handicrafts, creating opportunities for income generation and cultural preservation.

Innovations and Industrial Uses

Beyond traditional uses, bamboo and rattan are finding applications in modern industries. Bamboo fibers are processed into textiles, offering a sustainable alternative to synthetic fabrics. Bamboo-based bioenergy, including charcoal and ethanol, is being developed as a renewable energy source. Similarly, rattan is used in the production of veneers, composites, and other industrial materials. These innovations underscore the adaptability of bamboo and rattan, highlighting their potential to address contemporary challenges in sustainable development.

5.2 Economic and Ecological Significance of Bamboo and Rattan

The economic and ecological significance of bamboo and rattan extends far beyond their material uses, encompassing their role in livelihoods, biodiversity, and environmental sustainability.

Economic Contributions

Bamboo and rattan industries provide employment and income for millions of people, particularly in rural and forest-dependent regions. Small-scale enterprises based on these resources support artisans, carpenters, and traders, fostering economic development at the local level. For instance, the bamboo industry in China contributes significantly to the country's GDP, supporting a wide range of products from furniture to paper. Similarly, rattan exports from Southeast Asia generate substantial foreign exchange earnings, benefiting national economies. The global demand for sustainable materials has opened new markets for bamboo and rattan products, particularly in eco-conscious regions such as Europe and North America. Certified products, such as those labeled under the Forest Stewardship Council (FSC), command premium prices, incentivizing sustainable practices. The integration of bamboo and rattan into global value chains has also created opportunities for developing countries to diversify their economies and reduce poverty.

Ecological Importance

Bamboo and rattan play a crucial role in maintaining ecological balance. Bamboo's rapid growth and extensive root system make it highly effective in soil stabilization and erosion control, particularly in degraded or hilly terrains. It also acts as a carbon sink, sequestering significant amounts of carbon dioxide and contributing to climate change mitigation. Rattan, as a climber, supports forest regeneration by enhancing biodiversity and providing habitat for wildlife. The cultivation and sustainable management of bamboo and rattan contribute to forest conservation by reducing the pressure on timber resources. These plants can be grown in degraded lands and agroforestry systems, promoting ecosystem restoration and improving soil fertility. Additionally, bamboo and rattan-based products are biodegradable and renewable, aligning with global efforts to reduce plastic waste and transition to a circular economy.

5.3 Strategies for Sustainable Management and Cultivation

Ensuring the sustainability of bamboo and rattan resources is critical to meeting the growing demand for these materials while preserving their ecological and economic benefits. Strategies for sustainable management and cultivation involve a combination of policy support, community engagement, and technological innovation.

Sustainable Harvesting Practices Regulating harvesting practices is essential to prevent overexploitation and resource depletion. For bamboo, this includes selective harvesting, where mature culms are cut while younger shoots are left to grow. This approach ensures continuous regeneration and maintains the health of bamboo stands. Similarly, sustainable rattan harvesting involves rotational cutting and the preservation of seed-bearing plants to support natural regeneration.

Community-Based Management Engaging local communities in the management and cultivation of bamboo and rattan is crucial for ensuring their sustainable use. Community-based forestry initiatives provide local populations with the knowledge, tools, and incentives to manage these resources responsibly. Such programs often include capacity-building activities, such as training in sustainable harvesting techniques, value addition, and marketing.

Cultivation and Agroforestry Expanding the cultivation of bamboo and rattan in degraded lands and agroforestry systems can enhance their availability while supporting ecosystem restoration. Bamboo plantations, for instance, can be integrated with crops and livestock to create diversified farming systems that improve productivity and resilience. Rattan cultivation in secondary forests or mixed-species plantations can enhance biodiversity and provide additional income streams for farmers.

Market Development and Certification Improving market access and establishing certification systems are vital for promoting sustainable bamboo and rattan products. Certification schemes like FSC ensure that these resources are sourced responsibly, enhancing their value in international markets. Strengthening value chains, providing access to financing, and supporting small and medium-sized enterprises (SMEs) can further boost the economic potential of bamboo and rattan industries.

Research and Innovation Investing in research and innovation is critical for unlocking the full potential of bamboo and rattan. This includes developing high-yielding and disease-resistant varieties, improving processing technologies, and exploring new applications for these materials. Partnerships between research institutions, governments, and industries can drive innovation and ensure that bamboo and rattan remain competitive in evolving markets. The sustainable management and cultivation of bamboo and rattan not only addresses the growing demand for these resources but also contributes to environmental conservation, poverty alleviation, and sustainable development. By adopting integrated approaches that balance economic, social, and ecological considerations, bamboo and rattan can serve as models for renewable resource utilization in a rapidly changing world.

6. Essential Oils and Aromatics

Essential oils and aromatics are some of the most valuable and versatile minor forest products, deeply intertwined with traditional practices and modern industries. Derived from various parts of plants, including leaves, flowers, bark, and roots, these highly concentrated volatile compounds are prized for their therapeutic properties, aromatic qualities, and diverse applications. Essential oils form the cornerstone of several industries, including cosmetics, perfumery, and wellness, while also serving as raw materials for pharmaceuticals, food flavoring, and eco-friendly cleaning products. The growing demand for natural and sustainable products has significantly boosted the global market for essential oils, making their cultivation, extraction, and commercialization critical to both rural livelihoods and economic development. This section provides an in-depth exploration of extraction techniques and major sources, applications in key industries, and the immense opportunities in global markets.

6.1 Extraction Techniques and Major Sources

The extraction of essential oils is a highly specialized process that requires precision and expertise to preserve the delicate aromatic compounds present in plant materials. The choice of extraction technique depends on the type of plant material, the intended use of the oil, and the required purity and quality. The most commonly employed methods include steam distillation, cold pressing, and solvent extraction, each with unique advantages and limitations.

Steam Distillation Steam distillation is the most widely used method for extracting essential oils, particularly from aromatic plants like eucalyptus, lavender, and peppermint. In this process, steam is passed through the plant material, causing the volatile compounds to vaporize. The vapor is then condensed back into liquid form, separating the essential oil from the water. This method is favored for its ability to preserve the integrity of the oil's aromatic compounds, ensuring high quality and therapeutic efficacy. However, the process can be resource-intensive, requiring significant amounts of water and energy.

Cold-Pressing Cold pressing is typically used for citrus oils, such as orange, lemon, and bergamot, which are extracted from the peels of fruits. This mechanical process involves pressing the peels to release the oils, followed by filtration to remove impurities. Cold pressing is valued for its simplicity and ability to retain the oil's natural aroma and nutritional properties. However, the yield is often lower compared to other methods, making it less economical for large-scale production.

Solvent Extraction Solvent extraction is employed for delicate plant materials, such as jasmine and rose petals, which are sensitive to heat. In this method, a solvent is used to dissolve the aromatic compounds, which are then separated and purified to produce an absolute, a concentrated form of essential oil. While solvent extraction is effective in capturing the full spectrum of aromatic compounds, concerns about solvent residues and environmental impact have prompted the development of more sustainable alternatives, such as supercritical $CO2$ extraction. The sources of essential oils are as diverse as the plants from which they are derived, reflecting the ecological richness of forests and cultivated landscapes. Trees like sandalwood and cedarwood yield woody, grounding oils used in perfumery and aromatherapy. Herbs like basil, thyme, and oregano provide essential oils with potent antimicrobial properties, making them valuable in natural remedies and cleaning products. Floral oils, such as rose and jasmine, are cherished for their luxurious fragrance and skin-nourishing properties, while citrus oils are widely used for their uplifting scent and versatile applications. Forest ecosystems, in particular, are a treasure trove of essential oil sources, supporting the livelihoods of communities involved in their collection and processing.

6.2 Applications in Cosmetics, Perfumery, and Wellness Industries

The applications of essential oils span a wide range of industries, with cosmetics, perfumery, and wellness being among the most prominent sectors. The unique properties of essential oils, including their fragrance, therapeutic benefits, and natural composition, make them indispensable ingredients in these industries, catering to the growing demand for sustainable and plant-based products.

Cosmetics and Skincare In the cosmetics industry, essential oils are widely used for their skin-enhancing and aromatic properties. Oils like tea tree and lavender are renowned for their antimicrobial and soothing effects, making them common ingredients in acne treatments, moisturizers, and cleansers. Rosehip and argan oils are prized for their anti-aging properties, promoting skin elasticity and reducing the appearance of wrinkles. The natural origin of these oils aligns with consumer preferences for clean beauty products that are free from synthetic additives and harmful chemicals. Essential oils are also integral to the formulation of haircare products, such as shampoos, conditioners, and scalp treatments. Oils like peppermint and rosemary stimulate hair growth and improve scalp health, while chamomile and ylang-ylang provide conditioning and shine-enhancing effects. These oils not only enhance product efficacy but also contribute to the sensory experience of using cosmetic products.

Perfumery

The perfumery industry is perhaps the most iconic application of essential oils, where their complex and captivating aromas are used to create signature fragrances. Essential oils form the base, middle, and top notes of perfumes, contributing to their longevity, depth, and appeal. Woody oils like sandalwood and vetiver add warmth and earthiness to fragrances, while floral oils like rose and jasmine provide a romantic and luxurious touch. Citrus oils, such as bergamot and grapefruit, are valued for their fresh and uplifting scent, making them popular in both men's and women's perfumes. The use of essential oils in perfumery is not limited to high-end products; they are also used in natural and therapeutic fragrances designed for aromatherapy. These fragrances not only evoke emotions and memories but also offer relaxation, stress relief, and mood enhancement, aligning with the wellness-focused preferences of modern consumers.

Wellness and Aromatherapy

The wellness industry has embraced essential oils as powerful tools for promoting physical and mental well-being. Aromatherapy, the therapeutic use of essential oils, has gained widespread popularity for its ability to address stress, anxiety, insomnia, and other common ailments. Oils like eucalyptus and peppermint are used for respiratory health, while lavender and chamomile are known for their calming effects. Diffusers, massage oils, and bath products infused with essential oils are staples in wellness routines, offering a holistic approach to health and relaxation. Essential oils are also used in complementary and alternative medicine, where their antimicrobial, anti-inflammatory, and analgesic properties are leveraged to support healing. For example, clove oil is used as a natural remedy for toothache, while ginger oil alleviates muscle pain and digestive discomfort. The integration of essential oils into wellness practices reflects their enduring value as nature's pharmacy, addressing a wide range of health concerns safely and naturally.

6.3 Opportunities in Global Markets for Essential Oils

The global market for essential oils has witnessed exponential growth in recent years, driven by increasing consumer awareness of natural and sustainable products, advancements in extraction technologies, and expanding applications across industries. Essential oils are no longer niche commodities; they are now integral to mainstream markets, with significant opportunities for producers, exporters, and entrepreneurs.

Rising Demand for Natural Products

The shift towards natural and organic products has been a major driver of the essential oils market. Consumers are increasingly seeking alternatives to synthetic chemicals, favoring plant-based solutions for cosmetics, wellness, and household care. This trend has created opportunities for small-scale producers and forest-dependent communities to enter lucrative markets, provided they meet the quality and certification standards demanded by international buyers.

Export Potential

Countries with rich biodiversity and established essential oil industries, such as India, Indonesia, and Madagascar, are key players in the global market. Exports of essential oils like sandalwood, patchouli, and ylang-ylang contribute significantly to foreign exchange earnings and rural development. However, ensuring sustainable sourcing and fair trade practices is critical for maintaining market access and consumer trust.

Value Addition and Niche Markets

Value addition through processing and packaging can significantly enhance the profitability of essential oils. Products like essential oil blends, roll-ons, and diffuser kits cater to niche markets that prioritize convenience and experiential benefits. Furthermore, the growing interest in wellness tourism and aromatherapy workshops presents opportunities for experiential marketing, allowing producers to showcase the heritage and sustainability of their products.

7. Non-Timber Products in Livelihoods and Local Economies

Non-timber forest products (NTFPs) are a critical lifeline for millions of rural and forest-dependent communities worldwide. These products, encompassing a diverse range of resources such as fruits, nuts, honey, resins, medicinal plants, and fibers, contribute significantly to subsistence, income generation, and community development. Beyond their economic value, NTFPs also play a crucial role in gender empowerment, cultural preservation, and sustainable development.

7.1 Role in Supporting Subsistence and Income Generation

NTFPs are indispensable to the daily lives of rural and indigenous populations, providing essential resources for food, medicine, and shelter. These products often serve as a buffer during times of economic hardship, agricultural off-seasons, or natural disasters, ensuring food security and income stability for vulnerable households. Wild fruits, vegetables, nuts, and honey are among the most commonly collected NTFPs, offering vital nutrition while also generating income through local markets.

For many communities, NTFPs are a primary source of livelihood. Collecting and selling forest products like mushrooms, bamboo shoots, and medicinal herbs provide cash income, enabling families to meet their basic needs, invest in education, and access healthcare. Seasonal harvesting of high-value products, such as wild honey or tendu leaves (used in traditional cigarette production), contributes significantly to household earnings. These activities often require minimal initial investment and can be undertaken using traditional knowledge and tools, making them accessible even to marginalized groups.

The economic impact of NTFPs extends beyond individual households to the community level. Small-scale processing industries based on forest products, such as honey extraction units, basket weaving cooperatives, and herbal medicine preparation centers, create employment opportunities and stimulate local economies. In addition to direct income, these industries generate multiplier effects, fostering entrepreneurship and strengthening rural value chains. However, the sustainability of these benefits depends on equitable access to resources, effective management practices, and support for capacity-building initiatives that enhance the productivity and marketability of NTFPs.

7.2 Contribution of MFPs to Gender Empowerment and Community Well-Being

The collection, processing, and trade of NTFPs are often dominated by women, making these activities a powerful driver of gender empowerment and social inclusion. Women are key actors in the NTFP value chain, performing tasks such as harvesting, sorting, and selling products in local markets. This involvement not only provides them with independent income but also enhances their decision-making power within households and communities.

NTFP-related activities are particularly significant in empowering women in remote and forest-dependent regions, where formal employment opportunities are limited. Income generated from the sale of forest products enables women to invest in their children's education, improve family nutrition, and contribute to household expenses. In addition to economic empowerment, these activities often strengthen women's social networks and leadership roles within community organizations, fostering collective action and advocacy for resource rights.

The social benefits of NTFPs extend to community well-being, as these resources support traditional practices, cultural identity, and social cohesion. For example, the use of medicinal plants in traditional healing systems not only addresses healthcare needs but also preserves indigenous knowledge and cultural heritage. Similarly, the crafting of bamboo and rattan products sustains artisanal skills that are passed down through generations, enriching cultural diversity and community pride. Promoting gender equity and community well-being in the NTFP sector requires targeted interventions, such as capacity-building programs, access to credit, and market linkages for women-led enterprises. Policies that recognize and support the contributions of women and marginalized groups to the forest economy can enhance the inclusivity and sustainability of NTFP-based livelihoods.

7.3 Integration of MFPs in Local and Global Value Chains

The integration of NTFPs into local and global value chains represents a significant opportunity to enhance their economic impact and sustainability. Value chains for NTFPs encompass all stages of production, from collection and processing to distribution and marketing, and offer pathways for forest-dependent communities to connect with larger markets and benefit from fair trade practices. At the local level, NTFPs are often sold in informal markets or through direct transactions with traders. While these markets provide immediate income, they are typically characterized by low prices, limited bargaining power for producers, and vulnerability to exploitation by intermediaries. Strengthening local value chains through cooperatives, producer organizations, and capacity-building initiatives can improve the efficiency and equity of these transactions. For instance, training in value addition, such as drying, packaging, or processing, can enhance the quality and marketability of NTFPs, enabling producers to capture a greater share of the final value.

In the global context, the demand for sustainably sourced and ethically traded NTFPs has created new opportunities for forest-dependent communities. Products such as shea butter, wild honey, essential oils, and natural dyes are increasingly sought after in international markets for cosmetics, wellness, and food industries. Certification schemes like FairWild and Forest Stewardship Council (FSC) have played a critical role in facilitating access to these markets by ensuring that NTFPs meet environmental and social standards. These certifications not only add value to forest products but also promote sustainable harvesting practices and equitable benefit-sharing.

Integrating NTFPs into global value chains requires addressing key challenges, such as quality control, market access, and infrastructure development. Investments in transport, storage, and processing facilities can reduce post-harvest losses and enhance the competitiveness of NTFPs. Strengthening market linkages through partnerships with private companies, exporters, and e-commerce platforms can expand market reach and create long-term opportunities for producers. Additionally, fostering innovation and research in the NTFP sector can uncover new applications and value streams, driving growth and sustainability. By recognizing the economic, social, and environmental value of NTFPs and addressing the challenges associated with their integration into value chains, these products can continue to support livelihoods, empower communities, and contribute to sustainable development at local and global levels.

8. Sustainability and Conservation of Minor Forest Products

The sustainability and conservation of minor forest products (MFPs) are critical to ensuring the long-term viability of forest ecosystems and the communities that depend on them. These products, ranging from medicinal plants and fibers to resins and honey, provide essential resources for rural livelihoods and global markets. However, unsustainable harvesting, over-exploitation, and habitat destruction pose significant threats to their availability and ecological integrity. Addressing these challenges requires a holistic approach that integrates sustainable harvesting practices, leverages indigenous knowledge systems, and mitigates the impacts of over-exploitation and habitat loss.

8.1 Sustainable Harvesting Practices for Long-Term Resource Availability

Sustainable harvesting practices are essential for maintaining the availability of MFPs while preserving the health and productivity of forest ecosystems. These practices aim to minimize ecological disruption, ensure the regeneration of resources, and balance the needs of current and future generations.

Selective and Rotational Harvesting

Selective harvesting involves the careful collection of specific parts of plants, such as fruits, leaves, or bark, while leaving the rest of the plant intact to continue growing and reproducing. For example, in the case of medicinal plants, only mature leaves or flowers are harvested, ensuring that the plant retains its ability to regenerate. Rotational harvesting further enhances sustainability by allowing areas or species to rest and recover between harvesting cycles. This approach prevents over-harvesting and supports the natural growth and reproductive cycles of forest species.

Community-Based Harvesting Guidelines

In many forest-dependent communities, the adoption of community-based guidelines for harvesting ensures that practices are aligned with ecological principles. These guidelines often specify the timing, quantity, and methods of collection, reducing the risk of resource depletion. For instance, honey collectors may use non-destructive methods to extract honey, leaving enough for bees to sustain their colonies. Similarly, resin tappers may adopt techniques that minimize wounding of trees, promoting continued resin production and tree health.

Certification and Sustainable Trade

Certification schemes, such as FairWild and the Forest Stewardship Council (FSC), promote sustainable harvesting practices by setting standards for the collection, processing, and trade of MFPs. Certified products often command higher prices in markets, providing economic incentives for producers to adopt sustainable practices. These certifications also support traceability, ensuring that consumers can make informed choices and contribute to conservation efforts through their purchasing decisions.

Capacity Building and Training

Training programs for harvesters and communities play a vital role in disseminating knowledge about sustainable harvesting techniques. These programs often include practical demonstrations, workshops, and tools that help communities optimize resource use while protecting forest ecosystems. Capacity-building initiatives also enhance the economic value of MFPs by introducing value addition techniques, such as drying, grinding, or packaging, which increase the profitability of these resources.

8.2 Role of Indigenous Knowledge in MFP Management

Indigenous knowledge systems are invaluable in the sustainable management and conservation of MFPs. Rooted in centuries of observation and interaction with forest ecosystems, this knowledge encompasses practices, beliefs, and skills that contribute to the responsible use of forest resources.

Traditional Ecological Knowledge

Indigenous communities possess a deep understanding of the ecological relationships that sustain forest resources. This knowledge often includes detailed information about the life cycles, habitat requirements, and medicinal properties of specific plants and animals. For example, indigenous healers are adept at identifying medicinal plants, understanding their optimal harvesting times, and preparing remedies that maximize their efficacy. Similarly, traditional practices for collecting resins, fibers, and wild foods often prioritize resource sustainability and ecosystem health.

Sacred Groves and Cultural Practices

Many indigenous communities protect specific areas of forests as sacred groves, where harvesting is restricted or prohibited. These groves often serve as reservoirs of biodiversity, preserving rare or endangered species and maintaining the ecological balance of the surrounding landscape. Cultural practices, such as seasonal festivals and rituals, also play a role in regulating resource use, aligning harvesting activities with ecological cycles and fostering community-wide stewardship.

Integration with Modern Conservation Efforts

Indigenous knowledge is increasingly being recognized and integrated into modern conservation efforts. Collaborative approaches, such as participatory forest management, bring together indigenous communities, conservationists, and policymakers to develop strategies that combine traditional practices with scientific methods. This integration enhances the effectiveness of conservation initiatives while empowering indigenous communities to maintain their cultural heritage and livelihoods.

8.3 Addressing Over-Exploitation and Habitat Loss

Over-exploitation and habitat loss are among the most significant threats to the sustainability of MFPs, driven by factors such as commercial demand, population pressure, and deforestation. Addressing these challenges requires targeted interventions that balance resource utilization with conservation.

Monitoring and Regulation

Establishing monitoring systems to track the extraction and trade of MFPs is essential for preventing over-exploitation. These systems can include community-based monitoring programs, where local groups assess the availability and health of resources, as well as government-led initiatives that enforce quotas and licensing for commercial harvesting. Regulations that limit the collection of rare or endangered species, combined with penalties for illegal harvesting, can also deter unsustainable practices.

Habitat Restoration and Protection

Restoring degraded habitats and protecting critical ecosystems are vital for maintaining the availability of MFPs. Reforestation programs, agroforestry systems, and the establishment of protected areas help preserve the habitats of key species while enhancing the overall resilience of forest ecosystems. For example, planting host trees for lac insects or medicinal plants can simultaneously restore degraded lands and provide economic benefits for local communities.

Alternative Livelihoods

Providing alternative livelihood options for forest-dependent populations can reduce the pressure on MFPs and their habitats. Initiatives such as ecotourism, agroforestry, and small-scale manufacturing offer sustainable income sources while promoting conservation. For instance, community-based ecotourism projects that showcase the cultural and ecological value of MFPs create economic incentives for conservation, benefiting both people and nature.

Climate Change Mitigation

Climate change exacerbates the challenges of over-exploitation and habitat loss by altering the distribution, growth, and productivity of forest resources. Mitigating these impacts requires integrating climate resilience strategies into forest management plans, such as selecting climate-adaptive species for reforestation and diversifying resource use. Research and innovation can also play a role in developing climate-smart harvesting techniques and improving the resilience of forest-dependent communities.

Conclusion

Minor forest products (MFPs), though often overlooked in comparison to major forest resources like timber, are indispensable to both local economies and global industries. They serve as a vital link between human well-being and ecological health, supporting millions of rural and indigenous livelihoods while playing a crucial role in biodiversity conservation. Their diverse applications, ranging from food and medicine to industrial raw materials, demonstrate their significance as a renewable, versatile, and sustainable resource base. The importance of MFPs lies not only in their economic potential but also in their ability to bridge traditional knowledge and modern markets. Rural and indigenous communities have long depended on these products for subsistence and income generation, often preserving unique practices for their collection and processing. This deep-rooted knowledge is invaluable for sustainable harvesting, as it ensures the longevity of resources while respecting the balance of forest ecosystems. Additionally, MFPs contribute to cultural heritage, reflecting the intricate connections between people, forests, and their environment.

Plant-based MFPs such as fruits, nuts, seeds, medicinal plants, and fibers highlight the diverse contributions of forests to human life. These products provide essential nutrients, raw materials for industries, and therapeutic compounds for traditional and modern medicine. Their economic value is significant, supporting local economies and fostering international trade in sectors such as pharmaceuticals, cosmetics, and natural foods. Similarly, fibers and dyes derived from forest plants continue to play a key role in handicrafts and industrial applications, blending traditional practices with contemporary innovations. Animal-derived MFPs, including honey, beeswax, silk, and lac, further emphasize the multifaceted utility of forests. These products not only sustain industries but also support critical ecological processes such as pollination, which is vital for global food security. The careful integration of wildlife-based products into forest management plans ensures the dual benefit of conserving biodiversity while creating sustainable livelihoods. However, these practices must be balanced with robust conservation efforts to prevent over-exploitation and protect vulnerable species.

Resins, gums, and latex represent a specialized segment of MFPs with extensive industrial applications, from adhesives and varnishes to pharmaceuticals. While these products are highly valuable, their sustainable harvesting poses significant challenges. The development of efficient tapping techniques, coupled with policies promoting fair market access, can unlock their potential while safeguarding forest ecosystems. The focus on market development, combined with resource management, ensures that these products contribute to economic growth without depleting natural reserves.

Bamboo and rattan, as fast-growing and renewable resources, underscore the ecological and economic significance of MFPs. Their widespread use in construction, furniture, and crafts illustrates their versatility and importance in both rural and global markets. Sustainable management strategies for these resources are critical for maximizing their benefits while maintaining their role in carbon sequestration and soil stabilization. Promoting value-added products through branding and processing can further enhance the market potential of bamboo and rattan. Essential oils and aromatic compounds extracted from forest plants represent another thriving sector of MFPs. Their applications in wellness, perfumery, and cosmetics industries highlight their growing relevance in global markets. The increasing demand for natural and organic products provides opportunities for forest-dependent communities to integrate these resources into international value chains. Sustainable practices and innovations in extraction and processing are vital for scaling up their production while minimizing environmental impact.

Despite their vast potential, MFPs face significant challenges, including limited market access, lack of fair pricing for local collectors, and the adverse effects of climate change. Addressing these issues requires policy interventions, capacity building, and investment in research and infrastructure. Indigenous knowledge, combined with scientific advancements, can play a pivotal role in overcoming these challenges, ensuring that MFPs contribute to both local resilience and global sustainability. The future of MFPs lies in their integration into sustainable development frameworks, aligning with global goals such as the United Nations Sustainable Development Goals (SDGs). Enhancing value addition through processing and branding, fostering research and innovation, and promoting sustainable harvesting practices are essential steps in unlocking their full potential. By addressing challenges such as over-exploitation and habitat loss, MFPs can become a cornerstone of sustainable forest management, balancing economic growth with ecological preservation.

Chapter - 15

Medicinal Plants and Their Applications

1. Introduction to Medicinal Plants

Medicinal plants have been a cornerstone of human health and healing for millennia. From ancient civilizations to modern-day pharmaceutical laboratories, these plants have provided the raw materials for treatments and remedies used to prevent, manage, and cure diseases. The significance of medicinal plants lies not only in their therapeutic properties but also in their deeply intertwined role within the cultural, ecological, and economic spheres of societies worldwide. With the rise of industrialization and the increasing reliance on synthetic drugs, the role of medicinal plants has evolved, yet they continue to serve as a valuable resource in both traditional and modern medicine. Medicinal plants are those plant species that possess natural chemical compounds that provide therapeutic effects on the human body. These plants have been used throughout history for a wide range of health-related purposes, from treating minor ailments to chronic diseases. The importance of medicinal plants lies in their diverse range of biologically active compounds, such as alkaloids, terpenoids, flavonoids, and glycosides, which can exert powerful pharmacological effects. In addition to their therapeutic potential, medicinal plants often play a role in providing cost-effective and accessible healthcare options for communities, particularly in rural areas where access to modern pharmaceuticals may be limited. Beyond their practical applications in healthcare, medicinal plants are essential for their cultural and spiritual significance. Many indigenous cultures around the world have developed elaborate systems of plant-based medicine, often rooted in centuries of experiential knowledge passed down through generations. The wisdom embedded in these traditions continues to shape contemporary approaches to natural healing. Furthermore, medicinal plants contribute to global biodiversity, fostering rich ecosystems that support numerous other species, including those essential to human life. The environmental and ecological value of medicinal plants cannot be overstated, as they form integral parts of the natural world that humans depend upon for food, shelter, and other necessities.

Throughout history, the utilization of medicinal plants has been deeply rooted in the traditions of many civilizations. From the ancient Egyptians and Greeks to the indigenous peoples of the Americas and Africa, plants have been revered not only for their medicinal properties but also for their role in spiritual and ritualistic practices. The historical significance of medicinal plants is vast, with written records from various ancient cultures providing evidence of the longstanding use of plant-based remedies. The ancient Egyptians, for instance, documented the use of myrrh, frankincense,

DOI: 10.1201/9781042011063-15

and aloe vera for healing purposes. Similarly, in traditional Chinese medicine, herbs like ginseng and ginger have been used for thousands of years to treat various ailments. These historical uses laid the groundwork for modern pharmacology, where many drugs are derived from plants, continuing the legacy of plant-based medicine into the present day.

The global distribution and biodiversity of medicinal plants are extraordinary. From the tropical rainforests of South America to the arid deserts of Africa, plants with medicinal properties are found across the globe. Biodiversity hotspots, such as the Amazon rainforest, the Himalayas, and Southeast Asia, are home to a vast array of plant species that have not yet been fully explored for their potential medicinal uses. This diversity is not only a testament to the richness of life on Earth but also a source of immense potential for discovering new drugs and therapies. For instance, the search for anticancer, anti-inflammatory, and antiviral compounds continues to yield promising results from plant-based sources. However, with the increasing loss of biodiversity due to habitat destruction and climate change, the preservation of medicinal plant species has become a critical issue.

The therapeutic properties of medicinal plants are largely attributed to the bioactive compounds they contain. For example, alkaloids, found in plants such as morphine and quinine, have potent effects on pain relief and the treatment of malaria, respectively. Flavonoids, commonly found in fruits like citrus and berries, possess antioxidant properties that contribute to cardiovascular health. Terpenoids, found in essential oils, are known for their anti-inflammatory, antimicrobial, and anticancer effects. These compounds, and many others, form the basis of the pharmacological applications of medicinal plants. In modern healthcare, medicinal plants continue to play a crucial role in the treatment of a wide range of conditions, from chronic diseases like diabetes and cancer to more common ailments such as colds and headaches.

The role of medicinal plants in modern pharmaceuticals is undeniable. Many of the drugs used today have been derived from plant compounds, with aspirin, digitalis, and taxol being notable examples of plant-based medications that have revolutionized healthcare. In recent years, there has been a resurgence of interest in the potential of medicinal plants, driven by the need for natural alternatives to synthetic drugs. This has led to increased research and development in the field of phytomedicine, which seeks to harness the healing properties of plants while ensuring their safety and efficacy. Biotechnology and drug discovery are also areas where medicinal plants are playing an increasingly important role, with the advent of technologies such as genomics and metabolomics enabling researchers to explore the molecular mechanisms behind plant-based medicines. In addition to their medicinal properties, the cultivation, trade, and use of medicinal plants contribute significantly to local and global economies. In many developing countries, the harvesting and sale of medicinal plants form an important part of the local economy, providing livelihoods for thousands of people. These plants are often cultivated by small-scale farmers or harvested from wild habitats, with the products being sold in local markets or exported internationally. The growing demand for herbal medicines and natural products has further expanded the market for medicinal plants, creating new opportunities for economic development. This economic significance is especially important in rural areas, where medicinal plant cultivation can provide a sustainable source of income and contribute to community development.

The increasing demand for medicinal plants has led to concerns about sustainability and conservation. Over-harvesting, habitat destruction, and the commercialization of wild plant species have resulted in the decline of many medicinal plant populations. Some plants, such as ginseng, echinacea, and goldenseal, have become endangered due to unsustainable harvesting practices. The loss of these plants not only threatens biodiversity but also diminishes the potential for discovering new medicines. Conservation efforts are therefore critical to ensure that medicinal plants continue to be available for future generations. Sustainable harvesting methods, cultivation practices, and the development of policies to regulate the trade in medicinal plants are essential for the preservation of these valuable resources. Furthermore, ensuring the quality and safety of medicinal plant products is a major concern. While medicinal plants are often perceived as safer alternatives to synthetic drugs, they are not without risks. Contamination with pesticides, heavy metals, and other harmful substances can compromise the safety of plant-based medicines. Additionally, variability in the potency and efficacy of plant products, often due to differences in cultivation conditions, harvesting times, and processing methods, can lead to inconsistent therapeutic outcomes. The standardization and quality control of medicinal plant products are therefore crucial to ensure their safety and effectiveness. This has prompted increased regulation and oversight of the herbal medicine industry, with many countries implementing guidelines for the production and sale of medicinal plant-based products.

1.1 Definition and Importance of Medicinal Plants

Medicinal plants are defined as plant species that contain bioactive compounds with therapeutic properties, used to treat, prevent, or manage diseases. These plants may produce a wide variety of secondary metabolites, such as alkaloids, flavonoids, glycosides, tannins, terpenoids, and essential oils, which exhibit biological activity beneficial for human health. These compounds are synthesized by plants as part of their defense mechanisms against pathogens, herbivores, and environmental stressors, but are harnessed by humans for their medicinal value. Medicinal plants are found across diverse ecosystems, from tropical rainforests and temperate woodlands to arid deserts, reflecting their vast distribution and adaptability.

The importance of medicinal plants extends across multiple dimensions, encompassing healthcare, economic development, cultural heritage, and biodiversity conservation. In terms of healthcare, medicinal plants are often the first line of defense for millions of people, particularly in rural and remote areas with limited access to modern medical facilities. According to the World Health Organization (WHO), around 80% of the global population relies on traditional medicine, much of which is plant-based, for primary healthcare. For instance, neem, turmeric, and aloe vera are widely used in traditional remedies for their antibacterial, anti-inflammatory, and wound-healing properties. Economically, medicinal plants contribute significantly to the pharmaceutical and herbal product industries, generating billions of dollars annually. Plants like cinchona, which produces quinine for malaria treatment, and yew trees, a source of the anticancer drug paclitaxel, highlight the economic potential of medicinal plants as raw materials for drug manufacturing.

From an ecological perspective, medicinal plants play a critical role in maintaining biodiversity and ecosystem stability. Their sustainable cultivation and harvesting promote the conservation of natural habitats, prevent the loss of genetic diversity, and support the coexistence of human and natural systems. However, the overexploitation of medicinal plants for commercial purposes, coupled with habitat destruction and climate change, poses significant threats to their availability and sustainability. These challenges necessitate the adoption of conservation strategies, such as in-situ and ex-situ conservation, and the development of guidelines for sustainable harvesting practices. The cultural significance of medicinal plants cannot be overlooked. For many indigenous communities, these plants are deeply intertwined with their spiritual and cultural identity, forming the basis of rituals, ceremonies, and traditional healing practices. The knowledge associated with medicinal plants is often passed down through generations, representing a rich repository of ethnobotanical wisdom. Efforts to document, preserve, and integrate this traditional knowledge into modern healthcare systems are essential for fostering a holistic approach to medicine and promoting respect for cultural diversity.

1.2 Historical Significance and Traditional Uses of Medicinal Plants

The historical significance of medicinal plants is profound, as they have been used by human societies for thousands of years to address a wide range of health issues. Archaeological evidence suggests that the use of medicinal plants dates back to prehistoric times, with ancient civilizations such as the Egyptians, Greeks, Chinese, and Indians developing sophisticated systems of plant-based medicine. These early practices laid the foundation for modern pharmacology and herbal medicine, demonstrating the enduring relevance of medicinal plants in human history. In ancient Egypt, medicinal plants were extensively used for therapeutic purposes, as evidenced by the Ebers Papyrus, one of the oldest medical texts. This document, dating back to 1500 BCE, contains detailed descriptions of over 700 plant-based remedies, including the use of aloe for skin conditions, garlic for infections, and myrrh for wound healing. Similarly, ancient Greek physicians like Hippocrates and Dioscorides documented the medicinal properties of plants in their texts, emphasizing the role of natural remedies in maintaining health and treating diseases.

Traditional Chinese Medicine (TCM) has a rich history of using medicinal plants, with texts like the "Shennong Bencao Jing" (The Divine Farmer's Materia Medica) highlighting the therapeutic properties of plants such as ginseng, licorice, and cinnamon. These plants are still widely used in TCM today, demonstrating the enduring relevance of ancient practices. Similarly, Ayurveda, the traditional system of medicine in India, is based on the use of medicinal plants to restore balance and harmony in the body. Key Ayurvedic herbs like ashwagandha, tulsi, and brahmi have gained global recognition for their adaptogenic, immune-boosting, and cognitive-enhancing properties.

Indigenous communities around the world have also relied on medicinal plants for centuries to address health challenges and maintain well-being. For example, Native American tribes use plants like echinacea and sage for their immune-boosting and antimicrobial properties, while Amazonian tribes rely on plants like ayahuasca for spiritual and therapeutic purposes. These traditional practices are often deeply rooted in an understanding of local ecosystems, reflecting a symbiotic relationship between humans and nature. The traditional uses of medicinal plants extend beyond their therapeutic applications to include food, cosmetics, and cultural practices. Many plants used in traditional medicine also serve as sources of nutrition, providing essential vitamins, minerals, and antioxidants. For instance, moringa leaves are consumed for their high nutrient content, while turmeric is used as both a spice and a natural remedy for inflammation and digestive disorders. Similarly, the use of plant-based oils, extracts, and powders in traditional cosmetics, such as henna for hair and skin care, highlights the versatility of medicinal plants.

While traditional practices have provided valuable insights into the medicinal properties of plants, they have also faced challenges in terms of documentation, standardization, and integration into modern healthcare systems. The erosion of traditional knowledge due to globalization, urbanization, and cultural assimilation poses a significant threat to the preservation of ethnobotanical wisdom. Efforts to document and validate traditional uses of medicinal plants through scientific research are essential for bridging the gap between traditional and modern medicine.

1.3 Role of Medicinal Plants in Modern Healthcare and Pharmaceuticals

The role of medicinal plants in modern healthcare and pharmaceuticals is both foundational and transformative. These plants serve as sources of novel drugs, templates for synthetic compounds, and complementary therapies, highlighting their critical importance in addressing global health challenges. Advances in pharmacological research and biotechnology have expanded the scope of medicinal plant utilization, enabling the discovery and development of life-saving medicines. One of the most significant contributions of medicinal plants to modern healthcare is their role in drug discovery. Many of the drugs used today were originally derived from or inspired by plant compounds. For example, the pain-relieving drug aspirin is based on salicin, a compound found in willow bark, while the anticancer drug paclitaxel is extracted from the bark of the Pacific yew tree. Similarly, the alkaloids quinine and artemisinin, derived from cinchona and Artemisia annua, respectively, have revolutionized the treatment of malaria. These examples underscore the potential of medicinal plants as sources of bioactive compounds with diverse therapeutic applications. Medicinal plants also play a critical role in the development of herbal medicines and dietary supplements, which are widely used as complementary and alternative therapies. These products, often standardized for specific bioactive compounds, address a range of health issues, from stress and fatigue to chronic conditions like arthritis and diabetes. For instance, ginkgo biloba is commonly used to improve cognitive function, while St. John's wort is used to alleviate mild depression. The popularity of these products reflects a growing consumer preference for natural and holistic approaches to health.

In addition to their direct therapeutic applications, medicinal plants contribute to modern healthcare through their use in research and education. They serve as model systems for studying plant biochemistry, pharmacology, and molecular biology, advancing our understanding of disease mechanisms and drug interactions. For example, the study of alkaloids from medicinal plants has provided valuable insights into the nervous system, leading to the development of drugs for conditions like Parkinson's disease and hypertension. The integration of medicinal plants into modern healthcare systems also raises important considerations for quality control, standardization, and safety. Ensuring the efficacy and safety of plant-based medicines requires rigorous testing, including clinical trials, to validate their therapeutic claims. Standardization of active compounds, good manufacturing practices (GMP), and regulatory frameworks are essential for maintaining the quality and reliability of herbal products. Moreover, addressing challenges such as adulteration, contamination, and sustainability in the supply chain is critical for building trust and confidence among consumers and healthcare professionals. The global resurgence of interest in medicinal plants has also sparked initiatives to conserve these valuable resources and ensure their equitable use. Sustainable cultivation, bioprospecting, and benefit-sharing agreements are important strategies for balancing the economic and ecological aspects of medicinal plant utilization. Collaborations between traditional healers, researchers, and pharmaceutical companies can facilitate the development of innovative therapies while respecting the cultural and intellectual property rights of indigenous communities.

2. Diversity of Medicinal Plants

Medicinal plants represent one of the most diverse and invaluable components of global biodiversity, forming the basis of traditional medicine and contributing significantly to modern healthcare systems. Their diversity reflects the richness of ecosystems worldwide, encompassing an array of species with unique therapeutic properties. The widespread use of medicinal plants spans cultures and regions, with particular attention drawn to their role in biodiversity hotspots where the density of these plants is highest. However, this diversity is under threat due to habitat destruction, overexploitation, and climate change, necessitating conservation efforts to preserve these vital resources.

2.1 Global Distribution and Biodiversity Hotspots for Medicinal Plants

The distribution of medicinal plants is global, with species occurring across all continents and ecosystems, from tropical rainforests to temperate woodlands and arid deserts. However, certain regions, known as biodiversity hotspots, are particularly rich in medicinal plant diversity due to their unique climatic, geological, and ecological conditions. These hotspots not only harbor a high density of medicinal plant species but also support a significant number of endemic species that are found nowhere else in the world.

Tropical Rainforests

Tropical rainforests, such as the Amazon, Congo Basin, and Southeast Asian forests, are among the most important repositories of medicinal plant diversity. The Amazon Rainforest alone is estimated to contain tens of thousands of plant species, many of which have therapeutic potential. Plants such as cinchona, which yields quinine for malaria treatment, and guarana, used for its stimulant properties, are native to this region. The dense biodiversity of tropical rainforests has made them a focus of bioprospecting efforts, aiming to discover novel compounds for pharmaceutical development.

Temperate Forests and Grasslands

Temperate regions, including North America, Europe, and parts of Asia, also contribute significantly to medicinal plant diversity. These regions are home to species such as ginseng (Panax spp.), echinacea (Echinacea spp.), and valerian (Valeriana officinalis), which are widely used in traditional and modern medicine. The unique seasonal cycles and soil conditions in these areas influence the growth and biochemical composition of medicinal plants, enhancing their therapeutic value.

Arid and Semi-Arid Ecosystems

Medicinal plants in arid and semi-arid regions, such as the Sahel, Middle East, and parts of Australia, have adapted to extreme environmental conditions, often producing potent secondary metabolites as defense mechanisms. Species like aloe vera and frankincense (Boswellia spp.) are well-known examples, valued for their anti-inflammatory, antimicrobial, and skin-healing properties. These plants demonstrate the ecological resilience and pharmacological richness of arid environments.

Biodiversity Hotspots

Global biodiversity hotspots, as defined by Conservation International, include regions like the Himalayas, Indo-Burma, and the Mediterranean Basin, which are particularly rich in medicinal plant species. The Himalayas, for instance, are a treasure trove of medicinal plants such as ashwagandha (*Withania somnifera*) and Himalayan yew (*Taxus wallichiana*), which produce taxol, an anticancer compound. These hotspots are also home to traditional medicine systems like Ayurveda and Traditional Chinese Medicine (TCM), which rely heavily on local plant species. Despite their richness, these regions face significant threats from deforestation, urbanization, and agricultural expansion, underscoring the need for targeted conservation efforts. Protecting these biodiversity hotspots is not only essential for preserving medicinal plant diversity but also for maintaining ecosystem services that benefit local and global communities.

2.2 Notable Medicinal Plants and Their Therapeutic Properties

Medicinal plants are renowned for their bioactive compounds, which provide a wide range of therapeutic effects. These compounds, including alkaloids, flavonoids, terpenoids, and phenolics, have been harnessed to develop treatments for various health conditions. Several medicinal plants have gained prominence for their well-documented therapeutic properties and their use in both traditional and modern medicine.

Cinchona (*Cinchona spp.*)

Native to South America, cinchona is one of the most famous medicinal plants, known for producing quinine, a compound used to treat malaria. The discovery of quinine revolutionized the treatment of this disease and highlighted the potential of plant-based medicines to combat global health challenges. Today, derivatives of quinine continue to play a role in antimalarial therapies.

Neem (*Azadirachta indica*)

Neem, native to the Indian subcontinent, is a versatile medicinal plant with antibacterial, antifungal, and anti-inflammatory properties. Its leaves, bark, and seeds are used to treat skin conditions, infections, and dental problems. Neem oil is also widely used in cosmetics and as a natural pesticide, reflecting its diverse applications.

Ginseng (*Panax spp.*)

Ginseng is a globally recognized adaptogenic herb, known for its ability to enhance physical and mental performance, boost immunity, and reduce stress. Species such as *Panax ginseng* (Asian ginseng) and Panax quinquefolius (American ginseng) are widely cultivated and used in traditional medicine systems as well as modern health supplements.

Turmeric (*Curcuma longa*)

Turmeric, a staple in Ayurvedic medicine, is valued for its active compound curcumin, which exhibits potent anti-inflammatory, antioxidant, and anticancer properties. Turmeric is used to manage conditions such as arthritis, digestive disorders, and cardiovascular diseases, and its popularity has surged globally as a dietary supplement and functional food ingredient.

Aloe Vera (*Aloe barbadensis miller*)

Aloe vera is a widely used medicinal plant known for its skin-soothing and wound-healing properties. Its gel is applied topically to treat burns, cuts, and other skin conditions, while its juice is consumed to improve digestion and detoxify the body. Aloe vera's versatility has made it a staple in the cosmetic, pharmaceutical, and wellness industries.

Taxus (*Taxus spp.*)

The yew tree, particularly *Taxus brevifolia* and *Taxus wallichiana*, is the source of the anticancer drug paclitaxel (Taxol). This compound inhibits the growth of cancer cells and has become a cornerstone of chemotherapy for various cancers, including breast and ovarian cancer. The discovery of paclitaxel underscores the critical role of medicinal plants in advancing modern medicine.

2.3 Endangered Medicinal Plants and Conservation Efforts

Despite their importance, many medicinal plants face the threat of extinction due to overexploitation, habitat destruction, and climate change. Endangered species include those with high commercial value or limited distribution, making their conservation a priority for sustainable healthcare and biodiversity preservation.

Endangered Species

Species like the Himalayan yew, overharvested for paclitaxel, and wild ginseng, which has been depleted due to high demand in traditional medicine, are prime examples of endangered medicinal plants. Other species, such as the African cherry (*Prunus africana*) and goldenseal (*Hydrastis canadensis*), face similar threats, driven by global trade and unsustainable harvesting practices.

Conservation Strategies

Conservation efforts for medicinal plants involve a combination of in-situ and ex-situ approaches. In-situ conservation focuses on protecting natural habitats through the establishment of protected areas, biodiversity reserves, and community-managed forests. Ex-situ conservation includes cultivating medicinal plants in botanical gardens, seed banks, and plantations to reduce pressure on wild populations.

Sustainable Harvesting and Trade

Sustainable harvesting practices, such as rotational collection and quota systems, ensure that medicinal plants can regenerate and maintain their populations. Certification schemes, such as FairWild, promote ethical and sustainable trade, enabling consumers to support conservation efforts through their purchases. These initiatives also create economic incentives for communities to engage in sustainable resource management.

Research and Community Involvement

Research on cultivation techniques, propagation methods, and alternative sources of bioactive compounds is critical for reducing the dependency on wild populations. Engaging local communities in conservation efforts, such as through participatory forest management and benefit-sharing agreements, fosters a sense of ownership and responsibility for preserving medicinal plant diversity.

3. Pharmacological Applications of Medicinal Plants

Medicinal plants have long been a cornerstone of pharmacology, offering a vast repository of bioactive compounds with diverse therapeutic applications. Their role in treating infectious diseases, managing chronic conditions, and alleviating pain has made them invaluable to both traditional and modern medicine. The pharmacological properties of medicinal plants stem from their production of secondary metabolites, such as alkaloids, flavonoids, terpenoids, and phenolic compounds, which interact with biological systems to produce therapeutic effects.

3.1 Antimicrobial and Antiviral Properties

Medicinal plants have been used for centuries to combat infectious diseases caused by bacteria, viruses, fungi, and parasites. Their antimicrobial and antiviral properties are attributed to bioactive compounds that inhibit the growth or replication of pathogens. These properties make medicinal plants a critical resource in the fight against antibiotic resistance and emerging infectious diseases.

Antibacterial Properties

Plants like neem (*Azadirachta indica*), garlic (*Allium sativum*), and turmeric (*Curcuma longa*) are well-known for their antibacterial properties. Neem leaves and extracts contain compounds such as nimbidin and azadirachtin, which exhibit potent antibacterial activity against Gram-positive and Gram-negative bacteria. Neem has been used to treat skin infections, dental problems, and gastrointestinal disorders caused by bacterial pathogens. Similarly, garlic's active compound allicin has been shown to inhibit bacterial enzymes, making it effective against pathogens like *Escherichia coli, Staphylococcus aureus,* and *Salmonella* spp.

Antifungal Properties

Medicinal plants also play a significant role in antifungal therapies. Tea tree oil, derived from Melaleuca alternifolia, is widely used for its antifungal properties. It is effective against dermatophytes, which cause conditions like athlete's foot and ringworm. Other plants, such as calendula (*Calendula officinalis*) and aloe vera, have demonstrated antifungal activity against Candida species, providing natural alternatives for treating fungal infections.

Antiviral Properties

The antiviral properties of medicinal plants have garnered significant attention, particularly in the context of global viral outbreaks. Plants like licorice (*Glycyrrhiza glabra*) contain glycyrrhizin, a compound that inhibits viral replication by interfering with the attachment and penetration of viruses into host cells. Licorice has shown efficacy against viruses such as hepatitis B, hepatitis C, and herpes simplex virus. Similarly, Andrographis paniculata, commonly used in traditional medicine, has demonstrated antiviral activity against influenza and dengue viruses. These findings highlight the potential of medicinal plants in developing novel antiviral therapies.

Combating Antibiotic Resistance

The emergence of antibiotic-resistant bacteria has become a major public health concern, prompting the search for alternative treatments. Medicinal plants offer a promising solution, as their bioactive compounds often target multiple pathways in pathogens, reducing the likelihood of resistance development. For example, the combination of plant extracts with conventional antibiotics has been shown to enhance antimicrobial efficacy and overcome resistance mechanisms. This synergistic approach represents a novel strategy for addressing the global challenge of antibiotic resistance.

3.2 Role in Treating Chronic Diseases

Chronic diseases, including diabetes, cancer, and cardiovascular disorders, represent some of the most pressing global health challenges. Medicinal plants play a critical role in managing these conditions, offering natural remedies that are often safer and more accessible than synthetic drugs.

Diabetes Management

Plants like fenugreek (*Trigonella foenum-graecum*), bitter melon (*Momordica charantia*), and cinnamon (*Cinnamomum spp.*) have been extensively studied for their antidiabetic properties. Fenugreek seeds contain compounds such as 4-hydroxyisoleucine, which enhance insulin secretion and improve glucose uptake in cells. Bitter melon contains polypeptide-p, an insulin-like compound that lowers blood sugar levels. Cinnamon has been shown to improve insulin sensitivity and reduce fasting blood glucose levels in individuals with type 2 diabetes. These plants offer complementary approaches to conventional diabetes management, reducing the risk of complications associated with prolonged hyperglycemia.

Cancer Treatment

Medicinal plants have been a rich source of anticancer agents, with compounds such as taxol from the Pacific yew tree (*Taxus brevifolia*) and vincristine from the Madagascar periwinkle (*Catharanthus roseus*) playing pivotal roles in chemotherapy. Taxol inhibits the division of cancer cells by stabilizing microtubules, making it effective against ovarian, breast, and lung cancers. Vincristine disrupts cell division by binding to tubulin, and it is used to treat leukemia, lymphoma, and other cancers. Additionally, curcumin from turmeric has shown promise in inhibiting tumor growth and metastasis through its anti-inflammatory and antioxidant effects. These examples illustrate the potential of medicinal plants in developing targeted and effective cancer therapies.

Cardiovascular Health

Medicinal plants contribute significantly to cardiovascular health by lowering blood pressure, reducing cholesterol levels, and preventing atherosclerosis. Garlic is a well-known cardioprotective agent, with studies demonstrating its ability to lower LDL cholesterol and improve blood circulation. Hawthorn (*Crataegus spp.*) is used to strengthen the heart muscle, improve cardiac output, and alleviate symptoms of heart failure. Resveratrol, a compound found in grapes and berries, has been shown to protect against oxidative stress and inflammation, reducing the risk of cardiovascular diseases. These plants provide holistic approaches to maintaining heart health and preventing complications such as stroke and myocardial infarction.

3.3 Applications in Pain Management and Anti-Inflammatory Therapies

Pain and inflammation are common symptoms of various health conditions, and medicinal plants have been used for centuries to address these issues. Their bioactive compounds provide natural alternatives to synthetic painkillers and anti-inflammatory drugs, which often come with adverse side effects.

Pain Management

Medicinal plants like willow bark (*Salix spp.*) and opium poppy (*Papaver somniferum*) are among the most well-known sources of natural analgesics. Willow bark contains salicin, a precursor to aspirin, which provides pain relief by inhibiting the production of prostaglandins. Opium poppy produces morphine and codeine, potent opioids used to manage severe pain. These compounds interact with the central nervous system to block pain signals, making them invaluable in palliative care and postoperative recovery.

Anti-Inflammatory Therapies

Inflammation is a natural immune response, but chronic inflammation can lead to conditions such as arthritis, asthma, and inflammatory bowel disease. Medicinal plants like turmeric, ginger (*Zingiber officinale*), and boswellia (*Boswellia serrata*) have demonstrated significant anti-inflammatory properties. Curcumin, the active compound in turmeric, inhibits inflammatory pathways and reduces the production of pro-inflammatory cytokines. Gingerol, found in ginger, has similar effects, alleviating pain and inflammation in conditions like osteoarthritis. Boswellic acids, derived from boswellia resin, are used to treat rheumatoid arthritis and other inflammatory disorders by targeting enzymes involved in the inflammatory process.

Adjunct Therapies

In addition to their primary effects, medicinal plants often provide adjunct benefits, such as reducing oxidative stress and supporting tissue repair. For example, arnica (*Arnica montana*) is used topically to reduce bruising and swelling, while chamomile (*Matricaria chamomilla*) soothes muscle pain and promotes relaxation. These properties make medicinal plants versatile tools in comprehensive pain and inflammation management strategies.

4. Traditional and Indigenous Uses

Medicinal plants have been deeply intertwined with traditional and indigenous knowledge systems for millennia, serving as the cornerstone of healthcare practices across cultures. These plants, revered for their healing properties, are central to ethnomedicine and traditional healing practices, reflecting a profound understanding of the natural world. Indigenous communities have developed intricate systems of plant-based medicine that not only address physical ailments but also encompass spiritual, social, and cultural dimensions of health.

4.1 Ethnomedicine

Ethnomedicine, the study of traditional medical practices among indigenous communities, highlights the deep relationship between humans and their natural environments. Medicinal plants are an integral part of these systems, used to treat a wide range of ailments and maintain overall well-being. Indigenous knowledge about these plants is often passed down orally through generations, forming a repository of wisdom that combines empirical observation, experimentation, and cultural beliefs.

Indigenous communities possess a remarkable understanding of the medicinal properties of plants in their local environments. For example, the Amazonian tribes have identified thousands of plant species with therapeutic uses, such as the bark of the cinchona tree for malaria and the leaves of coca plants for pain relief and energy enhancement. Similarly, the San people of Southern Africa use the hoodia plant (*Hoodia gordonii*) as an appetite suppressant during long hunting expeditions, showcasing their knowledge of plant-based remedies tailored to their ecological and lifestyle needs.

Ethnomedicine often employs a holistic approach, recognizing the interconnectedness of physical, emotional, and spiritual health. Medicinal plants are used not only for their pharmacological properties but also as part of rituals and ceremonies aimed at restoring balance and harmony. For instance, the use of sacred herbs like sage (Salvia spp.) in smudging rituals among Native American tribes reflects the belief in cleansing both physical and spiritual spaces. These practices underscore the multifunctional roles of medicinal plants in indigenous healthcare systems. The study of ethnomedicine has significant implications for modern pharmacology and conservation. Documenting and validating traditional knowledge can lead to the discovery of novel drugs and therapies while promoting the sustainable use of plant resources. However, the erosion of indigenous knowledge due to cultural assimilation, urbanization, and environmental degradation poses a significant threat to this invaluable heritage. Collaborative efforts between researchers, policymakers, and indigenous communities are essential to preserve and integrate ethnomedical practices into contemporary healthcare frameworks.

4.2 Traditional Healing Practices and their Cultural Importance

Traditional healing practices represent a rich tapestry of cultural beliefs, rituals, and techniques centered around the use of medicinal plants. These practices, rooted in the cultural identity and spirituality of communities, are often inseparable from their social and ecological contexts. Traditional healers, such as shamans, herbalists, and midwives, play a pivotal role in these systems, serving as custodians of knowledge and providers of care. In many traditional systems, healing practices extend beyond treating physical symptoms to address the underlying causes of illness, which may be attributed to spiritual imbalances, social conflicts, or environmental factors. Medicinal plants are used in conjunction with other elements, such as prayers, chants, and symbolic rituals, to create a holistic healing experience. For example, the use of ayahuasca, a psychoactive brew made from plants like *Banisteriopsis caapi*, in Amazonian shamanic rituals is believed to facilitate spiritual healing and insight, demonstrating the integration of medicinal plants into broader cultural practices.

The cultural importance of traditional healing practices is evident in their role in maintaining community cohesion and identity. These practices often involve collective participation, fostering a sense of belonging and mutual support. For instance, the preparation and application of herbal remedies in Ayurvedic medicine are often accompanied by familial or community involvement, reinforcing social bonds and cultural continuity. Similarly, the use of traditional medicines during ceremonies and festivals, such as the Ugandan use of waragi infused with medicinal herbs, reflects their significance in cultural expression and resilience. Traditional healing practices also contribute to the preservation of biodiversity and sustainable resource management. Indigenous communities often follow guidelines for harvesting medicinal plants, such as taking only what is needed and leaving seed-bearing plants untouched, to ensure the availability of resources for future generations. These practices align with the principles of ecological stewardship, highlighting the interconnectedness of cultural and environmental sustainability. Despite their importance, traditional healing practices face challenges from the dominance of Western medicine, legal restrictions, and the commodification of traditional knowledge. Recognizing and supporting the role of traditional healers, promoting intercultural dialogue, and protecting intellectual property rights are critical steps toward preserving these practices and ensuring their integration into global health systems.

5. Extraction and Processing of Medicinal Compounds

The extraction and processing of medicinal compounds from plants are critical steps in transforming raw plant materials into therapeutic products. These processes ensure that the bioactive components responsible for the plant's medicinal properties are effectively isolated, standardized, and formulated for use in traditional medicine, pharmaceuticals, and health supplements. The techniques employed, quality control measures implemented, and value addition through herbal formulations play a significant role in ensuring the efficacy, safety, and marketability of medicinal plant products.

5.1 Techniques for Extracting Active Ingredients from Plants

The extraction of active ingredients from medicinal plants involves isolating bioactive compounds, such as alkaloids, flavonoids, terpenoids, and glycosides, which are responsible for the plant's therapeutic effects. Various extraction techniques have been developed to optimize the yield, purity, and bioavailability of these compounds. The choice of technique depends on factors such as the type of plant material, the nature of the active ingredient, and the intended application of the extract.

Traditional Extraction Methods

Traditional methods, such as decoction, infusion, and maceration, are widely used in traditional medicine systems. Decoction involves boiling plant materials in water to extract water-soluble compounds, commonly used for roots, bark, and seeds. Infusion, similar to brewing tea, is suitable for delicate plant parts like leaves and flowers. Maceration involves soaking plant materials in a solvent, typically water or alcohol, at room temperature to extract bioactive compounds. These methods are simple, cost-effective, and require minimal equipment, making them accessible for small-scale production and traditional healers.

Modern Extraction Techniques

Modern extraction techniques, including steam distillation, solvent extraction, supercritical fluid extraction, and ultrasound-assisted extraction, offer higher efficiency and precision compared to traditional methods.

- **Steam Distillation**: This technique is widely used for extracting essential oils from aromatic plants such as lavender, eucalyptus, and peppermint. Steam passes through the plant material, vaporizing the volatile compounds, which are then condensed and separated from water. Steam distillation preserves the integrity of heat-sensitive compounds and is favored for its simplicity and scalability.
- **Solvent Extraction**: Solvents like ethanol, methanol, and hexane are used to dissolve bioactive compounds from plant materials. This method is effective for extracting lipophilic compounds, such as alkaloids and flavonoids. The choice of solvent depends on the target compound and its solubility properties. However, solvent residues must be carefully removed to ensure the safety of the extract.
- **Supercritical Fluid Extraction (SFE)**: SFE uses supercritical carbon dioxide (CO2) as a solvent to extract bioactive compounds. This technique is highly efficient and environmentally friendly, producing solvent-free extracts suitable for pharmaceuticals and food products. SFE

is commonly used for high-value extracts, such as curcumin from turmeric or carotenoids from marigold flowers.

» **Ultrasound-Assisted Extraction (UAE)**: UAE enhances the extraction process by using ultrasonic waves to disrupt plant cell walls, releasing bioactive compounds into the solvent. This technique reduces extraction time and increases yield, making it suitable for industrial-scale production of plant extracts.

Advances in Biotechnology

Biotechnological approaches, such as plant tissue culture and metabolic engineering, are emerging as innovative methods for producing bioactive compounds. Tissue culture techniques allow for the production of secondary metabolites in controlled environments, reducing the dependency on wild plant populations. Metabolic engineering involves modifying the biosynthetic pathways of plants to enhance the production of specific compounds, such as paclitaxel from Taxus spp. These advances offer sustainable and scalable alternatives for extracting medicinal compounds.

5.2 Standardization and Quality Control in Medicinal Plant Products

Standardization and quality control are essential to ensure the safety, efficacy, and consistency of medicinal plant products. Unlike synthetic drugs, which contain a single active ingredient, plant-based medicines are complex mixtures of multiple compounds, making standardization a critical challenge.

Importance of Standardization

Standardization involves defining specific parameters for plant materials and extracts, such as the concentration of active compounds, physical and chemical properties, and biological activity. These parameters ensure that the product meets predefined quality standards, regardless of variations in plant origin, cultivation conditions, or processing methods. For example, the standardization of ginseng extracts focuses on the concentration of ginsenosides, while turmeric extracts are standardized based on curcumin content.

Quality Control Measures

Quality control encompasses the entire production process, from sourcing raw materials to final product testing. Key measures include:

» **Authentication of Plant Materials**: Ensuring the correct identification and sourcing of plant species is the first step in quality control. Morphological, anatomical, and molecular techniques are used to verify the authenticity of plant materials and prevent adulteration.

» **Testing for Contaminants**: Medicinal plant products must be free from contaminants such as heavy metals, pesticides, microbial pathogens, and mycotoxins. Advanced analytical techniques, such as gas chromatography-mass spectrometry (GC-MS) and high-performance liquid chromatography (HPLC), are used for detecting and quantifying contaminants.

» **Stability Testing**: Stability tests evaluate the shelf life and storage conditions of medicinal plant products, ensuring that they retain their potency and safety over time. These tests assess factors such as temperature, humidity, and light exposure on the product's quality.

» **Good Manufacturing Practices (GMP)**: Adherence to GMP guidelines ensures that medicinal plant products are manufactured in clean, controlled environments, minimizing the risk of contamination and errors. GMP certification is a prerequisite for exporting herbal products to international markets.

Role of Regulatory Frameworks

National and international regulatory frameworks play a critical role in standardizing and controlling the quality of medicinal plant products. Organizations such as the World Health Organization (WHO) and the European Medicines Agency (EMA) provide guidelines for the quality assurance of herbal medicines. Compliance with these guidelines enhances consumer confidence and facilitates the global trade of medicinal plant products.

5.3 Value Addition through Formulation of Herbal Remedies

Value addition transforms raw plant materials into market-ready products with enhanced therapeutic efficacy, convenience, and consumer appeal. Formulation involves combining plant extracts with other ingredients to create standardized, easy-to-use products, such as tablets, capsules, teas, and topical preparations.

Traditional Formulations

Traditional medicine systems, such as Ayurveda and Traditional Chinese Medicine (TCM), offer a wealth of knowledge on herbal formulations. Ayurvedic preparations like chyawanprash and triphala are complex blends of multiple herbs, designed to balance the body's doshas and improve overall health. Similarly, TCM formulations, such as the use of ginseng and astragalus in tonics, are based on the principles of yin-yang balance and meridian theory. These traditional formulations serve as the foundation for modern herbal product development.

Modern Herbal Products

The global demand for herbal remedies has driven the development of innovative products that cater to modern lifestyles. Examples include:

» **Nutraceuticals**: Herbal supplements enriched with vitamins, minerals, and antioxidants, such as curcumin capsules for inflammation and spirulina tablets for nutritional support.

» **Functional Foods**: Products like herbal teas, energy bars, and fortified beverages incorporate medicinal plant extracts to promote health and wellness.

» **Cosmeceuticals**: Skincare and haircare products infused with plant-based ingredients, such as aloe vera gel for hydration and neem oil for acne treatment.

Nanotechnology in Herbal Formulations

Nanotechnology has emerged as a cutting-edge approach to enhance the bioavailability and efficacy of medicinal plant compounds. Nano-encapsulation of plant extracts protects active ingredients from degradation and improves their absorption in the body. For example, curcumin-loaded nanoparticles have shown superior anti-inflammatory and anticancer effects compared to conventional formulations. This technology represents a significant advancement in the formulation of herbal remedies.

Economic and Environmental Benefits

Value addition not only increases the economic value of medicinal plants but also promotes sustainable resource use. By creating high-value products, value addition reduces the need for large quantities of raw materials, alleviating pressure on wild plant populations. It also generates employment opportunities and supports rural economies by involving local communities in processing and packaging activities.

6. Medicinal Plants in Modern Pharmaceuticals

Medicinal plants have been a cornerstone of pharmaceutical development, providing invaluable compounds that serve as drugs or templates for synthetic derivatives. From ancient remedies to cutting-edge biotechnology, these plants have shaped the evolution of modern medicine. Their bioactive compounds, such as alkaloids, flavonoids, terpenoids, and phenolics, continue to inspire drug discovery and innovation in treating complex diseases. However, the integration of medicinal plants into pharmaceuticals comes with challenges, including scalability, standardization, and conservation concerns.

6.1 Development of Plant-Derived Drugs: Success Stories and Innovations

Plant-derived drugs are among the most significant achievements in pharmacology, offering groundbreaking treatments for various health conditions. These success stories highlight the enduring value of medicinal plants in developing effective and innovative therapies.

Historical Success Stories

Several plant-derived drugs have revolutionized modern medicine, demonstrating the therapeutic potential of natural compounds:

- **Quinine from Cinchona**: Quinine, an alkaloid extracted from the bark of the cinchona tree, was the first effective treatment for malaria. Its discovery in the 17th century marked a turning point in the fight against this deadly disease. Quinine's efficacy inspired the development of synthetic antimalarial drugs, such as chloroquine, which remain critical in global health initiatives.
- **Morphine from Opium Poppy**: The opium poppy (*Papaver somniferum*) has been a source of analgesics for millennia. Morphine, its most potent alkaloid, was isolated in the early 19th century and became the gold standard for pain management. Its derivatives, such as codeine and oxycodone, are widely used in modern medicine.
- **Aspirin from Willow Bark**: Salicin, derived from willow bark (*Salix spp.),* served as the precursor to aspirin. Synthesized in 1897, aspirin became one of the most widely used drugs for pain relief, inflammation, and cardiovascular protection.
- **Taxol from Pacific Yew**: Paclitaxel (Taxol), derived from the bark of the Pacific yew tree (Taxus brevifolia), is a powerful chemotherapeutic agent used to treat ovarian, breast, and lung cancers. Its discovery highlighted the importance of medicinal plants in oncology.

Recent Innovations

The discovery and development of plant-derived drugs continue to advance, driven by modern techniques in extraction, synthesis, and bioassay screening:

» **Artemisinin from Sweet Wormwood**: Artemisinin, isolated from Artemisia annua, is the foundation of artemisinin-based combination therapies (ACTs), the frontline treatment for malaria. Its rapid action and low toxicity have made it indispensable in global health strategies.

» **Cannabinoids from Cannabis**: Cannabinoids, such as cannabidiol (CBD) and tetrahydrocannabinol (THC), are extracted from Cannabis sativa and used to treat epilepsy, chronic pain, and multiple sclerosis. The legalization of medical cannabis has spurred extensive research into its therapeutic applications.

» **Resveratrol from Grapes**: Resveratrol, a polyphenol found in grapes, has garnered attention for its antioxidant and cardioprotective effects. It is being explored for potential applications in aging, neurodegenerative diseases, and cancer.

The success stories of these drugs underscore the immense potential of medicinal plants as sources of novel therapies. Advances in synthetic biology, metabolic engineering, and nanotechnology are further expanding the horizons of plant-derived pharmaceuticals.

6.2 Role of Medicinal Plants in Biotechnology and Drug Discovery

Biotechnology has revolutionized the exploration of medicinal plants, enabling the identification, isolation, and enhancement of bioactive compounds. Modern biotechnological tools, such as genomics, proteomics, and metabolic engineering, have accelerated drug discovery and improved the scalability of plant-derived pharmaceuticals.

Genomics and Metabolomics

The genomic sequencing of medicinal plants has opened new avenues for understanding their biosynthetic pathways and identifying genes responsible for the production of bioactive compounds. For example, the genome of *Catharanthus roseus* (Madagascar periwinkle) has been sequenced to study the production of anticancer alkaloids like vincristine and vinblastine. Metabolomics, the study of the chemical fingerprints of plants, complements genomics by profiling secondary metabolites and their pharmacological properties.

Plant Tissue Culture

Plant tissue culture is a powerful technique for producing secondary metabolites in controlled environments. It allows for the large-scale production of valuable compounds without relying on wild populations. For instance, tissue culture methods have been used to produce paclitaxel and shikonin, reducing the environmental impact of harvesting rare plants.

Synthetic Biology

Synthetic biology integrates biological engineering with plant science to enhance the production of bioactive compounds. By introducing plant genes into microbial hosts, such as yeast or bacteria, researchers can produce plant-derived compounds in industrial-scale bioreactors. Artemisinin production in genetically engineered yeast is a landmark example of this approach, enabling a steady supply of the antimalarial drug.

High-Throughput Screening and Bioassays

Advances in high-throughput screening technologies have accelerated the identification of bioactive compounds from plant extracts. Robotic systems and automated assays test thousands of compounds against disease targets, streamlining the drug discovery process. For example, screening libraries of plant-derived compounds has led to the identification of promising candidates for neurodegenerative diseases and antibiotic-resistant infections.

Applications in Personalized Medicine

The integration of medicinal plants with biotechnology is paving the way for personalized medicine. By tailoring plant-derived therapies to an individual's genetic profile, researchers aim to develop treatments with enhanced efficacy and minimal side effects. This approach is particularly relevant for complex diseases like cancer and autoimmune disorders.

6.3 Challenges in Scaling Up Production for Pharmaceutical Use

Despite the therapeutic potential of medicinal plants, scaling up their production for pharmaceutical use poses significant challenges. These include issues related to resource sustainability, standardization, and regulatory compliance.

Resource Limitations

Many medicinal plants are harvested from wild populations, leading to overexploitation and habitat loss. For example, the demand for paclitaxel initially threatened the survival of the Pacific yew tree. Sustainable sourcing, cultivation, and tissue culture techniques are critical to addressing resource limitations.

Variability in Plant Compounds

The concentration of bioactive compounds in medicinal plants can vary based on factors such as soil type, climate, and harvesting methods. This variability complicates standardization and quality control, which are essential for pharmaceutical applications. Advances in precision agriculture and controlled-environment farming offer potential solutions to this challenge.

Complexity of Plant Extracts

Unlike synthetic drugs with a single active ingredient, plant extracts are complex mixtures of compounds that may interact synergistically or antagonistically. This complexity poses challenges for isolating and characterizing the active components, as well as for ensuring consistency in pharmaceutical formulations.

Regulatory Hurdles

The pharmaceutical industry is highly regulated, with stringent requirements for safety, efficacy, and quality. Medicinal plant products often face additional scrutiny due to their variability and natural origin. Compliance with good manufacturing practices (GMP), clinical trials, and pharmacovigilance standards is essential but resource-intensive.

Intellectual Property and Benefit Sharing

Issues related to intellectual property rights and benefit-sharing agreements with indigenous communities pose ethical and legal challenges. The Nagoya Protocol, which governs access to genetic resources and the equitable sharing of benefits, aims to address these concerns, but its implementation varies across countries.

Cost and Scalability

The cost of extracting and purifying bioactive compounds from plants can be prohibitively high, particularly for compounds present in low concentrations. Biotechnology offers scalable solutions, but the initial investment in research and infrastructure remains a barrier for many pharmaceutical companies.

7. Economic and Social Significance

Medicinal plants are a cornerstone of economic development and social well-being, providing significant contributions to local livelihoods, global markets, and community empowerment. Their role extends beyond healthcare, encompassing trade, agriculture, and cultural preservation, making them invaluable resources in both developed and developing economies. The cultivation, processing, and trade of medicinal plants offer employment opportunities to millions of people worldwide, particularly in rural and forest-dependent regions. Additionally, medicinal plants serve as vehicles for community development, fostering social cohesion, gender empowerment, and sustainable development.

7.1 Contribution of Medicinal Plants to Local and Global Economies

Medicinal plants are a major driver of economic growth, contributing to both local economies in rural areas and global markets dominated by the pharmaceutical, nutraceutical, and herbal industries. The global trade in medicinal plants and their derivatives is valued at billions of dollars annually, reflecting their immense economic importance. From the production of herbal remedies and dietary supplements to their incorporation into modern pharmaceuticals, medicinal plants are at the center of a thriving global industry that continues to grow at a rapid pace.

At the local level, medicinal plants provide vital income sources for rural and forest-dependent communities. Small-scale farmers and collectors harvest a wide variety of medicinal plants, including roots, leaves, seeds, and bark, which are sold in local markets or to intermediaries. This trade supports local economies, particularly in developing countries where formal employment opportunities are limited. For instance, in India, which is one of the largest producers of medicinal plants, the trade of herbs such as ashwagandha, turmeric, and neem contributes significantly to the livelihoods of rural populations. Similarly, in African countries, plants like rooibos, hibiscus, and baobab are cultivated and exported for their medicinal and nutritional properties, generating substantial revenue for local economies.

Globally, the demand for medicinal plants has surged due to increasing consumer preferences for natural and plant-based products. The herbal medicine industry, which includes products such as teas, tinctures, and capsules, is one of the fastest-growing segments of the health and wellness market. This demand is driven by a growing awareness of the health benefits of medicinal plants, as well as a shift toward sustainable and eco-friendly lifestyles. Countries like China and India, with their rich traditions of herbal medicine, dominate the export of medicinal plants, supplying raw materials and finished products to markets in Europe, North America, and the Middle East.

The pharmaceutical industry also relies heavily on medicinal plants for the development of drugs. Many life-saving medications, such as paclitaxel, artemisinin, and morphine, are derived from plant sources, highlighting the critical role of medicinal plants in healthcare. The integration of medicinal plants into pharmaceutical research and development not only drives economic growth but also stimulates innovation, leading to the discovery of new drugs and therapies. Additionally, the commercialization of plant-based cosmetics and personal care products, such as aloe vera gels, neem shampoos, and tea tree oil creams, has further expanded the economic footprint of medicinal plants. Despite their economic potential, the trade of medicinal plants faces challenges such as market volatility, regulatory barriers, and unsustainable harvesting practices. Addressing these issues requires coordinated efforts to establish fair trade systems, promote value addition, and ensure the sustainability of medicinal plant resources. Governments, non-governmental organizations (NGOs), and private enterprises must collaborate to create supportive policies and infrastructure that enable producers and traders to access domestic and international markets effectively.

7.2 Livelihood Opportunities in Medicinal Plant Cultivation and Trade

The cultivation and trade of medicinal plants offer significant livelihood opportunities, particularly for marginalized and rural populations. These activities provide income and employment for millions of people worldwide, creating pathways for economic independence and social mobility. Medicinal plant cultivation is often labor-intensive, requiring local knowledge, traditional expertise, and sustainable practices, which align well with the skillsets of rural communities. One of the most notable aspects of medicinal plant cultivation is its adaptability to small-scale farming. Unlike large-scale commercial agriculture, medicinal plants can be grown on small plots of land or integrated into agroforestry systems, making them accessible to farmers with limited resources. For example, farmers in India cultivate plants like aloe vera, tulsi, and brahmi on small farms, generating steady income with relatively low investment. Similarly, in Kenya, farmers grow medicinal herbs such as Artemisia annua for the production of artemisinin, which is used in malaria treatment. These activities not only enhance household incomes but also reduce reliance on monoculture farming, promoting biodiversity and soil health.

The collection of wild medicinal plants also plays a significant role in supporting livelihoods, particularly for indigenous and forest-dependent communities. Wild harvesting involves the sustainable collection of plants from natural habitats, often guided by traditional knowledge about seasonal cycles, growth patterns, and ecological balance. For instance, in Nepal, the collection of yarsagumba (Cordyceps sinensis), a valuable medicinal fungus, provides substantial income for local communities during the harvesting season. Similarly, in the Amazon rainforest, the collection of copaiba resin and açaí berries supports both livelihoods and forest conservation.

The trade of medicinal plants, whether in local markets or global supply chains, creates additional employment opportunities in processing, packaging, and distribution. Small and medium enterprises (SMEs) involved in value addition, such as the production of herbal teas, oils, and ointments, generate jobs for women, youth, and other marginalized groups. For example, women in Morocco are heavily involved in the production of argan oil, which is extracted from the seeds of the argan tree and used in cosmetics and health products. This activity not only empowers women economically but also promotes social equity and cultural preservation. To maximize the livelihood potential of medicinal plants, it is essential to invest in capacity-building programs, such as training in sustainable cultivation techniques, quality assurance, and market access. Establishing cooperatives and producer organizations can further enhance the bargaining power of farmers and collectors, enabling them to negotiate fair prices and secure long-term contracts with buyers. Additionally, promoting certification schemes, such as organic and FairWild labels, can add value to medicinal plant products, ensuring that producers receive a fair share of the benefits while adhering to sustainable practices.

7.3 Role of Medicinal Plants in Community Development and Empowerment

Medicinal plants play a transformative role in community development, fostering economic resilience, social cohesion, and environmental sustainability. Their cultivation, trade, and use are deeply embedded in the cultural fabric of many communities, providing not only economic benefits but also opportunities for empowerment and collective action. One of the most significant ways medicinal plants contribute to community development is through gender empowerment. Women, who are often the primary custodians of traditional knowledge about medicinal plants, play a central role in their cultivation, collection, and processing. For instance, women in the Himalayas are actively involved in harvesting medicinal herbs like jatamansi (Nardostachys jatamansi) and preparing herbal remedies for local markets. By generating income from these activities, women gain financial independence and enhanced decision-making power within their households and communities. Women-led cooperatives, such as those producing essential oils or herbal teas, further exemplify the potential of medicinal plants to drive gender-inclusive development.

Medicinal plants also promote social cohesion and cultural preservation by reinforcing traditional knowledge systems and practices. Indigenous communities often view these plants as integral to their identity, health, and spirituality. For example, the use of sacred herbs in rituals and ceremonies strengthens cultural heritage and intergenerational knowledge transfer. Community-based initiatives, such as the establishment of ethnobotanical gardens and herbal medicine training programs, create platforms for sharing and preserving this knowledge while fostering a sense of pride and ownership among participants. Environmental sustainability is another critical aspect of community development linked to medicinal plants. Sustainable harvesting and cultivation practices, such as agroforestry and rotational cropping, help conserve biodiversity, protect ecosystems, and mitigate climate change. Communities engaged in these activities often become stewards of their natural resources, advocating for the protection of forests, watersheds, and wildlife. For instance, projects like the sustainable harvesting of rooibos in South Africa or the cultivation of endangered medicinal plants in community-managed forests in Nepal demonstrate the synergistic relationship between ecological conservation and community empowerment. To fully harness the potential of medicinal plants for community development, it is essential to address challenges such as resource depletion, inequitable benefit-sharing, and lack of market access. Policies that recognize and support the contributions of local communities, such as access to land, credit, and technical assistance, can create enabling environments for sustainable development. Partnerships between governments, NGOs, and private sector actors can further amplify the impact of medicinal plants on community well-being. The economic and social significance of medicinal plants underscores their potential to transform lives and landscapes. By integrating traditional knowledge with modern innovations, promoting equitable trade, and fostering community-driven initiatives, medicinal plants can serve as powerful catalysts for sustainable development and empowerment.

8. Sustainability and Conservation of Medicinal Plants

The sustainability and conservation of medicinal plants are critical for preserving their ecological, cultural, and economic value. These plants, which provide essential healthcare solutions and livelihoods to millions, are increasingly under threat due to over-harvesting, habitat loss, and climate change. Addressing these challenges requires a multifaceted approach that integrates sustainable harvesting, conservation strategies, and supportive policy frameworks.

8.1 Threats to Medicinal Plant Resources: Over-Harvesting and Habitat Loss

Medicinal plants face numerous threats that jeopardize their availability and ecological stability. Among the most significant challenges are over-harvesting and habitat loss, both of which are exacerbated by growing global demand for natural remedies and traditional medicine.

Over-Harvesting of Medicinal Plants

Over-harvesting is a major threat to medicinal plant populations, particularly for species with high commercial value or those that grow slowly. Many medicinal plants, such as ginseng (*Panax spp.*), yew (*Taxus spp.*), and frankincense (*Boswellia spp.*), are harvested unsustainably to meet the demands of pharmaceutical and herbal industries. In many cases, entire plants are uprooted or harvested excessively, leaving little opportunity for regeneration. This unsustainable practice often leads to local extinction, reduced genetic diversity, and disrupted ecosystems. For instance, the Himalayan yew (*Taxus wallichiana*) is heavily harvested for its anticancer compound paclitaxel, causing significant population declines in its native range. Similarly, wild ginseng has become critically endangered in many regions due to over-collection for traditional medicine and global trade. The depletion of these plants not only affects their availability but also undermines the ecosystems in which they play critical roles.

Habitat Loss

Habitat loss is another pressing issue affecting medicinal plants, driven by deforestation, agricultural expansion, urbanization, and infrastructure development. Forest ecosystems, which harbor a vast majority of medicinal plant species, are particularly vulnerable. The conversion of forested land into agricultural fields or urban spaces destroys the natural habitats of many plants, reducing their populations and limiting their potential for recovery. Climate change further compounds the problem by altering the distribution and growth patterns of medicinal plants. Changes in temperature, precipitation, and seasonal cycles affect the flowering, fruiting, and regeneration of these plants, making them more susceptible to decline. For example, alpine and montane medicinal plants, such as Rhodiola (Rhodiola rosea), face habitat shrinkage due to rising temperatures and melting glaciers. The combined effects of over-harvesting and habitat loss pose significant challenges for the conservation of medicinal plants. Addressing these threats requires urgent and coordinated efforts to promote sustainable practices, protect natural habitats, and mitigate the impacts of climate change.

8.2 Strategies for Sustainable Harvesting and Cultivation

Sustainable harvesting and cultivation are essential strategies for conserving medicinal plants while ensuring their continued availability for traditional and modern healthcare systems. These approaches emphasize responsible resource use, habitat restoration, and community engagement.

Sustainable Harvesting Practices

Sustainable harvesting involves collecting plant materials in ways that allow for natural regeneration and minimize ecological disruption. Key practices include:

» **Selective Harvesting**: Harvesting specific parts of the plant, such as leaves, fruits, or bark, without damaging the entire plant. For example, neem (Azadirachta indica) leaves and seeds are harvested without harming the tree, allowing it to continue producing valuable resources.

» **Rotational Harvesting**: Rotating harvesting sites or species to prevent over-exploitation of a single population. This practice ensures that plants have sufficient time to recover between harvesting cycles.

» **Seasonal Harvesting**: Timing the collection of plant materials to coincide with optimal periods for regeneration. For instance, harvesting medicinal roots after seed dispersal ensures that the plant can reproduce before being collected.

» **Community Monitoring**: Involving local communities in monitoring and managing medicinal plant resources. Indigenous and local knowledge about plant ecology and growth cycles can guide sustainable harvesting practices.

Cultivation of Medicinal Plants

Cultivating medicinal plants in controlled environments reduces pressure on wild populations and provides a reliable supply of raw materials for the pharmaceutical and herbal industries. Key aspects of sustainable cultivation include:

» **Agroforestry Systems**: Integrating medicinal plants into agroforestry systems, where they are grown alongside food crops and timber species. This approach enhances biodiversity, soil fertility, and farmers' incomes.

» **Domestication of Wild Species**: Developing cultivation protocols for high-demand wild species, such as ginseng, sandalwood, and saffron, to reduce reliance on wild harvesting.

» **Organic and Sustainable Farming**: Adopting organic farming practices that minimize the use of chemical inputs and promote soil health. Certification schemes, such as organic and FairWild labels, add value to cultivated medicinal plants and ensure compliance with sustainability standards.

Habitat Restoration

Restoring degraded habitats is critical for the conservation of medicinal plants and their ecosystems. Initiatives such as reforestation, invasive species removal, and soil rehabilitation can create favorable conditions for the growth and regeneration of medicinal plants. Community-led conservation projects, such as the establishment of herbal gardens and seed banks, further contribute to habitat restoration and the preservation of plant diversity.

Conclusion

In conclusion, medicinal plants represent a critical component of human health and well-being, with their importance extending from ancient traditions to modern scientific applications. Throughout history, these plants have been integral to the healing practices of diverse cultures, providing remedies for a wide array of ailments. The depth of knowledge embedded in indigenous and traditional medicine systems continues to shape the landscape of contemporary healthcare, emphasizing the enduring relevance of plant-based therapeutics.

The diversity of medicinal plants across the globe is staggering, with biodiversity hotspots rich in plant species offering a wealth of therapeutic potential. From tropical rainforests to temperate woodlands, medicinal plants have evolved unique properties that support health and resilience. However, the rapid depletion of these resources due to over-harvesting, habitat destruction, and climate change presents a significant challenge for future generations. Conservation efforts are essential to protect these invaluable plants, particularly endangered species that have not yet been fully studied for their medicinal potential.

Pharmacologically, medicinal plants are increasingly recognized for their antimicrobial, antiviral, and anti-inflammatory properties, alongside their ability to manage chronic diseases such as diabetes, cancer, and cardiovascular disorders. These plants not only provide natural alternatives to synthetic drugs but also contribute to the development of novel pharmaceuticals, offering treatments that are more affordable and accessible to underserved populations. The role of medicinal plants in pain management and in supporting overall well-being through anti-inflammatory therapies further demonstrates their expansive therapeutic value.

The integration of medicinal plants into modern healthcare, particularly in biotechnology and drug discovery, highlights their potential to revolutionize treatment paradigms. Numerous plant-derived compounds have been isolated and synthesized, contributing to the development of life-saving medications. However, challenges remain in scaling up production, ensuring quality control, and addressing the regulatory hurdles associated with herbal medicines. Overcoming these barriers requires concerted efforts from the scientific, pharmaceutical, and regulatory sectors to ensure that the efficacy and safety of medicinal plant products meet the highest standards.

Beyond their medical value, medicinal plants play a crucial role in the economies of many regions, providing livelihoods for millions of people involved in their cultivation, trade, and use. They form the backbone of local economies, especially in rural areas where traditional knowledge about plant-based remedies is passed down through generations. The social impact of medicinal plant cultivation and trade is profound, empowering communities, promoting gender equality, and enhancing economic development. As such, the promotion of medicinal plant industries can foster sustainable development and poverty alleviation, providing opportunities for marginalized groups and contributing to broader societal welfare.

Sustainability and conservation are central to the future of medicinal plants. As demand for these plants continues to grow, it is essential to adopt sustainable harvesting practices and prioritize the cultivation of medicinal plants in a way that supports ecosystem health. Incorporating these practices into policy frameworks and promoting ethical trade standards are vital steps in ensuring that the global supply of medicinal plants remains secure and equitable. The protection of these resources will also require international collaboration and the integration of traditional knowledge into modern conservation strategies.

Despite the many advantages of medicinal plants, challenges remain in terms of regulation, intellectual property rights, and ensuring access to traditional knowledge. The balance between protecting the intellectual property of indigenous communities and promoting open access to plant-based medicines will require nuanced, ethical, and legal frameworks that respect both traditional and modern knowledge systems. As the field of medicinal plants continues to evolve, ongoing research and technological advancements will play a critical role in bridging gaps between traditional practices and modern healthcare systems.

Looking to the future, medicinal plants will undoubtedly continue to play a pivotal role in achieving global health goals. Advances in biotechnology, research, and drug discovery hold the promise of unlocking new therapeutic potentials, while integrative medicine approaches will facilitate the blending of traditional and modern health practices. By focusing on sustainability, innovation, and community empowerment, medicinal plants will remain an indispensable resource in both healing and societal development for generations to come.

Chapter - 16

Overview of Forest-Based Industries

1. Introduction to Forest-Based Industries

Forest-based industries are essential components of the global economy, deeply rooted in the history of human civilization and continuing to shape modern economic, environmental, and social landscapes. These industries encompass a wide array of sectors, including timber-based industries, non-timber forest product (NTFP) industries, bioenergy, and eco-tourism, among others. They utilize resources derived from forests to produce goods and services that meet the needs of local and global markets. The importance of forest-based industries extends beyond their economic contributions, as they also play a critical role in sustainable development, employment generation, and ecological conservation.

Defined broadly, forest-based industries involve the extraction, processing, and commercialization of products sourced from forest ecosystems. These include timber products like sawmilled wood, plywood, and furniture, as well as non-timber resources such as resins, essential oils, bamboo, and medicinal plants. These industries also contribute to renewable energy production, chemical manufacturing, and the growing eco-tourism sector. Forest-based industries are vital in supporting rural economies, especially in forest-rich regions, and they form the backbone of many global supply chains, from construction materials to consumer goods. Historically, forest-based industries have been integral to the survival and progress of human societies. Early civilizations relied on forest resources for building shelters, producing tools, and obtaining fuel. Over time, advancements in technology and market demands led to the evolution of these industries into organized economic sectors. The industrial revolution marked a significant turning point, with the mechanization of wood processing and the rise of industries such as sawmilling, paper production, and furniture manufacturing. Today, forest-based industries continue to evolve, leveraging modern technologies to meet the demands of an increasingly resource-conscious and environmentally aware global market.

Economically, forest-based industries contribute significantly to both local and global economies. Timber-based industries are foundational to sectors like construction, infrastructure, and furniture manufacturing. Sawmilling and wood processing provide essential materials for building and manufacturing, while value-added products like plywood, veneers, and engineered wood products meet the demands of modern construction and design. Furniture manufacturing, which began as a local craft in many cultures, has grown into a global industry, blending traditional techniques with modern

DOI: 10.1201/9781042011063-16

design and mass production. These industries not only generate substantial revenue but also create millions of jobs, particularly in rural areas where forest resources are abundant. Non-timber forest product (NTFP)-based industries, while distinct from timber-based sectors, are equally important. NTFPs such as resins, gums, essential oils, bamboo, rattan, and edible forest products form the basis of industries that cater to diverse markets, from cosmetics and wellness to food and beverages. These products are often harvested sustainably, allowing local communities to benefit economically while preserving forest ecosystems. The rising consumer demand for natural and eco-friendly products has further propelled the growth of NTFP-based industries, making them a vital component of sustainable forest economies.

The bamboo and rattan industries, in particular, have emerged as significant contributors to forest-based economies. Bamboo, known for its rapid growth and versatility, is used in construction, furniture production, paper manufacturing, and even textile industries. Rattan, similarly, is highly valued for its use in furniture making and handicrafts. These industries not only provide economic opportunities but also promote sustainable resource use, as bamboo and rattan are renewable materials with minimal environmental impact when harvested responsibly. Innovations in processing and product development continue to enhance the value and applications of these resources.

Energy and biofuel industries derived from forest resources have gained prominence in recent years, particularly in the context of the global transition toward renewable energy. Traditional sources like charcoal and firewood remain vital in many developing countries, while modern biofuel production, including wood pellets and biomass energy, has become an important part of the global energy mix. These industries utilize forest residues such as branches, sawdust, and other byproducts, contributing to both energy security and waste reduction. The integration of forest-based bioenergy into national and global energy strategies underscores the potential of forests as sustainable sources of renewable energy.

Forest-based chemical industries also play a significant role in producing materials like tannins, dyes, and cellulose derivatives. These chemicals are used in a variety of applications, including leather processing, textiles, and pharmaceuticals. Advances in biotechnology have paved the way for the development of bioplastics and forest-derived polymers, which offer eco-friendly alternatives to conventional petroleum-based plastics. The growing emphasis on sustainability and circular economy models has further spurred innovation in forest-based chemical industries, positioning them as key players in reducing environmental impacts. Eco-tourism, as a segment of forest-based industries, capitalizes on the natural beauty, biodiversity, and cultural heritage associated with forests. This sector provides economic benefits to local communities while promoting forest conservation and environmental education. Activities such as guided forest tours, wildlife observation, and cultural experiences create sustainable livelihoods for local populations and generate funding for conservation efforts. However, the rapid growth of eco-tourism also poses challenges, such as habitat degradation and over-tourism, requiring careful management to ensure that the benefits of eco-tourism are not outweighed by its environmental costs. Despite their contributions, forest-based industries face significant challenges. Resource depletion and over-exploitation, particularly in timber industries, have led to deforestation and biodiversity loss. Unsustainable logging practices, combined with increasing global demand for forest products, place immense pressure on forest ecosystems. Environmental concerns such as carbon emissions, soil degradation, and water pollution associated with industrial activities further complicate the sustainability of these industries. Additionally, economic challenges, including market volatility, competition from synthetic alternatives, and regulatory constraints, create barriers to growth and innovation. The future of forest-based industries lies in embracing sustainability and innovation. The adoption of certification programs like the Forest Stewardship Council (FSC) and the Programme for the Endorsement of Forest Certification (PEFC) promotes responsible sourcing and sustainable practices in forest management. Integrating these industries into circular economy models, where resources are reused and waste is minimized, can further enhance their sustainability. Technological advancements in forest management, processing, and product development will be critical in addressing environmental and economic challenges while unlocking new opportunities.

1.1 Definition and Scope of Forest-Based Industries

Forest-based industries are defined as industries that rely on forest resources, either directly or indirectly, for raw materials, processing, and production. These industries encompass a broad spectrum of activities, ranging from the extraction of timber and non-timber products to the manufacturing and marketing of finished goods derived from forest resources. They are distinguished by their reliance on the ecological, economic, and social services provided by forests. The scope of forest-based industries is vast and diverse, reflecting the multiple uses of forest resources. The primary categories within this sector include:

» **Timber-Based Industries:** These industries focus on the harvesting, processing, and utilization of wood for construction, furniture, flooring, and other applications. Sawmills, plywood factories, and joineries are integral components of timber-based industries.

» **Pulp and Paper Industries**: Utilizing wood pulp as a raw material, these industries produce paper, cardboard, and packaging materials. They play a crucial role in the global supply chain, supporting sectors like education, publishing, and e-commerce.

» **Bioenergy and Biomass Industries**: Forests serve as a source of renewable energy, with industries producing biofuels, charcoal, and wood pellets for heating and electricity generation. This sub-sector contributes to reducing reliance on fossil fuels and mitigating climate change.

» **Non-Timber Forest Product (NTFP) Industries**: This category includes industries that process and market products such as resins, gums, essential oils, medicinal plants, and forest-based food items like honey, nuts, and berries. These industries often overlap with traditional practices and small-scale enterprises.

» **Eco-Tourism and Recreational Industries**: Although not directly linked to material extraction, these industries depend on forest landscapes and biodiversity for their services, catering to recreational, educational, and conservation-focused tourism.

The scope of forest-based industries extends beyond economic activities, encompassing social, environmental, and cultural dimensions. They provide livelihoods for millions of people, particularly in rural and forest-dependent communities, while also contributing to ecosystem services such as carbon sequestration, soil conservation, and water regulation. Furthermore, the integration of technology and innovation has expanded the scope of forest-based industries, leading to the development of advanced products like engineered wood, bio-composites, and bioplastics.

1.2 Historical Development and Evolution of Forest Industries

The evolution of forest-based industries is closely tied to the history of human civilization and the changing relationship between societies and natural resources. From the early use of wood for shelter and tools to the modern era of industrial forestry, forest industries have undergone significant transformations driven by technological advancements, economic demands, and environmental considerations.

Pre-Industrial Era

In the pre-industrial era, forests were primarily used for subsistence purposes, with timber being the main resource for construction, fuel, and tool-making. Traditional industries, such as shipbuilding and carpentry, relied heavily on local forests for materials. The use of non-timber products, such as resins, gums, and medicinal plants, also dates back to ancient times, forming the basis of traditional medicine and trade.

Industrial Revolution

The industrial revolution marked a turning point for forest-based industries, as technological innovations enabled large-scale extraction and processing of forest resources. The advent of steam engines and mechanized sawmills increased the efficiency of timber harvesting, while the development of the paper-making process created a new demand for wood pulp. During this period, forest industries became a driving force behind urbanization and economic expansion, but they also led to widespread deforestation and ecological degradation.

Post-Industrial Era and Modernization

In the 20th century, forest-based industries underwent further modernization, with advancements in machinery, chemical processing, and logistics enabling higher productivity and diversification. The rise of global trade and industrial forestry practices facilitated the establishment of large-scale plantations, particularly for fast-growing tree species like eucalyptus and pine. These developments, however, raised concerns about biodiversity loss and unsustainable resource use.

Emergence of Sustainability

The late 20th and early 21st centuries have seen a growing emphasis on sustainable forestry and the integration of environmental and social considerations into forest-based industries. Certification schemes like the Forest Stewardship Council (FSC) and Programme for the Endorsement of Forest Certification (PEFC) emerged to promote responsible forestry practices and ensure traceability in supply chains. Innovations in bioeconomy, circular economy models, and renewable energy have further transformed forest industries, aligning them with global sustainability goals.

Digital Transformation

The current era is characterized by the digital transformation of forest industries, with technologies like remote sensing, GIS, and blockchain enhancing resource management, monitoring, and transparency. Smart manufacturing and automation have improved efficiency and reduced waste, while e-commerce has created new opportunities for forest-based products in global markets.

1.3 Role of Forest-Based Industries in Global and Local Economies

Forest-based industries are a vital component of both global and local economies, contributing significantly to employment, trade, and economic development. Their role extends across multiple sectors, influencing not only industrial output but also rural livelihoods, environmental conservation, and cultural heritage.

Global Economic Contributions

At the global level, forest-based industries account for a substantial share of industrial production and trade. The paper and pulp sector alone generates billions of dollars annually, supporting industries like publishing, packaging, and printing. Similarly, the timber sector contributes to global construction and furniture markets, with exports of sawn wood, plywood, and engineered wood products driving international trade. The bioenergy sector is also growing rapidly, driven by increasing demand for renewable energy sources.

Countries with significant forest resources, such as Canada, Russia, Brazil, and Indonesia, derive considerable economic benefits from forest-based industries. These countries are major exporters of timber, pulp, and other forest products, contributing to their GDP and foreign exchange earnings. Additionally, the global market for NTFPs, such as essential oils, medicinal plants, and forest-based food products, is expanding, offering opportunities for small-scale producers and entrepreneurs.

Local Economic Impacts

At the local level, forest-based industries provide critical livelihoods for rural and forest-dependent communities. Small-scale enterprises, such as sawmills, craft workshops, and herbal medicine producers, contribute to local economies and create employment opportunities. For example, the processing of bamboo and rattan in Southeast Asia supports millions of livelihoods, while the collection and sale of wild honey and medicinal herbs sustain forest-dwelling communities in Africa and South Asia. In addition to direct economic benefits, forest-based industries play a role in community development by fostering skills, infrastructure, and market access. Many initiatives, such as community forestry and cooperatives, integrate local knowledge and participation into industrial activities, promoting equitable growth and social inclusion.

Environmental and Social Contributions

Beyond their economic impact, forest-based industries contribute to environmental conservation and sustainable development. Certified forest products support responsible resource management, while bioenergy and bioproducts reduce dependence on fossil fuels. Forest industries also promote ecosystem services, such as carbon sequestration, water regulation, and biodiversity preservation, which are critical for climate resilience and human well-being.

2. Timber-Based Industries

Timber-based industries are among the most significant components of forest-based economies, forming a critical link between natural resources and modern industrial applications. These industries encompass a wide range of activities that utilize wood, a versatile and renewable resource, for producing essential products such as construction materials, furniture, veneers, and composite woods. Timber-based industries serve as the backbone of rural and global economies, offering employment opportunities, generating revenue, and contributing to sustainable development. Moreover, they play a pivotal role in reducing the carbon footprint of industries by offering environmentally friendly alternatives to synthetic materials. From sawmilling and wood processing to the production of plywood and composite wood, and further to the furniture manufacturing sector, timber-based industries reflect the multifaceted uses of forests in modern life.

2.1 Sawmilling and Wood Processing Industries

The sawmilling and wood processing industries are foundational to the timber sector, forming the initial stage of converting raw forest resources into usable materials for construction, manufacturing, and energy production. Sawmills are critical infrastructure in this industry, as they are responsible for transforming harvested logs into lumber and other primary wood products. These industries supply raw materials to secondary and tertiary timber-based industries, creating a value chain that impacts various aspects of the economy and environment. Historically, sawmilling began as a small-scale manual operation, often serving local communities. However, with the advent of the industrial revolution and technological advancements, sawmilling has transformed into a highly mechanized and efficient industry capable of processing large volumes of wood while minimizing waste.

The sawmilling process begins with log transportation to the mill, where they are debarked to remove the outer layer. This step not only prepares the wood for further processing but also generates byproducts like bark, which can be used for mulch, bioenergy, or other applications. Once debarked, the logs are subjected to primary sawing using band saws, circular saws, or gang saws to produce rough lumber. The rough lumber is then sorted and resawn into finer dimensions based on its intended use, such as construction beams, planks, or boards for furniture manufacturing. The processed wood undergoes drying, either through natural air drying or kiln drying, to reduce its moisture content. This step is crucial for ensuring the wood's stability, durability, and resistance to warping or cracking during use. The final stage involves grading and sorting the lumber based on its quality, dimensions, and intended application.

Modern sawmills employ advanced technologies such as computer-controlled saws, automated log scanners, and optimization software to maximize yield and efficiency. These innovations ensure precise cutting, minimal waste, and better utilization of the raw material. For example, computer-aided sawmills can analyze the shape and size of logs to determine the most efficient way to cut them, maximizing the recovery of usable lumber. Additionally, sawmilling operations have embraced the circular economy model by utilizing byproducts such as sawdust, wood chips, and shavings for bioenergy, particleboard production, and other purposes. This holistic approach not only reduces waste but also contributes to the industry's sustainability. Sawmilling industries are vital to local economies, especially in rural areas where they provide employment and infrastructure development. They also play a significant role in global markets, supplying raw materials to major timber-exporting nations like Canada, Russia, and the United States. The industry's adherence to sustainable practices, such as sourcing timber from certified forests, further enhances its ecological contribution. Certification schemes like the Forest Stewardship Council (FSC) and the Programme for the Endorsement of Forest Certification (PEFC) ensure that sawmills operate responsibly, balancing economic growth with environmental stewardship.

2.2 Plywood, Veneer, and Composite Wood Production

The production of plywood, veneer, and composite wood represents a significant segment of timber-based industries, characterized by its focus on value addition and material innovation. These products are integral to construction, interior design, and furniture manufacturing, offering strength, durability, and versatility at a relatively low cost. The transition from solid wood to engineered wood products like plywood and composites has revolutionized the timber industry, making efficient use of raw materials and enabling the development of advanced applications.

Plywood is one of the most widely used engineered wood products, consisting of thin layers or veneers of wood glued together with their grains arranged at right angles. This cross-grain arrangement imparts exceptional strength, dimensional stability, and resistance to splitting, making plywood ideal for structural applications such as flooring, roofing, and wall sheathing. The production process begins with peeling logs into thin veneers using rotary lathes or slicing machines. These veneers are dried to a specific moisture content to ensure proper adhesion and are then coated with adhesive resins like phenol formaldehyde or urea formaldehyde. The veneers are stacked in alternating grain orientations and pressed under high heat and pressure to create sheets of plywood. After pressing, the sheets are trimmed, sanded, and graded for specific applications. Veneer, the thin sheets of wood obtained during plywood production, is also used independently for decorative purposes. High-quality veneers from species like oak, walnut, and mahogany are applied to furniture, cabinetry, and paneling to enhance aesthetics while reducing the demand for solid wood. The use of veneers not only adds value to the timber industry but also promotes sustainability by utilizing every part of the log efficiently.

Composite wood products, including medium-density fiberboard (MDF), particleboard, and oriented strand board (OSB), are manufactured by combining wood fibers, particles, or strands with adhesives under heat and pressure. These products offer cost-effective and versatile alternatives to solid wood, with applications ranging from furniture and cabinetry to flooring and packaging. For example, MDF is known for its uniform density and machinability, making it ideal for intricate designs and finishes in furniture production. OSB, on the other hand, is widely used in structural applications for its strength and affordability. The production of plywood, veneer, and composite wood has embraced sustainability and innovation, incorporating recycled wood fibers, low-emission adhesives, and energy-efficient processes. Advanced products like cross-laminated timber (CLT) and laminated veneer lumber (LVL) have gained prominence in the construction industry for their ability to replace traditional materials like steel and concrete. These engineered wood products offer environmental benefits such as reduced carbon footprint, improved insulation, and renewable sourcing, aligning with global efforts toward sustainable development.

2.3 Furniture Manufacturing

Furniture manufacturing is a prominent sector within timber-based industries, reflecting the transformation of raw wood into functional and artistic products. This industry spans a broad spectrum, from traditional artisanal craftsmanship to large-scale industrial production for global markets. Furniture manufacturing not only represents a high-value application of timber but also serves as a cultural and economic symbol, showcasing the interplay between design, craftsmanship, and industrial innovation.

Historically, furniture making was a localized craft, deeply rooted in cultural traditions and reliant on the skill of artisans. Local wood species were used to create furniture that met both functional and aesthetic needs, often incorporating intricate carvings and designs that reflected regional identities. Artisanal workshops, passed down through generations, produced handcrafted furniture that was valued for its uniqueness, durability, and cultural significance. Examples of such traditions include Chippendale furniture from England, Scandinavian minimalist designs, and intricate wooden carvings from Asia. The Industrial Revolution marked a significant shift in furniture manufacturing, introducing mechanization and standardization. The use of powered machinery enabled mass production, making furniture more affordable and accessible. Over time, technological advancements such as computer-aided design (CAD), automated cutting, and robotic assembly have further transformed the industry, allowing for precision manufacturing and customization at scale. Modern factories can produce large volumes of furniture while maintaining consistent quality and reducing production costs.

Furniture manufacturing has also evolved into a global trade powerhouse, driven by international demand and cross-border supply chains. Countries like China, Vietnam, and Poland are major exporters of furniture, supplying markets in Europe, North America, and the Middle East. The integration of e-commerce platforms has further expanded the reach of furniture manufacturers, enabling direct-to-consumer sales and personalized designs. At the same time, sustainable practices such as using certified wood, incorporating recycled materials, and adopting energy-efficient production methods have gained traction, aligning the industry with consumer preferences for environmentally responsible products. Despite its industrialization, furniture manufacturing continues to preserve elements of artisanal craftsmanship, particularly in high-end and bespoke markets. Custom-made furniture that blends traditional techniques with modern designs remains in demand, reflecting a growing appreciation for authenticity and heritage. Initiatives to promote artisanal furniture, such as government-backed craft cooperatives and trade fairs, have helped preserve cultural traditions while creating economic opportunities for skilled craftsmen.

The furniture manufacturing sector also contributes to local economies by generating employment, fostering entrepreneurship, and supporting related industries such as textiles, metals, and logistics. Community-based initiatives that integrate traditional knowledge and sustainable practices into furniture production offer pathways for inclusive development. For example, projects that promote bamboo and rattan furniture in Southeast Asia have empowered rural artisans while contributing to forest conservation. As the industry moves forward, challenges such as resource depletion, competition from synthetic alternatives, and shifting consumer preferences must be addressed through innovation and collaboration. Furniture manufacturing, as a cornerstone of timber-based industries, exemplifies the potential of sustainable design and production to create economic value while preserving the ecological and cultural significance of forests.

3. Pulp and Paper Industries

The pulp and paper industries are integral components of the global economy, providing essential products such as writing paper, packaging materials, and hygiene products. These industries represent one of the largest sectors within forest-based industries, relying heavily on renewable raw materials like wood, agricultural residues, and recycled fibers. The global demand for paper and paperboard has evolved significantly over the years, influenced by technological advancements, changing consumer preferences, and environmental concerns. Despite their economic importance, the pulp and paper industries face challenges related to resource use, environmental sustainability, and waste management.

3.1 Production Processes and Raw Material Sources

The production of pulp and paper involves complex processes that transform raw materials into various paper products. These processes are highly industrialized, relying on mechanical, chemical, and hybrid methods to produce pulp, the primary material for paper manufacturing. The choice of raw materials and production techniques significantly impacts the quality of the final product, production costs, and environmental footprint.

Raw Material Sources

The primary raw material for the pulp and paper industry is wood, sourced from forests and plantations. Softwoods like pine, spruce, and fir are preferred for their long fibers, which impart strength and durability to paper. Hardwoods such as birch, eucalyptus, and poplar are also used, particularly for producing fine paper and coated grades, due to their shorter fibers that enhance smoothness and printability. In addition to wood, non-wood fibers from agricultural residues like straw, bagasse (sugarcane residue), and hemp serve as alternative raw materials in regions where wood resources are scarce. Recycled paper is another significant raw material, contributing to the circular economy and reducing dependence on virgin fibers. Recycled fibers are obtained from post-consumer paper products, such as newspapers, cardboard, and office paper, which are reprocessed into new paper products.

Pulp Production Processes

The production of pulp, the fibrous material used to make paper, involves separating cellulose fibers from lignin and other non-fibrous components in raw materials. The main pulping processes include:

1. **Mechanical Pulping**: This method involves grinding wood logs or chips to separate fibers. While mechanical pulping is energy-intensive, it preserves most of the wood's lignin, resulting in high yields. The pulp produced is suitable for newsprint and lightweight paper but has lower strength and brightness.
2. **Chemical Pulping**: In this method, chemicals such as sodium hydroxide and sodium sulfide (used in the kraft process) or sulfurous acid (used in the sulfite process) are employed to dissolve lignin and separate cellulose fibers. Chemical pulping produces high-strength and high-quality pulp, suitable for premium paper products and packaging materials. However, it generates lower yields compared to mechanical pulping.
3. **Semi-Chemical Pulping**: Combining mechanical and chemical methods, this process softens wood with chemicals before mechanical refining. It offers a balance between yield and quality, making it suitable for producing corrugating medium and linerboard.
4. **Recycled Pulping**: This process involves de-inking and cleaning recovered paper to produce pulp suitable for various applications. Recycled pulping contributes to waste reduction and resource efficiency.

Paper Manufacturing

Once pulp is produced, it is processed into paper through a series of steps that include refining, blending, pressing, and drying. Additives such as fillers, sizing agents, and dyes are incorporated to enhance the paper's properties, such as smoothness, opacity, and water resistance. The finished paper is then cut, rolled, or packaged according to market requirements.

3.2 Global Paper Demand and Market Dynamics

The global paper and paperboard market has undergone significant shifts over the past few decades, driven by changes in technology, consumer behavior, and environmental awareness. While the demand for certain paper products has declined due to digitalization, other segments, such as packaging materials, have experienced robust growth.

Changing Demand Patterns

The advent of digital communication and the decline of print media have reduced the demand for traditional printing and writing paper in many regions. Newspapers, magazines, and office paper have seen declining consumption as electronic devices and digital platforms replace physical media. However, the packaging sector has emerged as a major growth driver for the paper industry. The rise of e-commerce, along with increased consumer preference for sustainable packaging, has fueled demand for corrugated boxes, paperboard cartons, and kraft paper. Hygiene and tissue products, including toilet paper, facial tissues, and paper towels, have also witnessed steady demand, driven by rising disposable incomes, urbanization, and increased health awareness. These products are particularly important in developing markets, where improving living standards are creating new opportunities for the pulp and paper industry.

Regional Market Dynamics

The paper industry is characterized by significant regional variations in production and consumption. Developed regions like North America and Europe have well-established paper industries but face declining demand for certain products due to digitalization and environmental regulations. In contrast, emerging markets in Asia, Africa, and Latin America are experiencing growth, supported by industrialization, urbanization, and expanding middle-class populations. China is the largest producer and consumer of paper globally, driven by its extensive manufacturing sector and domestic consumption. India, Indonesia, and Vietnam are also emerging as key players, benefiting from abundant raw materials, competitive labor costs, and growing domestic markets.

Global Trade and Competition

The paper industry is highly competitive, with major players like International Paper, Stora Enso, and UPM operating on a global scale. Trade in paper products is influenced by factors such as raw material availability, production costs, and environmental regulations. Countries with abundant forest resources and advanced infrastructure, such as Canada, Finland, and Brazil, are major exporters of pulp and paper products. Meanwhile, net importers like India and the Middle East rely on imports to meet their growing demand.

3.3 Environmental Impacts and Advances in Recycling Technologies

The pulp and paper industry has long been associated with significant environmental impacts, including deforestation, water and energy consumption, pollution, and waste generation. However, growing awareness and technological advancements have led to substantial improvements in the industry's environmental performance.

Environmental Challenges

- **Deforestation**: Unsustainable logging practices to meet raw material demand can lead to deforestation, habitat loss, and biodiversity decline. While plantation forestry has reduced pressure on natural forests, the conversion of native ecosystems into monoculture plantations remains a concern.
- **Water and Energy Use**: Pulp and paper production is water-intensive, with large volumes required for pulping, washing, and bleaching. Energy consumption is also high, particularly for mechanical pulping and drying processes, contributing to greenhouse gas emissions.
- **Pollution**: The industry generates significant amounts of wastewater containing lignin, chlorinated compounds, and other pollutants. Air emissions, including carbon dioxide, sulfur dioxide, and particulate matter, also impact local environments and contribute to climate change.

Advances in Recycling Technologies

Recycling has emerged as a key strategy for mitigating the environmental impacts of the pulp and paper industry. Technological advancements have improved the efficiency and quality of recycled fiber production, enabling greater use of recovered paper in various applications.

- **De-Inking Processes**: Modern de-inking technologies use chemical and mechanical methods to remove ink and contaminants from recovered paper, producing clean pulp suitable for printing and packaging products.
- **Energy-Efficient Recycling**: Advances in energy-efficient pulping and drying systems have reduced the carbon footprint of recycled paper production. For example, the use of steam recovery systems and energy-efficient dryers minimizes energy consumption.
- **Circular Economy Models**: The adoption of circular economy principles, such as designing products for recyclability and establishing robust collection systems, has increased the recovery rate of paper products. In some regions, recycling rates exceed 70%, demonstrating the effectiveness of these models.

Sustainable Practices and Innovations

The industry has embraced sustainability initiatives, including certification schemes, renewable energy adoption, and waste reduction. For instance, mills are increasingly using biomass residues as a source of renewable energy, reducing reliance on fossil fuels. Non-wood fibers and agricultural residues are being explored as alternative raw materials, further diversifying the industry's resource base.

4. Non-Timber Forest Product (NTFP)-Based Industries

Non-Timber Forest Products (NTFPs) form the backbone of numerous industries that rely on forest-derived resources other than timber. These industries represent an essential segment of the forest-based economy, offering diverse applications in sectors such as food, cosmetics, pharmaceuticals, and industrial manufacturing. NTFP-based industries not only contribute to economic growth but also provide livelihoods to millions of people, particularly in rural and forest-dependent communities. Unlike timber extraction, which often involves large-scale operations, NTFP-based industries are characterized by their localized, community-oriented nature, making them crucial for sustainable development and biodiversity conservation.

4.1 Industries Focused on Resin, Gum, and Latex Extraction

Industries centered on resin, gum, and latex extraction represent some of the oldest and most enduring uses of forest resources. These natural exudates, obtained from a wide variety of trees and plants, are integral to numerous industrial and commercial applications. Resins, gums, and latex are valued for their unique chemical properties, which make them indispensable in sectors such as adhesives, varnishes, pharmaceuticals, and manufacturing. The process of extraction, refinement, and commercialization of these materials underscores their economic and environmental significance. Resins, which are hydrocarbon-rich substances secreted by trees, are used extensively in industries producing varnishes, adhesives, and incense. Notable examples include pine resin, which is distilled into turpentine and rosin, both of which are widely used in the chemical and manufacturing industries. Turpentine serves as a solvent in paints and varnishes, while rosin is a critical ingredient in adhesives, sealants, and even stringed musical instrument bows. The extraction of resin typically involves tapping trees like pines, where the exudate is collected and processed without harming the tree, making it a sustainable practice when managed correctly.

Gums, such as gum arabic, guar gum, and tragacanth, are water-soluble polysaccharides used in food, pharmaceuticals, and cosmetics. Gum arabic, derived from Acacia species, is an emulsifier and stabilizer in beverages, confectionery, and pharmaceuticals. Guar gum, extracted from guar beans, is a thickening agent in food products and a key component in oil drilling fluids. These gums are prized for their versatility, with demand driven by their applications in both industrial and consumer products.

Latex, a milky fluid secreted by plants like rubber trees (Hevea brasiliensis), forms the basis of the rubber industry. Natural rubber, obtained by coagulating latex, is a critical raw material for producing tires, footwear, and medical devices. The cultivation of rubber trees in regions such as Southeast Asia, Africa, and South America has created vast plantation economies, supporting millions of livelihoods. Latex is also used in the production of gloves, balloons, and adhesives, underscoring its wide-ranging industrial applications. The extraction of resin, gum, and latex has significant socio-economic and environmental implications. These industries provide income to rural communities engaged in the collection, processing, and trade of these products. Sustainable harvesting practices, such as rotational tapping and ensuring tree health, are crucial for maintaining the ecological balance of forests. Furthermore, advancements in biochemistry and material science have led to the development of synthetic alternatives and improved processing methods, enhancing the efficiency and sustainability of these industries.

4.2 Essential Oils and Aromatics: Perfumes, Cosmetics, and Wellness

The essential oils and aromatics industry represents one of the most dynamic and rapidly growing segments of NTFP-based industries. Essential oils, extracted from the leaves, flowers, bark, and roots of aromatic plants, are highly concentrated plant extracts known for their distinctive fragrances and therapeutic properties. These oils form the backbone of several high-value industries, including perfumes, cosmetics, and wellness products, contributing significantly to global trade and economic development.

Essential oils such as lavender, eucalyptus, peppermint, and tea tree are widely used in aromatherapy, a holistic healing practice that leverages the therapeutic properties of plant-derived oils to promote physical and mental well-being. Aromatherapy products, including diffusers, massage oils, and bath salts, are part of a burgeoning wellness industry that has gained popularity worldwide. For instance, tea tree oil is renowned for its antimicrobial properties, making it a staple in skincare products for treating acne and infections, while lavender oil is prized for its calming effects, commonly used in stress relief and sleep aids.

The perfume and fragrance industry is another major consumer of essential oils and aromatics. Natural extracts like rose oil, sandalwood, and jasmine are blended with synthetic compounds to create perfumes and colognes. These high-value products cater to luxury markets, where demand for premium and natural ingredients continues to grow. Essential oils are also integral to the production of scented candles, air fresheners, and household cleaning products, demonstrating their versatility and widespread appeal.

In the cosmetics industry, essential oils are valued for their multifunctional properties, including anti-aging, moisturizing, and antimicrobial effects. Products such as facial serums, hair oils, and body lotions often incorporate essential oils as active ingredients to enhance efficacy and consumer appeal. For example, argan oil, derived from the kernels of the argan tree, is widely used in hair and skincare products for its nourishing and restorative properties. Similarly, frankincense oil, extracted from the resin of Boswellia trees, is a popular ingredient in anti-aging creams due to its ability to reduce wrinkles and promote skin regeneration.

The production of essential oils involves extraction techniques such as steam distillation, cold pressing, and solvent extraction. Steam distillation, the most common method, involves passing steam through plant material to release volatile compounds, which are then condensed and separated from water. Advances in distillation technology have improved the yield and quality of essential oils, enabling producers to meet growing demand efficiently.

The essential oils and aromatics industry plays a vital role in supporting rural economies, particularly in developing countries where aromatic plants are cultivated. Small-scale farmers and cooperatives benefit from the cultivation and processing of essential oil crops, such as lavender in Provence, France, and eucalyptus in Australia. Sustainable sourcing practices, including organic farming and fair trade certification, are increasingly prioritized to ensure the long-term viability of aromatic plant resources and equitable benefit-sharing with local communities.

4.3 Edible Forest Products: Food, Beverages, and Nutraceuticals

Edible forest products constitute a vital component of NTFP-based industries, providing food, beverages, and nutraceuticals that support nutrition, health, and economic livelihoods. These products, derived from fruits, nuts, seeds, mushrooms, and other forest resources, are increasingly recognized for their role in promoting sustainable diets and contributing to global food security. Edible forest products also serve as raw materials for value-added industries, ranging from gourmet foods and beverages to dietary supplements and functional foods.

Fruits such as wild berries, baobab, and açaí are popular edible forest products with high nutritional value and growing market demand. Açaí, harvested from the Amazon rainforest, is a superfood rich in antioxidants, omega fatty acids, and vitamins, widely used in smoothie bowls, beverages, and health supplements. Similarly, baobab fruit, sourced from Africa, is celebrated for its high vitamin C content and dietary fiber, making it a sought-after ingredient in the nutraceutical and functional food industries.

Nuts and seeds, including Brazil nuts, pine nuts, and shea nuts, are staple products of forest ecosystems, supporting both subsistence and commercial economies. Brazil nuts, collected from wild forests in the Amazon, are an important export commodity, known for their rich selenium content and flavor. Shea nuts, processed into shea butter, are used not only in food but also in cosmetics and skincare products, exemplifying the multifunctionality of edible forest products. Mushrooms, particularly wild and medicinal varieties, are another significant category of edible forest products. Species like truffles, chanterelles, and shiitake are prized for their culinary applications, while reishi and lion's mane mushrooms are valued for their health benefits, such as boosting immunity and enhancing cognitive function. The cultivation and foraging of mushrooms provide income to rural communities and promote sustainable forest management practices.

Beverages derived from forest products, such as coffee, tea, and herbal infusions, are integral to global trade and consumption. Specialty coffees, such as shade-grown varieties, emphasize sustainability and biodiversity conservation. Herbal teas made from ingredients like hibiscus, lemongrass, and forest herbs are celebrated for their health-promoting properties, offering consumers a natural alternative to synthetic beverages.

The nutraceutical industry has emerged as a major consumer of edible forest products, incorporating them into dietary supplements and functional foods that address specific health needs. Products such as omega-3-rich chia seeds, antioxidant-packed açaí powder, and adaptogenic mushroom extracts exemplify the growing integration of forest resources into modern health and wellness trends. Edible forest products contribute significantly to rural development by supporting livelihoods, enhancing food security, and promoting sustainable land use. Community-based initiatives, such as forest food cooperatives and agroforestry systems, integrate the collection and cultivation of edible products with conservation objectives. Certifications like organic and fair trade add value to these products, ensuring market access and equitable income distribution for producers.

5. Bamboo and Rattan Industries

Bamboo and rattan industries represent some of the most versatile and sustainable sectors within forest-based economies. These industries utilize two of the most abundant and renewable natural resources to produce a wide array of products that span construction, furniture, handicrafts, and industrial applications. Bamboo, often referred to as "poor man's timber," is a fast-growing grass with incredible strength, flexibility, and durability, making it an ideal material for structural and aesthetic applications. Similarly, rattan, a climbing palm native to tropical regions, is prized for its pliability, light weight, and decorative appeal. Together, these resources have supported traditional livelihoods for centuries while emerging as key players in modern, sustainable development initiatives. The bamboo and rattan industries are deeply intertwined with cultural heritage, economic growth, and environmental conservation, offering immense potential for innovative applications and value addition.

5.1 Applications in Construction, Furniture, and Handicrafts

Bamboo and rattan have long been integral to construction, furniture making, and handicrafts, providing functional and aesthetically pleasing products for daily life and cultural expression. In construction, bamboo stands out as one of the most eco-friendly and cost-effective building materials, with applications ranging from scaffolding to load-bearing structures. Its remarkable tensile strength, which rivals that of steel, makes it a preferred material for building frameworks, bridges, and roofing in many parts of Asia, Africa, and Latin America. Bamboo is also widely used in earthquake-resistant construction due to its flexibility and ability to absorb seismic shocks. Modern architectural projects have embraced bamboo for its sustainable properties, incorporating it into eco-friendly buildings, resorts, and pavilions that showcase its structural and aesthetic versatility. Laminated bamboo panels and composite boards are increasingly being used in flooring, wall paneling, and ceilings, combining the natural beauty of bamboo with enhanced durability and design flexibility.

In the furniture industry, bamboo and rattan are highly valued for their lightweight yet sturdy characteristics, making them ideal for crafting chairs, tables, beds, and storage units. Traditional bamboo furniture, known for its rustic charm and durability, has been a staple in rural households for generations. Meanwhile, rattan furniture, with its smooth finish and intricate weaving patterns, has gained international recognition for its elegance and functionality. The global demand for bamboo and rattan furniture has surged, driven by trends in sustainable living and minimalist interior design. Designers and manufacturers are exploring innovative techniques to blend bamboo and rattan with modern materials such as metal and glass, creating hybrid furniture pieces that cater to contemporary aesthetics.

Handicrafts made from bamboo and rattan hold immense cultural and economic significance, particularly in regions where these resources are abundant. Artisans skillfully transform bamboo and rattan into decorative items, kitchenware, musical instruments, and woven baskets, showcasing the rich craftsmanship and heritage of local communities. These handicrafts are not only prized for their artistic value but also play a vital role in supporting rural livelihoods. Bamboo and rattan craft traditions are often passed down through generations, preserving cultural identities while adapting to modern market demands. Export-oriented industries have emerged around bamboo and rattan handicrafts, catering to global markets with an appreciation for handmade, sustainable, and ethically sourced products.

5.2 Sustainable Management and Economic Potential of Bamboo and Rattan

The sustainable management of bamboo and rattan resources is crucial for ensuring their long-term availability and maximizing their economic potential. Both bamboo and rattan are renewable resources with rapid growth rates, making them excellent alternatives to timber in mitigating deforestation and promoting sustainable forestry practices. Bamboo, in particular, can grow up to several feet per day, depending on the species, and can be harvested within three to five years without replanting, as its root system continues to produce new shoots. This regenerative ability makes bamboo an ideal crop for degraded land restoration, soil erosion control, and carbon sequestration. Similarly, rattan, which thrives in tropical rainforests, contributes to forest health by supporting biodiversity and stabilizing ecosystems.

The economic potential of bamboo and rattan industries is vast, with applications spanning multiple sectors and value chains. Bamboo alone supports a global market worth billions of dollars, encompassing construction materials, furniture, handicrafts, textiles, bioenergy, and even food products. For example, bamboo shoots are a popular delicacy in Asian cuisine, offering nutritional benefits and market opportunities for farmers. Rattan, though less widely cultivated than bamboo, commands significant value in furniture and handicrafts, particularly in export markets where handcrafted products are highly sought after. Countries like China, India, Indonesia, and Vietnam have capitalized on their abundant bamboo and rattan resources, developing thriving industries that contribute to rural development, employment generation, and export earnings.

Sustainable management practices are essential to unlocking the full economic potential of bamboo and rattan industries. Agroforestry systems that integrate bamboo and rattan cultivation with other crops offer a viable solution for smallholder farmers, providing diversified income streams and enhancing food security. Certification schemes such as the Forest Stewardship Council (FSC) and the International Network for Bamboo and Rattan (INBAR) promote responsible harvesting, traceability, and fair trade, ensuring that bamboo and rattan products meet international sustainability standards. Community-based initiatives, such as cooperatives and producer associations, play a vital role in managing bamboo and rattan resources equitably, empowering local communities to benefit from their natural assets while conserving them for future generations.

5.3 Innovations in Bamboo Processing and Product Development

The bamboo and rattan industries have witnessed significant advancements in processing technologies and product development, driven by growing demand for sustainable materials and innovative applications. These innovations have expanded the utility of bamboo and rattan beyond traditional uses, opening new avenues for industrial production, high-value products, and eco-friendly alternatives to conventional materials. In bamboo processing, technological innovations have focused on improving efficiency, reducing waste, and enhancing product quality. For instance, advanced machinery for splitting, slicing, and laminating bamboo has enabled the production of uniform, high-quality panels, boards, and veneers. These engineered bamboo products are used in modern construction, furniture, and interior design, offering superior strength, durability, and aesthetic appeal. Cross-laminated bamboo, a sustainable alternative to cross-laminated timber, has emerged as a key material in green building projects, providing structural integrity and environmental benefits. Bamboo fibers are also being used to produce textiles, paper, and biodegradable packaging, addressing the need for sustainable solutions in fashion, stationery, and e-commerce industries.

Rattan processing has similarly benefited from technological advancements, particularly in weaving and finishing techniques. Innovations in rattan treatment, such as chemical-free curing and dyeing processes, have improved the durability and eco-friendliness of rattan products. Designers and manufacturers are experimenting with combining rattan with other materials, such as leather, metal, and fabric, to create hybrid products that appeal to modern consumers. Rattan's flexibility and aesthetic versatility make it a preferred material for creating luxury furniture and bespoke interior décor items, catering to high-end markets. Bamboo and rattan are also making inroads into cutting-edge industries such as bioplastics, composites, and renewable energy. Bamboo fibers are being incorporated into bio-composites used in automotive and aerospace manufacturing, providing lightweight, durable, and sustainable alternatives to synthetic materials. Bamboo charcoal, derived from pyrolyzing bamboo, is gaining popularity as a renewable energy source and a component in water filtration, air purification, and soil improvement. The development of bamboo-based biofuels and bioplastics further highlights the potential of bamboo as a sustainable resource for reducing reliance on fossil fuels and addressing plastic pollution.

The integration of digital technologies, such as computer-aided design (CAD), 3D printing, and blockchain, is transforming the bamboo and rattan industries, enhancing product innovation, supply chain transparency, and consumer engagement. CAD tools enable designers to create intricate patterns and prototypes for bamboo and rattan products, while 3D printing facilitates the production of customized and complex structures. Blockchain technology ensures traceability and authenticity in bamboo and rattan supply chains, allowing consumers to verify the sustainability and ethical sourcing of their purchases. The future of bamboo and rattan industries lies in their ability to combine traditional knowledge with modern innovation, fostering a balance between cultural heritage, economic development, and environmental sustainability. By leveraging their versatility, renewability, and low ecological footprint, bamboo and rattan can continue to serve as cornerstones of sustainable industries, contributing to a greener, more inclusive global economy. These industries not only offer solutions to pressing challenges such as deforestation, climate change, and poverty but also exemplify the transformative potential of forest-based resources in driving sustainable development.

6. Energy and Biofuel Industries

The energy and biofuel industries that rely on forest resources are integral to both traditional and modern energy systems. Forests have historically provided charcoal and firewood as primary energy sources, especially in rural and developing regions. These traditional uses of wood fuel continue to sustain millions of livelihoods and provide essential energy for cooking and heating. At the same time, advancements in bioenergy technology have transformed forest residues and biomass into significant contributors to renewable energy. Modern biofuels, such as wood pellets and other biomass products, are emerging as sustainable alternatives to fossil fuels, aligning with global efforts to achieve carbon neutrality and renewable energy goals. The integration of forest residues into energy systems not only reduces waste but also fosters a circular economy, creating value from materials that would otherwise remain unused.

6.1 Charcoal and Firewood as Traditional Energy Sources

Charcoal and firewood have been fundamental to energy systems for centuries, particularly in regions where access to modern energy infrastructure is limited. As the earliest forms of bioenergy, these resources have provided essential fuel for cooking, heating, and industrial processes. Despite the global transition toward modern energy systems, charcoal and firewood remain indispensable in many parts of the world, especially in rural areas of Africa, Asia, and Latin America.

Firewood, the most basic form of forest-based energy, is typically collected from fallen branches, deadwood, or purposefully harvested trees. It serves as a critical energy source for domestic cooking and heating, especially in developing regions where it constitutes up to 90% of household energy consumption. Firewood's accessibility and affordability make it a lifeline for low-income communities, even as concerns over deforestation and indoor air pollution highlight the need for improved practices. Charcoal, produced by the slow pyrolysis of wood in low-oxygen conditions, offers a more concentrated and portable energy source compared to firewood. Its higher energy density and cleaner combustion make it a preferred fuel for urban households and small-scale industries, such as brick kilns, iron smelting, and pottery making. Charcoal production, often carried out in traditional earth kilns or more advanced retort systems, supports millions of livelihoods worldwide. However, the environmental and social challenges associated with traditional charcoal production, such as forest degradation, inefficient conversion processes, and unsustainable harvesting, necessitate the adoption of more sustainable practices.

Efforts to improve the sustainability of charcoal and firewood use have focused on promoting efficient cooking technologies, such as improved stoves that reduce fuel consumption and indoor air pollution. Community-based forest management programs also play a crucial role in ensuring sustainable wood fuel harvesting, fostering a balance between energy needs and forest conservation. These initiatives not only mitigate environmental impacts but also enhance energy security for rural and peri-urban communities.

6.2 Modern Biofuel Production: Wood Pellets and Biomass Energy

The shift toward renewable energy has positioned modern biofuels, such as wood pellets and biomass energy, as pivotal components of sustainable energy systems. These innovations leverage forest resources and residues to produce clean, efficient, and renewable energy alternatives to fossil fuels. Modern biofuels represent a significant advancement in the utilization of forest biomass, offering scalable solutions for both domestic and industrial energy needs.

Wood Pellets

Wood pellets are a highly efficient and standardized form of biomass fuel made from compacted sawdust, wood chips, and other forest residues. Their uniform size, high energy density, and low moisture content make them ideal for use in heating systems, power plants, and industrial boilers. Pellets are commonly used in residential pellet stoves for home heating and in large-scale power stations as a substitute for coal, reducing greenhouse gas emissions. The production of wood pellets has grown rapidly in recent years, driven by increasing demand in Europe, North America, and Asia. The manufacturing process of wood pellets involves grinding forest residues into fine particles, drying them to reduce moisture, and compressing them under high pressure to form dense, cylindrical pellets. The lack of chemical additives in pellet production ensures that they remain a clean and sustainable energy source. Pellet production not only creates value from byproducts of the timber and sawmill industries but also reduces waste, aligning with the principles of a circular economy.

Biomass Energy

Biomass energy encompasses a broader category of renewable energy derived from organic materials, including wood, agricultural residues, and municipal waste. Forest biomass, in the form of logs, chips, or residues, is a primary feedstock for biomass power plants, which generate electricity and heat through combustion or gasification. Biomass energy systems are versatile, catering to both on-grid and off-grid applications, and are increasingly integrated into district heating networks and industrial processes. The environmental benefits of biomass energy lie in its potential for carbon neutrality. The carbon dioxide released during biomass combustion is offset by the carbon absorbed by trees during their growth, creating a closed carbon cycle. However, ensuring the sustainability of biomass energy requires responsible sourcing practices, efficient conversion technologies, and adherence to certification standards, such as the Sustainable Biomass Program (SBP).

6.3 Contribution of Forest Residues to Renewable Energy Goals

The utilization of forest residues for renewable energy production is a critical strategy for achieving global climate goals, enhancing energy security, and promoting sustainable resource use. Forest residues, including branches, bark, sawdust, and logging slash, represent a vast and underutilized resource with significant potential for energy generation. Their conversion into biofuels and biomass energy not only reduces waste but also supports the transition to a low-carbon economy.

Role in Climate Mitigation

Forest residues contribute to renewable energy goals by providing a low-carbon alternative to fossil fuels, helping to mitigate greenhouse gas emissions. Biomass-based energy systems are particularly valuable in decarbonizing sectors that are difficult to electrify, such as heavy industry, aviation, and maritime transport. Advanced biofuels, derived from lignocellulosic biomass, offer sustainable solutions for these sectors, reducing their reliance on conventional petroleum-based fuels.

Economic and Rural Development

The collection, processing, and utilization of forest residues create significant economic opportunities, particularly in rural areas where forestry activities are concentrated. Forest-based bioenergy projects generate employment in biomass collection, pellet production, and energy plant operations, contributing to rural development and poverty alleviation. Community-led bioenergy initiatives, such as small-scale biogas plants or biomass heating systems, enhance energy access and resilience in off-grid regions.

Circular Economy and Waste Management

The integration of forest residues into energy systems exemplifies the principles of a circular economy, where waste materials are repurposed into valuable products. By utilizing residues that would otherwise decompose or be burned in open fields, bioenergy production reduces air pollution, prevents resource wastage, and adds value to forestry operations. Innovations in bioconversion technologies, such as torrefaction and pyrolysis, are further enhancing the efficiency and feasibility of utilizing forest residues for energy.

Policy and Global Goals

The role of forest residues in renewable energy is reinforced by supportive policies and international commitments. Renewable energy targets set by organizations like the United Nations Framework Convention on Climate Change (UNFCCC) and the European Union's Renewable Energy Directive highlight the importance of bioenergy in meeting global sustainability goals. National bioenergy strategies, incentives for biomass projects, and investments in research and development are driving the growth of forest-based renewable energy.

7. Forest-Based Chemical Industries

The forest-based chemical industries are a vital subset of forest utilization, leveraging the chemical properties of wood and plant derivatives to produce a wide range of products essential for modern life. From traditional uses like tannins and dyes to innovative developments in bioplastics and forest-based polymers, this sector has continually evolved to meet industrial and consumer demands. By utilizing renewable resources, forest-based chemical industries contribute to sustainability, economic growth, and the reduction of reliance on fossil fuels.

7.1 Production of Chemicals from Wood and Plant Derivatives

The production of chemicals from wood and plant derivatives is a cornerstone of forest-based industries. The unique chemical composition of wood, comprising cellulose, hemicellulose, lignin, and extractives, offers a diverse array of opportunities for chemical production. These components serve as raw materials for a range of industries, including paper and pulp, textiles, pharmaceuticals, and energy.

Cellulose and Its Derivatives

Cellulose, the primary structural component of wood and plant fibers, is a versatile raw material for producing various chemicals and products. It is widely used in the manufacture of cellulose derivatives, such as cellulose acetate, carboxymethyl cellulose (CMC), and methylcellulose. These derivatives find applications in textiles, food additives, pharmaceuticals, and cosmetics. For example, cellulose acetate is a key material in the production of photographic films, cigarette filters, and eyeglass frames, while CMC is used as a thickening agent in food and personal care products.

Lignin Utilization

Lignin, the complex polymer that provides rigidity to wood, has traditionally been considered a byproduct of the paper and pulp industry. However, advances in lignin chemistry have unlocked its potential as a valuable feedstock for producing chemicals such as vanillin, phenols, and adhesives. Lignin-based products are increasingly being used in bioplastics, carbon fibers, and sustainable construction materials, reducing dependence on petroleum-derived chemicals.

Extractives and Resins

Plant extractives, including essential oils, gums, and resins, are another important source of forest-based chemicals. Turpentine, a byproduct of pine resin, is used in solvents, paints, and cleaning products, while rosin serves as a base material for adhesives, printing inks, and chewing gum. Other extractives, such as polyphenols and flavonoids, are being explored for their antioxidant and antimicrobial properties in nutraceuticals and pharmaceuticals. The production of chemicals from wood and plant derivatives exemplifies the value of forest resources beyond traditional uses, contributing to a circular economy by utilizing every part of the tree. Innovations in biorefinery technologies, which integrate the processing of biomass into multiple products, are further enhancing the efficiency and sustainability of forest-based chemical production.

7.2 Industrial Applications of Tannins, Dyes, and Cellulose Derivatives

The industrial applications of tannins, dyes, and cellulose derivatives underscore the diverse uses of forest-based chemicals across sectors such as textiles, leather, pharmaceuticals, and construction. These natural compounds, extracted from wood and plant materials, offer sustainable alternatives to synthetic chemicals, aligning with growing consumer and regulatory demand for eco-friendly solutions.

Tannins

Tannins, water-soluble polyphenolic compounds extracted from the bark, leaves, and fruits of trees, are widely used in leather tanning, adhesives, and water treatment. In the leather industry, tannins are essential for converting animal hides into durable and flexible leather. Natural tannins, such as those derived from oak and chestnut, are valued for their ability to produce high-quality leather with minimal environmental impact. In adhesives, tannins are used to manufacture wood adhesives and binders for particleboard and plywood, offering a renewable alternative to petroleum-based resins. Additionally, tannins are employed in water treatment processes to remove heavy metals and contaminants, highlighting their role in environmental sustainability.

Natural Dyes

Natural dyes, extracted from plants, trees, and other natural sources, have been used for centuries to impart color to textiles, paper, and cosmetics. The resurgence of interest in natural dyes is driven by concerns over the environmental and health impacts of synthetic dyes. Forest-derived dyes, such as indigo, catechu, and logwood, are valued for their non-toxic properties and rich hues. These dyes are used in fashion and home textiles, catering to niche markets for sustainable and artisanal products. Advances in dye extraction and stabilization techniques have enhanced the usability of natural dyes in industrial applications, expanding their appeal to larger markets.

Cellulose Derivatives

Cellulose derivatives are among the most versatile forest-based chemicals, finding applications across diverse industries. In construction, methylcellulose is used as a binder and thickener in cement and plaster, improving workability and strength. In the food industry, cellulose derivatives such as CMC are used as stabilizers and thickeners in products like ice cream, sauces, and baked goods. Pharmaceuticals also rely on cellulose derivatives for tablet binding, drug delivery systems, and wound dressings. The biocompatibility and biodegradability of cellulose derivatives make them ideal for applications requiring sustainability and safety. The industrial applications of tannins, dyes, and cellulose derivatives illustrate the adaptability of forest-based chemicals in meeting the demands of modern industries while supporting environmental goals. By replacing synthetic counterparts, these natural compounds reduce the ecological footprint of industrial processes and contribute to sustainable development.

7.3 Emerging Trends in Bioplastics and Forest-Based Polymers

The growing concern over plastic pollution and the environmental impact of petroleum-based materials has spurred the development of bioplastics and forest-based polymers. These innovative materials, derived from renewable forest resources, offer sustainable alternatives for packaging, construction, automotive, and consumer goods industries. Emerging trends in this sector highlight the potential of forests to provide solutions to some of the most pressing environmental challenges of our time.

Bioplastics from Cellulose and Hemicellulose

Cellulose and hemicellulose, abundant in wood and plant biomass, are key feedstocks for producing bioplastics. Cellulose-based bioplastics, such as cellulose acetate and cellulose propionate, are used in packaging, films, and coatings. These materials are biodegradable and compostable, addressing the growing demand for sustainable packaging solutions. Advances in cellulose chemistry have enabled the production of transparent, flexible, and water-resistant bioplastics, expanding their applications in food packaging and electronics. Hemicellulose, a complex carbohydrate found alongside cellulose, is being explored for its potential in bioplastic production. Hemicellulose-derived films and coatings offer barrier properties that enhance the shelf life of packaged goods, making them suitable for food and beverage industries.

Lignin-Based Polymers

Lignin, traditionally considered a waste product in the pulp and paper industry, has emerged as a valuable resource for polymer production. Lignin-based thermoplastics and thermosetting resins are being developed for use in construction, automotive components, and electronics. These polymers offer strength, durability, and heat resistance while reducing the reliance on fossil fuels. Lignin's ability to replace phenols in adhesives and epoxy resins further underscores its versatility and environmental benefits.

Nanocellulose

Nanocellulose, derived from cellulose fibers, represents a breakthrough in forest-based polymers. This material, characterized by its high strength, low weight, and biodegradability, has applications in composites, coatings, electronics, and biomedical devices. Nanocellulose-reinforced composites are being used to manufacture lightweight and durable automotive parts, reducing vehicle weight and improving fuel efficiency. In the medical field, nanocellulose is used in wound dressings, drug delivery systems, and tissue engineering due to its biocompatibility and antimicrobial properties.

Sustainability and Market Potential

The development of bioplastics and forest-based polymers aligns with global sustainability goals, addressing issues such as waste reduction, resource efficiency, and climate change mitigation. These materials have the potential to replace a significant portion of conventional plastics, reducing the environmental impact of packaging, construction, and consumer goods industries. Governments and corporations are increasingly investing in bioplastics, with policies promoting renewable materials and bans on single-use plastics driving market growth. The forest-based chemical industries are at the forefront of innovation, transforming renewable resources into sustainable solutions for a wide range of applications. By harnessing the potential of wood and plant derivatives, these industries not only contribute to economic growth but also address critical environmental challenges, paving the way for a greener future.

8. Eco-Tourism and Recreation Industries

Forests play a vital role in supporting eco-tourism and recreation industries, serving as natural destinations for millions of visitors seeking respite, adventure, and a connection with nature. Eco-tourism leverages the ecological, cultural, and aesthetic values of forests to create sustainable tourism experiences that benefit both local communities and the environment. The global expansion of eco-tourism highlights its potential to generate economic growth while fostering conservation and environmental awareness. However, as tourism grows, challenges arise in balancing development with the preservation of fragile ecosystems.

8.1 Role of Forests in Supporting Tourism and Recreational Activities

Forests are among the most sought-after destinations for tourism and recreation, offering a wide range of activities that cater to diverse interests and preferences. From hiking, camping, and wildlife observation to spiritual retreats and adventure sports, forests provide a unique environment for recreation that combines natural beauty, biodiversity, and cultural heritage. Forests are home to iconic landscapes, such as national parks, wildlife sanctuaries, and biosphere reserves, that attract tourists from around the world. These areas often serve as the backbone of eco-tourism, showcasing the rich biodiversity and ecological significance of forest ecosystems. For example, the Amazon Rainforest, the Sundarbans Mangroves, and the Black Forest are internationally renowned destinations that draw millions of visitors annually. Forests also host unique geological features, such as waterfalls, caves, and cliffs, which enhance their appeal as natural attractions. Recreational activities in forests range from low-impact pursuits like birdwatching, photography, and nature walks to high-adrenaline adventures such as zip-lining, rock climbing, and mountain biking. Many forests also serve as spiritual and cultural retreats, offering spaces for meditation, yoga, and traditional healing practices. Sacred groves and forest temples hold cultural and religious significance for local communities, further enriching the tourism experience.

The role of forests in supporting recreational activities extends beyond physical landscapes to include their sensory and emotional appeal. The tranquility, fresh air, and immersive experiences provided by forests contribute to physical and mental well-being, making them ideal destinations for health and wellness tourism. Forest bathing, or shinrin-yoku, a Japanese practice of immersing oneself in the forest environment, has gained popularity as a therapeutic activity that reduces stress and enhances mental clarity. Forests also serve as venues for environmental education and awareness, offering opportunities for tourists to learn about ecosystems, wildlife, and conservation efforts. Educational programs, guided tours, and interpretive centers in forested areas help foster a deeper appreciation for nature and inspire responsible behavior among visitors. By connecting people with nature, forests play a crucial role in promoting sustainable lifestyles and conservation ethics.

8.2 Economic and Conservation Benefits of Eco-Tourism

Eco-tourism, which emphasizes sustainable travel and responsible interaction with natural environments, offers significant economic and conservation benefits. This form of tourism not only generates revenue and employment but also supports biodiversity conservation, community development, and cultural preservation.

Economic Benefits

Eco-tourism is a major driver of local and regional economies, particularly in rural and remote areas where other economic opportunities may be limited. By attracting visitors to forested destinations, eco-tourism creates income streams for various stakeholders, including tour operators, accommodation providers, local artisans, and food vendors. For example, national parks and wildlife reserves often generate substantial revenue from entrance fees, guided tours, and eco-lodges, which are reinvested into conservation efforts and infrastructure development. The employment opportunities created by eco-tourism extend across multiple sectors, from hospitality and transportation to craft production and nature guiding. Local communities, especially those residing near forested areas, benefit directly from these opportunities, which enhance their livelihoods and reduce dependence on unsustainable practices like poaching or illegal logging. In many cases, eco-tourism projects actively involve indigenous and marginalized communities, empowering them through skill development, entrepreneurship, and equitable benefit-sharing.

Conservation Benefits

One of the most significant advantages of eco-tourism is its role in funding and promoting conservation efforts. Revenue generated from eco-tourism is often used to support activities such as habitat restoration, anti-poaching patrols, and wildlife monitoring. By creating economic value from the preservation of natural resources, eco-tourism provides a strong incentive for governments, communities, and private stakeholders to protect forest ecosystems.

Eco-tourism also raises awareness about conservation issues, encouraging visitors to become advocates for environmental protection. Educational programs and interpretive experiences in eco-tourism destinations help visitors understand the ecological importance of forests and the threats they face, fostering a sense of responsibility and stewardship. Cultural preservation is another key benefit of eco-tourism. Many forested areas are home to indigenous communities whose traditional knowledge and practices are deeply intertwined with the local environment. By showcasing their cultural heritage through eco-tourism, these communities can preserve their traditions, languages, and crafts while gaining recognition and respect from the global audience.

Sustainable Development Goals

Eco-tourism aligns with several United Nations Sustainable Development Goals (SDGs), including those related to poverty alleviation, gender equality, and climate action. By integrating economic, social, and environmental objectives, eco-tourism serves as a model for sustainable development, offering a pathway to balance human needs with ecological integrity.

8.3 Challenges in Balancing Tourism Development with Forest Preservation

Despite its benefits, eco-tourism faces several challenges in balancing tourism development with the preservation of forest ecosystems. The increasing popularity of eco-tourism destinations has led to concerns about over-tourism, habitat degradation, and cultural commodification, highlighting the need for careful planning and management.

Environmental Impacts

Unregulated or poorly managed tourism can lead to significant environmental impacts, including soil erosion, water pollution, and habitat disturbance. The construction of infrastructure such as roads, hotels, and parking lots often involves clearing forested land, fragmenting habitats, and disrupting wildlife corridors. Activities such as off-road driving, camping, and trekking can damage fragile ecosystems, while littering and waste generation pose additional threats to forest health. Wildlife disturbance is a critical issue in forest-based tourism. Frequent human presence, noise, and artificial lighting can disrupt animal behavior, migration patterns, and breeding cycles. In extreme cases, close encounters with tourists can lead to stress, aggression, or habituation in wildlife, undermining conservation objectives.

Social and Cultural Challenges

The influx of tourists into forested areas can have social and cultural repercussions for local communities. While eco-tourism provides economic opportunities, it can also lead to the commodification of cultural practices and traditions, reducing their authenticity and value. The integration of global tourism norms may overshadow local identities and disrupt traditional ways of life. Conflicts over resource use are another concern, as tourism activities may compete with local needs for water, land, and energy. In some cases, the benefits of eco-tourism are unevenly distributed, with external operators or corporations reaping most of the profits while local communities receive limited rewards.

Carrying Capacity and Over-Tourism

Determining the carrying capacity of eco-tourism destinations is essential to prevent over-tourism, which can strain ecosystems and infrastructure. Exceeding the carrying capacity can lead to overcrowding, environmental degradation, and a decline in the quality of visitor experiences. For instance, popular destinations like national parks often face challenges in managing visitor numbers during peak seasons, resulting in congestion, resource depletion, and increased waste.

Solutions and Strategies

Addressing these challenges requires a multi-faceted approach that integrates stakeholder collaboration, regulatory frameworks, and community participation. Effective solutions include:

» **Sustainable Planning**: Developing eco-tourism infrastructure that minimizes environmental impact, such as eco-lodges, renewable energy systems, and low-impact transportation.

» **Visitor Management**: Implementing measures such as visitor quotas, timed entry, and designated trails to regulate tourist activities and reduce pressure on ecosystems.

» **Community Engagement**: Ensuring that local communities are actively involved in decision-making, benefit-sharing, and conservation initiatives, fostering a sense of ownership and stewardship.

» **Education and Awareness**: Promoting responsible tourism practices through visitor education, codes of conduct, and certification schemes such as the Global Sustainable Tourism Council (GSTC).

Conclusion

Forest-based industries are integral to the global economy, environment, and human livelihoods, offering a wide range of products and services that meet critical societal needs. Their contribution spans timber-based industries, non-timber forest product (NTFP)-based industries, renewable energy, and eco-tourism. These industries not only generate significant economic activity but also serve as a cornerstone for sustainable development, employment, and cultural preservation. However, their long-term viability depends on balancing economic growth with environmental stewardship and equitable resource management.

The economic significance of forest-based industries is evident in their contributions to GDP, employment, and trade. Timber-based industries, such as sawmilling, furniture manufacturing, and plywood production, form the backbone of the construction, infrastructure, and design sectors worldwide. Similarly, NTFP-based industries, which utilize resources like resin, gums, essential oils, bamboo, and medicinal plants, support rural livelihoods while catering to global markets demanding natural and sustainable products. The growing popularity of these industries highlights the evolving consumer preference for eco-friendly solutions and the potential of forests to provide diverse economic opportunities.

Bamboo and rattan industries are prime examples of how renewable forest resources can drive economic growth while promoting sustainability. Bamboo's rapid growth and versatility make it ideal for applications ranging from construction to textiles, while rattan's use in furniture and handicrafts creates value in both local and global markets. These industries underscore the potential of innovative product development and sustainable management practices in maximizing the economic and ecological benefits of forest resources. Energy and biofuel industries derived from forests demonstrate the sector's critical role in supporting renewable energy transitions. While traditional energy sources like charcoal and firewood remain vital in developing regions, modern biofuels such as wood pellets and biomass energy have gained prominence in global energy strategies. These industries utilize forest residues and byproducts, reducing waste and contributing to energy security while minimizing environmental impacts. The integration of bioenergy into national policies highlights the potential of forest-based industries in addressing global energy challenges.

Eco-tourism further exemplifies the multifaceted value of forest-based industries by combining conservation and economic development. Forest-based tourism activities generate income for local communities, support conservation initiatives, and raise awareness of the importance of preserving natural habitats. However, the rapid expansion of eco-tourism must be managed responsibly to prevent adverse effects such as habitat degradation and over-tourism, ensuring that its benefits remain sustainable. Despite their contributions, forest-based industries face significant challenges that threaten their sustainability. Resource depletion, unsustainable logging practices, and biodiversity loss are critical concerns, particularly for timber-based industries. Environmental impacts, including deforestation, carbon emissions, and soil erosion, exacerbate the pressure on forest ecosystems. Economic challenges such as market volatility, competition from synthetic alternatives, and regulatory complexities further hinder the growth and stability of these industries. Addressing these challenges requires innovative approaches and a commitment to sustainable practices.

The future of forest-based industries depends on integrating sustainability into their core operations. Certification programs like FSC and PEFC, coupled with circular economy models, provide frameworks for responsible resource use and waste reduction. Technological advancements in processing, monitoring, and product innovation can enhance efficiency and reduce environmental impacts. Collaboration among stakeholders, including governments, industries, and local communities, is crucial to creating equitable and sustainable forest economies.

References

Agrios GN. 2005. *Plant Pathology.* Elsevier Acad. Press. Singapore.

AK and Gibson GL. 1997. *Forest Genetics and Tree Breeding.* CBS.

Context (Ed.). Springer, Verlag Berlin Heidelberg.

Cosasalter C and C Pye-Smith. 2003. *Fast Wood Forestry – Myths and Realities.* CIFOR. Bogor, Indonesia. 50p.

Dube HC. 2014. *Modern Plant Pathology*, Second Edition. Agribios, Jodhpur (India).

Evane JW. 1989. *Insect Pest and their Control.* Samir BooK Center, New Delhi (India).

Gonthia P and Nicolotti G. 2013. *Infectious Forest Diseases.* CABI, UK. Guy Watson., 2013,Tree Pests and Diseases.

Harsh NS. 2012. *Disease Resistance in Genetic Material in Tree Improvement Programme.* Lambert Acad. Publications.

Heybroek HM, Stephan BR and Weissenberg KV. 1990. *Resistance to Diseases and Pests in Forest Trees.* IBD, Dehra Dun (India).

Kozlowski TT. 1971. *Growth and Development of Trees. Vol. I.* Academic Press.

Kramer PJ and Kozlowshi TT. 1979. *Physiology of Woody Plants.*Academic Press.

Ksenzhek OS and Volkov AG. 1998. *Plant Energetics.* Academic Press, New York.

Lack AJ and Evans DE. 2001. *Plant Biology- Instant Notes.* Vina Books Pvt. Ltd., New Delhi.

Larcher W. 2003. *Physiological Plant Ecology.* 4th edn, Springer-Verlag, Germany.

Linskens HF and Jackson JF. 1991. *Essential Oils and Waxes* (Ed.). Springer-Verlag Berlin Heidelberg.

Luttge U. 2008. *Physiological Ecology of Tropical Plants.* Springer-Verlag, Germany Mandal

Mathe A. 2015. *Medicinal and Aromatic Plants of the World-Scientific, Production, Commercial and Utilization Aspects.* Springer Netherlands.

Mehta T. 1981. *A Hand Book of Forest Utilization.* International Book Distributors, Dehradun.

Nair KSS, Sharma JK and Varma RV. 1996. *Impact of Diseases and Insect Pest in Forest Trees.*

Nair PKR. 1993. *An Introduction to Agroforestry.* Kluwer Academic publishers.

Panda H. 2005. *Hand Book on Specialty Gums, Adhesive, Oils, Rosin And Derivatives, Resins, Oleoresins, Katha, Chemicals with others Natural Products.* Asia Pacific business press. Inc.

Panshin AJ, Harrer ES and Bethel JS. *Forest Products, their Sources, Production and Utilization.*

Parker J. 2008. *Molecular Aspects of Plant Disease Resistance.* Ann. Pl. Rev., 34. Blackwell Publications UK.

Parthiban KT, Umarani R, Umesh Kanna S, Sekar I, Rajendran P and Durairasu P. 2014.

Pathak H, Maru S, Satya HN and Silawat SC. 2015. *Fungal Diseases of Trees in Forest Nurseries of Indore, India.* J Plant Pathol Microb.

Raghavendra AS. 1991. *Physiology of Trees.* John Wiley & Sons.

Ross Wylie F and Martin R Speight. 2012. *Insect Pests in Tropical Forestry* (2nd Ed.). CABI *Tropical Forests.*

Shackleton S, Shackleton C and Shanley P. 2011. *Non-Timber Forest Products in the Global*

Sinclair W and Howard HL. 2005. *Diseases of Trees and Shrubs.*

Speight MR. 2000. *Insect Pest in Tropical Forestry.* Rose Willey Publications

Taiz L and Zeiger E. 2007. *Plant Physiology* 4th Ed. Sinauer Associates Inc. Publishers, Sunderland.

Tejwani KG. 1994. *Agroforestry in India.* Oxford and IBH publishing Co., New Delhi.

Van der Plank JE. 1982. *Host Pathogen Interactions in Plant Disease.* Academic Press Inc., New York.

Van der Plank JE. 1984. *Disease Resistance in Plants.* Academic Press Inc., New York.

Willium M Ciesla. 2010. *Forest Entomology-A Global Perspective.* Wiley-Blackwell.

Zimmerman RH. 1972. *Juvenility and Flowering in Woody Plants: A Review.* Hort. Science 7(5): 447-455.

For Product Safety Concerns and Information please contact our EU
representative GPSR@taylorandfrancis.com
Taylor & Francis Verlag GmbH, Kaufingerstraße 24, 80331 München, Germany

www.ingramcontent.com/pod-product-compliance
Ingram Content Group UK Ltd.
Pitfield, Milton Keynes, MK11 3LW, UK
UKHW052224190926
475702UK00007B/1193
* 9 7 8 1 0 4 1 3 7 2 0 8 0 *